Lipoproteins: Current Concepts

Lipoproteins:
Current Concepts

Edited by **Caroline Gardner**

R CALLISTO REFERENCE

New York

Published by Callisto Reference,
106 Park Avenue, Suite 200,
New York, NY 10016, USA
www.callistoreference.com

Lipoproteins: Current Concepts
Edited by Caroline Gardner

International Standard Book Number: 978-1-63239-451-4 (Hardback)

Contents

Preface

This book elucidates the current concepts evolving in the area of lipoproteins. Searching for the word "lipoprotein" in the databases of Medline or PubMed, one gets more than 100,000 results which show the prevailing interest in the topic. This book will be valuable to new investigators in the field to get familiar with the general topic of lipoprotein research and will guide scientists interested in this domain. The sections highlight the important issue of lipid oxidation, role of lipoproteins in the development of atherosclerotic diseases, involvement of lipids and lipoproteins in the development of cancer.

This book unites the global concepts and researches in an organized manner for a comprehensive understanding of the subject. It is a ripe text for all researchers, students, scientists or anyone else who is interested in acquiring a better knowledge of this dynamic field.

I extend my sincere thanks to the contributors for such eloquent research chapters. Finally, I thank my family for being a source of support and help.

Editor

Lipid Oxidation and Anti-Oxidants

Pathophysiology of Lipoprotein Oxidation

Vikram Jairam, Koji Uchida and Vasanthy Narayanaswami

Additional information is available at the end of the chapter

1. Introduction

Lipoproteins are large lipid/protein complexes that play a major role in transport of lipids and lipophilic molecules in the plasma and central nervous system (CNS). Plasma lipoproteins represent a dynamic continuum of particles that are constantly undergoing re-modulation in their lipid and protein components leading to re-structuring of the particle under normal physiological conditions. The re-modulation is the result of lipid transfer and metabolism mediated by non-enzymatic and enzymatic processes, and of lipid association/dissociation behavior of apolipoproteins. Variations in protein and lipid components and composition arise because of changes in the feeding, metabolic and hormonal states, and due to differences in age, gender and disease states.

In this chapter, we will focus on oxidation of lipoproteins, paying attention to sources of oxidative stress, oxidation products of specific lipid and protein components, and the pathophysiology of oxidized lipoproteins in various disease states. The disease states that are addressed in this chapter include cardiovascular disease (CVD)/atherosclerosis, Alzheimer's disease and diabetes.

2. Lipoproteins

2.1. Classes of lipoproteins in plasma and CNS

Lipoproteins are classified based on their density, ranging from high density lipoproteins (HDL), low density lipoproteins (LDL), intermediate density lipoproteins (IDL), very low density lipoproteins (VLDL) and, chylomicrons (CM), Figure 1. They can also be classified based on their mobility during electrophoresis on an agarose gel as α-, pre-β- and β-lipoproteins, which correspond to HDL, VLDL and LDL, respectively. Lipoproteins vary significantly in particle diameter and density, protein and lipid components and composition. In general, the particle diameter is inversely related to the density.

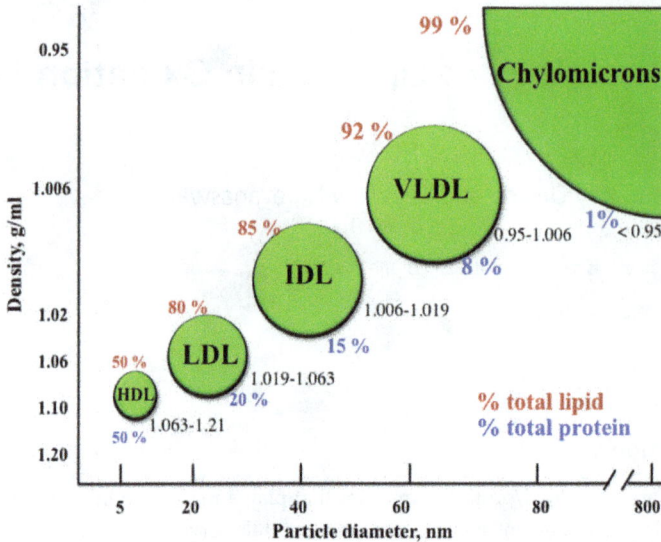

Figure 1. Lipoprotein classes. The classification of the major types of lipoproteins is based on their densities obtained by flotation ultracentrifugation analysis. The density range for each class is shown, in addition to the lipid (red) and protein (blue) content. The diagram is not to scale.

One of the main functions of lipoproteins is to transport hydrophobic factors in the highly aqueous vascular system. The CM is composed of lipids of dietary origin and is synthesized by the intestines. The VLDL is synthesized and secreted by the liver, and plays a role in distribution of lipids synthesized by the liver to the peripheral tissues. In the blood stream, the VLDL undergoes particle re-modulation to IDL and LDL, during which process the LDL become enriched in cholesterylesters. LDL plays a predominant role in delivery of cholesterol to the peripheral tissues and to the liver, with the cell surface-localized LDL receptor (LDLr) family of proteins playing a role in the cellular uptake and internalization of lipoproteins in target cells. The HDL may be synthesized by the liver and intestines or derived from other lipoproteins. In addition, peripheral tissues such as the macrophages are an important source of HDL, which are formed from cellular cholesterol efflux, and eventually transported to the liver. This process is called reverse cholesterol transport (RCT), which is an important mechanism for removal of peripheral cholesterol to the liver for eventual disposal.

Much less is known about lipoprotein metabolism in the CNS. Studies on lipoproteins secreted by astrocytes and those isolated from the cerebrospinal fluid (CSF) indicate the presence of only HDL-sized particles; large triglyceride-containing lipoprotein particles have not been detected in the CNS. The CNS maintains autonomy in terms of cholesterol synthesis and metabolism, right from the time when the blood brain barrier is established during development. One of the main functions of HDL secreted by astrocytes appears to be cholesterol delivery to the neurons via the LDLr family of proteins, for eventual use in the process of synaptogenesis.

2.2. Lipid and protein components of lipoproteins

In general lipoproteins are spherical in shape with a monolayer of amphipathic lipids (for example phospholipids, cholesterol and sphingolipids) and proteins encircling a core of neutral lipids (such as triglycerides and cholesterylesters), **Figure 2**. The lipid composition and content vary significantly in the different lipoproteins: in general, the larger lipoproteins (CM, VLDL and IDL) are enriched in triglycerides, while the smaller lipoproteins (LDL and HDL) are enriched in cholesterylesters and cholesterol. Unesterified or free cholesterol is esterified to cholesterylester by the action of lecithin-cholesterol acyltransferase (LCAT) on HDL. The phospholipids serve as the donor of the fatty acyl chains (especially 18:1 or 18:2 fatty acids) utilized in the esterification process. The fatty acid composition of triglycerides in the fasted state is dominated by 16:0 and 18:1 fatty acids.

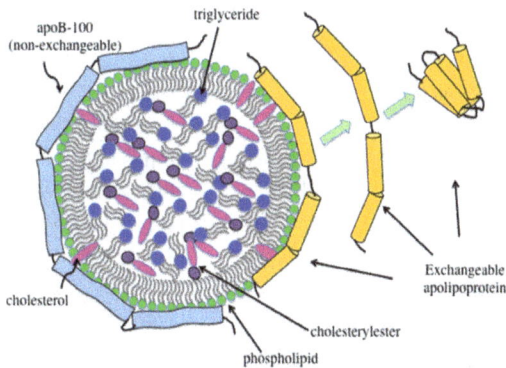

Figure 2. Schematic representation of a generic lipoprotein particle. Lipoproteins have a spherical geometry with a monolayer of amphipathic lipids and proteins encircling a core of neutral lipids. ApoB-100 (pale blue) is a single large polypeptide and is the non-exchangeable component of lipoproteins such as VLDL and LDL. The molecular mass of exchangeable lipoproteins (yellow cylinders) varies from 8-50 kDa. The exchangeable apolipoproteins have the ability to exist in lipid-free and lipid-bound states.

The protein composition and content of the lipoproteins also vary significantly from one particle to another. There are two types of apolipoproteins: non-exchangeable and exchangeable. The non-exchangeable apolipoproteins, apolipoprotein B-100 (apoB-100) and apoB-48, are present as a single copy per lipoprotein particle: apoB-100 on VLDL, IDL or LDL and apoB-48 on CM and CM remnants. ApoB-100 is 4536 residues long that is synthesized in the liver, while apoB-48, represents the N-terminal 48% of apoB-100, and is synthesized in the intestines. Both apoB-100 and apoB-48 are integral to the structure and stability of the lipoprotein particle. There are different types of exchangeable apolipoproteins, Table 1, which undergo a reversible association with lipoproteins depending on the metabolic state.

Apolipoprotein	Lipoprotein	Function
ApoAI	HDL	activates LCAT; promotes ABCA1-mediated cholesterol efflux in RCT
ApoAII	HDL	Inhibits LCAT
ApoAIV	Chylomicrons, HDL	Activates LCAT, cholesterol clearance and transport
ApoB-48	Chylomicron and chylomicron remnants	Cholesterol clearance and transport; lacks LDLr binding sites
ApoB-100	VLDL, IDL and LDL	Binds to LDLr
ApoCI	VLDL, HDL	Activates LCAT
ApoCII	VLDL, IDL, chylomicrons	Activates lipoprotein lipase
ApoCIII	VLDL, IDL, chylomicrons	Inhibits lipoprotein lipase
ApoD	HDL	Carrier proteins family (lipocalins)
ApoE2	VLDL, IDL, chylomicrons, chylomicron remnants	Poor LDLr binding activity; associated with Type III hyperlipoproteinemia and CVD
ApoE3	HDL, VLDL, IDL, chylomicron remnants (higher binding preference for HDL over VLDL)	Binds to LDLr family of proteins with high affinity; significant role in cholesterol efflux and RCT in atherosclerosis
ApoE4	VLDL, HDL, IDL, chylomicron remnants (higher binding preference for VLDL over HDL)	Binds to LDLr family of proteins with high affinity; associated with CVD and Alzheimer's disease
ApoM	HDL	Transports sphingosine-1-phosphate
Apo(a)	Lipoprotein(a) (Lp(a))	Linked to apoB-100 via disulfide bond; similar to plasminogen

[1] Adapted with permission from the AOCS Lipid Library, "Plasma Lipoproteins: Composition, Structure and Biochemistry," Table 3, 'The main properties of apoproteins, in http://lipidlibrary.aocs.org/lipids/lipoprot/index.htm, accessed June 12, 2012

Table 1. Major apolipoprotein components of lipoproteins and associated functions[1]

The exchangeable apolipoproteins have the ability to exist in lipid-free and lipid-bound states, and undergo a large conformational change upon transitioning from one state to the other. Exchangeable apolipoproteins are characterized by an abundance of amphipathic α-helices that are folded into a helix bundle. For example, in the case of apoE, the protein is composed of a series of α-helices that are folded into an N-terminal (NT) and a C-terminal (CT) domain. The NT domain is comprised of a 4-helix bundle bearing the LDLr binding sites on helix 4, which has an abundance of basic residues, while the CT domain harbors high-affinity lipid binding sites and apoE self-association sites. A similar arrangement was noted for apoAI; however, apoAI does not have the capability to interact with LDLr.

In summary, lipoprotein particles offer several targets that are vulnerable to attack by oxidative species. Oxidative modification is expected to compromise the structure and function of the protein and lipid components.

3. Oxidative stress

Reactive oxygen species (ROS) generally refer to oxygen free radicals with one or more unpaired electrons, and other highly reactive oxygen-containing molecules. They may be generated by the cells as products of normal cellular metabolism, or derived from exogenous sources including environmental pollutants, which could in turn, trigger further release of ROS. Whereas the ROS may be beneficial at low concentrations and play key physiological roles, their deleterious effect at high concentrations contributes to the etiology of several disease states. Organisms have developed an exquisite arsenal of defense mechanisms to combat the harmful effects of ROS, thereby maintaining a redox homeostasis. However, when the ROS overcome the cellular defense mechanism, there is a dysregulation of the redox balance, which leads to the state of oxidative stress (Esterbauer et al., 1991).

3.1. ROS

The high reactivity of free radicals is attributed to the unpaired electron(s) on the oxygen molecule. The primary ROS is the superoxide anion, which is generated by the addition of one electron to molecular oxygen during mitochondrial electron transport (Equation 1). This in turn gives rise to hydrogen peroxide following the action of antioxidant enzymes, superoxide dismutases (SOD) (Equation 2). The superoxide anions also give rise to the hydroxyl radicals in the presence of trace amounts of Fe^{2+} (Fenton reaction) (Equation 3); the hydroxyl radical is an extremely deleterious species that is capable of causing indiscriminate damage in the immediate vicinity of its formation.

$$O_2 + e^- \longrightarrow \bullet O_2^- \qquad \text{Superoxide anion} \qquad (1)$$

$$\bullet O_2^- + 2H^+ \xrightarrow{SOD} H_2O_2 + O_2 \qquad \text{Hydrogen peroxide} \qquad (2)$$

$$\bullet O_2^- + H_2O_2 \xrightarrow{Fe^{2+}} \bullet OH + OH^- + O_2 \quad \text{Hydroxyl radical} \qquad (3)$$

SOD catalyzes the dismutation of superoxide anion into oxygen and hydrogen peroxide, thereby affording protection to the cell. Three forms of SOD exist in humans and other mammals: SOD1 is located in the cytoplasm, SOD2 in the mitochondria, while SOD3 is extracellular. Mutations in SOD lead to diseases commonly associated with high oxidative stress such as familial amyotrophic lateral sclerosis, Parkinson's Disease, and cardiovascular disease (Fukai et al., 2002; Noor et al., 2002; Tórsdóttir et al., 2006). Overexpression of SOD inhibits LDL oxidation by endothelial cells (Fang et al., 1998). Higher SOD levels were demonstrated to be protective against LDL oxidation *in vitro* (Laukkanen et al., 2000).

Red blood cells (RBC) are particularly susceptible to ROS and free radical stress due to its constant interactions with oxygen. Despite their lack of mitochondria, RBC are under tremendous oxidative stress, due to the abundance of hemoglobin. Hemoglobin is prone to oxidation through either exogenous or endogenous sources, which results in superoxide production. Therefore, the RBC contains many SOD enzymes to convert the superoxides to hydrogen peroxide. Furthermore, hydrogen peroxide can combine with hemoglobin to form ferrylhemoglobin, a strong oxidizing agent (Rifkind et al, 2002). The RBCs are also equipped with other antioxidant enzymes such as catalase and glutathione peroxidase to overcome the harmful effects of ROS.

3.2. Reactive aldehydes

It is well established that polyunsaturated fatty acids (PUFA) undergo lipid peroxidation that is initiated by ROS, which generate a 'spectrum' of reactive aldehyde species. The reactive aldehydes formed in biological systems are more complex than those formed in simple systems like purified lipid preparations and their physiological concentrations are believed to be lower. They play a crucial role in amplifying the free radical-initiated reaction by generating a complex array of toxic end products. Although considered as end products of oxidative damage of lipids, the aldehydic products display reactivity with a wide variety of biological molecules under cellular conditions, thereby enhancing the pathogenesis of the diseases. They are therefore considered as toxic second messengers of oxidative stress and lipid peroxidation. Protein-bound aldehydes have been proposed as potential markers of oxidative stress as evidenced by immunohistochemical analysis of atherosclerotic lesions (Uchida et al., 1998a). Numerous α,β-unsaturated aldehydes have been reported in literature: we will focus on acrolein, 4-hydroxynonenal (4-HNE) and malondialdehyde (MDA), **Figure 3.**

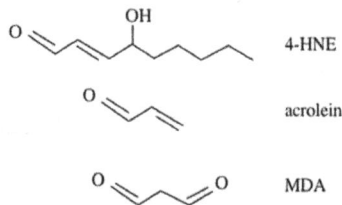

Figure 3. Major aldehydic products of lipid peroxidation

3.2.1. Acrolein

Acrolein is an acrid smelling environmental pollutant that is formed during combustion of organic and plastic substances (Beauchamp et al., 1985). It is present as one of the major components in the gaseous phase of tobacco smoke (up to 140 µg/cigarette) (Witz, 1989). It is also generated as a natural metabolite during oxidative stress mediated lipid peroxidation

(Uchida et al., 1998b). In addition, oxidation of threonines by myeloperoxidase (MPO) gives rise to acrolein (Anderson, et al., 1997; Savenkova et al., 1994). Acrolein is the most reactive of all α,β-unsaturated aldehydes; it causes oxidative modification of proteins by reacting with the sulfhydryls of cysteines, ε-amino groups of lysines and imidazole group of histidines (Witz, 1989; Esterbauer et al., 1991). Some possible products formed upon reaction of acrolein with lysine side chains include aldimine adducts (Schiff base formation), propanal adducts (Michel addition) and Nε-(3-formyl-3,4-dehydropiperidino)lysine (FDP-lysine), **Figure 4**.

3.2.2. 4-HNE

Discovered more than 50 years ago, alkenals have been under intense scrutiny in terms of their chemistry, biochemistry, toxicology and as oxidative stress agents. Amongst these alkenals, 4-HNE was determined to be the most toxic. It was discovered to be a product of lipid peroxidation, particularly of n-6 PUFA such as linoleic acid and arachidonic acid. Since then, 4-HNE has been implicated in a number of diseases, including atherosclerosis, Alzheimer's Disease, Parkinson's Disease, liver cirrhosis, and cancer (Zarkovic, 2003).

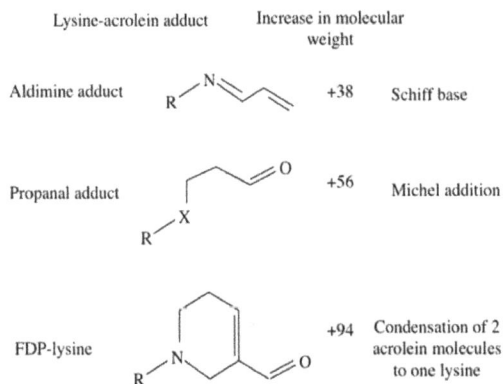

Figure 4. Modifications of lysine side chains by acrolein. The major lysine modification products of acrolein are shown, along with the expected increments in molecular weight during mass spectral analysis. Adapted from Furuhata et al, 2003.

From a chemical perspective, the reactivity of 4-HNE is conferred by the carbonyl group at C1 position, and a C=C bond between the second and third carbons which provide a partial positive charge to C3 (Schaur, 2003). This results in 4-HNE being highly reactive towards thiol and amino groups of proteins. 4-HNE reacts with thiols via a Michael addition, in which a nucleophile such as the sulfhydryl of cysteine or glutathione reacts with the C3 of HNE to form a covalent adduct. In addition, amino compounds such as lysine, ethanolamine, guanine, and the imidazole group of histidine are capable of forming Michael addition adducts with 4-HNE. 4-HNE can also undergo reactions involving a reduction or an epoxidation of the double bond. 4-HNE has been shown to react *in vivo* with common biomolecules, and can often have an inhibitory role with enzymes (Schaur, 2003; Korotchkina et al, 2001). Cross-reactivity of anti-4-HNE antibody with oxidized LDL (Ox-LDL) suggests that 4-HNE may contribute to the overall process of atherosclerotic plaque formation (Uchida et al., 1993).

3.2.3. MDA

In fresh samples, MDA is formed mainly from PUFA such as arachidonic acid (Esterbauer et al., 1991). Initially determined by the thiobarbituric acid (TBA) assay as TBA reactive substances (TBARS), the presence of pre-existing MDA or protein-bound MDA needs to be confirmed by other sensitive assays as well. Under physiological conditions, MDA readily modifies protein side chains forming stable cross-linked adducts with the ε-amino groups of lysines. Other side chains that may be modified include those of histidine, tyrosine, arginine and methionine.

While the levels of acrolein, 4-HNE, MDA and other aldehydic compounds *per se* are thought to be too low for detection, their role in mediating oxidative damage has been confirmed by the presence of autoantibodies against aldehyde-modified proteins. It provides evidence for the *in vivo* occurrence of lipid peroxidation products and their role in disease progression (Steinberg et al, 1989). We have a better understanding of the association between lipid peroxidative products, oxidative stress and disease progression, in part due to the development of antibodies directed against oxidized lipids and protein-bound aldehydes in the past two decades, Table 2. Currently, we have a repertoire of antibodies directed against different types of oxidatively modified lipids and proteins, **Table 2**, that serve as powerful tools for ELISA, immunohistochemical and immunoblotting analyses.

3.3. Sources of ROS

In addition to the mitochondrial electron transport, there are enzymatic sources of ROS that contribute to the redox status in biological systems. Located in the blood stream, they are of relevance to lipoprotein oxidation.

3.3.1. MPO

MPO is an enzyme found abundantly in neutrophil granulocytes and to some extent in macrophages. It is a lysosomal protein stored in the azurophilic granules in neutrophils and can be identified by its characteristic green heme pigment. During the respiratory burst of the neutrophil, MPO catalyses the production of hypochlorous acid (HOCl) from hydrogen peroxide and chloride anion (Equation 4). MPO is also responsible for generation of 3-chlorotyrosine and 3-nitrotyrosine; individuals with coronary artery disease show elevated levels of these two products in their blood and HDL. MPO has been shown to play a role in promoting atherosclerotic lesions through oxidative modification of apoAI (Shao et al., 2012) (described under apoAI).

$$H_2O_2 + Cl^- + H^+ \rightarrow HOCl + H_2O \qquad\qquad (4)$$

Lipid peroxidation products	mAb	Epitope
Oxidized lipids		
[a]9-Hydroxy-10E,12Z-octadecadienoic acid (9-HODE)	9H2	9-HODE
[b]13-Hydroxy-9Z,11E-octadecadienoic acid (13-HODE)	13H1	13-HODE
[a]12-Hydroxyeicosatetraenoic aci (12-HETE)	12H8	12-HETE
[a]Leukotoxin (epoxylinoleic acid)	21D1	Leukotoxin
[a]7-Ketocholesterol (7-KC)	35A-8	7-KC
[c]15-deoxy-$\Delta^{12,14}$-prostaglandin J$_2$ (15d-PGJ$_2$)	11G2	15d-PGJ$_2$
Protein-bound aldehydes		
[d]Acrolein (ACR)	5F6	ACR-lysine
[e]Crotonaldehyde (CRA)	82D3	CRA-lysine
[a]2-Hexenal (HE)	CT5	HE-lysine
[f]2-Nonenal (NE)	27Q4	NE-lysine
[g]Malondialdehyde (MDA)	1F83	MDA-lysine
[h]4-Hydroxy-2-hexenal (HHE)	53	HHE-histidine
[i]2-Hydroxyheptanal (2HH)	3C8	2-HH-lysine
[j]4-Hydroxy-2-nonenal (HNE)	HNEJ2	HNE-histidine
[k]4-Hydroxy-2-nonenal (HNE)	2C12	HNE-lysine
[l]4R-4-Hydroxy-2-nonenal ((R)-HNE)	R310	(R)-HNE-histidine
[m]4S-4-Hydroxy-2-nonenal ((S)-HNE)	S412	(S)-HNE-histidine
[n]4-Hydroperoxy-2-nonenal (HPNE)	PM9	HPNE-lysine
[o]4-Oxo-2-nonenal (ONE)	9K3	ONE-lysine
Protein-bound core aldehyde		
[p]9-Oxononanoylcholesterol (9-ONC)	2Λ81	9-ONC-lysine

[a]Unpublished
[b]Shibata et al. (2009) Acta Histochem. Cytochem. 42, 197; [c]Shibata et al. (2011) JBC 277, 10459; [d]Uchida et al. (1998) PNAS 95, 4882; [e]Ichihashi et al. (2002) JBC 276, 23903; [f]Ishino et al. (2010) JBC 285, 15302; [g]Yamada et al. (2001) JLR 42, 1187; [h]Yamada et al. (2004) JLR 45, 626; [i]Itakura et al. (2003) BBRC 308, 452; [j]Toyokuni et al. (1995) FEBS Lett. 359, 189; [k]Itakura et al. (2000) FEBS Lett. 473, 249; [l]Hashimoto et al. (2003) JBC 278, 5044; [m]Hashimoto et al. (1995) JBC 278, 5044; [n]Shimozu et al. (2011) JBC 286, 29313; [o]Shibata et al. (2011) JBC 286, 19943; [p]Kawai et al. (2003) JBC 278, 21040; [q]Kawai et al. (2003) JBC 278, 50346

Table 2. Lipid peroxidation-specific monoclonal antibodies

3.3.2. NADPH oxidase

NADPH oxidase is an enzyme complex that is normally found in the plasma membranes of neutrophils and monocytes. Like MPO, it is activated during the respiratory burst of the neutrophil. Its main role is to generate superoxide by transferring electrons from NADPH to molecular oxygen (Equation 5). The superoxide is then used to destroy phagocytosed bacteria or pathogens. NADPH oxidase and its product, superoxide, are major contributory factors for foam cell formation in atherosclerosis (Meyer & Schmitt, 2000).

$$2O_2 + NADPH \rightarrow 2 \bullet O_2^- + NADP^+ + H^+ \qquad (5)$$

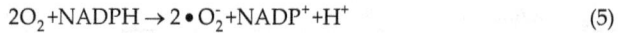

3.3.3. Lipoxygenase

Lipoxygenases are iron-containing enzymes that catalyze the dioxygenation of PUFA in lipids. Specifically, they form hydroperoxides from fatty acids and molecular oxygen. 15-lipoxygenase is the main protein in this family; it is involved in the metabolism of eicosanoids (prostaglandins, leukotrienes), which function as secondary messengers. 15-lipoxygenase has been shown to be involved in LDL oxidation (Bailey et al., 1995), with considerably greater 15-lipoxygenase activities in atherosclerotic compared to normal aortas (Hiltunen et al., 1995).

4. Lipoprotein oxidation

We now turn to the specifics of lipoprotein oxidation with reference to the different lipid and protein components.

4.1. Oxidation of lipid components

Of the different lipid components that may be potentially oxidized in various lipoprotein particles, we will focus on the oxidation of fatty acyl chains, including those on phospholipids, cholesterylester and triglycerides, and on cholesterol derivatives. The reader is referred to a comprehensive review by Subbaiah and colleagues (Levitan et al., 2010) for an introduction to the role of ceramides in Ox-LDL and of sphingosine 1-phosphate in HDL

4.1.1. Oxidation of fatty acyl chain

One group of biological targets that are highly vulnerable to attack by ROS and aldehydes are the lipids: their abundance in lipoproteins and the ease with which their unsaturated bonds are oxidatively modified make them susceptible to damage. Products of lipid peroxidation have been associated with the pathophysiology of numerous disease states, including atherosclerosis, diabetes and cancer. Although found in low concentrations in normal healthy tissues, they are found to be enriched in pathological cells and tissues, including macrophage foam cells and atherosclerotic lesions (Olkkonnen, 2008; Brown & Jessup, 1999; Olkkonen & Lehto, 2004; Javitt, 2008; Tsimikas et al., 2005; Berliner & Watson, 2005). PUFA peroxidation products cause further damage to proteins by oxidative modification of amino acid side chains and formation of protein carbonyl groups (Refsgaard et al., 2000).

Briefly, lipid peroxidation is initiated by the hydroxyl radical abstracting a hydrogen from the methylene group adjacent to a double bond of fatty acids, **Figure 5**. The fatty acid may also be part of the phospholipid at the *sn*-2 position or esterified to the –OH group of cholesterol in a lipoprotein. This process gives rise to an unstable lipid radical, which undergoes rearrangement of double bonds and addition of oxygen to form a peroxyl radical.

The lipid radical or the peroxyl radical could react with a neighbouring fatty acyl chain as well, thereby propagating the peroxidation process. The chain reaction may be terminated by eventual formation of a lipid hydroperoxide.

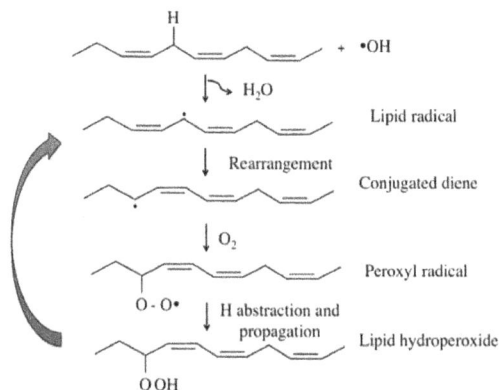

Figure 5. Lipid peroxidation. The lipid peroxidation process may be divided into the initiation, propagation and termination steps; only the double bond containing segment of a fatty acid is shown.

In the context of a lipoprotein, it was postulated that LDL oxidation was initiated through cell-generated ROS formation, with the involvement of the lipoxygenase pathway (Cyrus et al, 1999). Both apoE-null and LDLr-null mice genetically deficient in the 12/15-lipoxygenase were found to have significantly less atherosclerosis. Other groups developed mice overexpressing the 12/15-lipoxygenase, and found spontaneous aortic fatty streak lesions on a chow diet (Reilly et al., 2004). Finally, it was shown that the endothelial cells within the vessels required 12/15-lipoxygenase to generate oxidized phospholipids. It was also shown that the MPO pathway contributed to Ox-LDL in human atherosclerotic lesions (Savenkova et al., 1994).

LDL oxidation may occur within the arterial endothelial cells, which have high levels of precursor molecules like linoleic acid and arachidonic acid - fatty acids that are involved in producing eicosanoids. Hydroperoxide reaction with linoleic acid produces 13(S)-hydroperoxy-octadecadienoic acid (13(S)-HPODE), and reaction with arachidonic acid produces 15(S)-hydroperoxy-eicosatetraenoic acid (15(S)-HPETE). These molecules are present as part of LDL surface phospholipids and trigger further oxidation of phospholipids with arachidonic acid. These are early events occurring prior to apoB-100 modification and constitute the 'minimally-modified' LDL (Navab et al., 2001). Another consequence of phospholipid oxidation is fragmentation of the fatty acyl group at sn-2 position resulting in short chain fatty acids, which structurally and functionally resemble platelet activating factors with chemotactic activity.

Although less understood, *in vitro* oxidation of HDL is also shown to generate oxidized lipids and proteins. It is likely that HDL lipids may be oxidized initially even before LDL lipids, upon exposure of human plasma to peroxyl radicals, Cu^{2+} ions or lipoxygenase

(Garner et al., 1998). These studies also show that Met residues are oxidized in apoAI and apoAII. Further, ^{1}H and ^{31}P NMR analysis indicate a loss of the unsaturated system with appearance of epoxides on fatty acyl chains and 5,6-epoxide derivatives of cholesterol indicating significant modification of HDL (Bradamante et al., 1992).

4.1.2. Oxysterols

Oxysterols are molecules that are formed from the oxidation of cholesterol, which can occur at several sites (Vaya & Schipper, 2007). They display higher water solubility compared to cholesterol. The major oxysterols isolated from plasma LDL include 7α-OH and 7β-OH cholesterol. In addition, 7-keto cholesterol, which impairs cholesterol efflux and reduces cell membrane fluidity, and 5,6-epoxide derivatives of cholesterol, have also been identified in Ox-LDL (Levitan et al., 2010). Other types of oxysterols have been localized in atherosclerotic lesions, including those wherein the side chains of cholesterol are oxidized (for example, 27-hydroxycholesterol). The cytochrome P-450 system is largely responsible for generating the hydroxylated derivatives. They also function as transcriptional effectors and attenuate the Liver X Receptor (LXR). A majority of PUFA in LDL is esterified as cholesterylester; thus, hydroperoxide and hydroxide derivatives of cholesterylester are found in abundance in human atherosclerotic lesions.

In the CNS, the neurons convert cholesterol to 24S-hydroxycholesterol (also known as cerebrosterol), **Figure 6**; the conversion facilitates its movement out of the CNS, since cholesterol as such does not cross the blood brain barrier. Almost all circulating 24S-hydroxycholesterol originates from the brain (Lutjohann et al., 1996; Lutjohann et al., 2000; Bjorkhem et al., 1998) and may reflect CNS cholesterol turnover.

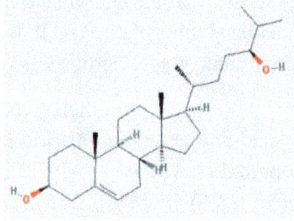

Figure 6. 24S-hydroxycholesterol (obtained from: http://pubchem.ncbi.nlm.nih.gov)

4.2. Oxidation of protein components

4.2.1. ApoB-100

From a historical perspective, apoB-100 was the earliest apolipoprotein to be identified as being a target for oxidative modification on lipoproteins (Steinberg et al., 1989). It is implied in the discussions involving Ox-LDL hypothesis, along with oxidation of the lipid moieties in LDL. The term Ox-LDL is used to identify LDL that has been modified to an extent that it is not recognized by the LDLr anymore; instead it becomes a ligand for the scavenger receptors family of proteins. On the other hand, the term 'minimally-modified' LDL has been adopted to encompass the different preparations that have been modified enough to be chemically distinguishable from, but recapitulates the LDLr binding feature of, unmodified LDL. ApoB-100 is one of the oxidizable targets on lipoproteins: one of the earliest *in vitro* studies demonstrate that incubation of human LDL with 4-HNE (Haberland et al., 1984; Jurgens et al., 1986), results in modification of 45 lysines, 7 histidines, 23 serines and 51 tyrosine residues on apoB-100.

4-HNE forms covalent adducts with lysine residues on apoB-100, thereby blocking its ability to recognize the macrophage LDLr. Evidence for *in vivo* LDL oxidation was provided by immunocytochemical staining and immunoblot analysis of extracts from atherosclerotic lesions of LDLr-deficient rabbits using antibodies against Ox-LDL, MDA-lysine or 4-HNE-lysine, (Palinski et al., 1990). This study also demonstrated higher titers of autoantibodies against MDA-conjugated LDL in the human and rabbit antisera. 4-HNE-modified LDL is an efficient ligand for scavenger receptors (Hoff & O'Neil, et al, 1993; Rosenfeld et al., 1990). ApoB-100 can also be modified by MPO, which leads to formation of chlorotyrosine and nitrotyrosine derivatives.

4.2.2. ApoE

ApoE is a 299 residue, 34 kDa protein that is commonly associated with VLDL, CM remnants and a sub-class of HDL. It is a major cholesterol transport protein in the plasma and the CNS (Hatters et al., 2006). In humans, apoE is polymorphic; variation in the *APOE* gene results in three major alleles, ε2, ε3 and ε4, occurring at frequencies of 8%, 77% and 15%, respectively in the population. The products of the three alleles are the isoforms, apoE2, apoE3 and apoE4, which differ in the amino acids at positions 112 and 158: apoE2 has Cys while apoE4 has Arg at these locations; apoE3 has a Cys and Arg at these locations, respectively. ApoE3 is considered an anti-atherogenic protein; individuals homozygous for the *APOE* ε2 allele are prone to developing familial type III hyperlipoproteinemia and premature atherosclerosis. The inheritance of one or more of the *APOE* ε4 alleles predisposes the bearer to hypercholesterolemia, as well as Alzheimer's disease, affecting both the age of onset and the severity of these diseases.

In vitro oxidative modification of the receptor-binding domain of apoE3 by acrolein generates epitopes recognized by an antibody specific for acrolein-lysine adducts, mAb5F6 (Tamamizu-Kato et al., 2007) with formation of both intra- and inter-molecular cross-linked products. This modification resulted in severe impairment of three major functions of apoE3: (i) its ability to interact with the LDLr, a function mediated by specific lysines and arginines located on helix 4 of the receptor-binding domain; (ii) its ability to bind heparin, which is facilitated predominantly by two specific lysines (K143 and K146), **Figure 7**; (iii) its ability to bind lipids. These studies indicate that acrolein either directly modifies the lysines in helix 4 that are involved in LDLr and heparin binding or that modification of lysines elsewhere on apoE3 alters the conformation of lysines in helix 4, thereby disrupting its binding. Further evidence was provided by direct exposure of VLDL isolated from human plasma to acrolein or Cu^{2+}, which disrupted its ability to bind and internalize the lipoprotein particle via LDLr, LDLr-related protein or HSPG on hepatocytes. (Arai et al., 1999; 2005). Taken together, oxidative modification appears to compromise the functional integrity of apoE3.

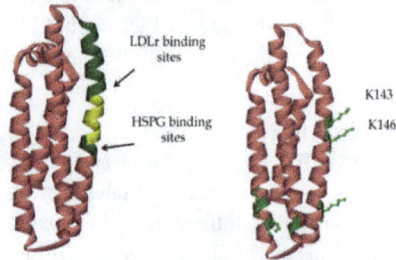

Figure 7. Lysines relevant in LDLr and HSPG binding function of apoE3. Ribbon diagram of apoE3 receptor-binding domain is shown with the LDLr- and HSPG-binding sites represented in green and yellow, respectively.

From a physiological perspective, the loss of apoE function by oxidative modification has direct implications in CVD: (i) decreased uptake and internalization of apoE-containing lipoproteins leading to their accumulation in the blood; (ii) decreased ability of apoE to interact with heparan sulfate proteoglycans (HSPG) lining the blood vessels and cell surfaces, where apoE is believed to be stored; and (iii) impaired ability of apoE to interact with lipoproteins, which is an essential prerequisite for the receptor-binding domain to elicit LDLr binding.

In the CNS, where apoE is the predominant apolipoprotein that has been identified so far, oxidative modification is expected to have serious implications in progression of neurological diseases such as Alzheimer's disease in an isoform-specific manner. About 65% of individuals with late-onset familial and sporadic Alzheimer's disease bear the *APOE* ε4 allele (Huang et al., 2004). The precise mechanism by which apoE4 is associated with Alzheimer's disease remains a contentious issue. While the role of apoE4 in aggravating the beta amyloid toxicity has received widespread attention, the inherent propensity of apoE4 to misfold noted under *in vitro* physiological conditions requires further scrutiny *in vivo*. Further, since oxidative damage plays a significant role in the pathogenesis associated with Alzheimer's disease (Perry et al., 2002), it is likely that oxidative modification of apoE4 further exacerbates its role in the etiology of the disease. Indeed, cerebrospinal fluid obtained by lumbar puncture in a limited number of Alzheimer's disease patients homozygous for *APOE3* or *APOE4*, and age-matched controls with or without dementia display a 50 kDa apoE-immunoreactive protein co-migrating with proteins immunoreactive for 4-HNE and MDA adducts (Montine et al., 1996; Bassett et al., 1999).

A similar 50 kDa apoE-immunoreactive protein was also reported in P19 neuroglial cultures differentiated into neurons and astrocytes subjected to oxidative stress. Interestingly, apoE3 appeared to be cross-linked to a greater extent than apoE4. The cross-linking has been attributed to the susceptible site provided by apoE3 in the form of Cys112, and the known reactivity of 4-HNE with sulfhydryl groups (Esterbauer et al., 1991). *In vitro* modification of apoE3 or apoE4 (isolated from plasma) by 4-HNE yielded cross-linked products with apparent molecular weights corresponding to dimeric and trimeric apoE (Montine et al., 1996). A similar trend was noted with MDA. It is possible that the greater susceptibility of apoE3 to cross-linking and oxidation than apoE4 is a reflection of its greater potency as an antioxidant.

4.2.3. ApoAI

ApoAI is a 243 residue, 28 kDa exchangeable apolipoprotein that is a major component of HDL. Like apoE, it is composed predominantly of amphipathic α-helices, with an N-terminal domain 4-helix bundle. Under normal physiological conditions, apoAI plays a critical role in promoting ATP Binding Cassette Transporter A1 (ABCA1)-mediated cholesterol efflux from macrophages. This aids in mobilizing cholesterol and phospholipids from peripheral tissues to the liver by the RCT process, for eventual disposal by biliary secretion. In atherosclerotic lesions, apoE plays a dominant role in RCT by virtue of the fact that cholesterol-laden macrophages secrete lipid-poor apoE, which in turn in promotes ABCA1-mediated cholesterol efflux (Huang et al, 1995).

With both apoAI and apoE, a nascent discoidal form of HDL is generated that is composed of a bilayer of phospholipids and cholesterol circumscribed by the α-helices of the protein. The discoidal HDL is an excellent substrate for LCAT, the enzyme that catalyzes the transfer of a fatty acyl chain from phospholipids to the free hydroxyl group of cholesterol to form cholesterylesters. The conversion of the amphipathic free cholesterol to the hydrophobic cholesterylesters promotes its transition to the core of the lipoprotein particle, thereby generating a spherical HDL containing a cholesterylester core. The HDL is targeted to the liver and steroidogenic tissues where they are recognized by the scavenger receptor class B Type 1 (SR-B1), which mediate selective uptake of cholesterylesters into the cells.

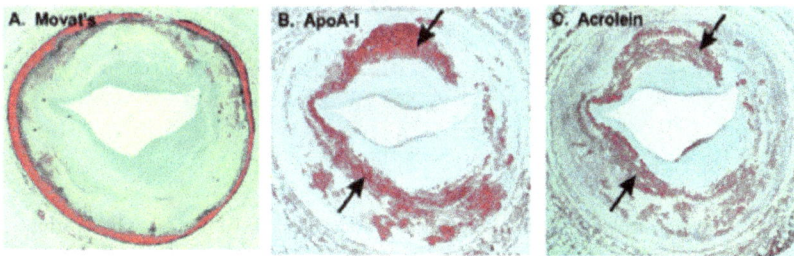

Figure 8. Immunohistochemical co-localization of apoAI and acrolein adducts in human atherosclerotic lesions. This research was originally published in The Journal of Biological Chemistry. Shao, B., Fu, X., McDonald, T. O., Green, P. S., Uchida, K., O'Brien, K. D., Oram, J. F. & Heinecke, J. W. Acrolein impairs ATP Binding Cassette Transporter A1-dependent cholesterol export from cells through site-specific modification of apolipoprotein A-I. J. Biol. Chem. (2005) Vol. 280, No. 43, pp. 36386-36396 © the American Society for Biochemistry and Molecular Biology.

There is strong evidence that MPO oxidizes HDL *in vivo* (Daugherty et al., 1994; Bergt et al., 2004; Pennathur et al., 2004; Zheng et al., 2004). In addition, aldehyde modification of HDL is also associated with the loss in ability of HDL to activate LCAT (McCall et al., 1995). Acrolein modifies apoAI specifically at Lys226 in helix 10 converting it to N^{ε}-(3-methylpyridinium)lysine (MP-Lys). A corresponding decrease in ABCA1-mediated cholesterol efflux was also noted. In addition, immunohistochemical analysis demonstrated co-localization of acrolein-adducts and apoAI in human atherosclerotic lesions, **Figure 8**, confirming previous studies (Uchida et al., 1998b). Further chlorination of Tyr192 and oxidation of specific Met residues in apoAI via the MPO pathway impairs its ability to promote cholesterol efflux (Shao et al., 2010). Taken together, it appears that oxidative modification of HDL apoAI may be one of the contributory factors to atherogenesis.

Figure 9 provides a simplified overview of the roles of apoB-100, apoAI and apoE in lipid distribution between liver and peripheral tissues. It also shows potential functional sites that are likely to be affected because of oxidative modification of lipoproteins.

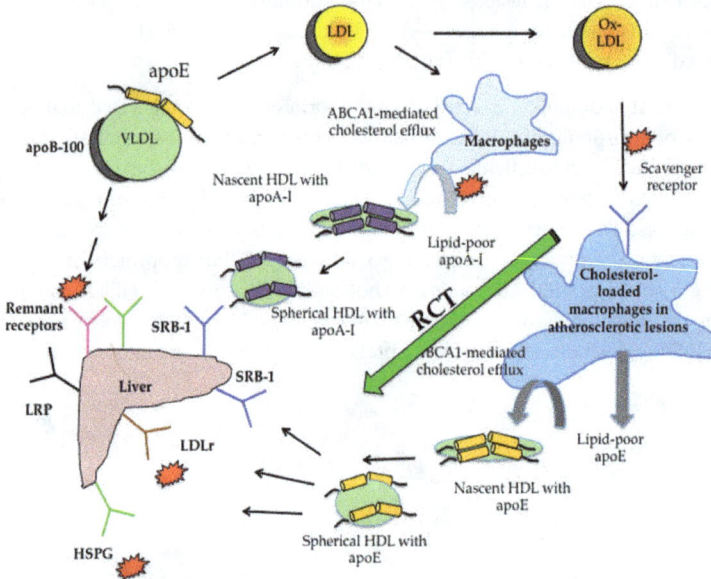

Figure 9. Distribution cholesterol mediated by apoB-100, apoE and apoAI. The green block arrow shows the general direction of the RCT process. The red stars draw attention to processes that are affected by oxidative modification of protein and lipid components of lipoproteins.

5. Lipoprotein oxidation and disease states

5.1. Atherosclerosis

Atherosclerosis is one of the leading causes of death worldwide, and is commonly associated with coronary and cerebrovascular diseases (Rocha & Libby, 2009; Moore & Tabas, 2011). Originally considered purely a lipid-storage disease, it is now recognized that atherosclerosis is intrinsically linked to inflammation, particularly with respect to involvement of innate and adaptive immunity.

Atherosclerosis is a progressive disorder marked by several stages with varying extent of lesions marking each stage. It is initiated by accumulation of lipoproteins in the sub-endothelium, typically at arterial branch points, which tend to have impaired laminar flow. The physiological response to this retention includes: (i) chemical modification (due to the production of ROS and other oxidative factors from intimal macrophages) and aggregation of the lipoproteins, (ii) early inflammatory response such as T cell recruitment, (iii) cytokine secretion, and, (iv) endothelial alterations. In response to the chemokines, the circulating monocytes enter the arterial wall and eventually differentiate to macrophages under the influence of macrophage colony stimulating factors.

The macrophages internalize oxidized and modified lipoproteins via several types of scavenger receptors, including class A and B scavenger receptors (e.g. CD36) and lectin-like oxidized LDLr-1 (LOX-1). Mice lacking CD36, platelet-activating-factor receptor, and toll-like receptors 2 and 4 show decreased atherosclerosis. The internalized cholesterol is stored in the cytoplasm as cholesterylester lipid droplets, surrounded by a monolayer of phospholipids. When viewed by electron microscopy, these cells have a foamy appearance, and are therefore called foam cells. Foam cell formation marks the earliest pathological lesion in atherosclerosis called 'fatty streaks'.

As the fatty streaks progress, they induce migration of smooth muscle cells from media to the intima of the arterial wall, which is accompanied by secretion of collagen and matrix proteins and macrophage proliferation. As lipid accumulation continues to occur in macrophages, smooth muscle cells also take up lipids. Collectively, these processes give rise to fibrous lesions. As the lesion progresses, foam cells die and release lipids, which aggregate with lipoproteins trapped in the matrix, eventually leading to advanced lesions with calcification and hemorrhage.

Ox-LDL also plays a role in modulating smooth muscle function, by increasing their adhesion to macrophages and foam cells in the plaques. At low concentrations, Ox-LDL stimulates proliferation of smooth muscle cells, whereas at higher concentrations, they cause smooth muscle apoptosis by up-regulating levels of the pro-apoptotic lipid ceramide. During this process, a necrotic core is formed in the vessel lumen. The necrotic plaque is unstable and subject to disintegration or rupture, leading to formation of a thrombus. Ox-LDL are key stimulators of the coagulation pathway, and promote the secretion of tissue factor from endothelial cells, which is required to form a clot. At various stages of the lesions unstable angina, heart attack or stroke may occur.

In summary, our understanding of the atherogenesis process has progressed rapidly since the deleterious nature of Ox-LDL was first pointed out about 3 decades back. We now recognize it as a multifactorial disease with inflammation and oxidative stress playing key roles. The last decade has seen the additional role of HDL in mitigating the severity of atherosclerosis; in this context, the role of HDL (not HDL levels *per se*, but the robustness of HDL function) is currently under intense scrutiny.

5.2. Alzheimer's disease

Alzheimer's disease is a neurodegenerative disorder that is clinically characterized by progressive cognitive decline and dementia. The major neuropathological hallmarks of this age-related disease are the extra cellular accumulation of amyloid beta peptide (Aβ), the intracellular presence of neurofibrillary tangles composed of hyperphosphorylated tau and the loss of cholinergic neurons in the brain.

Several hypotheses have been put forward to explain the pathogenesis of Alzheimer's disease, of which the oxidative stress hypothesis is of relevance here. In general, the brain of Alzheimer's disease subjects is believed to be in a heightened state of oxidative stress and is characterized by higher levels of lipid peroxidation products (acrolein, 4-HNE and MDA) (Arlt et al., 2002; Butterfield et al., 2010). The brain is particularly susceptible to oxidative stress due to the high oxygen and metal content, lipid levels, and paucity of antioxidants (such as vitamins C and E) compared to normal tissues (Schippling et al., 2000). The Alzheimer's disease brain is also rich in isoprostanes (a stable marker of peroxidation of arachidonic acid) and neuroprostanes (Lovell et al., 2001).

Accumulation of soluble and insoluble assemblies of Aβ (a peptide composed of 39-43 residues) in the brain parenchyma and in the cerebral vasculature is considered the primary event in Alzheimer's disease pathogenesis by the amyloid hypothesis. Aβ is derived from the ubiquitously expressed transmembrane protein amyloid precursor protein by regulated intramembranous proteolysis. The oligomeric form of Aβ is considered as the most toxic species; one of the reasons for its toxicity is its ability to act as an oxidative stress agent in a lipid environment. The oxidative nature of Aβ has been attributed to the presence of Met35, which is capable of undergoing one-electron oxidation to form a sulfuranyl radical cation (S•+). This in turn is able to abstract a H-atom from a fatty acyl chain as shown in **Figure 7**, and initiate a series of free radical-mediated events in a lipid milieu such as a membrane bilayer or a lipoprotein particle (Butterfield et al., 2005).

In the CNS, apoE is localized on HDL-sized spherical and discoidal particles. Lipoproteins isolated from CSF of Alzheimer's disease patients were shown to have a higher sensitivity to *in vitro* oxidation compared to those from normal subjects. It has also been shown that CSF lipoproteins can be oxidized through transition metal ions such as Cu^{2+} or Fe^{3+}. These metal ions are present in CSF as complexes with metal-binding proteins like ceruloplasmin or transferrin (Schippling et al., 2000); under pathological conditions, they are released in a catalytically active form. Furthermore, Aβ itself can induce oxidation through interactions with metal ions. Aβ contains three histidines and one tyrosine, all of which can chelate transition metal ions. This promotes Aβ aggregation and a pro-oxidative state of Aβ through reduction of the transition metal (Artl, et al., 2002). The peptide can then produce ROS or induce lipid peroxidation, both of which strongly affect lipoprotein stability.

Conversion of cholesterol to 24S-hydroxycholesterol is believed to be a mechanism by which the brain maintains cholesterol homeostasis. Plasma concentrations of 24Shydroxycholesterol are utilized as a biomarker and a diagnostic tool for neurological disorders. Neuronal damage is accompanied by destruction of neuronal membranes, which causes more cholesterol to be converted into 24S-hydroxycholesterol. In agreement significantly higher peripheral concentrations of 24S-hydroxycholesterol were found in Alzheimer's disease and vascular demented patients (Lutjohann at al., 2000). Importantly, the latter study showed that the apoE genotype does not contribute significantly to the elevated plasma levels of 24S-hydroxycholesterol in Alzheimer's disease patients (Bretillon, 2000).

The CSF levels of 24S-hydroxycholesterol appears to be sensitive to changes in the brain (possibly because they are not affected by hepatic clearance rates of this oxysterol) and may represent better markers both for neurodegenerative diseases and for disturbances in the blood brain barrier (Leoni & Caccia, 2011). In early stages of Alzheimer's disease, there are significantly higher CSF concentrations of 24S-hydroxycholesterol suggesting increased cholesterol turnover in the CNS during degeneration (Leoni & Caccia, 2011). These levels decrease as the disease advances, possibly reflecting the loss of cells expressing cholesterol 24S-hydroxylase, the enzyme responsible for the conversion of brain cholesterol into 24S-hydroxycholesterol. In Alzheimer's disease and mild cognitive impairment, but not in normal individuals, the levels of 24S-hydroxycholesterol significantly correlate with CSF levels of apoE (Shafaati et al., 2007).

The elevation of 24S-hydroxycholesterol in CSF is consistent with a significant role for this oxysterol as a signaling molecule during neuronal degeneration. It has been shown that 24S-hydroxycholesterol is able to induce expression of apoE and ABC transporters in astrocytes through activation of LXR, and to stimulate cellular cholesterol efflux (Baldan et al., 2009).

5.3. Diabetes

Diabetes mellitus is a disease characterized by hyperglycemia and insufficiency or resistance to insulin. In general, diabetes is associated with increased generation of free radicals, heightened state of oxidative stress, and attenuated antioxidant response. Studies have revealed greater levels of TBARS in the plasma of diabetic patients compared to controls (Kawamura et al., 1994). Plasma lipoproteins isolated from diabetic rats have been shown to be cytotoxic *in vitro* in cultured cells, suggesting they may have been oxidatively modified *in vivo* (Morel & Chisolm, 1989). Other studies have shown that glucose autoxidation and protein glycation can result in the formation of radicals like superoxide anions that promote lipoprotein oxidation.

Hyperglycemia promotes glycation of the protein and lipid components of lipoproteins leading to generation of advanced glycation end-products (AGE) (Sun et al., 2009). Glycation is the process whereby glucose attaches to the ε-amino group of lysines or the α-amino group of an N-terminal amino acid in a non-enzymatic manner (Maritim et al., 2003). Biochemically, glucose attachment to the protein results in the formation of an unstable Schiff base, which then rearranges to an Amadori product. These Amadori products then undergo dehydration reactions and rearrange themselves to finally form AGE, which are responsible for many of the irreversible pathological effects seen in diabetes. The receptors for AGE (RAGE) are abundantly expressed on vascular endothelial cells, smooth muscle cells and macrophages, which are enhanced in atherosclerotic lesions in diabetes. Since Ox-LDL has AGE epitopes, it binds RAGE on macrophages and enhances macrophage proliferation and oxidative stress.

Like oxidation, glycation of LDL prevents LDLr-mediated cellular uptake of lipoproteins and promotes scavenger receptor-mediated uptake. *In vivo*, small, dense, which is more atherogenic than large buoyant LDL, appears to be preferentially glycated; also, *in vitro* studies suggest that it is more susceptible to glycation (Soran & Durrington, 2011). Diabetic individuals display higher plasma concentrations of glycated LDL than non-diabetic individuals.

Type 1 diabetes subjects have lipid disorders (diabetic dyslipidemia), with a pro-atherogenic lipid profile: increased concentration of TG and LDL cholesterol, low HDL levels (Verges, 2009). In addition, they display increased cholesterol-triglyceride ratio within their VLDL, increased triglyceride in their LDL and HDL, glycation of apolipoproteins, increased oxidation of LDL and an increase in small dense LDL (relatively more atherogenic). HDL from Type 1 diabetes subjects is less effective in promoting cholesterol efflux and has reduced antioxidant properties.

Subjects with Type 2 diabetes also have a proatherogenic lipid profile with quantitative and qualitative differences in their lipoproteins (Verges, 2005). Typically, they have increased triglyceride levels, increased VLDL production and decreased VLDL catabolism, and decreased HDL cholesterol levels. They have large VLDL particles that are richer in triglyceride, small dense LDL particles, increase in triglyceride content of LDL and HDL, Ox- and glycated-LDL, glycation of apolipoproteins and increased susceptibility of LDL to oxidation.

6. Concluding remarks

In conclusion, this chapter has provided a broad overview of the role of oxidative stress, ROS, and lipoprotein oxidation in the pathophysiology of disease states such as atherosclerosis, Alzheimer's disease and diabetes. While Ox-LDL and inflammation seem to be bona fide factors in the development of atherosclerosis, the role of dysfunctional HDL in these disease states is not known at this point. There are several points of uncertainty regarding the *in vivo* source of ROS and oxidative stress, the physiological behavior of oxidized lipoproteins, particularly in Alzheimer's disease and diabetes, and the line-up of antioxidants and autoantibodies in response to the oxidized factors. It is anticipated that the next decade will provide more insights into the molecular and mechanistic basis of the effect of oxidative damage on lipoprotein in disease states. This would pave the way for new therapeutic options for preventing and treating these diseases.

Author details

Vikram Jairam
Yale University School of Medicine, USA

Koji Uchida
Nagoya University, Japan

Vasanthy Narayanaswami
California State University Long Beach, USA

Acknowledgement

The authors were funded by the Tobacco Related Disease Research Program (TRDRP 17RT-0165), the Alzheimer's Association, the National Institutes of Health (NIH-HL096365), CSUPERB Faculty Development Grant and the Drake Family Trust.

7. References

Anderson, M. M., Hazen, S. L., Hsu, F. F. & Heinecke, J. W. (1997). Human neutrophils employ the myeloperoxidase-hydrogen peroxide-chloride system to convert hydroxy-amino acids into glycolaldehyde, 2-hydroxypropanal, and acrolein. A mechanism for the generation of highly reactive alpha-hydroxy and alpha,beta-unsaturated aldehydes by phagocytes at sites of inflammation. *J. Clin. Invest.*, Vol. 99, No. 3, (Feb 1997), pp. 424-432.

Arai, H., Kashiwagi, S., Nagasaka, Y., Uchida, K., Hoshii, Y. & Nakamura, K. (1999). Oxidative modification of apolipoprotein E in human very-low-density lipoprotein and its inhibition by glycosaminoglycans. *Arch. Biochem. Biophys.*, Vol. 367, No. 1, (Jul 1999), pp. 1-8.

Arai, H., Uchida, K. & Nakamura, K. (2005). Effect of ascorbate on acrolein modification of very low density lipoprotein and uptake of oxidized apolipoprotein e by hepatocytes. *Biosci., Biotechnol. Biochem.*, Vol. 69, No. 9, (Sep 2005), pp. 1760-1762.

Arlt, S., Beisiegel, U. & Kontush, A. (2002). Lipid Peroxidation in Neurodegeneration: New Insights into Alzheimer's Disease. *Curr. Opin. Lipidol.*, Vol. 13, No. 3, (June 2002), pp. 289-294, 0957-9672

Bailey, J. M., Makheja, A. M., Lee, R. & Simon, T. H. (1995). Systemic Activation of 15-lipoxygenase in Heart, Lung, and Vascular Tissues by Hypercholesterolemia: Relationship to Lipoprotein Oxidation and Atherogenesis. *Atherosclerosis*, Vol. 113, No. 2, (Mar 1995), pp. 247-258.

Baldán, Á., Bojanic, D. D., & Edwards, P. A. (2009) The ABCs of sterol transport. *J Lipid Res.* 2009 April; 50(Supplement): S80–S85

Bassett, C. N., Neely, M. D., Sidell, K. R., Markesbery, W. R., Swift, L. L. & Montine, T. J. (1999). Cerebrospinal Fluid Lipoproteins Are More Vulnerable to Oxidation in

Alzheimer's Disease and Are Neurotoxic When Oxidized Ex Vivo. *Lipids*, Vol. 34, No. 12, (Dec 1999), pp. 1273-1280.

Beauchamp, R. O., Jr., Andjelkovich, D. A., Kligerman, A. D., Morgan, K. T. & Heck, H. D. (1985). A critical review of the literature on acrolein toxicity. *Crit. Rev. Toxicol.*, Vol. 14, No. 4, (1985), pp. 309-380.

Bergt, C., Pennathur, S., Fu, X., Byun, J., O'Brien, K., McDonald, T. O., Singh, P., Anantharamaiah, G. M., Chait, A., Brunzell, J., Geary, R. L., Oram, J. F. & Heinecke, J. W. (2004). The myeloperoxidase product hypochlorous acid oxidizes HDL in the human artery wall and impairs ABCA1-dependent cholesterol transport. *Proc. Natl. Acad. Sci. (USA)*, Vol. 101, No. 35, (Aug 2004), pp. 13032-13037.

Berliner, J. A. & Watson, A. D. (2005). A Role for Oxidized Phospholipids in Atherosclerosis. *New Engl. J. Med.*, Vol. 353, No. 1, (Jul 2005), pp. 9-11.

Björkhem, I., Lütjohann, D., Diczfalusy, U., Ståhle, L., Ahlborg, G. & Wahren, J. (1998). Cholesterol homeostasis in human brain: turnover of 24S-hydroxycholesterol and evidence for a cerebral origin of most of this oxysterol in the circulation. *J. Lipid Res.*, Vol. 39, No. 8, (Aug 1998), pp. 1594-1600.

Bretillon, L., Sidén, A., Wahlund, L. O., Lütjohann, D., Minthon, L., Crisby, M., Hillert, J., Groth, C. G., Diczfalusy, U. & Björkhem, I. (2000). Plasma levels of 24S-hydroxycholesterol in patients with neurological diseases. *Neurosci. Lett.*, Vol. 293, No. 2, (Oct 2000), pp. 87-90.

Bradamante, S., Barenghi, L., Giudici, G. A. & Vergani, C. (1992). Free radicals promote modifications in plasma high-density lipoprotein: nuclear magnetic resonance analysis. *Free Rad. Biol. Med.*, Vol. 12, No. 3, (1992), pp. 193-203.

Brown, A. J. & Jessup, W. (1999). Oxysterols and atherosclerosis. *Atherosclerosis*, Vol. 142, No. 1, (Jan 1999), pp. 1-28.

Butterfield, D. A. & Boyd-Kimball, D. (2005). The critical role of methionine 35 in Alzheimer's amyloid beta-peptide (1-42)-induced oxidative stress and neurotoxicity. *Biochim. Biophys. Acta*, Vol. 1703, No. 2, (Jan 2005), pp. 149-156.

Butterfield, D. A., Bader Lange, M. L. & Sultana, R. (2010). Involvements of the lipid peroxidation product, HNE, in the pathogenesis and progression of Alzheimer's disease. *Biochim. Biophys. Acta*, Vol. 1801, No. 8, (Aug 2010), pp. 924-929.

Cyrus, T., Witztum, J. L., Rader, D. J., Tangirala, R., Fazio, S., Linton, M. F. & Funk, C. D. (1999). Disruption of the 12/15-lipoxygenase gene diminishes atherosclerosis in apo E-deficient mice. *J. Clin. Invest.*, Vol. 103, No. 11, (Jun 1999), pp. 1597-1604.

Daugherty, A., Dunn, J. L., Rateri, D. L. & Heinecke, J. W. (1994). Myeloperoxidase, a catalyst for lipoprotein oxidation, is expressed in human atherosclerotic lesions. *J. Clin. Invest.*, Vol. 94, No. 1, (Jul 1994), pp. 437-444.

Esterbauer, H., Schaur, R. & Zollner, H. (1991). Chemistry and Biochemistry of 4-hydroxynonenal, Malonaldehyde and Related Aldehydes. *Free Rad. Biol. Med.*, Vol. 11, No. 1, pp. 81-128.

Fang, X., Weintraub, N. L., Rios, C. D., Chappell, D. A., Zwacka, R. M., Engelhardt, J. F., Oberley, L. W., Yan, T., Heistad, D. D. & Spector, A.A. (1998). Overexpression of

Human Superoxide Dismutase Inhibits Oxidation of Low-Density Lipoprotein by Endothelial Cells. *Circ. Res.*, Vol. 82, No. 12, pp. 1289-1297, 1524-4571.

Furuhata, A., Ishii, T., Kumazawa, S., Yamada, T., Nakayama, T. & Uchida, K. (2003). N(epsilon)-(3-methylpyridinium)lysine, a major antigenic adduct generated in acrolein-modified protein. *J. Biol. Chem.*, Vol. 278, No. 49, (Dec 2003), pp. 48658-48665.

Fukai, T., Folz, R. J., Landmesser, U. & Harrison, D. G. (2002). Extracellular superoxide dismutase and cardiovascular disease. *Cardiovasc. Res.*, Vol. 55, No. 2, (Aug 2002), pp. 239-249.

Garner, B., Witting, P. K., Waldeck, A. R, Christison, J. K, Raftery, M. & Stocker, R. (1998). Oxidation of high density lipoproteins. I. Formation of methionine sulfoxide in apolipoproteins AI and AII is an early event that accompanies lipid peroxidation and can be enhanced by alpha-tocopherol. *J. Biol. Chem.*, Vol. 273, No. 11, (Mar 1998), pp. 6080-6087.

Haberland, M. E., Olch, C. L., Folgelman, A. M. (1984) Role of lysines in mediating interaction of modified low density lipoproteins with the scavenger receptor of human monocyte macrophages. *J. Biol. Chem.* 1984 Sep 25;259(18):11305-11311.

Hatters, D.M., Peters-Libeu, C.A., & Weisgraber, K. H. (2006). Apolipoprotein E structure: insights into function. *Trends Biochem. Sci.* 31, 445-454.

Hiltunen, T., Luoma, J., Nikkari, T. & Ylä-Herttuala, S. (1995). Induction of 15-lipoxygenase mRNA and protein in early atherosclerotic lesions. *Circulation*, Vol. 92, No. 11, (Dec 1995), pp. 3297-3303.

Hoff, H. F. & O'Neil, J. (1993). Structural and Functional Changes in LDL after Modification with Both 4-hydroxynonenal and Malondialdehyde. *J. Lipid Res.*, Vol. 34, No. 7, (Jul 1993), pp. 1209-1217.

Huang, Y., von Eckardstein, A., Wu, S., Assmann, G. (1995) Effects of the apolipoprotein E polymorphism on uptake and transfer of cell-derived cholesterol in plasma. *J. Clin. Invest.* 96, 2693-701.

Huang, Y., Weisgraber, K. H., Mucke, L. & Mahley, R. W. (2004) Apolipoprotein E: diversity of cellular origins, structural and biophysical properties, and effects in Alzheimer's disease. *J. Mol. Neurosci.* 23, 189-204.

Javitt, N. B. (2008). Oxysterols: novel biologic roles for the 21st century. *Steroids*, Vol. 73, No. 2, (Feb 2008), pp. 149-157.

Jürgens, G., Hoff, H. F., Chisolm, G. M. 3rd, & Esterbauer, H. (1987) *Chem. Phys. Lipids.* Vol. 45 (2-4, (Nov-Dec 1987), 315-336.

Kawamura, M., Heinecke, J. W. & Chait, A. (1994). Pathophysiological Concentrations of Glucose Promote Oxidative Modification of Low Density Lipoprotein by a Superoxide-dependent Pathway. *J. Clin. Invest.*, Vol. 94, No. 2, (Aug 1994), pp. 771-778.

Korotchkina, L. G., Yang, H., Tirosh, O., Packer, L. & Patel, MS. (2001). Protection by Thiols of the Mitochondrial Complexes from 4-hydroxy-2-nonenal. *Free Rad. Biol. Med.*, Vol. 30, No. 9, (May 2001), pp. 992-999.

Laukkanen, M. O., Lehtolainen, P., Turunen, P., Aittomäki, S., Oikari, P., Marklund, S. L., Ylä-Herttuala, S. (2000). Rabbit extracellular superoxide dismutase: expression and effect on LDL oxidation. *Gene*, Vol. 254, No. 1-2, (Aug 2000), pp. 173-179.

Leoni, V. & Caccia, C. (2011). Oxysterols as Biomarkers in Neurodegenerative Diseases. *Chem. Phys. Lipids*, Vol. 164, No. 6, (Sept 2011), pp. 515-524.

Levitan, I., Volkov, S. & Subbaiah, P. V. (2010). Oxidized LDL: diversity, patterns of recognition, and pathophysiology. *Antioxid. & Redox Signaling*, Vol. 13, No. 1, (Jul 2010), pp. 39-75.

Lovell, M. A., Xie, C. & Markesbery, W. R. (2001). Acrolein is increased in Alzheimer's disease brain and is toxic to primary hippocampal cultures. *Neurobiol. Aging*, Vol. 22, No. 2, (Mar-Apr 2001), pp. 187-194.

Lütjohann, D., Breuer, O., Ahlborg, G., Nennesmo, I., Sidén, A., Diczfalusy, U. & Björkhem, I. (1996). Cholesterol homeostasis in human brain: evidence for an age-dependent flux of 24S-hydroxycholesterol from the brain into the circulation. *Proc. Natl. Acad. Sci. (USA)*, Vol. 93, No. 18, (Sept 1996), pp. 9799-9804.

Maritim, A. C., Sanders, R. A., & Watkins, J. B. (2003). Diabetes, Oxidative Stress, and Antioxidants: A Review. *J. Biochem. Mol. Toxicol.*, Vol. 17, No. 1, (Feb 2003), pp. 24-38.

McCall, M. R., Tang, J. Y., Bielicki, J. K. & Forte, T. M. (1995). Inhibition of Lecithin-Cholesterol Acyltransferase and Modification of HDL Apolipoproteins by Aldehydes. *Arterioscl. Thromb. Vasc. Biol.*, Vol. 15, No. 10, (Oct 1995), pp. 1599-1606.

Meyer, J. & Schmitt, M.E. (2000). A Central Role for the Endothelial NADPH Oxidase in Atherosclerosis. *FEBS Lett.*, Vol. 472, No. 1, (Apr 2000), pp. 1-4.

Montine, T. J., Huang, D. Y., Valentine, W. M., Amarnath, V., Saunders, A., Weisgraber, K. H., Graham, D. G. & Strittmatter, W. J. (1996). Crosslinking of apolipoprotein E by products of lipid peroxidation. *J. Neuropathol. Exp. Neurol.*, Vol. 55, No. 2, (Feb 1996), pp. 202-210.

Moore, K. J. & Tabas, I. (2011) Macrophages in the pathogenesis of atherosclerosis. *Cell*, 2011 Vol. 145, No 3, (Apr 29), pp. 341-355.

Morel, D. W. & Chisolm, G. M. (1989). Antioxidant treatment of diabetic rats inhibits lipoprotein oxidation and cytotoxicity. *J. Lipid Res.*, Vol. 30, No. 12, (Dec 1989), pp. 1827-1834.

Navab, M., Berliner, J. A., Subbanagounder, G., Hama, S., Lusis, A. J., Castellani, L. W., Reddy, S., Shih, D., Shi, W., Watson, A. D., Van Lenten, B. J., Vora, D. & Fogelman, A. M. (2001). HDL and the inflammatory response induced by LDL-derived oxidized phospholipids. *Arterioscl. Thromb. Vasc. Biol.*, Vol. 21, No. 4, (Apr 2001), pp. 481-488.

Noor, R., Mittal, S. & Iqbal, J. (2002). Superoxide dismutase--applications and relevance to human diseases. *Medical Science Monitor: International Medical Journal of Experimental and Clinical Research*, Vol. 8, No. 9, (Sep 2002), pp. 210-215.

Olkkonen, V. M. & Lehto, M. (2004). Oxysterols and oxysterol binding proteins: role in lipid metabolism and atherosclerosis. *Ann. Med.*, Vol. 36, No. 8, (2004), pp. 562-572.

Palinski, W., Ylä-Herttuala, S., Rosenfeld, M. E., Butler, S. W., Socher, S. A., Parthasarathy, S., Curtiss, L. K., Witztum, J. L. (1990) Antisera and monoclonal antibodies specific for epitopes generated during oxidative modification of low density lipoprotein, *Arteriosclerosis*, Vol. 10, No 10 (3) (May-Jun 1990), 325-335.

Perry, G., Cash, A. D. & Smith, M. A. (2002). Alzheimer Disease and Oxidative Stress. *J. Biomed. Biotech.*, Vol. 2, No. 3, (2002), pp. 120-123.

Refsgaard, H. H., Tsai, L. & Stadtman, E. R. (2000). Modifications of proteins by polyunsaturated fatty acid peroxidation products. *Proc. Natl. Acad. Sci. (USA)*, Vol. 97, No. 2, (Jan 2000), pp. 611-616.

Reilly, K. B., Srinivasan, S., Hatley, M. E., Patricia, M. K., Lannigan, J., Bolick, D. T., Vandenhoff, G., Pei, H., Natarajan, R., Nadler, J. L. & Hedrick, C. C. (2004). 12/15-Lipoxygenase activity mediates inflammatory monocyte/endothelial interactions and atherosclerosis in vivo. *J. Biol. Chem.*, Vol. 279, No. 10, (Mar 2004), pp. 9440-9450.

Rifkind, J. M., Abugo, O. O., Nagababu, E., Ramasamy, S., Demehin, A., Jayakumar, R. (2002). Aging and the red cell. *Adv. Cell Aging & Geront.*, Vol. 11, (2002), pp. 283-307.

Rocha, V. Z. & Libby, P. (2009). Obesity, Inflammation, and Atherosclerosis. *Nature Reviews Cardiology*, Vol. 6, No. 6, (June 2009), pp. 399-409.

Rosenfeld, M. E., Palinski, W., Ylä-Herttuala, S., Butler, S., & Witztum, J. L. (1990) Distribution of oxidation specific lipid-protein adducts and apolipoprotein B in atherosclerotic lesions of varying severity from WHHL rabbits, *Arteriosclerosis*, Vol. 10, No. 10 (3) (May-Jun 1990), 336-349

Savenkova, M. L., Mueller, D. M. & Heinecke, J. W. (1994). Tyrosyl radical generated by myeloperoxidase is a physiological catalyst for the initiation of lipid peroxidation in low density lipoprotein. *J. Biol. Chem.*, Vol. 269, No. 32, (Aug 1994), pp. 20394-20400.

Schaur, R. (2003). Basic Aspects of the Biochemical Reactivity of 4-hydroxynonenal. *Mol. Asp. Med.*, Vol. 24, No. 4-5 (Aug-Oct 2003), pp. 149-159.

Schippling, S., Kontush, A., Arlt, S., Buhmann, C., Sturenberg, H., Mann, U., Muller-Thomsen, T. & Beisiegel, U. (2000). Increased Lipoprotein Oxidation in Alzheimer's Disease. *Free Rad. Biol. Med.*, Vol. 28, No. 3, (Feb 2000), pp. 351-360.

Shao, B., Fu, X., McDonald, T. O., green, P. S., Uchida, K., O'Brien, K. D., Oram, J. F. & Heinecke, J. W. (2005). Acrolein impairs ATP Binding Cassette Transporter A1-dependent cholesterol export from cells through site-specific modification of apolipoprotein A-I. *J. Biol. Chem.*, Vol. 280, No. 43, (Oct 2005), pp. 36386-36396.

Shao, B., Pennathur, S. & Heinecke, J. W. (2012). Myeloperoxidase Targets apoA-I, the Major High Density Lipoprotein Protein, for Site-Specific Oxidation in Human Atherosclerotic Lesions. *J. Biol. Chem.*, Vol. 287, No. 9, (Feb 2012), pp. 6375-6386.

Shafaati, M., Solomon, A., Kivipelto, M., Björkhem, I. & Leoni, V. (2007). Levels of ApoE in cerebrospinal fluid are correlated with Tau and 24S-hydroxycholesterol in patients with cognitive disorders. *Neurosci. Lett.*, Vol. 425, No. 2, (Sep 2007), pp. 78-82.

Soran, H. & Durrington, P. N. (2011). Susceptibility of LDL and its subfractions to glycation. *Curr. Opinion in Lipidology*, Vol. 22, No. 4, (Aug 2011), pp. 254-261.

Steinberg, D., Parthasarathy, S., Carew, T. E., Khoo, J. C. & Witztum, J. L. (1989). Beyond
    cholesterol. Modifications of low-density lipoprotein that increase its atherogenicity.
    New Eng. J. Med., Vol. 320, No. 14, (Apr 1989), pp. 915-924.
Sun, L., Ishida, T., Yasuda, T., Kojima, Y., Honjo, T., Yamamoto, Y., Yamamoto, H.,
    Ishibashi, S., Hirata, K. & Hayashi, Y. (2009). RAGE mediates oxidized LDL-induced
    pro-inflammatory effects and atherosclerosis in non-diabetic LDL receptor-deficient
    mice. *Cardiovasc. Res.*, Vol. 82, No. 2, (May 2009), pp. 371-381.
Tamamizu-Kato, S., Wong, J. Y., Jairam, V., Uchida, K., Raussens, V., Kato, H.,
    Ruysschaert, J. M, Narayanaswami, V. (2007). Modification by acrolein, a
    component of tobacco smoke and age-related oxidative stress, mediates functional
    impairment of human apolipoprotein E. *Biochemistry*, Vol. 46, No. 28, (Jul 2007), pp.
    8392-8400.
Tórsdóttir, G., Sveinbjörnsdóttir, S., Kristinsson, J., Snaedal, J. & Jóhannesson, T. (2006).
    Ceruloplasmin and superoxide dismutase (SOD1) in Parkinson's disease: a follow-up
    study. *J. Neurol. Sci. .*, Vol. 241, No. 1-2, (Feb 2006), pp. 53-58.
Uchida, K.,Szweda, L. I., Chae, H-Z., & Stadtman, E. R. (1993), Immunochemical detection of
    4-hydroxynonenal protein adducts in oxidized hepatocytes. *Proc. Natl. Acad. Sci. (USA)*,
    Vol. 90, No. 18, (Sept 1993), pp. 8742-8746.
Uchida, K., Kanematsu, M., Sakai, K., Matsuda, T., Hattori, N., Mizuno, Y., Suzuki, D.,
    Miyata, T., Noguchi, N., Niki, E. & Osawa, T. (1998a). Protein-bound acrolein: potential
    markers for oxidative stress. *Proc. Natl. Acad. Sci. (USA)*, Vol. 95, No. 9, (Apr 1998), pp.
    4882-4887.
Uchida, K., Kanematsu, M., Morimitsu, Y., Osawa, T., Noguchi, N. & Niki, E. (1998b).
    Acrolein is a product of lipid peroxidation reaction. Formation of free acrolein and its
    conjugate with lysine residues in oxidized low density lipoproteins. *J. Biol. Chem.*, Vol.
    273, No. 26, (Jun 1998), pp. 16058-16066.
Vaya, J. & Schipper, H. M. (2007). Oxysterols, Cholesterol Homeostasis, and Alzheimer
    Disease. *J. Neurochem.*, Vol. 102, No. 6, (Sep 2007), pp. 1727-1737.
Vergès B. (2005). New insight into the pathophysiology of lipid abnormalities in type 2
    diabetes. *Diab. Metab.*, Vol. 31, No. 5, (Nov 2005), pp. 429-439.
Vergès B. (2009). Lipid disorders in type 1 diabetes. *Diab. Metab.*, Vol. 35, No. 5, (Nov 2009a),
    pp. 353-360.
Witz, G. (1989). Biological interactions of alpha, beta-unsaturated aldehydes. *Free Rad. Biol.
    Med.*, Vol. 7, No. 3, (1989), pp. 333-349.
Zarkovic, N. (2003). 4-Hydroxynonenal as a bioactive marker of pathophysiological
    processes. *Mol. Asp. Med.*, Vol. 24, No. 4-5, pp. 281-291.
Zheng, L., Nukuna, B., Brennan, M. L., Sun, M., Goormastic, M., Settle, M., Schmitt, D., Fu,
    X., Thomson, L., Fox, P. L., Ischiropoulos, H., Smith, J. D., Kinter, M. & Hazen, S. L.
    (2004). Apolipoprotein A-I is a selective target for myeloperoxidase-catalyzed oxidation
    and functional impairment in subjects with cardiovascular disease. *J. Clin. Invest.*, Vol.
    114, No. 4 (Aug 2004), pp. 529-541.

Chapter 2

Oxidized Phospholipids:
Introduction and Biological Significance

Mohammad Z. Ashraf and Swati Srivastava

Additional information is available at the end of the chapter

1. Introduction

Phospholipids containing polyunsaturated fatty acids are highly prone to modification by reactive oxygen species. They tend to undergo lipid peroxidation to form OxPLs which induce cytotoxicity and apoptosis and plays a significant role in inflammation. There are reports that provide insights for involvement of OxPLs in interleukin transcription, phenotype switching of smooth muscle cells and apoptotic mechanisms of the modified phospholipids. Thus peroxidation greatly alters the physiochemical properties of membrane lipid bilayers and consequently induces signaling depending upon the formation or reorganization of membrane domains or specific molecular binding (Deigner et al, 2008). Distinct OxPLs species may interact with specific binding sites and receptors leading to the activation of individual signaling pathways. The most prevalent human coronary atherosclerosis is a chronic inflammatory disease that occurs due to lipid abnormalities. Pro-inflammatory oxidized low-density lipoprotein (OxLDL) has been suggested to be a link between lipid accumulation and inflammation in vessel walls. Increased levels of phospholipids' oxidation products have been detected in different organs and pathological states, including atherosclerotic vessels (Watson et al 1997, Subbanagounder et al 2000), inflamed lung (Yoshimi et al 2005, Nakamura et al 1998), non-alcoholic liver disease (Ikura et al 2006), plasma of patients with coronary artery disease (Tsimikas et al 2005), as well as in apoptotic cells (Huber et al 2002, Chang et al 2004), virus-infected cells (Van Lenten et al 2004) and cells stimulated with inflammatory agonists (Subbanagounder et al 2002). Moreover, studies have been done on two HDL-associated enzymes, serum paraoxonase (PON1) and PAF-acetylhydrolase (PAF-AH), which are responsible for hydrolysis of plasma oxidized phospholipids (Forte et al 2002) thereby providing evidence for their role in atherosclerosis. Another important marker of oxidative stress is the association of OxPLs with the apolipoprotein B-100 particle (OxPLs/apoB) of

LDL. Increased levels of OxPLs/apoB are implicated in coronary artery disease, progression of carotid and femoral atherosclerosis and the prediction of cardiovascular events (Tsimikas et al 2005).

2. Formation of OxPLs

OxPLs are generated by the oxidation of polyunsaturated fatty acid residues, which are usually present in the phospholipids at the sn-2 position. Oxidation of phospholipids is initiated either enzymatically by lipoxygenases or by reactive oxygen species and propagates *via* the classical mechanism of lipid peroxidation chain reaction. This implies that the production of OxPLs cannot be regulated by adjusting the amount or activity of enzymes. Hence there is a probability of the uncontrolled generation of OxPLs during oxidative stress. Several evidences suggest that OxPLs are formed from Poly Unsaturated Fatty Acids (PUFAs) at the sn-2 position (Bochkov et al 2007, Podrez et al 2002). Bioactive oxidized phospholipids may contain fragmentation products of PUFA, such as 1-palmitoyl-2-oxovaleroyl-sn-glycero-3-phosphorylcholine and 9-keto-10-dodecendioic acid ester of 2-lyso-phosphatidyl choline (KOdiA-PC); prostaglandins, such as 15 deoxy-delta 12, 14 prostaglandin I2 (PGI2) and 1-palmitoyl-2-(5,6-epoxyisoprostane E2)-sn-glycero-3-phosphoryl choline (PEIPC); and levuglandins. These molecules exhibit different biological activities. Chromatographic separation of many products formed by oxidation of 1-palmitoyl-2-arachidonoyl-sn-glycero-3-phosphorylcholine (PAPC) led to the identification of 1-palmitoyl-2-(5-oxovaleroyl)-sn-glycero-3-phosphatidylcholine (POVPC), 1-palmitoyl-2-glutaroyl-sn-glycero-3-phosphatidylcholine (PGPC) and 1-palmitoyl-2-(5,6-epoxyisopropane E2)-sn-glycero-3-phosphatidylcholine (PEIPC) as potent lipid mediators of inflammation. High structural variation may explain why OxPLs demonstrate a remarkable variety of biological activities (FIGURE-1).

Enzymatic and non-enzymatic reactions, free-radical, and radical-free processes are capable of initiating wide spectrum of reactions causing oxidation of PUFAs. Majority of these reactions produce identical primary oxidation products (i.e., peroxyl radicals and hydroperoxides). Subsequent oxidation of OxPLs is an enzyme-independent stochastic process producing a wide spectrum of OxPLs. Peroxidation products thus generated proceeds according to several mechanisms such as oxidation of PUFA residue, cyclization of peroxyl radical or oxidative fragmentation of esterified PUFAs generating either full-length residues incorporating several oxygen atoms, or shortened fatty acid residues. Introduction of additional oxygen atoms into PUFAs is a common mechanism that increases complexity of OxPLs mixtures however biological activities of poly-oxygenated PLs are still not characterized. On the other hand, cyclization of peroxyl radical produces cyclic peroxide, which undergoes re-arrangements yielding bicyclic endoperoxide, or oxidation introducing additional non-cyclic or cyclic peroxide group. Cyclization of peroxyl radical is only possible for FAs having three or more double bonds (Salomon et al 2005).

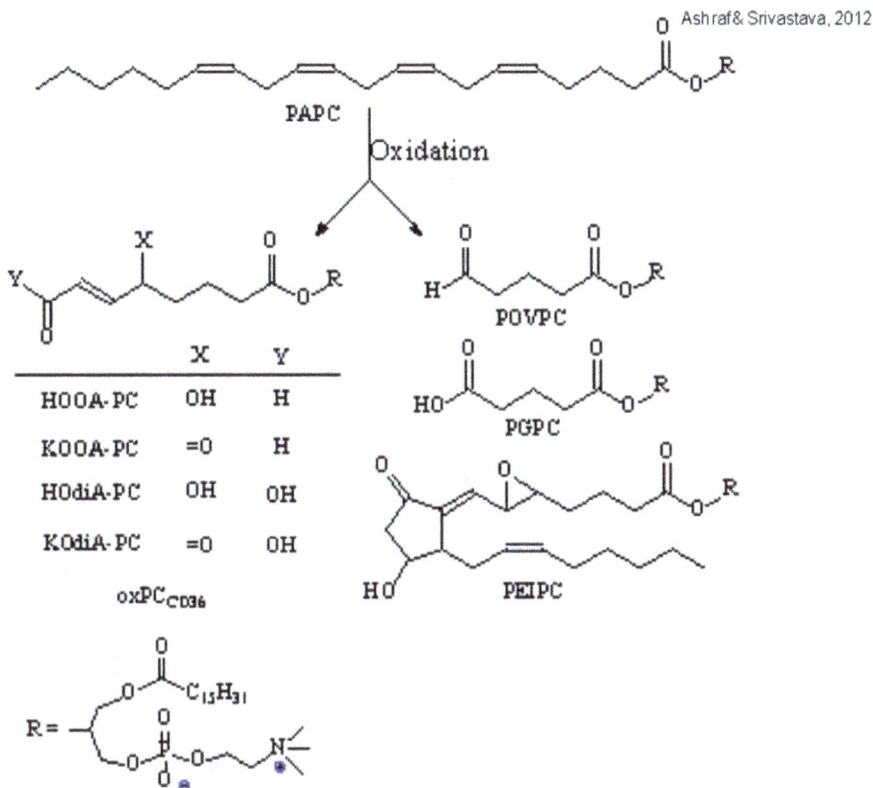

Figure 1. Representative chemical structures of oxidized phospholipids formed during oxidation of PAPC.

2.1. Oxidative cleavage and generation of fragmented OxPLs species

Peroxides/ peroxyls are transformed into advanced oxidation products by fragmentation of hydroperoxides. γ-Hydroxy (or oxo) a,b-unsaturated PLs with terminal aldehyde groups are produced from hydroperoxides via oxidation/fragmentation or polymerization/cleavage. Oxidative fragmentation of hydroperoxides occurs via several mechanisms including b-scission, Hock rearrangement, or cyclization of alkoxy radical produced from hydroperoxide (Gugiu et al 2006). γ-Hydroxy (or oxo)-α,β-unsaturated aldehyde PLs are highly reactive compounds, that are able to covalently link to amino groups of proteins, as well as thiol groups of biomolecules (Hoff et al 2003). On the other hand, peroxyl radical can cross-react with double bonds present in hydroperoxides yielding peroxydimers, these are unstable products and spontaneously break down forming either new radicals or α,β-unsaturated aldehydes (Schneider et al 2008). In addition to these products, saturated fragmented species containing terminal carbonyl groups are produced by oxidative fragmentation of PUFA-PLs, most common amongst which are oxononanoate and azelaoate formed from linoleic acid, oxovaleroate, and glutaroate generated from arachidonic acid, or oxobutyrate and succinate produced from docosahexaenoic acid (Gu et al 2003, Podrez et al 2002). Saturated fragmented OxPLs can be formed by further oxidation of γ-hydroxy (or oxo)-α, β-unsaturated PLs in addition to direct formation from hydroperoxides, (Podrez et al 2002). Saturated fragmented OxPLs l lack double bonds and hence they are resistant to further oxidation as the absence of double bonds within fragmented chains results in reduced reactivity of aldehyde containing saturated OxPLs as compared to α,β-unsaturated fragmented OxPLs.

2.2. Non-enzymatic oxidation of PL-PUFAs

This process is initiated by free radicals or non-radical reactive oxygen species (ROS). Free radical-mediated chain reaction is initiated by the formation of carbon-centered radicals and/or hydroperoxides of PUFAs (peroxidation of PUFAs). Due to the presence of methylene groups located between double bonds (bisallylic methylene groups), PUFAs are more susceptible to oxidation as compared to saturated FAs. As a result they are characterized by weakened hydrogen-carbon bonds. Free radicals can abstract hydrogen from bisallylic methylene leading to the formation of carbon-centered radicals within PUFAs. Now occurs the initiation step of lipid peroxidation, Carbon-centered radicals rapidly react with molecular oxygen, producing peroxyl radicals. These Peroxyl radicals react with bisallylic methylene groups in other PUFA molecules, leading to the transformation of peroxyl radicals to hydroperoxides and generation of new carbon-centered radicals. Thus, additional cycles of peroxidation are initiated. PUFA hydroperoxides in turn produce reactive alkoxyl and hydroxyl radicals via iron or copper-catalyzed Fenton-like reactions, further propagating the chain reaction (Bochkov et al 2010).

2.3. Enzymatic oxidation of PL-PUFAs

1, 4-pentadiene motifs are recognized within unsaturated fatty acids by lipoxygenases (LOXs) and molecular oxygen with high stereoselectivity is introduced. The majority of lipoxygenases oxidize only unesterified PUFAs. Only one group (12/15-LOX) amongst all known LOXs is capable of oxidizing PL-esterified fatty acids. This class of enzymes is present in different biological species and includes mouse, rat, rabbit, bovine, and porcine leukocyte-type 12- LOX, rabbit and human reticulocyte-type 15-LOX, and soybean LOX (Huang et al 2008, Wittwer et al 2007). Switching of activity of electron transport in mitochondria to peroxidation by cytochrome c (cyt c) has been suggested by Kagan et al (2005). This transformation begins when cyt c binds to negatively charged cardiolipin (CL), leading to conformational changes and subsequent release of PL-protein complex from mitochondria into cytosol. The complex of cyt c with CL activated by traces of PUFA-OOH or H_2O_2 acquires the ability to oxidize CL, PS, or PI, with formation of PL-OOH (Kagan et al 2009).

Alternatively, OxPLs are also generated by re-esterification of free oxidized PUFAs into lyso-PLs. Several types of OxPLs have been found to be generated by this mechanism both *in vivo* and *in vitro* (Arai et al 1997, Birkle et al 1984).

2.4. Detoxification of reactive OxPLs

Detoxification of OxPLs comprises the mechanisms that terminate peroxidation chain reaction and inactivate chemically reactive toxic groups produced by oxidation. Hydroxides are characterized by significantly lower chemical reactivity and therefore are considered to be stable and non-toxic compared to hydroperoxides (Spiteller et al 1997). Most commonly, the enzyme catalyzing the reduction of hydroperoxides to hydroxides is glutathione peroxidase (GPx). Lipid hydroperoxides are reduced in a reaction that involve selenocysteine residue of GPx and glutathione thus generating lipid hydroxide and oxidized glutathione. With respect to membrane-bound hydroperoxides of PL esterified PUFAs, PL glutathione peroxidase (GPx4) has the highest activity amongst GPx enzymes (Savaskan et al 2007).

A variety of products containing aldehyde and keto functional groups are formed upon oxidation of OxPLs which are further reduced by aldo-keto reductases to respective hydroxyl groups. Apart from playing physiological role in metabolism of sugar aldehydes, aldo-keto reductases also play a role in detoxification of toxic phospholipid aldehydes (Jin et al 2007).

Another aspect of detoxification is OxPLs cleavage. Platelet activating factor acetylhydrolase (PAF-AH) has been recognized for its ability to cleave and thus inactivate PAF (McIntyre et al 2009).The enzyme was shown to hydrolyze fragmented saturated OxPLs (Stremler et al 1991), as well as long-chain OxPLs, including esterified F2-isoprostanes, PC-hydroperoxides and PEIPC (Kriska et al 2007, Davis et al 2008).

3. Mechanism of action

Specific receptor binding of OxPLs is the subject of an ongoing debate. Available evidence suggests that OxPLs interact with various signal transduction receptors and pattern recognition receptors present on the cell surface. Most commonly known receptors include CD36, SRB1, EP2, VEGFR2 and the PAF receptor (Bochkov et al 2007, Zimman et al 2007). It has been demonstrated that when present in vesicles, truncated oxidized fatty acids at the sn-2 position move from the hydrophobic interior to the aqueous exterior of the vesicle. this would allow their recognition by cell surface receptors Earlier models of isoprostane-containing phospholipids have suggested that they are highly twisted and may distort membrane areas in which they are present (Morrow et al 1992). Moumtzi et al (2007) have shown that phospholipid oxidation products can integrate into lipid membranes of cells and lipoproteins; they can either act as ligands or may cause local membrane disruption. Besides, peroxidation of phospholipids leads to the accumulation of lysoforms as a result of both non-enzymatic decylation and enzymatic hydrolysis reactions catalyzed to a large extent by lipoprotein-as-associated phospholipase A2 (also known as PAF acetylhydrolase), which has high substrate selectivity toward polar phospholipids, including the oxidized forms (Zalewski et al 2005). Some lysophospholipids bind and activate G protein-coupled receptors (GPCR). Parhami et al (1993 & 1995) explained that oxidized phospholipids act by binding to a G protein-coupled receptor. These authors demonstrated that minimally modified LDL stimulated a putative Gs-coupled receptor, thus increasing cyclic AMP (cAMP) levels in endothelial cells. Lysophosphatidylcholine and lysophosphatidic acid triggered the activity of G2A and LPA1-LPA4 receptors respectively (Tomura et al 2005, Anliker et al 2004). In addition to GPCR, OxPLs also activate other classes of receptors such as peroxisome proliferator-activated receptors (PPAR). Thus, phospholipid peroxidation may induce the generation of lysophospholipids that are known to accumulate in LDL (OxLDL) and atherosclerotic lesions (Siess et al 2004, Tselepis et al 2002).

Prostaglandin receptors have been recently implicated into OxPLs-induced inflammation. OXPAPC and its component lipid PEIPC are able to stimulate prostaglandin E_2 and D_2 receptors (EP2 and DP respectively) and to compete with receptor binding of radio labeled prostaglandin E_2 (Li et al 2006). Previously, it was observed that POVPC binds to human macrophages via the PAF receptor (PAF-R). Occupancy of the PAF-R by the OxPLs modifies the transcription levels of pro-inflammatory genes such as IL-8 (Pegorier et al 2006).

Some effects of OxPLs are probably not mediated by signal transducing receptors. Modulation of cellular cholesterol depots has been suggested as a non-receptor mediated mechanism of OxPLs sensing by cells. It is well illustrated that OxPAPC induces depletion and re-distribution of cellular cholesterol reserves finally leading to the activation of a transcription factor SREBP, a well recognized sensor for cellular cholesterol contents. In turn, SREBP activates IL-8 production (Yeh et al 2004).The human aortic EC gene expression was found to be stimulated by PAPC. Furthermore, OxPAPC may bind to a 37KDa glycosylphosphatidylinositol anchored protein, which interacts with TLR4 to induce interleukin-8 (IL-8) transcription (Walton et al 2003). Leitinger et al (2003) and Watson et al (1997) have described a possible role of toll-like receptors (TLRs) in OxPLs-induced inflammation. Studies have confirmed that Asp299Gly-TLR4 polymorphism plays a protective role in attenuation of atherosclerosis.

Mitogen activated protein kinase phosphatase-1 (MKP-1) was reported to be involved in OxPAPC-induced MCP-1 production. Also activation of eNOS by OxPAPC is regulated via a phosphatidylinositol-3-kinase/Akt-mediated mechanism, OxPAPC-induced SREBP activation is significantly reduced with eNOS inhibition (Berliner and Gharavi, 2008).

Chen et al (2007) reported that LDL-associated phosphatidylcholine esterified with sn-2-azelaic acid at the sn-2 position is readily taken up by cells. This compound, one of the main phospholipid oxidation products in LDL, induces apoptosis of HL60 cells at low micromolar concentrations. Since the intact phospholipid is required for signaling, this effect can be prevented by over-expression of PAF acetyl hydrolase known for oxidizing phospholipids with polar residues at the sn-2 position.

Another biologically active phospholipid described is platelet activating factor (PAF) having various inflammatory actions such as platelet aggregation, hypotension, anaphylactic shock and increased vascular permeability (Prescott et al 2000). PAF is structurally identified as 1-0-alkyl-2-acetyl-sn-glycero-3-phosphocholine. Atherogenic effects are also induced by PAF by activating monocytes and stimulating smooth muscle cell growth. In contrast to the tightly regulated physiological generation of PAF, uncontrolled processes of free radical oxidation generate analogs of PAF *in vivo* and *in vitro*. As a result of this uncontrolled chemical reaction, fragmentation of the residue at sn-2 position occurs and these oxidatively generated PAF mimetics stimulate monocytes, leukocytes and platelets. They are found in atherosclerotic lesions and even in blood from individuals exposed to cigarette smoke (Heery et al 1995).

Other oxidized phospholipids such as POVPC and PGPC have also been shown to play major roles in activation of endothelial cells and induction of leukocyte binding. They are identified as abundant products in oxidized LDL. The effect of POVPC is protein kinase-A dependent leading to the stimulation of the cAMP-mediated pathway (Berliner and Gharavi, 2008).

OxPLs also induces autocrine mediators such as vascular endothelial growth factor (VEGF), which works through activation of transcription factor-4 (ATF4) (Oskolkova et al 2008).

4. OxPLs receptors

It has been shown that OxPLs stimulate a number of signal-transducing receptors located on the cell surface or in the nucleus, including G protein-coupled receptors, receptor tyrosine kinases, Toll-like receptors, receptors coupled to endocytosis, and nuclear ligand-activated transcription factors such as PPARs.

4.1. Prostaglandin receptors

OxPCs containing esterified PEIPC activate receptors recognizing prostaglandins E2 and D respectively (Li et al 2006). Activation of EP2 receptor on ECs results in activation of integrins and increased binding of monocytes.

4.2. Scavenger receptors

OxPLs comprise a major group of ligands for scavenger receptors. Different classes of Scavenger receptors range from Class A, B, D, E and F depending upon the nature and type of ligand (FIGURE-2). CD36 have been described as the major receptor expressed on macrophages and involved in the process of atherogenesis and apoptosis. The role of CD36 has been shown to be responsible for recognition of free oxidized phospholipids (Boullier et al 2000, Podrez et al 2000). Also Boullier et al (2000) and Watson et al (1997) have pointed out that oxidized phospholipid is covalently linked to apolipoprotein B-100 in extensively oxidized LDL (e.g. Cu^{2+}-oxLDL) and serve as ligand for CD36. Scavenger receptor- ligand interaction initiates signaling cascades that regulate macrophage activation, lipid metabolism and inflammatory pathways which may influence the development and stability of atherosclerotic plaque. Recent studies have demonstrated the expression of scavenger receptors especially CD36 and SR-BI on platelets suggesting their critical role in platelet hyper-reactivity in dyslipidemia and atheroprogression.

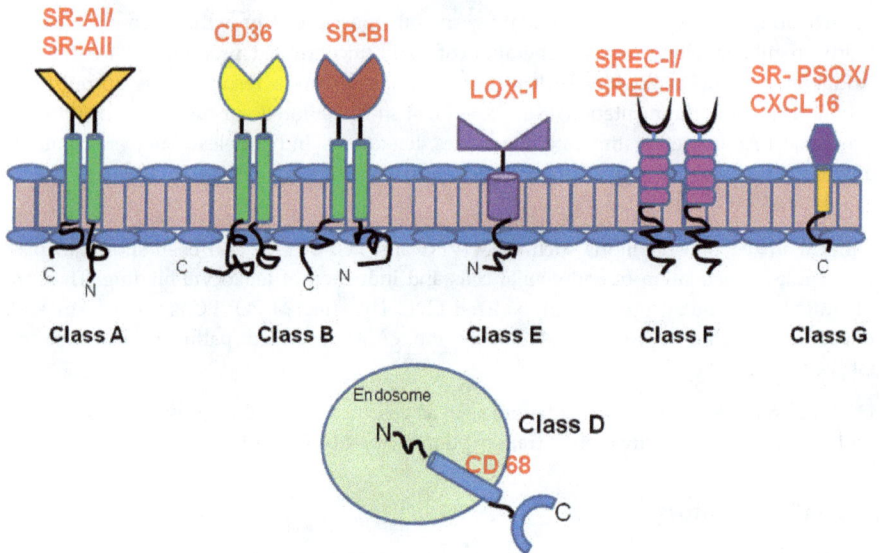

Ashraf & Srivastava, 2012

Figure 2. Schematic representations of different class of scavenger receptors involved in OxPLs binding.

4.3. PAF receptors

OxPLs initiate activation of receptor specific for PAF, which act as an important lipid mediator of inflammation and platelet aggregation. It recognizes alkyl-acyl-phosphatidylcholines specifically and contains an ether bond at the sn-1 position in combination with unusually short sn-2 acetyl residue. Oxidative fragmentation of sn-2 PUFAs in alkyl-PCs generates products such as 1-alkyl-2-butenoyl and 1-alkyl-2-butanoyl that are recognized by PAF receptor (Androulakis et al 2005, Marathe et al 1999). However, the role of the PAF receptor in the overall biological activity of OxPCs is not characterized.

4.4. VEGF receptors

It has been demonstrated that phosphorylation (activation) of VEGFR2 is enhanced within the first minutes of incubation with OxPAPC (Zimman et al 2007). They hypothesized that trans-activation of VEGFR2 in OxPAPC-treated cells was mediated by c-SRC.

4.5. Sphingosine-1-phosphate (S1P) receptor 1

It has shown that OxPAPC stimulates the recruitment of S1P1 to caveolin-enriched membrane microdomains, and induces its phosphorylation (activation) by AKT. Transactivation of S1P1 by OxPAPC plays a role in barrier-protective function of OxPLs.

4.6. Toll-like receptor 4

TLR4 plays a role in OxPAPC-mediated induction of IL-8 in HeLa cells. OxPAPC also induces lung injury and IL-6 production by mouse lung macrophages via the TLR4–TRIF–TRAF6 pathway (Imai et al 2008). On the other hand various classes of OxPLs do not influence the basal levels of E-selectin, ICAM-1, VCAM-1, TNFa, IL-6, IL-1a, IL-1b, and COX-2 in whole blood or individual cell types, including human umbilical vein ECs, blood monocytes, macrophage cell line, or fibroblasts (Bochkov et al 2002, Erridge et al 2008).

4.7. PPARα and PPARγ

Peroxisome proliferator-activated receptors (PPARs) are intracellular ligand-activated transcription factors. Diacyl-OxPLs stimulated a PPAR response element-driven reporter construct in transfected HAECs and the effect of OxPAPC, POVPC, and PGPC was mediated by PPARα as indicated by the activation of the ligand binding domain of PPARα, but not PPARγ or PPARδ (Lee et al 2000).

Second messengers up-regulated by OxPLs: Apart from the above described receptors , minimally modified Low Density Lipoproteins (MM-LDL) also induces elevation of Ca^{2+} in ECs (Honda et al 1999) and also OxPAPC was shown to induce rapid and reversible Ca₂þ-responses in ECs (Bochkov et al 2002). MM-LDL causes a saturable dose-dependent increase in cAMP levels in aortic ECs that may arise due to activation of Gs and inhibition of Gi heterotrimeric G-protein complexes (Parhami et al 1995).

5. Biological function

Many cellular events are initiated and modulated by biologically active oxidized phospholipids. OxPLs were initially characterized as an active principle of minimally modified LDL (MM-LDL), responsible for its ability to stimulate EC to bind the leukocytes (Watson et al 1995). MM-LDL and OxPLs has the characteristic feature of inflammatory agonist i.e., their ability to activate binding of monocytes but not neutrophils (Watson et al 1997). In contrast to lipopolysaccride (LPS), tumor necrosis factor α (TNFα), or interleukin 1 (IL-1), MM-LDL does not up-regulate the expression of ICAM-1, VCAM-1 and E-Selectin on EC (Kim et al 1994), but promotes surface deposition of CS-1-containing variant of fibronectin (CS-1 FN) serving as ligand for the α4β1 (VLA-4) integrin expressed on the surface of monocytes (Shih et al 1999). Similar to MM-LDL, OxPLs selectively stimulate adhesion of monocytes by CS-1 FN-dependent mechanism. Likewise other inflammatory agonists, OxPLs also stimulate the production of cyto- and chemokines. OxPLs are known to up-regulate expression IL-6, IL-8, MCP-1, GROα, MIP-1α, MIP-1β and CXCL3 (Subbanagounder et al 2002, Furnkranz et al 2005, Lee et al 2000, Reddy et al 2002, Kadl et al 2002, Gargalovic et al 2006, Huo et al 2001).

Expression of a number of genes related to angiogenesis, atherosclerosis, inflammation and wound healing are modulated by oxidized phospholipids in human aortic endothelial cells (Berliner and Gharavi, 2008; Gargalovic et al., 2006). Bochkov and colleagues (2002, 2007) have made known that OxPLs counteract the lipopolysaccride (LPS) pathway. Considering anti-inflammatory role of OxPLs, they reported that oxidized 1-palmitoyl-2arachidonoyl-*sn*-glycero-3-phosphocholine (OxPAPC) interfered with the ability of LPS to bind to the LPS-binding protein (LBP) and to CD-14, thus suppressing LPS-induced nuclear factor-κB (NF-κB)-mediated up-regulation of inflammatory genes.

Knapp and coworkers (2007) found that OxPAPC inhibits the interaction of LPS with LPS-binding protein and CD14. This also reduces phagocytotic activity of neutrophils and macrophages by a CD-14-independent mechanism. However, in these experiments, administration of OxPAPC rendered mice highly susceptible to *Escherichia coli* peritonitis, which may cause mortality during gram-negative sepsis *in vivo*. Thus the overall harmful profile of phospholipid oxidation products includes the impairment of host response to bacterial infections.

Recently, Gharavi and colleagues (2007) have reported the activation of JAK2/STAT3 pathway by phospholipids and implicated their role in atherogenesis. 1-Palmitoyl-2-epoxyisoprostane-*sn*-glycero-3-phosphocholine, an oxidation product of -1palmitoyl-2-arachidonoyl-*sn*-glycero-3-phosphocholine, induces c-Src kinase-dependent activation of JAK2 in endothelial cells and synthesis of chemotactic factors, such as interleukin (IL)-8. In turn, STAT3 activation and regulation of IL-8 transcription is dependent on JAK2 leading to the enhanced levels of STAT3 activity in inflammatory regions of human atherosclerotic lesions. Since STAT3 activation is involved in other chronic inflammatory diseases such as rheumatoid arthritis, psoriasis etc, it has been suggested that STAT3 activation by oxidized phospholipids could be an important interventional target for atherosclerosis and other diseases with inflammatory components.

5.1. Regulation of vascular cell function

OxPLs have multiple effects on endothelial cells. After 4h treatment with 50µg/ml of OxPAPC ~1000 genes are regulated amongst which ~600 are up-regulated and ~400 are down-regulated (Gargalovic et al 2006). Also, a major difference in responsiveness to specific effects of Ox-PAPC of endothelial cells from different human donors has been documented (Gargalovic et al 2006). The atherogenic pathways which were found to be upregulated include inflammation, cholesterol synthesis, coagulation and decrease in cell division. Some important effects of OxPAPC on endothelial cell function independent of gene regulation have been reported. OxPAPC has been shown to increase monocytes but not neutrophils binding by activating β-1 integrin (Berlin et al 2008, Leitinger et al 2005).

Many effects of OxPLs are mediated by its interaction with CD36. Several studies have indicated that LDL supplemented with OxPAPC or vesicles supplemented with fragmented α/β unsaturated fatty acids at the sn-2 position, such as KOdiA or HODA PC, bind to CD36 (Podrez et al 2002, Greenberg et al 2006). Another important phagocytic function of macrophages is the uptake of apoptotic cells, which are abundant in atherosclerotic plaques. OxPLs including oxidized phosphatidyl serine and phosphatidyl choline derivatives were shown to serve as ligands for macrophage uptake of apoptotic cells (Chou et al 2008, Greenberg et al 2006).

OxPLs also interact and bind with other recognition receptors in macrophages such as TLRs, CD14, LPS binding protein and C-reactive protein competing with negative ligands (Bochkov et al 2007, Bochkov et al 2002, Erridge et al 2008, Miller et al 2003). Thus, the formation of OxPLs during inflammation may represent an important feedback mechanism to limit further tissue damage. OxPLs have also been shown to activate macrophages. Currently conducted studies have revealed the role of OxPAPC in inducing lung injury and cytokine production by lung macrophages (Imai et al 2008).

The role of OxPLs in adaptive immune response can't be overlooked where they modulate the maturation process of dendritic cells (DCs). OxPLs also regulate innate immunity in human leprosy (Cruz et al 2008). In addition to the effects on DCs, OxPLs have also been shown to affect and induce T-cells (Seyerl et al 2008).

Phenotypic switching of smooth muscle cells (SMCs) involving increased proliferation; enhanced migration and down-regulation of SMC differentiation marker genes play a critical role in atherogenesis. Many studies have shown that OxPLs stimulate differentiation and cell division of SMCs (Heery et al 1995, Pidkovka et al 2007) while others have shown activation of apoptotic signaling pathways (Fruhwirth et al 2008).

5.2. Gene expression

OxPLs have profound effect on gene expression. OxPAPC have been shown to modulate the expression of approximately 1000 genes in human aortic ECs which include both up-regulated and down-regulated mRNAs (Gargalovic et al 2006). OxPLs regulate genes related to inflammation, lipid metabolism, cellular stress, proliferation, and differentiation. These include VEGF-A and IL-8, which are induced by OxPLs independent of their transcription factors.

5.3. Pathophysiological functions

Pathophysiologically OxPLs are involved in various proinflammatory and cardiovascular disorder; details are being described below (FIGURE-3).

5.4. Atherosclerosis

Quantification of OxPLs using liquid chromatography coupled with mass spectrometry has indicated that atherosclerotic vessels contain high concentrations of OxPCs. Different species of OxPCs were detected in atherosclerotic vessels including PL-hydroperoxides and hydroxides (Waddington et al 2001). In addition to elevated levels of OxPLs, atherosclerotic vessels express high amounts of proteins known to be induced by OxPLs in vitro. The latter includes MKP-1 (Reddy et al 2004), ATF3, ATF4 (Gargalovic et al 2006), SREBP-1 (Yeh et al 2004), HO-1 and IL-8 (Cheng et al 2009), MCP-1 and COX-2 (Ma et al 2008). OxPLs act on all major cell types involved in atherogenesis including monocytes, endothelial and vascular smooth muscle cells, lymphocytes, and platelets.

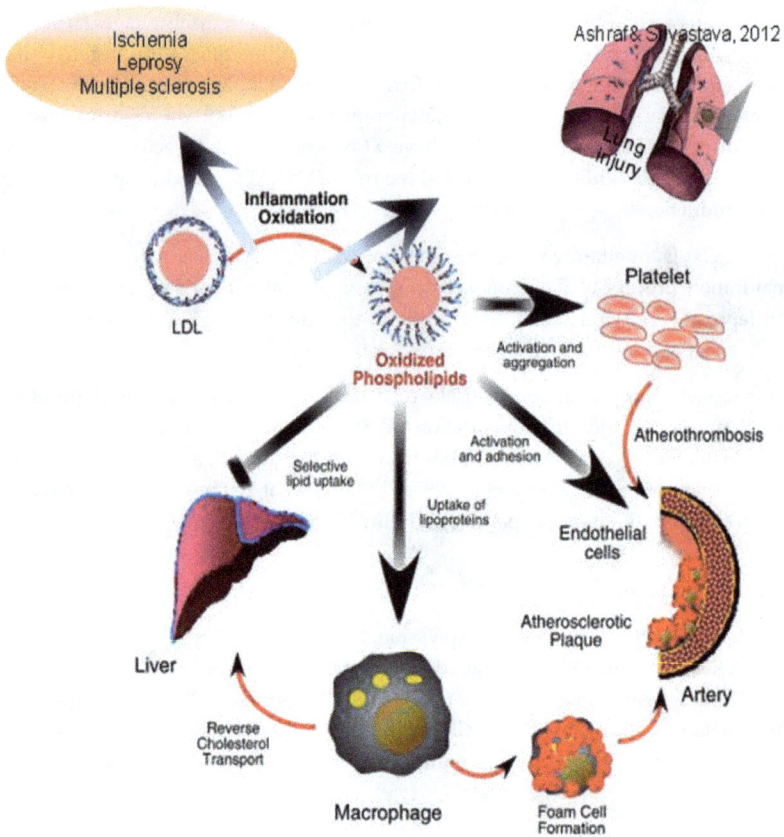

Figure 3. Oxidized phospholipids present in oxidized LDL induce various Diseases.

5.5. Lung injury

The epithelial lining pulmonary surfactant is permanently exposed to high concentrations of oxygen and other oxidants present in the air. Ozone gas also plays a role in generating oxidatively truncated PLs (Uhlson et al 2002). Under normal healthy conditions surfactant is protected from oxidation by maintaining low contents of PUFAs, antioxidant action of glutathione present in the lining fluid and surfactant proteins A and D (Kuzmenko et al 2004). However, the accumulation of biologically active OxPLs products occurs in pathological states due to the oxidation of surfactant PCs, membrane lipids and apoptosis of bronchial cells. Studies conducted with animal models have shown that OxPLs protect lungs from acute lung injury. Ma et al. (2004) showed that OxPAPC inhibits elevation of TNFa in mice upon intratracheal or systemic administration of LPS or CpG DNA. Hence the available data shows that OxPLs may induce either beneficial or detrimental effects on lungs. The action of OxPLs on the lungs may depend upon their concentrations, lower levels of OxPLs protect endothelial barrier whereas high concentrations of the same OxPLs induce disruptive effects (Birukov et al 2004, DeMaio et al 2006).

5.6. Ischemia

Ischemia/reperfusion results in elevated levels of OxPLs both in tissues and systemic levels. PAF like (alkyl-acyl) OxPLs were detected within the first minutes after reperfusion of kidneys after warm ischemia (Lloberas et al 2002). Plasma concentrations of fragmented OxPCs were increased in patients during the reperfusion period after coronary surgery with cardiopulmonary bypass (Frey et al 2000). Hence available data shows that ischemia/reperfusion is a pathological state characterized by elevated local and circulating levels of OxPLs.

5.7. Inflammation

Inflammation is characterized by a massive production of ROS. The elevation of circulating levels of OxLDL in response to inflammatory stimuli has already been shown. The OxPLs production in response to inflammation is induced by different cell types including leukocytes. Phorbol ester-stimulated neutrophils and monocytes incubated with PUFA-PCs produced mono- and bishydroperoxides of PC, as well as isoP–PC, thus suggesting that activated phagocytes can oxidize lipids in the surrounding medium (Jerlich et al 2003).

5.8. Radiation stress

Formation of OxPLs can be activated by visual and UV-light. OxPLs accumulating in retinas serve as ligands for CD36-dependent phagocytosis of shed photoreceptor outer segments by retinal pigment epithelium; this process is necessary for normal function of the retina (Sun et al 2006). Generation of OxPLs by light exposure has also been shown in skin cells. UVA-1-irradiated PAPC containing several OxPLs species induced expression of antioxidant and anti-inflammatory enzyme heme oxygenase-1 in dermal fibroblasts, keratinocytes, and in a three-dimensional epidermal equivalent model (Gruber et al 2007). Therefore, OxPLs are likely to play a protective role in UVA irradiated skin by inducing HO-1.

5.9. Leprosy

Oxidized PCs have been detected in lepromatous (disseminated) leprosy lesions, but not in tuberculous leprosy characterized by stronger host immune response and self-contained infection (Cruz et al 2008). Lepromatous leprosy lesions are characterized by the accumulation of OxPLs, which can counteract innate and specific immune responses, thereby promoting survival.

5.10. Multiple sclerosis

Multiple sclerosis (MS) is an autoimmune disease of the brain that causes neurodegeneration. Role of OxPLs in MS is supported by Qin et al. (2007), demonstrating the presence of OxPLs (alone and conjugated to a 15 KDa protein) in extracts of MS lesions directly by Western blot analyses using the E06 antibody. OxPLs might be promoting the inflammatory process in MS lesions.

6. Medical relevance

Increasing number of studies suggest the role of oxidized phospholipids in development of atherosclerosis by interacting with specific receptors as well as through their reactive groups that can bind covalently to proteins, forming lipid-protein adducts that become dysfunctional. It is a challenge to determine if therapeutic inhibition of the OxPLs interaction with vessel wall cells can inhibit atherosclerosis. Also it will be interesting to identify the lipid oxidation products that activate each response in the various cell types and the receptors or binding molecules and signal transduction pathways activated by these lipids.

Pro-inflammatory oxidized phospholipids are significant predictors of the presence of carotid and femoral atherosclerosis, development of new lesions and increased risk of cardiovascular events (Ashraf et al 2009). Hence oxidized phospholipids could serve as biomarker for diagnosis of coronary artery disease and they could also be used as potential targets for therapeutic intervention.

7. Conclusions

The inflammatory profile of OxPLs combines both pro- and anti-inflammatory effects. OxPLs may show detrimental as well as beneficial cellular effects. OxPLs exert pro-inflammatory effects on different cell types such as endothelium where they induce a shift from antithrombotic and anti-inflammatory state to procoagulant and inflammatory phenotype of EC. Although OxPLs stimulate a number of classical inflammation mechanisms, they are not capable of activating many signaling and adhesion events characteristic of acute inflammation, such as activation of the NFκB pathway, expression of ICAM-1 and E-selectin or adhesion of granulocytes. Several studies have provided evidence that OxPLs play an important role in atherosclerosis. In addition, OxPLs also up-regulate monocytes-specific chemokines and stimulate EC to bind monocytes, thus initiating monocytic inflammation. Thus it can be concluded that OxPLs can stimulate and inhibit inflammation depending upon the biological situation. Advancement in this field can be expected from studies that are based on well defined synthetic and labeled OxPLs species and the modern techniques of system biology. Also advances in the knowledge of signaling pathways and the interaction partners of oxidized phospholipid will increase our understanding of inflammatory processes and molecular mechanisms of various diseases such as atherosclerosis. These studies may also help in playing important role in future therapeutic diagnostics.

Abbreviations

Oxidized phospholipids (OxPL)
Oxidized low-density lipoprotein (OxLDL)
Serum paraoxonase (PON1)
PAF-acetylhydrolase (PAF-AH)

9-keto-10-dodecendioic acid ester of 2-lyso-phosphatidyl choline (KOdiA-PC)
15 deoxy-delta 12, 14 prostaglandin I2 (PGI2)
1-palmitoyl-2-(5,6-epoxyisoprostane E2)-sn-glycero-3-phosphoryl choline (PEIPC)
Reactive oxygen species (ROS)
Lipoxygenases (LOXs)
Glutathione peroxidase (GPx)
G protein-coupled receptors (GPCR)
Peroxisome proliferator-activated receptors (PPAR)
Toll-like receptors (TLRs)
Vascular endothelial growth factor (VEGF)
Lipopolysaccride (LPS)
Tumor necrosis factor α (TNFα)
Dendritic cells (DCs)
Smooth muscle cells (SMCs)
Multiple sclerosis (MS)

Author details

Mohammad Z. Ashraf and Swati Srivastava
Genomics Group, Defence Institute of Physiology & Allied Sciences, India

8. References

Androulakis N, Durand H, Ninio E, and Tsoukatos DC. Molecular and mechanistic characterization of platelet activating factor-like bioactivity produced upon LDL oxidation. J Lipid Res 2005; 46: 1923–1932.

Anliker B, Chun J. Cell surface receptors in lysophospholipids signaling. Semin cell Dev Biol 2004;92:1086-1094.

Arai M, Imai H, Metori A, and Nakagawa Y. Preferential esterification of endogenously formed 5-hydroxyeicosatetraenoic acid to phospholipids in activated polymorphonuclear leukocytes. Eur J Biochem 1997; 244: 513–519.

Ashraf M.Z., Kar N. S., Podrez E.A. Oxidized phospholipids: Biomarker for cardiovascular diseases. Int J Biochem Cell Biol 2009; 41: 1241-1244.

Berliner JA and NM Gharavi. Endothelial cell regulation by phospholipid oxidation products. Free Radic Biol Med 2008; 45:119-123.

Birkle DL and Bazan NG. Effect of K$\flat$ depolarization on the synthesis of prostaglandins and hydroxyeicosatetra- (5,8,11,14) enoic acids (HETE) in the rat retina. Evidence for esterification of 1 2-HETE in lipids. Biochim Biophys Acta 1984; 795: 564–573.

Birukov KG, Bochkov VN, Birukova AA, Kawkitinarong K, Rios A, Leitner A, Verin AD, Bokoch GM, Leitinger N, and Garcia JG. Epoxycyclopentenone-containing oxidized phospholipids restore endothelial barrier function via cdc42 and Rac. Circ Res 2004; 95: 892–901.

Bluml S, Rosc B, Lorincz A, Seyerl M, Kirchberger S, Oskolkova O, Bochkov VN, Majdic O, Ligeti E, and Stockl J. The oxidation state of phospholipids controls the oxidative burst in neutrophil granulocytes. J Immunol 2008; 181: 4347–4353.

Bochkov VN, Kadl A, Huber J, et al. Protective role of phospholipids oxidation products in endotoxin-induced tissue damage. Nature 2002; 419:77-81.

Bochkov VN, Mechtcheriakova D, Lucerna M, Huber J, Malli R, Graier WF, Hofer E, Binder BR, and Leitinger N. Oxidized phospholipids stimulate tissue factor expression in human endothelial cells via activation of ERK/EGR-1 and Ca(++)/NFAT. Blood 2002; 99: 199–206.

Bochkov VN, Oskolkova OV, Birukov KG, Levonen AL, CJ Binder, Stockl J. Generation and Biological Activities of Oxidized Phospholipids. Antioxidants & Redox Signaling. 2010; 12: , 1009-59.

Bochkov VN, Philippova M, Oskolkova O, Kadl A, Furnkranz A, Karabeg E, Afonyushkin T, Gruber F, Breuss J, Minchenko A, Mechtcheriakova D, Hohensinner P, Rychli K, Wojta J, Resink T, Erne P, Binder BR, and Leitinger N. Oxidized phospholipids stimulate angiogenesis via autocrine mechanisms, implicating a novel role for lipid oxidation in the evolution of atherosclerotic lesions. Circ Res 2006; 99: 900–908.

Bochkov VN. Inflammatory profile of oxidized phospholipids. Thromb Haemost 2007; 97:348-354.

Boullier A, Gillotte KL, Horkko S, Green SR, Friedman P, Dennis EA et al. The binding of oxidized low density lipoprotein to mouse CD36 is mediated in part by oxidized phospholipids that are associated with both the lipid and protein moieties of the lipoprotein. J Biol chem. 2000; 275:9163-9169.

Byun J, Mueller DM, Fabjan JS, and Heinecke JW. Nitrogen dioxide radical generated by the myeloperoxidasehydrogen peroxide-nitrite system promotes lipid peroxidation of low density lipoprotein. FEBS Lett 1999; 455: 243–246.

Carr AC, Winterbourn CC, and van den Berg JJ. Peroxidasemediated bromination of unsaturated fatty acids to form bromohydrins. Arch Biochem Biophys 1996; 327: 227–233.

Chang MK, Binder CJ, Miller YI, et al. Apoptotic cells with oxidized-specific epitopes are immunogenic and proinflammatory. J Exp Med 2004; 200: 1359-1370.

Chen R, Yang L, MsIntyre TM. Cytotoxic phospholipid oxidation products. Cell death from mitochondrial damage and the intrinsic caspase cascade. J Biol Chem 2007; 282:24842-24850.

Cheng C, Noordeloos AM, Jeney V, Soares MP, Moll F, Pasterkamp G, Serruys PW, and Duckers HJ. Heme oxygenase 1 determines atherosclerotic lesion progression into a vulnerable plaque. Circulation 2009; 119: 3017-3027.

Chou MY, K Hartvigsen, LF Hansen, L Fogelstrand, PX Shaw, A Boullier, CJ Binder and JL Witztum. Oxidation-specific epitopes are important targets of innate immunity. J Intern Med 2008; 263: 479-488.

Cruz D, AD Watson, CS Miller, D Montoya, MT Ochoa, PA Seiling, MA Gutierrez, M Navab, ST Reddy, JL Witztum, et al. Host derived oxidized phospholipids and HDL regulate innate immunity in human leprosy. J Clin Invest; 2008; 118:2917-2928.

Davis B, Koster G, Douet LJ, Scigelova M, Woffendin G, Ward JM, Smith A, Humphries J, Burnand KG, Macphee CH, and Postle AD. Electrospray ionization mass spectrometry identifies substrates and products of lipoproteinassociated phospholipase A2 in oxidized human low density lipoprotein. J Biol Chem 2008; 283: 6428–6437.

DeMaio L, Rouhanizadeh M, Reddy S, Sevanian A, Hwang J, and Hsiai TK. Oxidized phospholipids mediate occluding expression and phosphorylation in vascular endothelial cells. Am J Physiol Heart Circ Physiol 2006; 290: H674–H683.

Erridge C, S Kennedy, CM Spickett and DJ Webb. Oxidized phospholipid inhibition of toll-like receptor (TLR) signaling is restricted to TLR2 and TLR4: roles for CD14, LPS-binding protein and MD2 as targets for specificty of inhibition. J Biol Chem 2008; 283:24748-24759.

Forte TM, Subbanagounder G, Berliner JA, Blanche PJ, Clermont AO, Jia Z et al. Altered activities of anti-atherogenic enzymes LCAT, paraoxonase and platelet activating factor acetylhydrolase in atherosclerosis-susceptible mice. J Lipid Res 2002; 43: 477-485.

Frey B, Haupt R, Alms S, Holzmann G, Konig T, Kern H, Kox W, Rustow B, and Schlame M. Increase in fragmented phosphatidylcholine in blood plasma by oxidative stress. J Lipid Res 41: 1145–1153, 2000.

Fruhwirth GO and a Hermetter. Mediation of apoptosis by oxidized phospholipids. Subcell Biochem 2008; 49:351-367.

Furnkranz A, Schober A, Bochkov VN, et al. Oxidized phospholipids trigger atherogenic inflammation in murine arteries. Arterioscler Thromb Vasc Biol 2005; 25: 633–638.

Gargalovic PS, Gharavi NM, Clark MJ, et al. The unfolded protein response is an important regulator of inflammatory genes in endothelial cells. Arterioscler Thromb Vasc Biol 2006; 26: 2490–2496.

Gargalovic PS, Gharavi NM, Clark MJ, Pagnon J, Yang WP, He A, Truong A, Baruch-Oren T, Berliner JA, Kirchgessner TG, and Lusis AJ. The unfolded protein response is an important regulator of inflammatory genes in endothelial . Arterioscler Thromb Vasc Biol 2006; 26: 2490–2496.

Gargalovic PS, M Imura, B Zhang, NM Gharavi, MJ clark, J Pagnon, WP Yang, A He, A Troung, S Patel et al. Identification of imflammatory gene modules based on variations of human endothelial cell responses to oxidized lipids. Proc Natl Acad Sci USA 2006; 103:12741-12746.

Gharavi NM, Alva JA, Mouillesseaux KP et al. Role of the JAK/STAT pathway in the regulation of interleukin-8 transcription by oxidized phospholipids *in vitro* and in atherosclerosis *in vivo*. J Biol Chem 2007; 282: 31460-31468.

Girotti AW and Kriska T. Role of lipid hydroperoxides in photo-oxidative stress signaling. Antioxid Redox Signal 2004; 6: 301–310.

Greenberg ME, M Sun, R Zhang, M Febbraio, R. Silverstein and SL Hazen. Oxidized phosphatidylserine-CD36 interactions play an essential role in macrophage-dependent phagocytosis of apoptotic cells. J Exp Med 2006; 203:2613-2625.

Gruber F, Oskolkova O, Leitner A, Mildner M, Mlitz V, Lengauer B, Kadl A, Mrass P, Kronke G, Binder BR, Bochkov VN, Leitinger N, and Tschachler E. Photooxidation generates biologically active phospholipids that induce heme oxygenase-1 in skin cells. J Biol Chem 2007; 282: 16934–16941.

Gu X, Sun M, Gugiu B, Hazen S, Crabb JW, and Salomon RG. Oxidatively truncated docosahexaenoate phospholipids: Total synthesis, generation, and peptide adduction chemistry. J Org Chem 2003; 68: 3749–3761.

Gugiu BG, Mesaros CA, Sun M, Gu X, Crabb JW, and Salomon RG. Identification of oxidatively truncated ethanolamine phospholipids in retina and their generation from polyunsaturated phosphatidylethanolamines. Chem Res Toxicol 2006; 19: 262–271.

Heery JM, Kozak M, Stafforini DM, Jones DA, Zimmerman GA, McIntyre TM et al. Oxidatively modified LDL contains phospholipids with platelet-activating factor-like acitivity and stimulates the growth of smooth muscle cells. J clin Invest 1995; 96: 2322-2330.

Hoff HF, O'Neil J, Wu Z, Hoppe G, and Salomon RL. Phospholipid hydroxyalkenals: Biological and chemical properties of specific oxidized lipids present in atherosclerotic lesions. Arterioscler Thromb Vasc Biol 2003; 23: 275–282.

Honda HM, Leitinger N, Frankel M, Goldhaber JI, Natarajan R, Nadler JL, Weiss JN, and Berliner JA. Induction of monocyte binding to endothelial cells by MM-LDL: role of lipoxygenase metabolites. Arterioscler Thromb Vasc Biol 1999; 19: 680–686.

Huang LS, Kang JS, Kim MR, and Sok DE. Oxygenation of arachidonoyl lysophospholipids by lipoxygenases from soybean, porcine leukocyte, or rabbit reticulocyte. J Agric Food Chem 2008; 56: 1224–1232.

Huber J, Valves A, Mitulovic G, et al. Oxidized membrane vesicles and blebs from apoptotic cells contain biologically active oxidized phospholipids that induce monocyte-endothelial interactions. Arterioscler Thromb Vasc Biol 2002; 22: 101-107.

Huo Y, Weber C, Forlow SB, et al. The chemokine KC, but not monocyte chemoattractant protein-1, triggers monocyte arrest on early atherosclerotic endothelium. J Clin Invest 2001; 108: 1307–1314.

Ikura Y, Ohsawa M, Suekane T, et al. Localization of oxidized phosphatidylcholine in nonalcoholic fatty liver disease: impact on disease progression. Hepatology 2006; 43: 506-514.

Imai Y, Kuba K, Neely GG, Yaghubian-Malhami R, Perkmann T, van LG, Ermolaeva M, Veldhuizen R, Leung YH, Wang H, Liu H, Sun Y, Pasparakis M, Kopf M, Mech C, Bavari S, Peiris JS, Slutsky AS, Akira S, Hultqvist M, Holmdahl R, Nicholls J, Jiang C, Binder CJ, and Penninger JM. Identification of oxidative stress and Toll-like receptor 4 signaling as a key pathway of acute lung injury. Cell 2008; 133: 235–249.

Jerlich A, Schaur RJ, Pitt AR, and Spickett CM. The formation of phosphatidylcholine oxidation products by stimulated phagocytes. Free Radic Res 2003; 37: 645–653.

Jin Y and Penning TM. Aldo-keto reductases and bioactivation/detoxication. Annu Rev Pharmacol Toxicol 2007; 47: 263–292.

Kadl A, Huber J, Gruber F, et al. Analysis of inflammatory gene induction by oxidized phospholipids in vivo by quantitative real-time RT-PCR in comparison with effects of LPS. Vascul Pharmacol 2002; 38: 219–27.

Kagan VE, Bayir HA, Belikova NA, Kapralov O, Tyurina YY, Tyurin VA, Jiang J, Stoyanovsky DA, Wipf P, Kochanek PM, Greenberger JS, Pitt B, Shvedova AA, and Borisenko G. Cytochrome c/cardiolipin relations in mitochondria: A kiss of death. Free Radic Biol Med 2009; 46: 1439–1453.

Kagan VE, Tyurin VA, Jiang J, Tyurina YY, Ritov VB, Amoscato AA, Osipov AN, Belikova NA, Kapralov AA, Kini V, Vlasova II, Zhao Q, Zou M, Di P, Svistunenko DA, Kurnikov IV, and Borisenko GG. Cytochrome c acts as a cardiolipin oxygenase required for release of proapoptotic factors. Nat Chem Biol 2005; 1: 223–232.

Kim JA, Territo MC, Wayner E, et al. Partial characterization of leukocyte binding molecules on endothelial cells induced by minimally oxidized LDL. Arterioscler Thromb 1994; 14: 427–433.

Knapp S, Matt U, Leitinger N, et al. Oxidized phospholipids inhibit phagocytosis and impair outcome in gram-negative sepsis *in vivo*. J Immunol 2007; 178: 993-1001.

Kriska T, Marathe GK, Schmidt JC, McIntyre TM, and Girotti AW. Phospholipase action of platelet-activating factor acetylhydrolase, but not paraoxonase-1, on long fatty acyl chain phospholipid hydroperoxides. J Biol Chem 2007; 282: 100–108.

Kuzmenko AI, Wu H, Bridges JP, and McCormack FX. Surfactant lipid peroxidation damages surfactant protein A and inhibits interactions with phospholipid vesicles. J Lipid Res 2004; 45: 1061–1068.

Lee H, Shi W, Tontonoz P, et al. Role for peroxisome proliferator-activated receptor alpha in oxidized phospholipid-induced synthesis of monocyte chemotactic protein-1 and interleukin-8 by endothelial cells. Circ Res 2000; 87: 516–521.

Leitinger N, Tyner TR, Oslund L, et al. Structurally similar oxidized phospholipids differentially regulate endothelial binding of monocytes and neutrophils. Proc Natl Acad Sci USA 1999; 96: 12010–12015.

Leitinger N. Oxidized phospholipid as triggers of inflammation in atherosclerosis. Mol Nutr Food Res 2005; 49:1063-1071.

Leitinger N. Oxidized phospholipids as modulators of inflammation in atherosclerosis. Curr Opin Lipidol 2003; 14: 421-430.

Li R, Mouillesseaux KP, Montoya D, Cruz D, Gharavi N, Dun M, Koroniak L, and Berliner JA. Identification of prostaglandin E2 receptor subtype 2 as a receptor activated by OxPAPC. Circ Res 2006; 98: 642–650.

Lloberas N, Torras J, Herrero-Fresneda I, Cruzado JM, Riera M, Hurtado I, and Grinyo JM. Postischemic renal oxidative stress induces inflammatory response through PAF and oxidized phospholipids. Prevention by antioxidant treatment. FASEB J 2002; 16: 908–910.

Ma Y, Malbon CC, Williams DL, and Thorngate FE. Altered gene expression in early atherosclerosis is blocked by low level apolipoprotein E. PLoS One2008; 3: e2503.

Ma Z, Li J, Yang L, Mu Y, Xie W, Pitt B, and Li S. Inhibition of LPS- and CpG DNA-induced TN F-alpha response by oxidized phospholipids. Am J Physiol Lung Cell Mol Physiol 2004; 286: L808–L816.

Marathe GK, Davies SS, Harrison KA, Silva AR, Murphy RC, Castro-Faria Neto H, Prescott SM, Zimmerman GA, and McIntyre TM. Inflammatory platelet-activating factorlike phospholipids in oxidized low density lipoproteins are fragmented alkyl phosphatidylcholines. J Biol Chem 1999; 274: 28395–28404.

McEver RP and Cummings RD. Perspectives series: Cell adhesion in vascular biology. Role of PSGL-1 binding to selectins in leukocyte recruitment. J Clin Invest 1997; 100: 485– 491.

McIntyre TM, Prescott SM, and Stafforini DM. The emerging roles of PAF acetylhydrolase. J Lipid Res 2009; 50: S255– S259.

Miller YI, S Viriyakosol, CJ Binder, JR Feramisco, TN Kirkland and JL Witztum. Minimally modified LDL binds to CD14, induces macrophage spreading via TLR4/MD-2 and inhibits phagocytosis of apoptotic cells. J Biol Chem 2003; 278: 1561-1568.

Morrow JD, JA Awad, HJ Boss, IA Blair and LJ Roberts 2nd. Non-cyclooxygenase-derives prostanoids (F2-isoprostanes) are formed in situ on phospholipids. Proc Natl Acad Sci, USA 1992; 89:10721-10725.

Moumtzi A, M Trenker, K Flicker, E Zenzmaier, R Saf, and A Hermetter. Import and fate of fluorescent analogs of oxidized phospholipids in vascular smooth muscle cells. J Lipid Res 2007; 48:565-582.

Nakamura T, Henson PM, Murphy RC. Occurrence of oxidized metabolites of arachidonic acid esterified to phospholipids in murine lung tissue. Annal Biochem 1998; 262: 23-32.

O'Donnell VB, Eiserich JP, Chumley PH, Jablonsky MJ, Krishna NR, Kirk M, Barnes S, Darley–Usmar VM, and Freeman BA. Nitration of unsaturated fatty acids by nitric oxide-derived reactive nitrogen species peroxynitrite, nitrous acid, nitrogen dioxide, and nitronium ion. Chem Res Toxicol 1999; 12: 83–92.

Oskolkova OV, Afonyushkin T, Leitner A, von Schlieffen E, Gargalovic PS, Lusis AJ, et al. ATF4-dependent transcription is a key mechanism in VEGF up-regulation by oxidized phospholipids: critical role of oxidized sn-2 residues in activation of unfolded protein response. Blood 2008; 112: 330-339.

Parhami F, Fang ZT, Fogelman AM, et al. Minimally modified low density lipoprotein-induced inflammatory responses in endothelial cells are mediated by cyclic adenosine monophosphate. J Clin Invest 1993; 92: 471-478.

Parhami F, Fang ZT, Yang B, et al. Stimulation of Gs and inhibition of Gi protein functions by minimally oxidized LDL. Arteioscler Thromb Vasc Biol 1995; 15: 2019-2024.

Pegorier S, Stengel D, Durand H, et al. Oxidized phospholipid: POVPC binds to platelet-activating-factor on hman macrophages. Implications in atherosclerosis. Atherosclerosis 2006; 188: 433-443.

Pidkovka NA, OA Cherepanova, T Yoshida, MR Alexander, RA Deaton, JA Thomas, N Leitinger, and GK Owens. Oxidized phospholipids induce phenotypic swithcing of vascular smooth muscle cells in vivo and in vitro Circ Res 2007; 101:792-801.

Podrez E.A, E. Poliakov, Z Shen, R. Zhang, Y. Deng, M Sun, P J Finton, L. Shan, M Febbraio, D P Hajjar et al. A novel family of atherogenic oxidized phospholipids promotes macrophage foam cell formation via the scavenger receptor CD36 and is enriched in atherosclerotic lesions. J Biol Chem 2002; 277:38517-38523.

Podrez EA, Febbraio M, Sheibani N, Schmitt D, Silverstein RL, Hajjar DP, et al. Macrophage scavenger receptor CD36 is the major receptor for LDL modified by monocyte-generated reactive nitrogen species. J Clin Invest 2000; 105: 1095-1108.

Podrez EA, Poliakov E, Shen Z, Zhang R, Deng Y, Sun M, Finton PJ, Shan L, Gugiu B, Fox PL, Hoff HF, Salomon RG, and Hazen SL. Identification of a novel family of oxidized phospholipids that serve as ligands for the macrophage scavenger receptor CD36. J Biol Chem 2002; 277: 38503–38516.

Podrez EA, Poliakov E, Shen Z, Zhang R, Deng Y, Sun M, Finton PJ, Shan L, Gugiu B, Fox PL, Hoff HF, Salomon RG, and Hazen SL. Identification of a novel family of oxidized phospholipids that serve as ligands for the macrophage scavenger receptor CD36. J Biol Chem 2002; 277: 38503–38516.

Prescott SM, Zimmerman GA, Stafforini DM, MsIntyre TM. Platelet-activating factor and related lipid mediators. Annu Rev Biochem 2000; 69: 419-445.

Qin J, Goswami R, Balabanov R, and Dawson G. Oxidized phosphatidylcholine is a marker for neuroinflammation in multiple sclerosis brain. J Neurosci Res 2007; 85: 977–984.

Reddy ST, Grijalva V, Ng C, et al. Identification of genes induced by oxidized phospholipids in human aortic endothelial cells. Vascul Pharmacol 2002; 38: 211–218.

Reddy ST, Nguyen JT, Grijalva V, Hough G, Hama S, Navab M, and Fogelman AM. Potential role for mitogenactivated protein kinase phosphatase-1 in the development of atherosclerotic lesions in mouse models. Arterioscler Thromb Vasc Biol 2004; 24: 1676–1681.

Salomon RG. Levuglandins and isolevuglandins: Stealthy toxins of oxidative injury. Antioxid Redox Signal 2005; 7: 185–201.

Savaskan NE, Ufer C, Kuhn H, and Borchert A. Molecular biology of glutathione peroxidase 4: From genomic structure to developmental expression and neural function. Biol Chem 388: 1007–1017, 2007.

Schneider C, Porter NA, and Brash AR. Routes to 4- hydroxynonenal: fundamental issues in the mechanisms of lipid peroxidation. J Biol Chem 2008; 283: 15539–15543.

Seiss W, Tigyi G. Thrombogenic and artherogenic activities of lysophosphatidic acid. J Cell Biochem 2004; 92: 1086-1094.

Seyerl M, S Bluml, S. Kirchberger, VN Bochkov, O Oskolkova, O Majdic and J stockl. Oxidized phospholipid induce enery in human peripheral T cells. Eur J Immunol 2008; 38:778-787.

Shih PT, Elices MJ, Fang ZT, et al. Minimally modified low-density lipoprotein induces monocyte adhesion to endothelial connecting segment-1 by activating beta1 integrin. J Clin Invest 1999; 103: 613–625.

Spiteller P and Spiteller G. 9-Hydroxy-10,12-octadecadienoic acid (9-HODE) and 13-hydroxy-9,11-octadecadienoic acid (13-HODE): Excellent markers for lipid peroxidation. Chemistry and Physics of Lipids 1997; 89: 131–139.

Stremler KE, Stafforini DM, Prescott SM, and McIntyre TM. Human plasma platelet-activating factor acetylhydrolase. Oxidatively fragmented phospholipids as substrates. J Biol Chem 1991; 266: 11095–11103.

Subbanagounder G, Deng Y, Borromeo C, et al. Hydroxy alkenal phospholipids regulate inflammatory functions of endothelial cells. Vascul Pharmacol 2002; 38: 201–209.

Subbanagounder G, Leitinger N, Schwenke DC et al. Determinants of bioactivity of oxidized phospholipids. Specific oxidized fatty acyl groups at the sn-2 position. Arterioscler Thromb Vasc Biol 2000; 20: 2248-2254.

Subbanagounder G, Wong JW, Lee H, et al. Epoxyisoprostane and epoxycyclopentenone phospholipids regulate monocyte chemotactic protein-1 and interleukin-8 synthesis. Formation of these oxidized phospholipids in response to interleukin-1 beta. J Biol Chem 2002; 277: 7271-7281.

Sun M, Finnemann SC, Febbraio M, Shan L, Annangudi SP, Podrez EA, Hoppe G, Darrow R, Organisciak DT, Salomon RG, Silverstein RL, and Hazen SL. Light- induced oxidation of photoreceptor outer segment phospholipids generates ligands for CD36-mediated phagocytosis by retinal pigment epithelium: A potential mechanism for modulating outer segment phagocytosis under oxidant stress conditions. J Biol Chem 2006; 281: 4222–4230.

Tomura H, Mogi C, Sato K et al. proton-sensing and lysolipid-sensitive G-protein coupled receptors: a novel type of multi-functional receptors. Cell signal 2005; 17: 1466-1467.

Tselepis AD, John Chapman M, Inflammation, bioactive lipids and atherosclerosis: potential roles of a lipoprotein-associated phospholipase A2, platelet activating factor-acetylhydrolase. Atheroscler Suppl 2002; 3: 57-68.

Tsimikas S, Brilakis ES, Miller ER, et al. Oxidized phospholipid, Lp(a) lipoprotein, and coronary artery disease. N Engl J Med 2005; 353: 46-57.

Uhlson C, Harrison K, Allen CB, Ahmad S, White CW, and Murphy RC. Oxidized phospholipids derived from ozonetreated lung surfactant extract reduce macrophage and epithelial cell viability. Chem Res Toxicol 2002; 15: 896–906.

Van Lenten BJ, Wagner AC, Navab M, et al. D-4F, an apolipoprotein A-1 mimetic peptide, inhibits the inflammatory response induced by influenza A infection of human type II pneumocytes. Circulation 2004; 110:3252-3258.

Waddington E, Sienuarine K, Puddey I, and Croft K. Identification and quantitation of unique fatty acid oxidation products in human atherosclerotic plaque using highperformance liquid chromatography. Anal Biochem 2001;292: 234–244.

Walton KA, Hsieh X, Gharavi N, Wang S, Wang G, Yeh M, Cole AL, and Berliner JA. Receptors involved in the oxidized 1-palmitoyl 2-arachidonoyl-sn-glycero-3-phosphorylcholine-mediated synthesis of interleukin-8. A role for Toll-like receptor 4 and a glycosylphosphatidylinositol anchored protein. J Biol Chem 2003; 278: 29661–29666.

Watson AD, Berliner JA, Hama SY, et al. Protective effect of high density lipoprotein associated paraoxonase. Inhibition of the biological activity of minimally oxidized low density lipoprotein. J Clin Invest 1995; 96: 2882–2891.

Watson AD, Leitinger N, Navab M et al. Structural identification by mass spectrometry of oxidized phospholipids in minimally oxidized low density lipoprotein that induced monocyte/endothelial interactions and evidence for their presence *in vivo*. J Biol chem 1997; 272: 13597-13607.

Winterbourn CC, van den Berg JJ, Roitman E, and Kuypers FA. Chlorohydrin formation from unsaturated fatty acids reacted with hypochlorous acid. Arch Biochem Biophys 1992; 296: 547–555.

Wittwer J and Hersberger M. The two faces of the 15- lipoxygenase in atherosclerosis. Prostaglandins Leukot Essent Fatty Acids 2007;77: 67–77.

Yeh M, Cole AL, Choi J, Liu Y, Tulchinsky D, Qiao JH, Fishbein MC, Dooley AN, Hovnanian T, Mouilleseaux K, Vora DK, Yang WP, Gargalovic P, Kirchgessner T, Shyy JY, and Berliner JA. Role for sterol regulatory element-binding protein in activation of endothelial cells by phospholipid oxidation products. Circ Res 2004; 95: 780–788.

Yoshimi N, Ikura Y, Sugama Y et al. Oxidized phosphatidylcholine in alveolar macrophages in idiopathic interstitial pneumonias. Lung 2005; 183: 109-121.

Zalewski A, Macphee C. Role of lipoprotein-associated phospholipase A2 in atherosclerosis: biology, epidemiology and possible therapeutic target. Arterioscler Thromb Vasc Biol 2005; 25: 923-931.

Zimman A, K P Mouillesseaux, T Le, N M Gharavi, A Ryvkin, T G Graeber, T T Chen, A D Watson and J A Berliner. Vascular endothelial growth factor receptor 2 plays a role in the activation of aortic endothelial cells by oxidized phospholipids. Arterioscler Thromb Vasc Biol 2007; 27: 332-338.

Chapter 3

Antioxidant Complexes and Lipoprotein Metabolism – Experience of Grape Extracts Application Under Metabolic Syndrome and Neurogenic Stress

Andriy L. Zagayko, Anna B. Kravchenko,
Mykhaylo V. Voloshchenko and Oxana A. Krasilnikova

Additional information is available at the end of the chapter

1. Introduction

The oxidative hypothesis of atherosclerosis states that peroxide modification of LDL (or other lipoproteins) is important and probably required for the pathogenesis of arterial sclerotic disease; thus, there is an assumption that inhibition of LDL oxidation would increase or prevent atherosclerosis and its clinical consequences [1]. It is believed that the basis for the atherosclerotic plaque development is the foam cell formation from oxidized low-density lipoproteins (LDL) captured by monocytes and macrophages via scavenger-receptors.

Oxidation of LDL is also important for the healthy vessel functioning. High LDL concentrations can suppress the function of arteries in relation to release of nitric oxide from the endothelium, and many of such effects are mediated by the products of lipid oxidation [2]. Moreover, oxidized LDL inhibit the endothelium-dependent nitric oxide mediated relaxations in a rabbit isolated coronary arteries. Oxidized LDL induce apoptosis in the vascular cells, including macrophages, and this is prevented by nitric oxide [3].

One of the most important mechanisms of the inflammation proatherogenic effect is development of the systemic oxidative stress, and, as a consequence of proatherogenic abnormalities of the blood lipoprotein metabolism, there is appearance of antibodies to them, alterations of the main artery wall structure [4].

At the same time, on the one hand, a high atherogenicity of strongly oxidized LDL, especially tiny subfractions, has been confirmed; on the other hand, the oxidative stress is one of the causes of endothelial dysfunction.

Endothelium vascular wall cells are involved into the interaction with the pathogenic LDL [5]. While macrophages are being overloaded with esterified cholesterol, oxysterols and other biologically active substances, including powerful enzymes with a wide spectrum of action, a foam cell is formed from the macrophage. Yet so far to its apoptosis the foam cell secrets a wide complex of interleukins, enzymes, mediators. Many of them induce a local inflammatory process, destruction of the surrounding intercellular substance, damage of the fibrous structures and separate cells.

Many factors are considered as the most important factors for atherosclerosis development risk. Among such factors an important role belongs to the so-called proatherogenic states, including chronic stress and metabolic syndrome (MS) [6]. The proatherogenic character of stress is connected, first of all, with the activation of free radical oxidation and hyperlipidemia development. One of the principal statements of all contemporary conceptions of the atherosclerosis pathogenesis is thought to be the destruction of the cell membrane structure, which universal damage factor is peroxide oxidation of lipids (POL) [7].

It is well-known that free-radical processes play the leading role in atherosclerosis pathogenesis. So the antioxidants using in correction of proatherogenic states is fully explicable especially when we speak about natural antioxidants. Thus, the investigation of their biological effects under stress and metabolic syndrome is of grate interest and may be a perspective direction of research.

At the same time it is known that the enzymes associated with HDL, paraoxonase and PAF-acetyl hydrolase can hydrolyse biologically active lipids of mm-LDL, destroy monocyte aggregates and decrease the endothelial activation of mm-LDL [8]. HDL also contain a high concentration of tocopherol due to which they can be free radical scavengers as well.

Antioxidants protect LDL from peroxide oxidation and consequently from intensive uptake of LDL by macrophages decreasing the foam cell formation, the endothelium damage and possibility for lipids to infiltrate the intima. This condition supports the actuality of searching medicines for treating atherosclerosis, in which inhibition of the POL process plays an important part in the mechanism of their action [9]. Tocopherol, carotene, probucol, a number of plant medicines containing flavonoids are proposed as antioxidants.

The overwhelming majority of antioxidant substances used in pharmacotherapy are xenobiotics and so substrates of CYP system actvating ROS formation. Moreover some of them, such as probucol, leade to HDL-C decreasing.

Therefore, the substances of natural, in particular, plant origins that possess a complex activity draw attention of researchers.

Phenolic compounds are widely present in the world of plants; they are the most widespread product of the plant metabolism. Participation of polyphenols in redox processes to produce stable quinone structures by their phenolic forms reveals an antiradical direction of their action which provides their direct antioxidant activity. At present it has been proven that polyphenols as antiradical agents not only hinder the initiation of free radical oxidation, but also interrupt the chain of lipoperoxidation [10]. A great variety of studies carried out both *in vitro* and *in vivo* supports the ability of polyphenols to inactivate ("to bind", "to scavenge") the radicals that initiate chains of oxidation. First of all, it relates to the primary ROS - O_2 and OH [11].

There are some data that such natural polyphenols as catechins and procyanidins exposed to the human blood plasma produce certain complexes primarily with ApoA-1, i.e. with HDL.

One of the richest sources of polyphenols is *Vitis vinifera* and products of its processing, in particular wine.

Phenolic substances of grapes, including flavonoids and other polyphenols of grape, wine and grape seeds, are of a great interest due to their antioxidant properties and the ability to scavenge free radicals [12].

Studies *in vitro* have shown that grape, wine and grape seeds inhibit the oxidation of LDL. The activity of those substances as oxidation inhibitors in wine diluted 1,000 times markedly exceeded the analogous values for vitamins C and E [13]. It has been experimentally proven that red wine polyphenols slow down LDL oxidation processes and prevent platelet aggregation, thus preventing coronary heart diseases [14].

However, there is not a lot of research in this field yet. Arguments for anti-atherogenic properties of antioxidants are not enough. Results of convincing research are needed in order to decisively recommend antioxidants for treatment and prophylaxis of atherosclerosis.

2. Actuality

Taking into account the leading role of the free-radical processes in atherosclerosis pathogenesis one can make a conclusion about expediency of using natural antioxidants in prophylaxis and correction of this disease. [15]. Consequently, the study of the antioxidant influence on the development of stress-reactions and metabolic syndrome (MS) with the purpose of prevention of harmful complications for the cardiovascular system is of undoubted interest.

A number of studies also confirm the ability of a natural antioxidant α-tocopherol to reduce the risk of cardiovascular system diseases developed in patients with MS. It has been found that administration of α-tocopherol limits oxidation and cytotoxicity of LDL in the blood plasma significantly, supports the vascular endothelial function and reduces the intensity of systemic inflammation in the conditions of MS. The inhibiting effect of this antioxidant on aggregation and adhesion of platelets, adhesion of monocytes to endothelial cells and the smooth muscle cell proliferation has also been shown. However, numbers of experimental studies confirm that the single use of α-tocopherol is not enough for prevention of cardiovascular diseases in patients with MS [16]. It has been determined that the use of α-tocopherol in combination with ascorbic acid and aspirin (as a thrombolytic drug) is more effective [8]. In the ASAP study, the combination of vitamins E plus C was also tested, and this significantly decreased the intima-to-media progression rates in human. The ATBC clinical study used a combination of vitamin E and β-carotene in human as a secondary prevention strategy; however, no benefit on major coronary events has been found. The large MRC/BHF Heart Protection Study (HPS) for secondary prevention also examined the benefit of the antioxidant combination (vitamins E and C and β-carotene).

A strong dose-dependent effect of α-tocopherol administration is one of the unwanted effects. It is known that even a slight increase of the α-tocopherol dose could affect lipoprotein oxidation, the endothelium function and the degree of systemic inflammation [12].

The results of Cambridge Heart Antioxidant Study (CHAOS) of using antioxidants in cardiology published in 1996 give the opportunity to say that in patients with true (confirmed by angiography) coronary atherosclerosis vitamin E administration (a daily dose of 544-1088 mg (400-800 MU) reduces the risk of non-fatal myocardial infarction. The overall mortality from cardiovascular diseases in this case does not decrease. A favourable effect is revealed only after one-year administration of tocopherol.

At the same time in the Heart Outcomes Prevention Evaluation (HOPE) study, which was devoted to the study of the action of both ramipril and vitamin E (400 MU/daily dose), it was found that use of this antioxidant during approximately 4.5 years did not cause any effect on either the primary (myocardial infarction, insult and death from cardiovascular diseases) or any other end points of research. In another large-scale study on the primary prophylaxis of atherosclerotic diseases in people at least with one risk factors (hypertension, hypercholesterolemia, obesity, preliminary MI of the closest relative or advanced age) vitamin E (300 ME/daily dose) was used during 3.6 years and did not reveal any effect on any of the end points (the incidence of cardiovascular events and death). The vitamin E effectiveness was also not confirmed for various other cases (hypercholesterolemia, the level of sportsmen training, sexual potency, retardation of aging processes, etc.).

Empirically vitamin E is used in various diseases; however, the majority of the reports about tocopherol effectiveness is based on the single clinical observation and experiment data. Nowadays there are no reliable results on the role of vitamin E in prevention of tumour diseases, though the ability to reduce formation of nitrosamines (potentially carcinogenic substances being formed in the stomach), to decrease the formation of free radicals and have antitoxic effects when using chemotherapeutic remedies is well-known. In addition, the long-term intake of vitamin E in the doses from 11 to 800 mg does not cause side effects.

In HDL Atherosclerosis Treatment Study (HATS) there was the treatment of atherosclerosis depending on the high density lipoprotein cholesterol (HDL-C) level; in 160 patients with coronary heart disease with the confirmed coronary artery stenosis and the low HDL-C level the higher (800 MU/day) dose of vitamin E than in HOPE was used. The treatment combination also included 1000 mg of vitamin C, 25 mg of β-carotene and 100 mg of selenium. The study lasted 3 years and revealed that antioxidants had no influence on the HDL-C level, but in combination with hypocholesterolemic drugs they reduced their effect on LDL-C and especially – on HDL-C.

The dominant carotenoid revealed in blood and various tissues (such as liver, kidneys, adrenal glands, ovaries and prostate) is lycopene. Due to its structure and mechanism of action lycopene belongs to the group of antioxidants; a lycopene molecule contains 13 double bonds, which can interact with free radicals. Like β-carotene lycopene can serve as a precursor of vitamin A. However, the lycopene antioxidant activity is two times stronger than that of vitamin A.

Lycopene is recommended as an adjuvant in the treatment of the following diseases: idiopatic male infertility, chronic prostatitis, preeclampsia and intrauterine growth retardation (IUGR), mastopathia, diabetes mellitus, cardiovascular diseases, leucoplakia, age-related degeneration of yellow spots and cataract [17]. As with other oxidants, lycopene is administered in immunodeficiency states against chronic infections and to reduce the harmful action of unfavourable environmental factors.

The most representative evidence of the antioxidants' positive role in cardiovascular diseases prophylaxis was obtained in the multicultural European community multicentre study on antioxidants, myocardial infarction, and breast cancer (EURAMIC), during which the relationship between the antioxidant status and acute myocardial infarction in patients from 10 European countries was determined. The protective action was proven only for lycopene. In the Kuopio Ischemic Heart Disease Risk Factor Study (KIHD) the high level of blood plasma lycopene is associated with decreased risk of acute coronary syndrome and insult. In Erasmus Rotterdam Health Study (ERGO, also called "Rotterdam Study") it has been proven that lycopene prevents development and progression of atherosclerosis.

The meta-analysis of 72 epidemiological studies conducted concerning the connection between the tomato intake and cancer has determined the associative feed-back between the blood plasma lycopene level and the risk of cancer in 57 studies and 35 from 57 obtained associations were statistically significant [18].

Probucol (phenbutol) is a hypolipidemic medicine and belongs to butyl phenol derivatives. Probucol is the medicine that is similar in structure to hydroxytoluene – the compound with the potent antioxidant properties.

The hypolipidemic effect of Probucol is caused by activation of non-receptor ways of LDL extraction from the blood. It is believed that the prominent antioxidant activity of probucol prevents LDL oxidation.

Probucol decreases the total cholesterol content in plasma due to intensification of the LDL catabolism at the final stage of cholesterol elimination from the organism. It also inhibits the cholesterol biosynthesis at early stages and to a small extent slows down the food cholesterol absorption. It does not influence the triacylglycerol and the VLDL content, but significantly decreases the antiatherogenic HDL level in the blood. It is believed that decrease of the HDL-C level reflects improvement of cholesterol esters transfer with HDL on acceptor lipoproteins due to increase of the cholesteryl ester transfer protein (CETP) activity.

In spite of undesired decrease in the HDL-C concentration probucol causes regression of xanthelasma; this effect is revealed best of all in patients with the most dramatic HDL-C decrease. This important observation demonstrates that the low HDL-C content is not undoubtedly a negative phenomenon. The data obtained in experiments with animals indicate that probucol due to its antioxidant properties prevents lipid peroxidation and thereby inhibits the LDL uptake by macrophages, therefore, it inhibits atherogenesis. This allows suggesting that the therapeutic effect of the medicine may not be connected with its ability to decrease the LDL level. There is no clinical evidence of this hypothesis at the moment.

The medicine is absorbed slowly when taken internally, it is readily soluble in the adipose tissue releasing gradually into the bloodstream, and so its action is kept for a long time (up to 6 months after discontinuation of the treatment).

When using probucol in MultiVitamins and Probucol (MVP) research the renewal of the endothelium function in patients with IHD, decrease of restenosis cases after coronary angioplasty (when taking it at least 4 weeks before the procedure and further treatment during 6 months) was observed. Other antioxidants (α-tocopherol in high doses (700 mg per day), β-carotene and vitamin C) turned out to be ineffective.

Combined application of the endogenous antiradical antioxidants is of particular interest. In HPS (Heart Protection Study) along with the study of the simvastatin effectiveness the prophylactic action of antioxidants was investigated. The use of the vitamin complex (600 mg of vitamin E, 250 mg of vitamin C and 20 mg of β-carotene per day) lasted in average 5.5 years and did not reveal any differences in placebo groups and groups taking vitamins. Moreover, if the tendency exists, it reflects increasing of vascular events in the antioxidant intent-to-treat group. The action of antioxidants was compared with the effect of the combined use of simvastatin and nicotinic acid (niacin). Moreover, one of the groups received simvastatine+niacin and antioxidants. Angiographic and clinical data of this study were also disappointing with respect to the use of antioxidants.

Unfortunately, a great part of the compounds synthesized, which are used for pharmacocorrection of these states, are xenobiotics, so they can activate the free-radical formation process. Synthetic antioxidants, in particular probucol, can not be recommended for patient use because they decrease the HDL-C level.

The lack of antioxidant medicines popularity and the absence of traditions of their common use in practical medicine are caused a number of reasons: unsatisfactory previous study of this issue, complexity of adequate estimation of oxidation state parameters in the organism and the absence of the effective medicines with the antioxidant activity that are able to quickly reduce the consequences of the oxidative stress.

Therefore, the main indications for using antioxidants are excessively activated free-radical oxidation processes accompanying different pathologies. The choice of specific medicines, correct indications and contraindications for their use has not been developed yet and require further research.

3. Experiment design

In our experiments we studied the indicators of lipid and lipoprotein metabolism in the blood plasma and the liver under the experimental metabolic syndrome (MS) in Syrian hamsters of different sex and age.

In the experiments purebred male rats with 180-220 g of the body weight were used. The animals were kept in vivarium on a balanced diet. During 21 days the animals were given low alcoholic beverages from grapes of red and white grades *per os* daily. These beverages were introduced in the maximum effective doses of 9 mg of polyphenols/100 g the body weight. Taking into account the fact that the polyphenol content in the beverages investigated was quite low, the effective dose was introduced 3 times a day by 2 ml of liquids per 100 g of the animal's body weight. Control animals were introduced the corresponding volume of the saline solution. Ethanol was given in the corresponding dose.

Stress was caused by immobilization on the abdomen for 3 hours [19]. Animals were decapitated 3 hours after the immobilization. The blood was collected to get the serum. The liver was perfused by the cold extraction medium (0.25 M sucrose in 0.025 M tris-HCl, pH 7.5), homogenized in the Potter homogenizer with 2 ml of the extraction medium per 1 g of the liver. All manipulations with animals were held under chloralose-urethane anaesthesia.

To distribute the plasma lipoproteins the samples were centrifuged at 65,000 rpm (342,000 *g*) for 4 h at 4°C in the Optima XL-100K ultracentrifuge (Beckman Coulter) set at slow acceleration and deceleration [20]. Samples were fractionated within 1 h of centrifugation.

Lipids were extracted with chloroform and methanol (1:2 v/v) twice, as described by Bligh et al [21], and the supernatant was collected for determination of TG and FFA. TG and FFA were determined by enzymatic colorimetric methods with commercial kits (Zhongsheng, Beijing, China). The total cholesterol content was detected with the help of standard enzymatic cholesteroloxidase kits of "Boehringer Mannheim GmbH diagnostica" firm (Germany). The total lipid concentration was determined with the help of a standard kit "Eagle Diagnostics" (USA) – the reaction with vanillin reagent.

Determination of the lipid peroxide product quantity was performed in heptane-isopropanol extracts [22]. The optical density was measured at the wavelength of 220 nm (for compounds with isolated double bonds), 232 nm (for diene conjugates) and 278 nm – for ketodienes and conjugate trienes.

The TBA content was determined on the spectrophotometer with the help of the reaction with thiobarbituric acid [23].

A modified version of the high performance liquid chromatography (HPLC) procedure developed by Stacewicz-Sapuntzakis et al. [24] was used to measure vitamins E in the plasma. The HPLC system included a 150 × 3.9 mm Nova-pak C18 (4 microns) column with a guard-pak pre-column (both from Waters, Milford, MA), Waters Millipore TCM column heater, Waters 490 multi-wavelength detector, Hitachi 655–61 processor, Hitachi 655A-11 liquid chromatography, and BioRad autosampler AS-100.

The serum ascorbic acid concentrations were measured as described by using HPLC [25] with salicylsalicylic acid as a deproteinizing agent, metaphosphoric acid as a stabilizer.

The serum PON1 activity was measured by the rate of generation of p-nitrophenol determined at 405 nm according to MacKness B et al. [26].

The plasma cholesterol ester transfer protein (CETP) activity was examined using the modifications of Khosla et al. [44]. The CETP activity in duplicate10-μL aliquots of the plasma was determined after incubations with ^{3}H-cholesterol ester (CE)-labeled HDL$_3$ and LDL. Radioactivity transferred from ^{3}H-HDL$_3$ to LDL (measured in the supernatant after precipitation with heparin/MnCl^{2+}) was used to calculate the CETP activity (expressed as the percentage of radioactivity transferred from ^{3}H-HDL$_3$ to LDL per 16 h of incubation).

To measure endothelium-bound LPL, the perfusion solution was changed to buffer containing 1% fatty acid–free BSA and heparin (5 units/ml). The coronary effluent was collected in timed fractions over 10 min and assayed for the LPL activity by measuring the hydrolysis of a sonicated [3H]triolein substrate emulsion [27].

The plasma LCAT activity was measured by determination of the amount of radioactivity in each spot calculating the free cholesterol/ total cholesterol ratio in each plasma sample before and after the LCAT reaction and thus estimating the esterification rate [28]. The fractional esterification rate (% . h') expressed as the percentage of the free cholesterol esterified in the plasma sample per hour.

The HL activity was evaluated using the glycerol-stabilized emulsion of triolein and egg phosphatidylcholine containing glycerol-tri[9,10(n)-^{3}H] oleate by determination of the radioactivity amount during incubation [29].

Statistical analysis. All data were analyzed for statistical significance with SPSS 13.0 software. The data were presented as means ± standard deviation. Statistical analysis used one-way ANOVA. P<0.05 was considered to be statistically significant.

4. Discussion

The results of our studies suggest the existence of significant changes in the lipid metabolism, as well as sex and age differences in the lipid and lipoprotein metabolism both in healthy animals and in animals with MS.

In male hamsters fed with a high-calorie diet atherogenic dyslipidemia develops independently of age (Table 1). As it can be seen from the data obtained, increase of the total lipid content in the animal blood plasma is caused by increasing of the ApoB-containing lipoprotein (ApoB-LP) level since the HDL content is not changed. At the same time it has been found that the plasma TAG level in young (47%) and in adult animals (30%) increased in comparison with the intact group.

Increase of the TAG blood content in conditions of MS is considered to be a key factor for development of atherogenic dyslipidemia that is typical for this pathology [30]. A strong correlation between hypertriacylglycerolemia plus the HDL-C level decrease and accumulation of LDLB in the blood plasma has been demonstrated in many experiments and clinical studies [18].

It is assumed that atherogenic alterations occur as a result of lipoprotein disbalance in the blood plasma, i.e. because of predominance of the LDL and VLDL fractions over the antiatherogenic HDL fraction (especially when the values of the LDL+VLDL/HDL index are higher than 3.5).

Age	Group	Parameters			
		TAG, g/L	Total cholesterol, mmol/L	ApoB-LP, g/L	HDL, g/L
4 weeks	Intact	1.06±0.07	2.93±0.19	4.72±0.23	1.11±0.05
	MS	1.56±0.09*	3.56±0.10*	6.68±0.15*	0.98±0.07
20 weeks	Intact	1.57±0.22	2.84±0.15	5.66±0.34	1.01±0.02
	MS	2.00±0.13*	3.71±0.18*	6.68±0.21*	0.85±0.08
1 year	Intact	1.50±0.10	2.73±0.02	5.21±0.06	1.74±0.13
	MS	2.27±0.13*	3.15±0.08*	7.00±0.22*	2.32±0.13*

The data presented as mean±SD
* –p≤0.05 versus intact animals

Table 1. Some plasma lipid values in male Syrian golden hamsters with MS (in each group n=10).

As it is known, there are 2 phenotypes of LDL: LDLA and LDLB that differ by size, density, the lipid content and the atherogencity coefficient. LDLB have less size (d 25.5-25.75) comparing to LDLA (d > 25.75) and are characterized by a lower content of polar lipids, as well as a higher content of cholesterol esters. Lipoproteins of this subfraction are slowly removed from the bloodstream that is caused by their low affinity to B/E-receptors for LDL, higher sensitivity to glycosylation and oxidative damage [31]; they also have a high affinity to scavenger-receptors of macrophages [32].

All this features explain a high atherogenicity of LDLB subfraction. Numerous clinical and epidemiological studies have confirmed that accumulation of LDLB in the blood is an independent risk factor for atherosclerosis occurrence [33].

Normally, there are predominantly LDLA in the blood plasma, and LDLB are present in a small percent of the total LDL, but in MS and insulin resistance the LDLB content increases significantly.

It is well-known that in MS the key factor for TAG and ApoB-LP accumulation in the blood is the VLDL hyperproduction by the liver [34]. According to our data, accumulation of ApoB-LP in the blood occurs parallelly with increase in the content of this lipoprotein fraction in the liver (Table 2).

These results allow us to make a suggestion that VLDL formation is activated in the liver of the animals fed with a high-calorie diet in our experiment.

The mechanisms of the VLDL hyperproduction by the liver in the conditions of FFA intensive supply to hepatocytes have remained still unclear. The stimulation of VLDL formation can occur both by using the elevated uptake of the blood FFA and via activation of fatty acid biosynthesis *de novo* because of hyperglycemia.

Age	Group	Parameters				
		Total lipids, mg/g liver	ApoB-LP, mg/g liver	HDL, mg/g liver	G6PDH, nmol/mg protein/min	Lysosomal lipase, nmol/mg protein/min
Week 4	Intact	104.24±2.52	11.46±0.37	1.25±0.14	3.74±0.33	0.67±0.03
	MS	124.16±2.05*	15.16±0.54*	1.11±0.07	2.80±0.17*	1.09±0.07*
Week 20	Intact	112.62±2.66	13.03±0.50	0.94±0.10	4.44±0.28	0.54±0.03
	MS	143.59±2.65*	15.69±0.36*	1.10±0.20	3.13±0.28*	1.27±0.09*

The data presented as mean±SD

* –p≤0.05 versus intact animals

Table 2. Some liver lipid metabolism values in male Syrian golden hamsters with MS used in the current study (in the crude tissue, in each group n=10).

It is known that in insulin resistance FFA that come to hepatocytes from the blood are primarily used for the TAG re-synthesis. It leads to increase in the intracellular TAG content and correlates with the increase of the VLDL secretion rate into the bloodstream. The VLDL morphology, which is specified predominantly at the second stage of their formation, depends significantly on the intracellular TAG content and hepatocyte sensitivity to insulin [35]. More active phospholipase D-dependent pre-VLDL lipidation takes place in the elevated intercellular TAG content and insulin resistance of hepatocytes [36]. Insulin blocks the VLDL1 formation in the liver. In the conditions of insulin resistance this effect and the elevated intercellular TAG content stimulate formation and secretion predominantly of VLDL1 by the liver [37].

The VLDL1 secretion increase leads to significant changes in the lipid and lipoprotein metabolism in the blood: the increased TAG content and accumulation of LDL*B* with high atherogenicity in the blood. These changes are typical for MS and considered to be separate risk factors for development of atherosclerosis.

Metabolism of ApoB-LP in the blood plasma is tightly connected with metabolism of HDL performing a reverse cholesterol transport from peripheral tissues to the liver. The leading factors in the process of transformation of VLDL into LDL in the bloodstream and determination of the LDL morphology are the rate of cholesterol esters transfer from HDL to ApoB-LP mediated by cholesteryl ester transfer protein (CETP), and the rate of TAG hydrolysis in the ApoB-LP composition mediated by lipoprotein lipase (LPL) and hepatic lipase (HL) [38].

According to data of many clinical studies, increase of the CETP activity of the HDL composition in most cases leads to decrease of the HDL-C level and accumulation of LDL*B* in the blood plasma. Moreover, a degree of these modifications correlates with the blood TAG level.

We observed significant changes in the cholesterol and HDL metabolism in the blood plasma in animals fed with a high-calorie diet. These changes have expressed the proatherogenic character and could be one of the causes for the LDL*B* accumulation in the blood.

Our results suggest that increase of the total blood cholesterol level in hamsters fed with a high-calorie diet is obviously connected with increase of the cholesterol content in the ApoB-LP composition as its level in the HDL composition decreases (Table 3).

Age (at the beginning of the experiment)	Group	Parameters			
		HDL-C, mkmol/L	HDL-CE, mkmol/L	LCAT, mkmol/l/h	CETP, mkmol/l/h
Week 4	Intact	174.17±18.99	1028.33±12.76	54.92±0.58	20.42±1.76
	MS	80.83±9.17*	810.00±22.78*	49.00±2.50	33.83±1.56*
Week 20	Intact	138.00±8.00	770.00±32.56	45.50±2.55	59.50±5.39
	MS	164.50±9.97	512.50±0.01*	20.25±2.28*	116.88±9.43*

The data presented as mean±SD
* –p≤0.05 versus intact animals, ª –p≤0.05 versus intact animals 4 weeks.

Table 3. Plasma HDL-C and HDL-CE, cholesterol esterifying activity and CE transfer in Syrian golden hamsters with the experimental MS (in each group n=10).

Decrease in the HDL cholesterol level is apparently connected with increase of the transfer rate of cholesteryl esters from HDL to ApoB-LP. According to our data the rate of the cholesteryl esters transfer from HDL in the animals fed with a high-calorie diet grows to 166% and 199% compare to the values of young and adult intact animals, respectively (Table 3).

At the same time decrease in the free cholesterol and HDL esterified cholesterol levels was determined in young males, but in adult animals only the HDL esterified cholesterol content lowered. The cholesteryl ester transfer rate from HDL to ApoB-LP is activated when the TAG content increases in the blood, it is observed in the postprandial period, as well as in ApoB-LP metabolism abnormalities [39]. In both cases the cholesteryl ester transfer activation is a consequence of increasing the TAG-rich lipoproteins (TRL) in the bloodstream [40]. The latter is also confirmed by our data pertaining to the increase of the neutral lipids content in the ApoB-containing lipoproteins in hamsters with the experimental MS. These differences seem to be connected with the difference in the HDL free cholesterol esterification rate in males of various ages. This rate is primarily determined by the activity of LCAT – the enzyme associated with HDL [41].

The increase of the cholesteryl-ester transfer activity from HDL is mostly the result of the CETP activation. The increase of the CETP activity in MS was demonstrated in a great number of experiments [22]. It is known that the activation of CETP biosynthesis in the liver is primarily the cause for increasing the activity of this protein in the blood HDL composition, but mechanisms of CETP induction have been still unclear.

Thus, increase of the cholesteryl ester transfer rate from HDL on the background of hypertriacylglycerolemia, which is observed in our experiment in the animals fed with a high-calorie diet (Table 3), is atherogenic since the cholesteryl ester transfer predominantly to TAG-enriched lipoproteins leads to accumulation of CE-enriched VLDL1, which are major precursors of LDLB. Intensive TAG supply to HDL in exchange for cholesteryl esters results in accumulation of TAG-enriched HDL particles, which are the predominant substrate for hepatic lipase (HL), in the blood. So, HDL particles are rapidly removed from the bloodstream and it leads to decrease of the HDL-C content.

That is why changes in the enzymes activity, which hydrolyze lipoprotein lipids in the bloodstream, in particular – in LPL and HL activity, affect significantly the lipoprotein metabolism in MS.

TAG in the TAG-enriched lipoproteins (chylomicrons and VLDL) are the substrate for LPL. FFA, released after hydrolysis under the action of LPL, come to adipocytes and muscle cells where they are deposited as the TAG component or used as a source of energy. TAG hydrolysis in the VLDL composition increases availability of cholesterol for its transfer to HDL, therefore, in this way LPL mediates the reverse cholesterol transfer. The LPL activity is regulated by the influence on transcription, translation and enzyme transport from the cells. Insulin is known to activate LPL that results in decrease of the total blood TAG level and stimulation of cholesterol reverse transfer [35].

According to our data, the plasma LPL activity decreased in young male hamsters fed with a high-calorie diet (Table 4).

Age (at the beginning of the experiment)	Group	Parameters	
		LPL (U/ml)	HL (U/ml)
Week 4	Intact	8±2	51±4
	MS	4±1*	91±3*
Week 20	Intact	83±2	3±1
	MS	129±3*	2±1

The data presented as mean±SD
* –p≤0.05 versus intact animals

Table 4. Postheparin plasma lipase activities in Syrian golden hamsters with the experimental MS (in each group n=10).

The results obtained are in agreement with the literature data about the reduction of the LPL activity in obesity and insulin resistance [42]. The mechanisms of the LPL activity inhibition in these conditions are still unclear though a definite contribution could be made by development of insulin resistance.

The increase of the cholesteryl ester transfer rate from HDL on the background of hypertriacylglycerolemia, which was stated in our experiment both in animals fed with a high-calorie diet and in chronic stress, is an atherogenic factor for two reasons. Firstly, the cholesteryl ester transfer predominantly to the TAG-enriched lipoprotein fractions leads to accumulation of VLDL1 enriched with cholesteryl esters, which are the main LDL*B* precursors. Secondly, the intensive exchange of cholesteryl esters in HDL for TAGs results in accumulation of TAG-enriched HDL in the blood, which are the predominant substrates for HL, and they are rapidly removed from the bloodstream, and it, in turn, causes decrease in the HDL-C concentration. The activation of the lipoprotein secretion by the liver is also observed in the conditions of the acute chemical and emotional painful stress. This fact may be considered to be a sign of proatherogenesis since it is accompanied by hyperlipidemia development due to increase of atherogenic lipoprotein fractions.

As shown in our studies, decrease of the LPL activity in the blood plasma of young males fed with a high-calorie diet can be an additional factor for TAG accumulation in the blood and the HDL-C level reduction observed in our experiment.

HL mediates a selective transport of VLDL remnants to hepatocytes via LDL-receptors, takes part in reverse transport of cholesterol accelerating HDL coming into the liver via scavenger receptors (SRB1). Hydrolyzing TAG in the ApoB-LP composition HL plays a significant role in their re-modelling in the bloodstream. It is known that the HL activity specifies substantially the lipid composition, size and properties of LDL [43].

The HL activity is predominantly regulated at the transcriptional level under the influence of sex hormones, glucocorticoids and adipokines. The rate of the HL gene transcription is also dependent on the intercellular lipid content, primarily cholesterol in the hepatocytes [44].

In our experiment the blood plasma HL activity in male hamster fed with a high-calorie diet increased irrespective of age (Table 4), it corresponds to literature data. In a number of studies it has been shown that the HL activity increases in insulin resistance, obesity, and a high-calorie diet [45]. Moreover, it has been determined that increase of the HL mRNA content is observed when using a high-calorie diet; this is the evidence of the enzyme biosynthesis activation under these conditions. The authors associate this fact with the decrease in the blood plasma adiponectin level, which can inhibit the HL synthesis in hepatocytes.

Considering these data, as well as the data obtained in our studies about adiponectin decrease in the blood plasma in obesity (Fig. 1), we may suppose that one of the causes for the HL activity increasing when taking a high-calorie diet in our experiment is decrease of adiponectin secretion by the adipose tissue.

The HL activity increase is considered to be one of the key factors for the atherogenic dyslipidemia development in obesity and MS. In a number of works a clear correlation between the HL activity and the LDLB content in the blood plasma was demonstrated [19]. It is believed that namely HL activation results in the increased LDLB formation [33]. The latter occurs with increase of the TAG-enriched VLDL1 content in the blood and the CETP activation. Furthermore, the HL activity increase leads to decrease of the HDL cholesterol level [46]. This is associated with the fact that hydrolysis of TAG in the HDL3 composition results in their transformation into HDL2, which are rapidly removed from the bloodstream by the liver. Thus, the HDL-C level decrease that we determined in our experiment (Table 3) can be a consequence of the HL activity increase.

In the current study we have found that the blood FFA level increase is accompanied by the ApoB-LP synthesis activation in the liver of Syrian male hamsters fed with a high-calorie diet irrespective of age. This causes increase of the TAG and ApoB-LP level in the blood. Decrease of the HDL-C level is a consequence of the rate of cholesteryl ester exchange between HDL and LDL due to activation of CETP and HL. As a result of these changes the atherogenic dyslipidemia development, which is typical for MS, is observed.

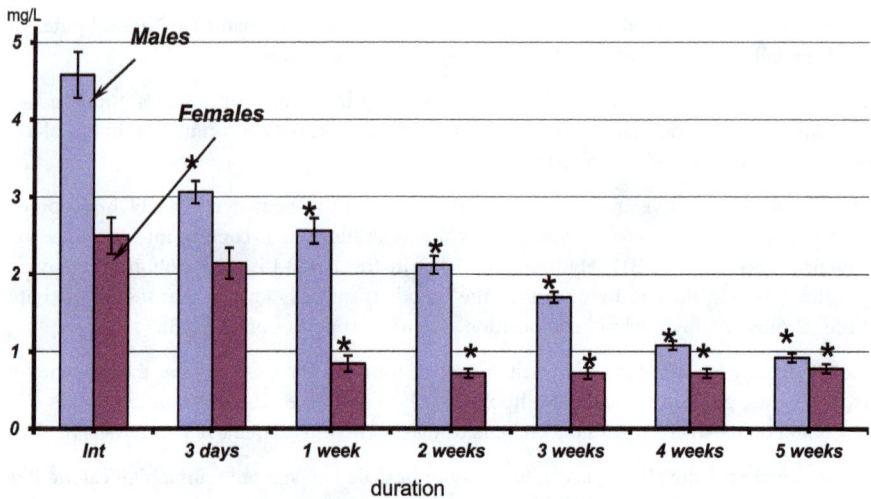

Figure 1. Plasma adiponectine level in Syrian golden hamsters with the experimental MS development (values are mean±SD; * –p≤0.05 versus intact animals, in each group n=10, * – p < 0.05 versus intact animals).

We have found the age differences of the lipid profile in the blood plasma of the normal male hamsters. So, in intact males (with age from 4 to 20 weeks) on the background of the constant content of total lipids and lipoproteins in the blood plasma there was increase of the FFA level (60% comparing to a 4-week intact), TAG (48%) and ApoB-LP (20%), and the HDL level showed a tendency to decrease. These results are the evidence of the lipidation increase with age. It has been also shown that in adult males the unesterified cholesterol and cholesteryl ester levels are lower than in young animals (20% and 25%, respectively), and the cholesteryl ester transfer rate from HDL in adult animals exceeds this index value in young animals (191%) (Table 3).

The data obtained correspond to the literature data about age-dependent changes in the lipid metabolism in males, which have the proatherogenic character [47]. It is known that with age the sex hormones level lowers in males and the glucocorticoid secretion level increases. The plasma lipid profile in males is determined, among other factors, by the secretion level of sex hormones possessing antiatherogenic properties. A lot of studies proved the presence of direct correlation between the blood testosterone plus the dehydrotestosterone level and the HDL-C content [48]. Moreover, the high level of sex hormones correlates with decrease of the TAG content and the total cholesterol in the blood. Thus, increase of the TAG level and decrease of the HDL cholesterol content in the blood plasma of males with age may be connected with reduction of the sex hormone secretion (Table 5). Changes of the lipid profile in the blood plasma of males with age may be also associated with increase of glucocorticoid secretion, which was observed in our experiment (Table 5).

Thus, with age the blood plasma lipid profile in males is subjected to unfavourable changes such as increase of the FFA and TAG content and decrease of HDL-C. The latter may be connected with decrease of the sex hormone level, and increase of the cortisol secretion.

However, despite the more favourable lipid profile in the blood plasma of young males comparing with normal adult animals, atherogenic dyslipidemia in obesity and insulin resistance develops irrespective of age.

In contrast with males, in females the atherogenic dyslipidemia development is significantly dependent on age (Table 6). In particular, while in males with age there are no significant changes in the liver ApoB-LP content, in females this index raises during maturation – in intact animals – to 20%, and in animals in the experimental MS – to 31%. That indicates intensification of lipolytic processes in the liver of females during ageing, and may serve as a manifestation of the lipid metabolism activation. The analogous changes in the total lipid content also proved this tendency (Table 6).

Sex	Parameter	Group	Age (at the beginning of the experiment)		
			4 weeks	20 weeks	1 year
Females	Estradiol, pmol/mL	Intact	0.55±0.05	0.64±0.06	0.54±0.05
		MS	0.63±0.06*	0.75±0.08	0.36±0.04*
	Cortisol, nmol/L	Intact	47.00±3.85	73.17±5.56	76.00±4.95
		MS	74.0±5.49*	87.5±4.45	117.00±2.63*
Males	Estradiol, pmol/mL	Intact	0.24±0.02	0.19±0.02	0.29±0.03
		MS	0.27±0.02	0.29±0.02*	0.20±0.03*
	Testosterone, pmol/mL	Intact	4.02±0.39	4.24±0.31	3.58±0.37
		MS	4.45±0.41*	3.51±0.40*	3.03±0.31*
	Cortisol, nmol/L	Intact	61.17±3.71	94.8±3.06	85.33±5.40
		MS	84.67±3.62*	132.00±7.88*	148.40±9.54*

The data presented as mean±SD
* –p≤0.05 versus intact animals

Table 5. Plasma sex hormones and cortisol levels in hamsters with the experimental MS (in each group n=16)

Age	Group	Parameters			
		Total lipids, mg/g	ApoB-LP, mg/g liver	HDL, mg/g	Lysosomal lipase, nmol/mg protein/min
4 weeks	Intact	117.67±4.72	8.87±0.24	1.27±0.08	0.34±0.03
	MS	144.34±5.00*	10.24±0.25*	0.65±0.05*	1.24±0.05*
10 weeks	Intact	137.54±3.91	10.65±0.46	0.89±0.07	0.83±0.04
	MS	179.22±3.44*	13.44±0.30*	0.46±0.06*	1.33±0.08*

The data presented as mean±SD
* –p≤0.05 versus intact animals

Table 6. Lipid metabolism parameters in the liver homogenate in Syrian golden female hamsters with MS (in the crude tissue, in each group n=16)

Oxidation of LDL and VLDL (i.e. ApoB-LP) is an alternative way of the lipoprotein catabolism, which leads to their uptake by macrophages via scavenger-receptors, and may lead to the transformation of these cells into "foam" ones. That is why it is one of the factors of atherogenesis in MS.

In our experiment we also observed the composition changes in lipoproteins and in particular HDL particle enrichment with lipids (Table 7). However, the cholesterol content of these lipoproteins decreased in contrast to the ApoB-LP cholesterol content that was increased.

Parameters	Group	
	Intact	MS
Total lipids, % of the total HDL composition	49.45±1.35	57.31±1.91*
Total cholesterol, % of the total HDL composition	14.97±0.23	11.21±0.76*
TAG, % % of the total HDL composition	1.75±0.07	3.08±0.15*
α-Tocopherol, mmol/L	8.02±0.39	5.70±0.35*
Isolated double bonds, U/ml	8.64±0.59	7.31±0.17*
Diene conjugates, mmol/L	18.88±2.10	31.68±1.65*
Ketodienes+conjugated trienes, U/ml	1.15±0.08	1.48±0.06*
Total hydroperoxides, mmol/L	69.04±3.46	78.31±1.33*

The data presented as mean±SD or percentage
* –$p \leq 0.05$ versus intact animals

Table 7. The plasma HDL composition in Syrian golden hamsters (1 year) with the experimental MS (in each group n=10).

There are several possible reasons for that phenomenon. One of them is a well-known fact that HDL contains high levels of both unsaturated fatty acids, which are rapidly utilized, and proteins, which hydrophilic properties compensate the lack of phospholipids, as well as α-tocopherol and enzymatic antioxidants, particularly paraoxonase, which protect these lipoproteins from peroxidation. There is no doubt that the changes in the cholesterol metabolism enzymes activity associated with HDL (CETP) are involved in this process (Table 3).

Nevertheless, the content decrease of compounds with isolated double bonds and accumulation of the lipoperoxidation products has been determined in the HDL fraction in MS (Table 7). Moreover, the data obtained have shown that the content of ketodienes and coupled trienes in the HDL fraction is 129% comparing to control; the content of diene conjugates – 168% and the content of the total hydroperoxides – 115%. It has been also found that there is decrease of the α-tocopherol content in HDL (41%) comparing to the control values (Table 7).

Thus, HDL can protect LDL from oxidation "providing" a cell with paraoxonase and PAF-acetyl hydrolase. However, this protective effect of HDL is reduced in response to induction of the stress acute phase in animal models [49].

As can be seen from our data (Table 8), the HDL-associated paraoxonase activity is generally decreased in experimental MS.

Sex	Age	Groups	Activity, nmol/mL/min
Males	4 week	Intact	80.78±3.69
		MS	67.06±3.70*
	20 week	Intact	62.19±2.63
		MS	37.29±3.33*
Females	4 week	Intact	104.41±2.95
		MS	75.45±2.21*
	20 week	Intact	127.27±2.95
		MS	121.93±3.05

The data presented as mean±SD
* –p≤0.05 versus intact animals

Table 8. The plasma HDL paraoxonase activity in Syrian golden hamsters with experimental MS (in each group n=10).

The data obtained show that application of antioxidant complexes for correction of unfavourable changes in proatherogenic states may be perspective since free radical oxidation activation is a common pathogenetic link of all those states; this link is not only involved in damage of cells and their components, but also as an alternative way of catabolism it accelerates the lipid recyclization.

At the same time, since a significant feature of proatherogenic states is the hormone status imbalance, polyphenolic antioxidants need special attention because these compounds along with the antioxidative activity also demonstrate phytoestrogen properties [50], and it may be an additional factor of the lipid metabolism regulation.

An important effect of flavonoids is scavenging of oxygen-derived free radicals. The experimental systems *in vitro* have also shown that flavonoids possess anti-inflammatory, antiallergic, antiviral, and anticarcinogenic properties. The so-called "Mediterranean diet" is thought to prevent cardiovascular diseases, as a consequence of its high content of antioxidants, which are crucial in ameliorating oxidative events implicated in many diseases. In addition to the antioxidant/antiradical activity, red wine polyphenols (RWPs) have been shown to possess many biological properties, including inhibition of platelet aggregation, the vasorelaxing activity, modulation of the lipid metabolism, and inhibition of the low-density lipoprotein oxidation.

In our research we have used wine, juice and polyphenolic extracts from grapes of different grades, and polyphenolic concentrates "Enoant" and "Polyphen" obtained from *Vitis Vinifera* grapes to correct the changes in the lipid metabolism in the conditions of the experimental metabolic syndrome, acute and chronical stress. All substances used in our research were developed in National Institute for Vine and Wine "Magarach" (Yalta, Ukraine). The studies carried out have specified that polyphenolic extracts and concentrates are quite active remedies that decrease negative effects in MS though the effectiveness of various substances administered are significantly different.

So, administration of any of the investigated substances has significantly decreased the total blood plasma lipoprotein content in hamsters with MS, but the use of "Cabernet" extract has the most pronounced effect (Fig. 2). The same tendency is observed in decreasing the ApoB-LP content, the total cholesterol and FFA level have also decreased, though practically no difference between the grape varieties investigated has found.

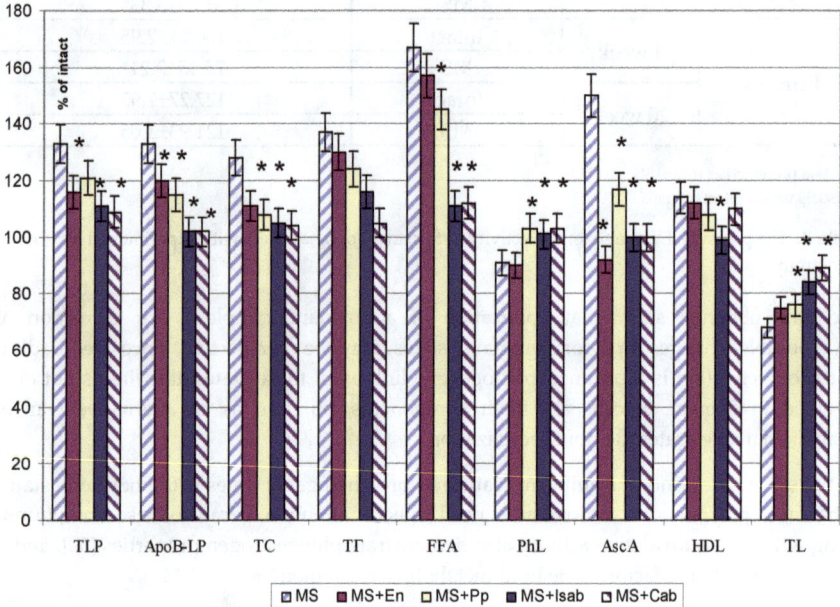

* – p≤0.05 versus intact animals.

Figure 2. The effect of *Vitis Vinifera* substances on some plasma lipid metabolism values in male Syrian golden hamsters (1 year old) with the experimental MS (in each group n= 7)

The non-enzyme antioxidant level (α-tocopherol, reduced glutathione and ascorbic acid) in the blood serum has also reached reference values under the influence of the polyphenolic extracts. This fact confirms the high antioxidant activity of the studied substances.

Normalization of the blood plasma phospholipid content under the influence of polyphenolic extracts arouses the interest. The phospholipid content returned to the intact level, which may be a result of their oxidation reduction, given that the unsaturated fatty acids in phospholipids are compounds that undergo oxidation by free radicals quickly and easily.

However, in spite of the quite favourable effect of "Isabella" extract, its administration also has negative consequences, particularly the HDL level decrease to the value observed in intact animals accompanied by the LDL content increase.

The investigated substances normalize also the blood lipoproteins composition. Thus, the total lipid and the total cholesterol content decrease in the ApoB-LP composition, moreover

"Enoant" lowers the cholesterol content in this atherogenic lipoprotein fraction even below the control level.

Generally, the TAG content is also normalized under the action of all the investigated substances, but taking into account the ratio – cholesterol/triacylglycerols, "Polyphen" has the most favourable effect.

The polyphenol extracts and concentrates have significantly improved the ApoB-LP oxidative status in animals with MS. The best results have been obtained when using "Cabernet", as well as for other indexes investigated (Table 9).

Parameter	Group				
	MS	MS +"Enoant"	MS +"Polyphen"	MS+ "Isabella"	MS+ "Cabernet"
Total lipids, % of the total ApoB-LP composition	88.87 ±0.71*	83.70 ±0.78 */**	83.12 ±0.37*/**	82.22 ±0.09*/**	81.00 ±0.19*
Total cholesterol, % of the total ApoB-LP composition	8.39 ±0.24	7.87 ±0.04*/**	8.13 ±0.04	8.06 ±0.12**	8.17 ±0.08*/**
TAG, % of the total ApoB-LP composition	37.55 ±1.89*/**	53.97 ±0.10*/**	53.02 ±0.14*/**	50.65 ±1.23*/**	49.57 ±0.40*/**
α - Tocopherol, mmol/L	2.68 ±0.08*	2.98 ±0.05*/**	3.06 ±0,04*/**	3.17 ±0,02*/**	3.19 ±0,05*/**
Isolated double bonds, U/ml	1.71 ±0.06*	1.84 ±0.03*/**	1.90 ±0.02*/**	1.97 ±0.03**	2.09 ±0.03**
Diene conjugates, mmol/L	37.25 ±1.50*	30.63 ±0.41*/**	29.54 ±0.34*/**	28.77 ±0.14*/**	26.68 ±1.94**
Ketodienes+conjugated trienes, U/ml	8.18 ±0.11*	7.49 ±0.04*/**	7.23 ±0.08*/**	7.17 ±0.08*/**	6.99 ±0.26**
Total hydroperoxides, mmol/L	108.25 ±1.39*	98.52 ±0.55*/**	94.97 ±0.15*/**	90,65 ±1.15*/**	89.30 ±1.06*/**

The data presented as mean±SD or percentage
* – p≤0.05 versus intact animals, ** – p≤0.05 versus model of MS

Table 9. The effect of *Vitis Vinifera* substances on the plasma ApoB-LP composition in male Syrian golden hamsters (1 year old) with the experimental MS (in each group n= 10)

The HDL composition in the blood is also affected by the substances studied. In these particles the total lipid content decreases and even reaches the level of intact animals when using "Cabernet" extract (Table 10).

The cholesterol level also changes: it decreases when using "Enoant" and increases under the action of "Isabella" and "Cabernet" extracts.

The HDL-C content decreases under the action of "Enoant" may occur due to peroxide processes inhibition since cholesterol accumulation in lipoprotein particles, as it was mentioned before, has a compensatory character in response to the phospholipid oxidation of the lipoprotein particle hydrophilic cover. The TAG content decreased under the action of all substances, and "Isabella" was the most effective substance. The TAG content decrease is probably mediated by the phytoestrogenic action of polyphenols directed to lipolysis inhibition in the adipose tissue.

Parameter	Group				
	MS	MS +"Enoant"	MS +"Polyphen"	MS+ "Isabella"	MS+ "Cabernet "
Total lipids, % of the total ApoB-LP composition	57.31 ±1.91*	54.91 ±0.21*/**	53.09 ±0.08*/**	51.74 ±0.74*/**	49.20 ±0.42**
Total cholesterol, % of the total ApoB-LP composition	11.21 ±0.76*	10.54 ±0.30*/**	11.14 ±0.04*	12.05 ±0.21*	12.45 ±0.34*/**
TAG, % of the total ApoB-LP composition	3.08 ±0.15*	2.90 ±0.09*/**	2.11 +0.12*/**	1.92 ±0.04*/**	1.96 ±0.03*/**
α - Tocopherol, mmol/L	5.70 ±0.35*	7.47 ±0.20*/**	7.19 ±0.17*/**	7.36 ±0.11*/**	8.13 ±0.06**
Isolated double bonds, U/ml	7.31 ±0.17*	7.67 ±0.08*/**	7.69 ±0.07*/**	7.99 ±0.05*/**	8.15 ±0.01*/**
Diene conjugates, mmol/L	31.68 ±1.65*	24.85 ±0.35*/**	23.44 ±0.40*/**	22.55 ±0.34*/**	21.88 ±0.23**
Ketodienes+conjugated trienes, U/ml	1.48 ±0.06*	1.24 ±0.03**	1.32 ±0.03*/**	1.25 ±0.03*/**	1.54 ±0.47*/**
Total hydroperoxides, mmol/L	78.31 ±1.33*	75.26 ±0.31*/**	75.62 ±0.54*/**	74.48 ±0.55*/**	73.41 ±0.39*/**

The data presented as mean±SD or percentage
* – p≤0.05 versus intact animals, ** – p≤0.05 versus model of MS

Table 10. The effect of *Vitis Vinifera* substances on the plasma HDL composition in male Syrian golden hamsters (1 year old) with the experimental MS (in each group n= 10)

The exact bimolecular mechanisms for this cardioprotection are unclear, but it is likely that actions mediated both through the estrogen receptors, such as the beneficial alteration in lipid profiles and upregulation of the low-density lipoprotein (LDL) receptor, and independently of the estrogen receptors, such as antioxidant action, contribute to the cardioprotective effects of phytoestrogens observed.

The potential role of phytoestrogens, including isoflavonoids, as cardioprotective agents has been extensively reviewed. The data obtained in our experiments showed that in male hamsters with the experimental MS the treatment with grape extracts reduced VLDL cholesterol (VLDL-C) and TG by 30 and 40 % compared with the control animals. Furthermore, golden Syrian hamsters fed with red wine phenolics had a significant decrease in the plasma apo B concentrations. Similar to our previous study, grape polyphenols may have altered hepatic secretion of TG-rich VLDL. This reduction is evident when observing the decreases in both plasma apo B and apo E concentrations. The significant decrease in apo E concentrations may have further reduced plasma TG concentrations. In general, apo E displaces apo C-II from the VLDL particle, thereby inhibiting the lipoprotein lipase (LPL) activity and overall lipolysis. Furthermore, Huang et al. [51] showed that adding apo C-II to transgenic apo-E3–enriched

VLDL increased the LPL activity in a dose-dependent manner. The reductions in apo E and TG concentrations suggest less displacement by apo E, thereby promoting the grape polyphenols activity and further reducing the TG concentrations in the plasma.

Due to decreases in TG concentrations, administration of "Cabernet" extract was shown to affect the overall lipoprotein metabolism. Decreased concentrations of the plasma TG altered substrate availability in the delipidation cascade, leading to the decrease observed in LDL-C concentrations. After a 3-week treatment period the grape polyphenols treatment induced a significant decrease in the cholesteryl ester transfer protein (CETP) activity as well. Such decrease in the CETP activity may be partially a result of the substantial decrease in substrate availability, including both the plasma TG and LDL-C.

It is evident that grape polyphenols modify the packaging of VLDL through alteration in the hepatic enzyme activity and apo B secretion. These modifications seem to decrease the overall secretion of the VLDL particles and therefore, decrease plasma TG and related apo concentrations. Due to decrease of the TG substrate, further modifications in the lipoprotein metabolism may occur.

The alteration in the TG metabolism may not be the single mechanism driving the hypocholesterolemic effects of grape polyphenols. When golden Syrian hamsters were treated with dealcoholized red wine, red wine, or grape juice, similar significant reductions in both TC and LDL-C concentrations were apparent in all treatment groups compared with the control [51]. Although there was a trend for decrease in TG concentrations in all treatment groups compared with the control, the differences were not significant. That study, along with others, suggests the presence of an additional mechanism by which grape polyphenols exert the cardioprotective effect. In Hep G-2 cells, dealcoholized red wine was shown to upregulate significantly the LDL receptor activity. This significant increase in activity was similar to the increase seen when Hep G-2 cells were treated with atorvastatin. Furthermore, when Hep G-2 cells were treated with increasing doses of red wine, LDL receptor mRNA abundance was significantly increased in a dose-responsive manner. The increase of the LDL receptor activity and abundance may be a result of the homeostatic intracellular cholesterol feedback loop. In general, decrease in the intracellular cholesterol will upregulate the LDL receptor expression and activity, whereas increase in the intracellular cholesterol will downregulate the receptor [48]. Grape polyphenols were shown to decrease hepatic cholesterol concentrations; therefore, the liver compensates for this deficiency by upregulating the LDL receptor and the overall decrease in the plasma LDL concentrations occurs.

One possible explanation of the anti-atherogenic activity of grape polyphenols is the well-known HDL cholesterol-increasing effect of polyphenols in various species, including transgenic mice [52].

In our experiments it has been found that the grape extract treatment induced slight (15%) increase in HDL cholesterol concentrations is possibly related to the significant decrease in the hepatic lipase activity (Table 11). The reductions observed in both hepatic and LPL activities by grape polyphenols treatment may prevent formation of small atherogenic VLDLB particles and may also decrease their uptake by the LDL receptor -related protein.

In addition to increases in HDL cholesterol concentrations, grape extracts also change the size and quality of HDL particles [53]. Although the mechanisms by which polyphenols influence the metabolism of HDL particles are not clear, changes in LPL and cholesteryl ester transfer protein (CETP) may play an important role.

Polyphenols treatment in humans is associated with decrease in the CETP content correlated with the concomitant increase in HDL cholesterol concentrations [54]. Consistent with our findings, grape extracts caused a significant increase in the postheparin LPL activity and HDL cholesterol concentrations in patients with moderate hypercholesterolemia and in hamsters [39]. However, the HDL cholesterol-increasing action of polyphenols in animals (mouse, hamster and rat) without CETP in some cases [52] suggests that this effect is may be independent of the CETP activity.

milliunits	Control (MS) (n=50)	Grape extract "Cabernet" (n=50)
LPL	356.0±53.2	258.6±57.3*
Hepatic lipase	232.6±25.9	216.2±34.7

*P<0.05 versus control animals.

Table 11. The plasma postheparin lipases activity in male hamsters with MS (in each group n=10)

The "Cabernet" extract appeared to be the most effective substance in relation to the HDL defence from peroxidation, though the other substances revealed the same but not so high activity. They decreased the content of products (diene conjugates, ketodienes+coupled trienes, total hydroperoxides) effectively and increased – substrates (compounds with isolated double bonds) of lipoperoxidation, prevented decrease of the antioxidant level (α-tocopherol).

It should be pointed out that the level of lipid peroxidation secondary products (ketodienes+coupled trienes) decreased more effectively under the influence of "Enoant" (to intact values).

Under the action of the studied substances the lipoprotein supply to the liver also decreases, evidenced by the decrease of the ApoB-LP content in the organ. Moreover, in the composition of these lipoproteins the TAG content normalizes, and it indicates normalization of the activity of lipases catalyzing the lipoprotein metabolism in the blood (Table 12).

The liver oxidative status is also improved: the antioxidant levels almost restore, the peroxidation products content decreases, the content of compounds with isolated double bonds increases (Tables 3, 5, 9, 12).

Testing of the "Enoant" action – one of the substances studied – in female hamsters of different age with the experimental MS proved the effectiveness of the antioxidant therapy of this pathology.

So, the total lipids, TAG and FFA contents decrease in the blood plasma of those animals under the action of "Enoant" (Table 12). In addition, in adult females "Enoant" causes decrease in the ApoB-LP and total cholesterol content, and it, in turn, reduces atherogenic changes in MS.

Parameter	Group				
	MS	MS + "Enoant"	MS + "Polyphen "	MS+ "Isabella"	MS+ "Cabernet"
Total cholesterol, % of the total ApoB-LP composition.	7.18±0.06*	8.23±0.26*/**	8.46±0.05**	8.76±0.05**	9.10±0.13**
TAG, % of the total ApoB-LP composition	42.00±1.29*	44.64±0.52**	44.42±0.43**	45.95±0.50**	45.41±0.73**
Isolated double bonds, U/g	2.13±0.06*	2.39±0.04*/**	2.71±0.03*/**	2.99±0.09**	2.77±0.16**
Total hydroperoxides, mmol/g	101.03±2.00*	90.55±1.54*/**	88.69±1.02*/**	80.46±0.77*/**	78.83±2.71*/**

The data presented as mean±SD or percentage
* – p≤0.05 versus intact animals, ** – p≤0.05 versus the model of MS

Table 12. The effect of *Vitis Vinifera* substances on the liver cytosol ApoB-LP composition in male Syrian golden hamsters (1 year old) with the experimental MS (in the crude tissue, in each group n= 10)

The increase of α-tocopherol (the main lipid-phase antioxidant) in the blood plasma of animals that received "Enoant" proved its antioxidant activity in our experiment (Table 13).

Furthermore, the significant decrease of the body weight was observed in hamsters that received "Enoant" along with a high-calorie diet compared to the animals on a high-calorie diet alone.

Based on these findings, we may conclude that introduction of grape polyphenolic extracts and concentrates in MS can prevent the increase of the total lipid and ApoB-LP content in the blood plasma, prevent the activation of free radical processes in the plasma lipoprotein particles, and normalize the liver lipid metabolism. The ability of the investigated substances to reduce negative consequences of MS such as atherosclerosis development has been proven.

The last suggestion is confirmed by our results concerning the aorta wall lipid composition in the experimental MS. The introduction of "Enoant" for prophylaxis and treatment reduces significantly atherogenesis manifestations in the aorta, decreasing the aorta media lipidation and the neutral lipid content (Fig. 3, 4).

Thus, from our data, we can conclude that antioxidants, particularly grape polyphenolic concentrates and extracts, which have pronounced antioxidant, phytoestrogenic and stress-protector properties, should be included into a complex therapy of MS to reduce its negative effects.

The next experiment was designed to investigate the action of grape wines and polyphenolic concentrates on development of proatherogenic effects of the emotional-painful stress. In our experiments we used purebred female rats because, as it was shown in previous studies, the acute stress response in females was more expressive than in males.

Age	Group	Parameter						
		Total lipids, mg/ml	ApoB-LP, mg/ml	Total cholesterol, mmol/L	TAG, mg/ml	FFA, mmol/L	Diene conjugates in ApoB-LP, nmol/ml	α - Tocopherol, nmol/ml
4 weeks	MS	4.52 ±0.17	4.00 ±0.16	2.04 ±0.08	1.08 ±0.49	1.17 ±0.06	24.82 ±1.46	6.67 ±0.22
	MS+ "Enoant"	3.54 ±0.16**	3.47 ±0.13	1.88 ±0.06	0.91 ±0.02**	0.95 ±0.02**	22.27 ±0.99	9.79 ±0.77**
20 weeks	MS	7.75 ±0.20	3.84 ±0.11	2.58 ±0.07	1.40 ±0.04	1.42 ±0.04	23.58 ±1.35	10.49 ±0.82
	MS+ "Enoant"	6.88 ±0.14**	3.30 ±0.08**	2.22 ±0.05**	1.18 ±0.03**	1.15 ±0.03**	21.10 ±1.14	12.87 ±0.36**

The data presented as mean±SD or percentage
** – $p \leq 0.05$ versus the model of MS

Table 13. The effect of polyphenol concentrate "Enoant" on some plasma lipid metabolism values in female Syrian golden hamsters with the experimental MS (in each group n= 10)

Age	Group	Parameters			
		Total lipids, mg/g	α - Tocopherol, nmol/g	Ascorbic acid, mkmol/g	TBA active substances, nmol/g
4 weeks	MS	140.75 ±9,15	21.01 ±1.47	3.73 ±0.14	1.97 ±0.06
	MS+ "Enoant"	111.53 ±4.08**	25.31 ±0.34**	4.08 ±0.10	1.74 ±0.09
20 weeks	MS	154.18 ±2.70	19.59 ±0.39	6.10 ±0.35	1.95 ±0.09
	MS+ "Enoant"	121.04 ±4.18**	26.11 ±1.03**	5.96 ±0.24	1.73 ±0.09

The data presented as mean±SD or percentage
** – $p \leq 0.05$ versus the model of MS

Table 14. The effect of polyphenol concentrate "Enoant" on some liver lipid metabolism values in female Syrian golden hamsters with the experimental MS (in each group n= 10)

During 21 days animals were daily given *per os* grape wines of "Cabernet" and "Rkatsiteli" grades in the doses that corresponded to 300 ml of wine for a human of 70 kg. Other animals were given alcohol in the dose corresponding to 30 ml of alcohol for a human of 70 kg, as well as polyphenolic concentrates "Enoant" and "Polyphen" in the doses of 0.05 ml/kg of the body weight. The grape wines and polyphenolic concentrates were produced by the National Institute of Grape and Wine "Magarach". Control animals were given the corresponding volume of the physiological solution.

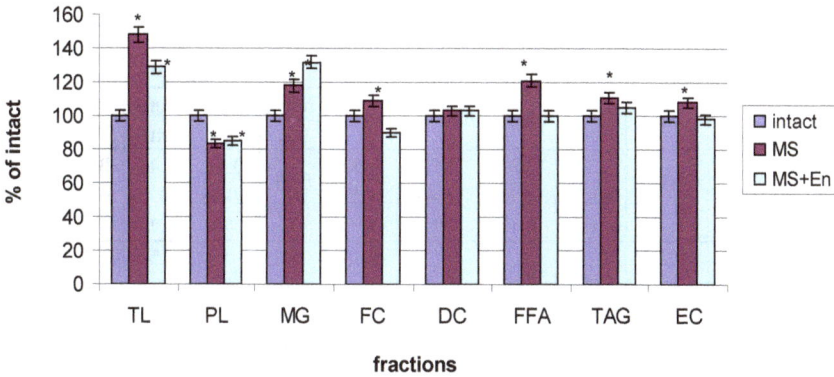

Figure 3. The lipid content in the aorta wall in male Syrian golden hamsters with the experimental MS and "Enoant" treatment (M±m, in each group n=10).

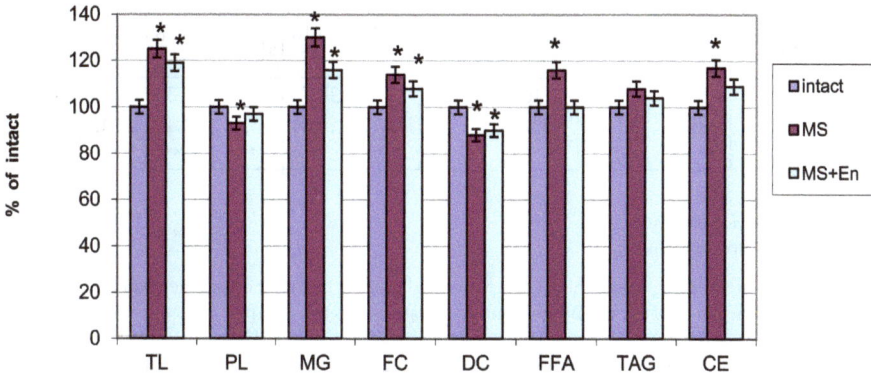

Figure 4. The lipid content in the aorta wall in female Syrian golden hamsters with the experimental MS and "Enoant" treatment (M±m, in each group n=10).

It was shown that all the substances investigated: polyphenolic concentrates "Enoant" and "Polyphen", 10% solution of ethanol, and grape wines "Cabernet" and "Rkatsiteli" possessed the stress-protective activity, which intensity was dependent on the substance used (Tables 15-17).

When introducing only "Enoant" and "Polyphen" to the animals these complexes did not cause any changes on the investigated indexes of the pro-oxidant and antioxidant status in the liver and it is an indication about safety of using these concentrates.

"Enoant" and "Polyphen" revealed the significant protective activity in the emotional-painful stress. It allows to use them as stress-protective, hepatoprotective and antiatherogenic remedies.

Parameter	Group									
	Stress + Enoant		Stress + Polyphen		Stress + ethanol		Stress + Cabernet		Stress + Rkatsiteli	
	str.	non-str.	str.	non-str.	str.	non-str.	str.	non-str.	str.	non-str.
TL, mg/ml	4.14 ±0.45*	3.98 ±0.31 **	3.81 ±0.03 */**	3.82 ±0.48 **	5.50 ±0.74 */**	5.58 ±0.37*	3.41 ±0.66 **	3.46 ±0.43 **	3.43 ±0.15 **	3.57 ±0.54 **
TAG, mg/ml	0.76 ±0.08 */**	0.78 ±0.09*	0.51 ±0.08	0.71 ±0.09*	0.66 ±0.12 */**	0.91 ±0.05 */**	0.50 ±0.05	0.58 ±0.04 */**	0.54 ±0.10	0.54 ±0.05 **
Total cholesterol, mg/ml	70.76 ±9.34*	54.90 ±6.92 */**	74.14 ±8.21*	82.93 ±9.48 **	85.71 ±5.71	69.10 ±7.37 **	56.71 ±7.32*	62.25 ±6.50 **	97.51 ±9.66 **	80.83 ±8.74 **
HDL, mg/ml	0.98 ±0.14	0.89 ±0.08	0.96 ±0.05	0.88 ±0.08	1.10 ±0.19	1.00 ±0.12	1.26 ±0.15	1.04 ±0.08	1.08 ±0.12	0.96 ±0.05
ApoB-LP, mg/ml	1.56 ±0.12 **	1.89 ±0.26*	1.59 ±0.16 **	1.54 ±0.22 **	1.85 ±0.19 **	1.67 ±0.18 **	2.07 ±0.29*	1.67 ±0.18 **	1.86 ±0.20 **	1.63 ±0.18 **
Corticosterone, nmol/l	5.50 ±0.99 */**	5.87 ±1.08 */**	7.10 ±0.80 */**	29.33 ±8.58	15.00 ±1.72 */**	38.75 ±5.64 **	32.75 ±5.17 **	26.33 ±4.40 */**	16.93 ±2.43 */**	40.60 ±6.38 **

The data presented as mean±SD

* - p≤0.05 versus intact animals

** - p≤0.05 versus stressed animals

Table 15. The effect of grape polyphenol complexes and grape wines on the lipid metabolism and the plasma corticosterone level in rats with the neurogenic stress (in each group n=10).

Moreover, we have found out that the stress-protective activity of grape wines is equal to the polyphenolic concentrates activity given in the similar dose.

Wines of "Cabernet" and "Rkatsiteli" grades normalized the total lipid content both in a liver homogenate and in the blood plasma in stress; in addition, TAG levels also reached the control values.

Grape wine components prevented the FFA content increase noted when introducing the solution of alcohol. This fact may prove the protective action of the components mediated by inhibition of fatty infiltration of organs.

The latter is confirmed by the absence of influence of grape wine introduction on the NADPH-generating dehydrogenases activity in the liver.

The cholesterol content decrease in the blood plasma when introducing grape wines has attracted our attention, as well as a favourable redistribution of cholesterol in the LP fractions – decrease of ApoB-containing lipoprotein level with the unchanged HDL content.

Parameter		Stress + Enoant		Stress + Polyphen		Stress + ethanol		Stress + Cabernet		Stress + Rkatsiteli	
		str.	non-str.	str.	non-str.	str.	non-str.	str.	non-str.	str.	non-str.
PON, nmol/ml/min		240.25 ±20.86 */**	216.00 ±16.24 */**	231.33 ±15.53 */**	231.25 ±15.52 **	197.50 ±14.37 */**	150.00 ±19.24 */**	239.25 ±24.18 */**	246.00 ±16.56 **	222.75 ±13.98**	215.25 ±13.78 **
Ascorbic acid, mkmol/ml		49.99 ±4.41*	43.20 ±4.16*	53.76 ±4.51 **	45.08 ±6.08 */**	38.04 ±4.12 */**	34.83 ±4.43*	48.07 ±2.13*	48.10 ±2.49 */**	55.87 ±6.65 **	53.41 ±5.42 */**
α-Tocopherol, nmol/ml		10.70 ±1.34 **	10.62 ±0.67 **	10.39 ±1.12 **	10.8 ±1.26 **	6.07 ±0.76 */**	5.35 ±0.78 */**	9.45 ±1.11	9.99 ±0.48 **	10.26 ±0.48 **	9.32 ±1.04
Oxidized ApoB-LP	Isolated double bonds, U/ml	2.76 ±0.36*	2.23 ±0.24*	3.14 ±0.46 **	3.51 ±0.42 */**	3.13 ±0.37 **	2.35 ±0.24*	3.25 ±0.38 **	2.64 ±0.28*	4.13 ±0.43 **	3.11 ±0.41*
	Diene conjugates, mmol/ml	22.47 ±3.20 */**	25.93 ±2.98 */**	21.54 ±1.22 */**	25.93 ±1.76 */**	30.44 ±2.20*	26.28 ±1.24*	23.05 ±4.48*	26.99 ±1.38*	19.54 ±0.93 **	21.56 ±0.99 */**
	Ketodienes+conjugated trienes, U/ml	2.31 ±0.14	2.58 ±0.29 */**	2.20 ±0.11 **	2.37 ±0.10*	3.01 ±0.50*	2.97 ±0.42*	2.29 ±0.16	2.60 ±0.31*	2.22 ±0.11	2.23 ±0.06 **

The data presented as mean±SD
* - p≤0.05 versus intact animals
** - p≤0.05 versus stressed animals

Table 16. The effect of grape polyphenol complexes and grape wines on the plasma oxidant/antioxidant status in rats with the neurogenic stress (in each group n=10).

Paraoxonase activity was normalized in the animals given wines and the antioxidant content both in the blood plasma and in the liver was significantly higher than in the control animals.

These effects together with much lower level of ApoB-LP oxidation in the animals given grape wines prove the high antiatherogenic potential of the wines investigated.

In addition, the grape wines have revealed a rather high level of the stress-protective activity and it is indicated by a significant decrease of the corticosterone content in the blood plasma in stressed animals given wines.

Since grape wines have shown a high level of the stress-protective activity we investigated how the ratio of wine components – polyphenols and ethanol – can influence the stress-protective activity of this complex.

As was shown in our experiments "Enoant" administration even in the combination with ethanol does not reduce the stress-protective action of it, but on the contrary – it intensifies this action preventing unfavourable effects of the alcohol. At the same time the TAG and FFA level in the liver tissue of rats given ethanol together with "Enoant" decreases even when using the lowest dose investigated (0.01 ml per 100 g) (Table 18). Since the content of TAG and FFA increases apparently due to the lipogenesis activation when using ethanol, which might lead to fatty infiltration of the liver, then reduction of this process activity could protect the liver.

Parameter	Group									
	Stress + Enoant		Stress + Polyphen		Stress + ethanol		Stress + Cabernet		Stress + Rkatsiteli	
	str.	non-str.	str.	non-str.	str.	non-str.	str.	non-str.	str.	non-str.
Total lipids, mg/g	149.17 ±33.14	151.84 ±19.86 **	130.98 ±23.66	201.93 ±33.71 **	193.32 ±25.61 */**	220.54 ±23.22 */**	171.39 ±28.66 **	180.31 ±15.98 **	164.93 ±16.93 **	190.03 ±18.21**
TAG, mg/g	4.65 ±0.78	3.77 ±0.47	6.00 ±0.32	5.94 ±1.08 **	7.86 ±0.25 **	6.89 ±0.49 **	4.90 ±1.08	3.88 ±1.08	5.74 ±0.61	4.93 ±0.68
ApoB-LP, mg/g	4.58 ±0.06 **	4.43 ±0.07 */**	3.72 ±0.34*	3.15 ±0.44*	5.00 ±0.82	4.10 ±0.49 */**	3.06 ±0.16 */**	3.15 ±0.16 */**	2.87 ±0.30 */**	3.02 ±0.36*
FFA, mg/g	1.06 ±0.12	1.00 ±0.11	1.08 ±0.16	1.26 ±0.08*	1.35 ±0.15*	1.45 ±0.13*	1.05 ±0.22	1.22 ±0.20	1.15 ±0.13	1.40 ±0.31

The data presented as mean±SD
* - p≤0.05 versus intact animals
** - p≤0.05 versus stressed animals

Table 17. The effect of grape polyphenol complexes and grape wines on the liver lipid metabolism in rats with the neurogenic stress, in the crude tissue (in each group n=10).

Parameter	Group									
	Stress + Enoant		Stress + Polyphen		Stress + ethanol		Stress + Cabernet		Stress + Rkatsiteli	
	str.	non-str.	str.	non-str.	str.	non-str.	str.	non-str.	str.	non-str.
GSH, mkmol/g	6.02 ±0.27 */**	3.70 ±0.42*	5.12 ±0.46	5.48 ±0.34	4.40 ±0.34	2.87 ±0.29 */**	4.64 ±0.70	4.38 ±0.35	4.47 ±0.51	3.59 ±0.50*
α-Tocopherol, nmol/g	27.82 ±2.86 */**	34.01 ±3.56 **	34.34 ±4.14 **	25.39 ±1.18 */**	6.55 ±0.61 */**	5.59 ±0.54 */**	26.95 ±1.32 */**	24.64 ±0.81 */**	29.24 ±1.82 */**	28.9 ±3.15 **
Ascorbic acid, mkmol/g	1.31 ±0.13 **	1.29 ±0.10 */**	1.17 ±0.23*	1.25 ±0.15 */**	0.59 ±0.07 */**	0.81 ±0.06*	1.04 ±0.21*	1.21 ±0.11 */**	1.24 ±0.20 */**	1.24 ±0.17 */**
Isolated double bonds, U/g	18.63 ±1.88 */**	15.99 ±1.50 */**	20.40 ±1.34 */**	18.66 ±1.52 */**	16.98 ±0.66*	13.89 ±1.73*	24.40 ±2.92 */**	16.15 ±1.76 */**	26.67 ±1.62 */**	18.82 ±1.90 */**
Diene conjugates, mmol/g	13.58 ±0.74 */**	12.87 ±0.54 */**	13.48 ±0.62 */**	12.85 ±0.26 */**	13.61 ±0.08 */**	13.20 ±0.58 */**	10.90 ±1.10 **	15.00 ±2.12*	10.66 ±1.88 **	14.32 ±1.62*
Ketodienes+conjugated trienes, U/g	13.57 ±0.54*	13.39 ±1.67	11.44 ±1.56 **	11.84 ±1.88 */**	12.68 ±1.91	13.04 ±1.31 **	14.20 ±2.02*	15.74 ±1.98	17.15 ±1.62*	16.04 ±1.78
TBA-active products, nmol/mg protein	0.21 ±0.02*	0.19 ±0.02 **	0.28 ±0.03*	0.22 ±0.03 **	0.55 ±0.08 */**	0.68 ±0.06 */**	0.15 ±0.02	0.28 ±0.04*	0.18 ±0.02	0.17 ±0.02 **

The data presented as mean±SD
* - $p \leq 0.05$ versus intact animals
** - $p \leq 0.05$ versus stressed animals

Table 18. The effect of grape polyphenol complexes and grape wines on the liver tissue oxidant/antioxidant status in rats with the neurogenic stress (in the crude tissue, in each group n=10).

It should be noted that the effect of high doses of "Enoant" (0.1 and 0.15 ml/100 g) was ambiguous. On the one hand, it caused α-tocopherol accumulation in the liver that might be an indicator of their protective action, but on the other hand, it probably revealed some prooxidative effect initiating the increase in the content of the POL final products – thiobarbituric acid-active products, and also activating ApoB-containing lipoproteins oxidation. In this case the secondary oxidative stress developed. A tendency to decrease the lipid content in the liver and to increase it in the blood plasma testifies about it.

Such effect is typical for high doses of many antioxidants capable to reveal the prooxidant action, including α-tocopherol. These data indicate the necessity of reasonable attitude to antioxidants therapy, including "Enoant".

Parameter	Group						
	Stress +Ethanol	Stress+Ethanol+ Enoant, ml per 100 g of the body weight:					
		0.01	0.03	0.05	0.07	0.1	0.15
Total lipids, mg/ml	5.89 ±0.08*	5.69 ±0.06*	5.30 ±0.05*	4.82 ±0.15*	3.51 ±0.08	3.44 ±0.09	3.58 ±0.15
TAG, mg/ml	0.91 ±0.07*	0.99 ±0.04*	0.73 ±0.04*	0.52 ±0.03	0.43 ±0.03	0.39 ±0.02*	0.50 ±0.02
Total cholesterol, mg/ml	0.49 ±0.07	0.46 ±0.05	0.54 ±0.04	0.40 ±0.01*	0.55 ±0.09	0.55 ±0.03	0.56 ±0.02
HDL, mg/ml	0.93 ±0.02	1.00 ±0.06	1.02 ±0.07	0.91 ±0.05	1.81 ±0.03*	1.23 ±0.06*	1.32 ±0.04*
ApoB-LP, mg/ml	1.73 ±0.05*	1.84 ±0.05*	1.48 ±0.06	1.44 ±0.03	1.18 ±0.05	1.41 ±0.02	1.64 ±0.04*
α-Tocopherol, nmol/мml	4.11 ±0.34*	4.88 ±0.17*	5.84 ±0.14*	6.68 ±0.24*	8.20 ±0.34*	9.23 ±0.35	8.80 ±0.46
Ascorbic acid, mkmol/L	33.81 ±1.73*	34.04 ±2.73*	39.04 ±1.60*	49.25 ±1.10*	55.01 ±1.67	56.98 ±2.03	49.96 ±3.43
Diene conjugates in ApoB-LP, mkmol/L	28.11 ±0.34*	28.99 ±0.14*	29.35 ±0.80*	28.81 ±1.30*	23.91 ±0.51	20.60 ±1.43	28,27 ±1.35*
TBA-active products, mkmol/L	2.39 ±0.55*	2.01 ±0.30*	1.48 ±0.16*	1.11 ±0.04	1.18 ±0.34	0.77 ±0.21	1.31 ±0.20*
Corticosterone, nmol/L	35.25 ±4.27	24.00 ±3.03	28.25 ±4.54	16.00 ±0.44*	25.50 ±0.50	17.50 ±2.33*	21.60 ±6.00

The data presented as mean±SD
* - p≤0.05 versus intact animals

Table 19. The effect of different doses of polyphenol concentrate "Enoant" in combination with ethanol on the plasma parameters of the stress response development in rats with the neurogenic stress (in each group n=10).

At the same time small doses of "Enoant" have a relatively low biological activity; they do not reduce negative effects of ethanol intake and do not inhibit the stress response significantly.

Therefore, we can conclude that the most effective doses of "Enoant" are 0.05-0.07 ml/100 g of the body weight because with these doses "Enoant" has not only high stress-protective, antiatherogenic and hepatoprotective activities, but practically neutralizes negative effects of ethanol.

Thus, our results suggest that grape wines have a high stress-protective, antiatherogenic and hepatoprotective activity that is equal to grape polyphenolic non-alcoholic concentrates characteristics, and the wine components in the doses studied have prevented negative effects of ethanol. Introduction of ethanol to animals in the human equivalent dose – 0.43 ml/kg of the body weight increases their tolerance to stress, but is an unfavourable factor that could result in MS development, fatty infiltration of organs and other pathologies. The polyphenolic concentrates "Enoant" and "Polyphen" in the human equivalent dose – 0.3 ml/kg of the body weight reveal a significant stress-protective, hepatoprotective and anti-atherogenic activity under the action of the emotional-painful stress. Grape wines from "Cabernet" and "Rkatsiteli" grades in the human equivalent dose – 4.3 ml/kg of the body weight also reveal a high stress-protective, antiatherogenic and hepatoprotective activity equal to grape polyphenolic non-alcoholic concentrates, and the wine components in the doses used prevented the negative effect of ethanol.

The highest activity has been shown by the combination of "Enoant" and ethanol that corresponds to the ratio of components in dry red wines, as well as the absence of significant difference in the protective effects of red and white wines, in spite of the difference in the polyphenol content [55]. Based on these results, in the second series of our experiments we decided to investigate "Cabernet" and "Rkatsiteli" wine effects on the development of stress-reaction proatherogenic consequences under the action of the emotional-painful stress in different periods of introduction.

It has been shown that "Cabernet" had a higher level of the anti-atherogenic activity than "Rkatsiteli"; in relation to the stress-protective activity the wines of these grades did not differ markedly. Such effect is likely connected with accumulation of polyphenols in the organism.

To examine the last supposition it was necessary to determine how different periods of introduction of the investigated wines influenced the stress-reaction development. We have carried out the study of wine intake influence on the development of proatherogenic consequences of the emotional and painful stress in different terms after consumption.

The data obtained in the experiments showed significant improvement of the antioxidant status both in the blood plasma and the liver tissue one day after the introduction of "Cabernet" wine (tables 20, 21).

At the same time "Rkatsiteli" wine did not reveal such activity. A similar condition persisted for 2-5 days of administration.

Periods of time		Total lipids, mg/g	TAG, mg/g	GSH, mkmol/g	α-Tocopherol, nmol/g	Diene conjugates, nmol/g	TBA-active products, nmol/g
		Parameter					
Day 1	C+Str	94.94±5.65*	3.80±0.11*	2.12±0.17*	18.88±0.79*	14.93±0.37*	1.68±0.10
	R+Str	103.16±4.63*	5.15±0.61*	2.37±0.28*	15.60±0.39*	15.94±0.39*	2.24±0.32
Day 2	C+Str	105.17±5.12*	3.58±0.23*	2.51±0.34*	26.04±2.17*	14.13±0.09*	1.07±0.24*
	R+Str	106.93±9.42*	5.67±0.34*	2.77±0.87*	15.65±0.92*	15.10±0.07*	1.91±0.15
Day 3	C+Str	115.38±11.65*	3.92±0.13*	3.82±0.37*	23.69±2.18*	13.85±0.48*	1.70±0.11
	R+Str	103.28±5.81*	6.47±0.28*	2.47±0.26*	16.89±0.71*	14.44±0.25*	2.29±0.20*
Day 5	C+Str	139.14±8.06*	5.57±0.31*	2.43±0.37*	26.96±2.12	13.19±0.34	1.35±0.15
	R+Str	116.66±3.60*	6.88±0.37*	2.43±0.42*	18.64±1.18*	14.22±0.16*	2.06±0.13
Day 8	C+Str	161.18±6.05*	7.27±0.15	4.11±0.21	32.12±0.85	12.28±0.52	1.64±0.13
	R+Str	122.20±7.07*	6.35±0.45	2.33±0.28*	23.08±2.08*	12.93±0.44	1.73±0.04
Day 10	C+Str	181.82±9.24	8.04±0.63	5.15±0.42	30.62±2.53	11.86±0.12	1.30±0.19*
	R+Str	153.99±5.30*	8.41±0.56	3.22±0.48*	26.31±1.26*	11.84±0.48	1.29±0.14*
Day 12	C+Str	174.72±6.15	8.34±0.55	3.74±0.25	32.55±5.58	12.42±0.36	1.47±0.27
	R+Str	164.4±8.03*	8.63±0.47	3.25±0.17*	28.28±2.26	11.41±0.22	1.45±0.31
Day 15	C+Str	162.16±12.81	7.50±0.43	4.61±0.22	38.16±2.06*	10.50±0.52*	1.34±0.06*
	R+Str	172.13±10.42	7.68±0.69	4.42±0.34	41.99±2.42	10.27±0.63*	1.24±0.17

The data presented as mean±SD
* - p≤0.05 versus intact animals

Table 20. The effect of prophylactic administration of grape wines of "Cabernet" (C) and "Rkatsiteli" (R) grades on the stress response development in the liver tissue in rats with the neurogenic stress in different periods of time, in the crude tissue (in each group n=10).

Periods of time		Parameter						
		Total lipids, mg/ml	TAG, mg/ml	Total cholesterol, mg/ml	ApoB-LP, mg/ml	α-Tocopherol, nmol/ml	Diene conjugates, nmol/ml	Corticosterone nmol/l
Day 1	C+Str	5.15±0.50*	0.78±0.07*	0.94±0.04	1.53±0.02*	8.52±0.27*	31.49±2.58*	75.00±8.66*
	R+Str	6.15±0.44*	0.93±0.07*	1.08±0.05ᶻ	1.68±0.03*	6.29±0.47*	38.35±1.56*	111.70±10.00
Day 2	C+Str	3.87±0.23	0.71±0.04*	0.89±0.03	1.52±0.02*	8.90±0.72*	31.65±1.92*	96.67±21.86*
	R+Str	4.85±0.16*	0.84±0.06*	0.86±0.04	1.70±0.04*	7.77±0.27	32.17±1.71*	81.54±16.50*
Day 3	C+Str	3.42±0.32	0.56±0.03	0.73±0.05*	1.54±0.04*	10.20±0.52	24.72±2.89	35.67±9.49
	R+Str	4.45±0.41	0.70±0.04*	0.89±0.06	1.66±0.04*	8.49±0.38*	22.24±1.84	43.00±8.50
Day 5	C+Str	3.54±0.27	0.49±0.05	0.72±0.02*	1.45±0.06	10.28±0.65	28.34±1.77*	41.50±18.50
	R+Str	3.56±0.23	0.58±0.05	0.88±0.02	1.66±0.04*	10.01±0.27	20.46±2.03	57.10±3.00*
Day 8	C+Str	3.47±0.39	0.49±0.08	0.77±0.02*	1.36±0.03	12.02±0.35	20.47±2.03	42.50±10.61
	R+Str	4.08±0.31	0.4±0.06	0.84±0.03	1.55±0.04	10.04±0.68	20.15±2.61	40.50±9.19
Day 10	C+Str	3.73±0.24	0.54±0.05	0.69±0.02*	1.33±0.02	11.03±0.89	22.68±2.46	34.00±18.38
	R+Str	3.60±0.35	0.55±0.09	0.76±0.04*	1.43±0.04	11.39±0.47	22.90±1.93	38.00±2.82
Day 12	C+Str	3.71±0.35	0.49±0.07	0.65±0.03*	1.26±0.04	11.18±1.01	21.72±1.57	25.50±2.12
	R+Str	3.55±0.45	0.054±0.06	0.75±0.03*	1.35±0.03	11.40±0.93	20.08±1.45	27.00±1.31
Day 15	C+Str	3.78±0.36	0.59±0.04	0.61±0.02*	1.33±0.02	11.04±1.32	20.99±0.92	34.00±7.00
	R+Str	4.22±0.57	0.51±0.09	0.69±0.03*	1.29±0.06	11.72±0.93	20.59±1.92	31.00±10.82

The data presented as mean±SD
* - p≤0.05 versus intact animals

Table 21. The effect of prophylactic administration of grape wines of "Cabernet" (C) and "Rkatsiteli" (R) grades on the plasma parameters of the stress response development in rats with the neurogenic stress (in each group n=10).

However, on day 8 of administration the antioxidant and stress-protective effects of these wines were almost similar, and on the day 10 –they practically did not differ.

On days 12 and 15 there were also no differences as to the antioxidant and stress-protective action of the wines studied, which significantly reduced activation of the free radical oxidation under the action of stress normalizing the most of the indexes investigated.

Thus, the investigated wines are characterized by the high level of the antioxidant and stress-protective activity, and in the first days of introduction "Cabernet" wine improved more effectively the antioxidant status in the blood and the liver tissue than "Rkatsiteli" wine, but by day 10 the effects of the studied wines had no substantial difference.

Probably, these results are dependent on polyphenol cumulation in the organism because it is known that the polyphenol content of "Cabernet" is 10 times more than of "Rkatsiteli".

Thus, the results suggest that "Cabernet" and "Rkatsiteli" wines have already revealed the high stress-protective, hepatoprotective and anti-atherogenic activity in the conditions of the emotional-painful stress on the 2-3 days after introduction, and practically normalized the oxidative status and the lipid metabolism under the action of stress in prophylactic administration within 10 days. This indicates that grape polyphenols possess a high total antioxidant activity. At the same time the last suggestion required further research.

In order to examine the effects of wine stocks and polyphenolic concentrates obtained from other grape grades on development of proatherogenic consequences of the emotional-painful stress we investigated the action of substances obtained from the grapes of hybrid grades "Krasen", "Golubok" and "Podarok Magaracha" produced by the National Institute of Grape and Wine "Magarach".

In the series of experiments we used purebred male rats that during 21 day were given daily, *per os*, table wine stocks of the grades "Podarok Magaracha", "Krasen" and "Golubok" in the human equivalent dose corresponding to 300 ml of wine for a human with 70 kg of the body weight. Other groups of animals were given ethanol in the human equivalent dose corresponding to 30 ml of ethanol for a human with 70 kg of the body weight taking into account the species sensitivity coefficients, as well as the table wine stocks of the grades mentioned in doses equivalent to the polyphenol content of the given wines calculated by the polyphenol content in active doses (AD – 9 mg of polyphenols/100 g of the body weight).

The results have demonstrated that not only polyphenolic concentrates, but the table wine stocks also revealed a substantial stress-protective activity to a different extent (Tables 22-25).

In fact, "Krasen" table wine stock revealed the highest activity; the stress-protective activity was almost 2.4 times more the ethanol activity in the dose studied. This product effectively prevented the activation of free radical oxidation both in the blood (increased the level of compounds with isolated double bonds in the atherogenic ApoB-LP, decreased the content of peroxidation products – diene conjugates – almost 3 times comparing to the stressed animals, and 15% - comparing to the intact animals), and the liver tissue (prevented the antioxidant content decrease, particularly the content of α-tocopherol and ascorbic acid returned practically to the intact level, and there was 40% decrease of the diene conjugates level). At the same time this table wine stock prevented hyperlipidemia and the shift of metabolism to the increased lipolysis, there was 60% decrease of the blood total lipid content comparing to the stressed animals, and 11% - comparing with the intact animals. At the same time the TAG content in the liver was equal to the intact level that also demonstrated the protective action of this table wine stock. Reduction of lipogenesis in the liver tissue under the action of this product is important, and it protects the organ from steatosis. It should be also mentioned that the given product normalized the cholesterol content in the blood plasma.

Group	Parameter				
	Total lipids, mg/g	TAG, mg/g	FFA, mmol/g	ApoB-LP, mg/g	Lysosomal lipase, nmol/mg protein/min
Str.+Con. Podarok Magaracha (AD)	149.03 ±2.59*,**	5.74 ±0.07**	4.46 ±0.04*	4.22 ±0.03*,**	0.45 ±0.03**
Str.+Con. Krasen (AD)	147.13 ±1.15*	4.27 ±0.02*,**	4.19 ±0.05*,**	4.61 ±0.01*,**	0.50 ±0.02
Srt.+Wine Podarok Magaracha	161.74 ±1.91*,**	6.33 ±0.05**	4.30 ±0.11*,**	4.43 ±0.03*,**	0.32 ±0.01*,**
Srt.+Wine Krasen	155.88 ±1.35*,**	6.04 ±0.16**	3.24 ±0.04**	4.50 ±0.07*,**	0.56 ±0.03**
Str.+Con. Podarok Magaracha (DW)	141.29 ±1.79*	4.95 ±0.15*,**	3.83 ±0.09**	3.18 ±0.03*	0.65 ±0.02**
Str.+Con. Krasen (DW)	145.87 ±3.19*	4.24 ±0.07*,**	4.23 ±0.07*,**	4.39 ±0.08*,**	0.55 ±0.03**
Wine Podarok Magaracha	182.4 ±3.08*	6.51 ±0.07**	2.70 ±0.08*,**	4.80 ±0.10*,**	0.37 ±0.01*
Wine Krasen	164.36 ±1.86	5.97 ±0.17	2.95 ±0.09*	4.93 ±0.18*	0.35 ±0.02*
Con. Podarok Magaracha (AD)	191.33 ±2.03*	6.47 ±0.04*	4.61 ±0.31*	4.33 ±0.08*	0.83 ±0.03*
Con. Krasen (AD)	170.4 ±2.09	6.15 ±0.14	3.22 ±0.05	4.97 ±0.11*	0.69 ±0.02#
Ethanol	229.76 ±3.39*	7.40 ±0.13*	4.57 ±0.13*	6.11 ±0.07*	0.38 ±0.02*

The data presented as mean±SD
* - p≤0.05 versus intact animals
** - p≤0.05 versus stressed animals

Table 22. The effect of grape polyphenol concentrates and grape wines on the liver lipid metabolism in rats with the neurogenic stress (in the crude tissue, in each group n=10).

It is also necessary to point out that the control intake of the investigated substances (Tables 22-25) did not reveal negative effects on the organisms of the experimental animals. Moreover, in addition to the antioxidant activity these substances revealed a significant hypocholesterolemic and anti-atherogenic action, which was more pronounced when using "Krasen" grade wine stock and the concentrate.

Group	Parameter				
	GSH, mkmol/g	α-Tocopherol, nmol/g	Ascorbic acid, mkmol/g	Diene conjugates, nmol/g	TBA-active products, nmol/ mg protein
Str.+Con. Podarok Magaracha (AD)	3.34 ±0.02*,**	24.01 ±0.47*,**	1.21 ±0.01*,**	14.96 ±0.22*,**	0.49 ±0,01
Str.+Con. Krasen (AD)	3.8 ±0.02*,**	26.71 ±0.46*,**	1.28 ±0.01*,**	15.03 ±0.11*,**	0,45 ±0.02
Srt.+Wine Podarok Magaracha	3.48 ±0.02*,**	26.71 ±0.34*,**	1.33 ±0.02*,**	13.81 ±0.18#,**	0.42 ±0.03**
Srt.+Wine Krasen	3.57 ±0.08*,**	28.46 ±0.75**	1.41 ±0.03*,**	13.36 ±0.40**	0.21 ±0.01*,**
Str.+Con. Podarok Magaracha (DW)	2.00 ±0.07*,**	21.72 ±0.51*,**	1.02 ±0.03*,**	16.05 ±0.11*,##	0.47 ±0.01
Str.+Con. Krasen (DW)	3.27 ±0.06*	21.12 ±0.39*,**	1.04 ±0.03*,**	16.72 ±0.16*,##	0.45 ±0.02
Wine Podarok Magaracha	4.68 ±0.10*,**	33.59 ±0.60	2.16 ±0.04*	10.42 ±0.53*	0.15 ±0.01*
Wine Krasen	4.56 ±0.24	35.48 ±0.78*	1.57 ±0.03	9.27 ±0.24*	0.13 ±0.01*
Con. Podarok Magaracha (AD)	4.46 ±0.13	35.29 ±0.45*	1.95 ±0.03*	10.87 ±0.41	0.20 ±0.01*
Con. Krasen (AD)	4.82 ±0.14*	27.84 ±0.39	2.00 ±0.04*	10.26 ±0.06*	0.15 ±0.01*
Ethanol	5.31 ±0.35*	24.16 ±1.40*	2.06 ±0.03*	12.62 ±0.60	0.46 ±0.02

The data presented as mean±SD
* - p≤0.05 versus intact animals
** - p≤0.05 versus stressed animals

Table 23. The effect of grape polyphenol concentrates and grape wines on the oxidant/antioxidant status in the liver tissue in rats with the neurogenic stress (in the crude tissue, in each group n=10).

Group	Parameter						
	Total lipides, mg/ml	TAG, g/ml	FFA, mmol/L	Total cholesterol, g/ml	HDL, mg/ml	ApoB-LP, mg/ml	Cortico-sterone, nmol/L
Str.+Con. Podarok Magaracha (AD)	3.89 ±0.08**	0.72 ±0.01*,**	1.40 ±0.02#,**	56.66 ±1.49*,**	0.79 ±0.02*,**	1.45 ±0.05*,**	47 ±2**
Str.+Con. Krasen (AD)	4.12 ±0.08#	0.67 ±0.01*,**	1.16 ±0.03*,**	58.70 ±1.77	0.82 ±0.03	0.,90 ±0.04*,**	47 ±3**
Srt.+Wine Podarok Magaracha	4.60 ±0.11*	0.65 ±0.02*,**	1.63 ±0.02*	63.32 ±1.01**	0.85 ±0.02**	1.21 ±0.02**	41 ±1*,**
Srt.+Wine Krasen	3.42 ±0.09*,**	0.52 ±0.03**	1.40 ±0.04**	64.69 ±1,70**	0.86 ±0.02	1.08 ±0.03#,**	42 ±1**
Str.+Con. Podarok Magaracha (DW)	5.25 ±0.07*	0.74 ±0.02*	0.74 ±0.02*,**	56.67 ±1.15*,**	0.87 ±0.03	1.75 ±0.07*,**	60 ±2
Str.+Con. Krasen (DW)	3.88 ±0.08**	0.67 ±0.01*,**	0.67 ±0.01*,**	55.58 ±1.21*,**	0.83 ±0.03**	1.06 ±0.03*,**	63 ±1*
Wine Podarok Magaracha	3.72 ±0.09	0.38 ±0.01*	1.14 ±0.02*	60.78 ±1.64	1.4 ±0.02*	1.11 ±0.02	54 ±2
Wine Krasen	3.76 ±0.03	0.36 ±0.01*	1.18 ±0.02*	57.21 ±0.81#	1.11 ±0.02*	1.14 ±0.01	44 ±4
Con. Podarok Magaracha (AD)	4.32 ±0.04*	0.56 ±0.02	1.23 ±0.01	68.70 ±1.22	1.04 ±0.02*	1.14 ±0.02	71 ±2*
Con. Krasen (AD)	3.78 ±0.04	0.39 ±0.02*	1.43 ±0.04*	60.89 ±1.67	1.16 ±0.02*	1.20 ±0.02	35 ±2*
Ethanol	4.05 ±0.09	0.74 ±0.03*	1.65 ±0.02*	55.80 ±1.46*	1.24 ±0.06*	1.32 ±0.02*	75 ±3*

The data presented as mean±SD
* - p≤0.05 versus intact animals
** - p≤0.05 versus stressed animals

Table 24. The effect of grape polyphenol concentrates and grape wines on the plasma lipid metabolism parameters and corticosterone level in rats with the neurogenic stress (in each group n=10).

Group	Parameter				
	PON, nmol/ ml×min	Ascorbic acid, mkmol/L	α-Tocopherol, nmol/ml	Isolated double bonds in ApoB-LP	Diene conjugates in ApoB-LP
Str.+Con. Podarok Magaracha (AD)	174 ±2*	37.26 ±1.60*,**	7.77 ±0.21*	2.07 ±0.09*	27.03 ±1.30*,#
Str.+Con. Krasen (AD)	194 ±3*,**	41.79 ±0.44*,**	8.25 ±0.66*,**	2.35 ±0.15*	26.86 ±0.43*
Srt.+Wine Podarok Magaracha	173 ±3*	40.08 ±1.72*,**	8.91 ±0.29**	1.92 ±0.25*	18.45 ±0.18*,**
Srt.+Wine Krasen	189 ±2*,**	54.30 ±0.97*,**	10.42 ±0.16*,**	2.47 ±0.03*,**	14.50 ±0.35*,**
Str.+Con. Podarok Magaracha (DW)	159 ±1*	35.46 ±1.10*	7.04 ±0.07*	1.52 ±0.06*	25.73 ±0.68*,**
Str.+Con. Krasen (DW)	179 ±2*,**	48.59 ±0.96*,**	8.03 ±0.06*,**	2.21 ±0.07*	27.68 ±0.60*,**
Wine Podarok Magaracha	223 ±6	75.95 ±0.74*	10.03 ±1.72	5.43 ±0.12*	15.67 ±0.15*
Wine Krasen	215 ±3*	74.79 ±0.34*	11.52 ±0.24*	5.55 ±0.15*	15.81 ±0.15*
Con. Podarok Magaracha (AD)	221 ±4	67.04 ±1.40	9.49 ±0.24	4.83 ±0.11	16.87 ±0.21
Con. Krasen (AD)	255 ±3	70.61 ±2.31	9.97 ±0.16	5.39 ±0.04*	15.47 ±0.34*
Ethanol	236 ±3	79.49 ±2.18*	10.22 ±0.36	5.75 ±0.17*	17.82 ±0.24

The data presented as mean±SD
* - p≤0.05 versus intact animals
** - p≤0.05 versus stressed animals

Table 25. The effect of grape polyphenol concentrates and grape wines on the plasma oxidant/antioxidant status in rats with the emotional-painful stress (in each group n=10).

5. Conclusion

Based on our findings, it is possible to state that antioxidant complexes, particularly polyphenol extracts and the concentrates obtained from *Vitis Vinifera*, which are safe and reveal the potent antioxidant and stress-protective activity, should be used for reduction of proaterogenic states consequences in the complex prophylactic and treatment of atherosclerosis as effective stress-protective remedies.

Thus, administration of *Vitis Vinifera* substances can prevent the increase of the total lipoprotein and ApoB-LP content in the blood, and prevent the free radical process activation in the plasma lipoprotein particles, and, in general, normalize the lipid and lipoprotein metabolism in the liver in metabolic syndrome. These results have proven the ability of the investigated complexes to reduce such negative consequence of metabolic syndrome as development of atherosclerosis.

In addition, according to obtained research data the polyphenolic concentrates possess a potent protective activity both in acute and chronical neurogenic stress.

Our studies suggest that multicomponent active substances with antioxidant properties are more effective in correction of the proatherogenic states caused by stress and metabolic syndrome negative effects in comparison with individual antioxidants (particularly, α-tocopherol). The research data suggest that the increased plasma antioxidant activity alone does not result in decreased foam cell formation, at least in the studied animal model. Moreover, *in vitro* studies have shown that α-tocopherol can be pro-oxidative rather than protective for lipids in isolated LDL. Similarly with vitamin E, vitamin C additives do not offer consistent benefit against atherosclerosis in animals.

The occurrence of tocopherol-mediated peroxidation and the mode of its prevention predicts that the balance of α-tocopherol and available coantioxidants, rather than α-tocopherol alone, determines whether LDL lipid peroxidation occurs in biological systems. Inhibition of the free radical process with the polyphenolic complexes administration can be associated with their ability to increase the level of antioxidants – α-tocopherol, ascorbic acid and reduced glutathione in the test animal liver tissue compared with the group of the stressed animals. The complexes obtained from *Vitis Vinifera*, in particular, polyphenolic concentrates "Enoant" and "Polyphen", as well as grape wines (particularly "Cabernet") with their moderate use revealed the potent antioxidant activity. The preliminary results also suggest that coantioxidants inhibit lipoprotein lipid peroxidation *in vivo*. Thus, if LDL oxidation causes atherosclerosis, the requirement for coantioxidants may explain why supplementation with individual antioxidants, particular vitamin E alone, overall has yielded inconclusive results in the controlled human and animal intervention studies.

In conclusion, our research results may be used for the development of the atherosclerosis prophylaxis strategy, and treatment of diabetes mellitus and metabolic syndrome because recent studies proved insufficient effectiveness of α-tocopherol and advantages of multicomponent antioxidant complexes administration. The high effectiveness of the polyphenolic complexes obtained from *Vitis Vinifera*, including polyphenolic concentrates

"Enoant" (from grape of "Cabernet" grade) and "Polyphen" (from grape of "Rkatsiteli" grade) produced by the National Institute of Grape and Wine "Magarach" has been proven. Our results also confirmed the high effectiveness of the antioxidant complexes from grapes in the correction of the endothelial dysfunction, thus, including these extracts in the treatment schemes is very reasonable. As it would be expected from our observations, increasing the antioxidant oxidant defense by antioxidant supplementation has the ability to restore the endothelial vasomotor function.

An important question to be asked is whether the polyphenol antioxidants exerted their inhibitory effect on lesion progression only because of their antioxidant properties or, possibly, because of additional biological properties, in particular – the phytoestrogen activity.

However, further studies, especially in humans, are required to validate the role of these antioxidants in inhibiting LDL oxidation.

Nevertheless, there are some limitations in the use of the concentrates produced from red grade grapes because of the uric acid content changes.

Author details

Andriy L. Zagayko* , Anna B. Kravchenko,
Mykhaylo V. Voloshchenko and Oxana A. Krasilnikova
Biochemistry Department, National University of Pharmacy, Kharkiv, Ukraine

6. References

[1] Navab M, Anantharamaiah GM, Fogelman AM (2005) The role of high-density lipoprotein in inflammation. Trends Cardiovasc Med. 15:158-161.

[2] Carvalho MD, Vendrame CM, Ketelhuth DF, Yamashiro-Kanashiro EH, Goto H, Gidlund M (2010) High-density lipoprotein inhibits the uptake of modified low- density lipoprotein and the expression of CD36 and FcgammaRI. J. Atheroscler. Thromb. 17: 844-857.

[3] Mumby S, Koh TW, Pepper JR, Gutteridge JM (2001) Risk of iron overload is decreased in beating heart coronary artery surgery compared to conventional bypass. Biochim Biophys Acta. 1537: 204-210.

[4] Park D, Kyung J, Kim D, Hwang SY, Choi EK, Kim YB (2012) Anti-hypercholesterolemic and anti-atherosclerotic effects of polarized-light therapy in rabbits fed a high-cholesterol diet. Lab. Anim Res. 28: 39-46

[5] Kang MK, Chang HJ, Kim YJ, Park AR, Park S, Jang Y, Chung N (2012) Prevalence and determinants of coronary artery disease in first-degree relatives of premature coronary artery disease. Coron. Artery Dis. 2:167-173.

---

* Corresponding Author

[6] Miyazaki Y, Glass L, Triplitt C, Wajcberg E, Mandarino LJ, DeFronzo RA (2002) Abdominal fat distribution and peripheral and hepatic insulin resistance in type 2 diabetes mellitus. Am. J. Physiol. Endocrinol. Metab. 283: E1135- E1143.

[7] Mayr M. (2006) Oxidized low-density lipoprotein autoantibodies, chronic infections, and carotid atherosclerosis in a population-based study. J. Am. Coll. Cardiol. 47: 2436-2443.

[8] Stocker R, Keaney JF (2004) Role of oxidative modifications in atherosclerosis. Physiol. Rev. 84: 1381 – 1478.

[9] Beckman JA, Creager MA, Libby P (2002) Diabetes and Atherosclerosis: Epidemiology, Pathophysiology, and Management. JAMA. 287: 2570-2570.

[10] Masson D, Jiang XC, Lagrost L, Tall AR (2009) The role of plasma lipid transfer proteins in lipoprotein metabolism and atherogenesis. J. Lipid Res. 50: S201- S206.

[11] Sun AY, Wang Q, Simonyi A, Sun GY (2008) Botanical phenolics and brain health. Neuromolecular Med. 10: 259-274.

[12] Iriti M, Faoro F (2009) Bioactivity of grape chemicals for human health. Nat. Prod. Commun. 4: 611-634.

[13] Walzem RL (2008) Wine and health: state of proofs and research needs. Inflammopharmacology. 16: 265-271.

[14] Goupy P, Bautista-Ortin AB, Fulcrand H, Dangles O (2009) Antioxidant activity of wine pigments derived from anthocyanins: hydrogen transfer reactions to the dpph radical and inhibition of the heme-induced peroxidation of linoleic acid. Agric. Food Chem. 57: 5762-5770.

[15] Goupy P, Bautista-Ortin AB, Fulcrand H, Dangles O (2009) Antioxidant activity of wine pigments derived from anthocyanins: hydrogen transfer reactions to the dpph radical and inhibition of the heme-induced peroxidation of linoleic acid. Agric. Food Chem. 57: 5762-5770.

[16] Dohadwala MM, Vita JA (2009) Grapes and cardiovascular disease. J. Nutr. 139:1788S-17893S.

[17] Chorell E, Svensson MB, Moritz T, Antti H (2012) Physical fitness level is reflected by alterations in the human plasma metabolome. Mol. Biosyst. 8: 1187-1196.

[18] Yamaguchi N, Mezaki Y, Miura M, Imai K, Morii M, Hebiguchi T, Yoshikawa K (2011) Antiproliferative and proapoptotic effects of tocopherol and tocol on activated hepatic stellate cells. J. Nutr. Sci. Vitaminol. 57: 317-325.

[19] Raghavamenon A, Garelnabi M, Babu S, Aldrich A, Litvinov D, Parthasarathy S (2009) Alpha-tocopherolis ineffective in preventing the decomposition of preformed lipid peroxides and may promote the accumulation of toxic aldehydes: a potential explanation for the failure of antioxidants to affect human atherosclerosis. Antioxid. Redox. Signal. 11: 1237-1248.

[20] Mein JR, Lian F, Wang XD. (2008) Biological activity of lycopene metabolites: implications for cancer prevention. Nutr. Rev. 66: 667-683.

[21] Bligh EG, Dyer WJ (1959) A rapid method of total lipid extraction and purification. Can. J. Biochem. Physiol. 37: 911–917.

[22] Makarova OP, Saperova MA, Skurupiy VA (2010) Lipid peroxidation in the liver and lungs in SiO(2)-induced granulomatosis. Bull. Exp. Biol. Med. 149: 702-705.

[23] Gao H, Zhou YW (2005) Anti-lipid peroxidation and protection of liver mitochondria against injuries by picroside II. World J. Gastroenterol. 11: 3671-3674.

[24] Stacewicz-Sapuntzakis M, Bowen PE, Kikendall JW, Burgess M (1987) Simultaneous determination of serum retinol and various carotinoids: their distribution in middle-age men and women. J. Micronutr. Anal. 3: 27-33.

[25] Cahill L, Corey P, El-Sohemy A (2009) Vitamin C deficiency in a population of young Canadian adults. Am. J. Epidemiol. 170: 464–471.

[26] MacKness B, Mackness MI, Durrington PN, Arrol S, Evans AE, McMaster D, Ferrières J, Ruidavets JB, Williams NR, Howard AN (2000) Paraoxonase activity in two healthy populations with differing rates of coronary heart disease. Eur. J. Clin. Invest. 30: 4–10.

[27] Idris CA, Sundram K (2002) Effect of dietary cholesterol, trans and saturated fatty acids on serum lipoproteins in non-human primates. Asia Pac J. Clin. Nutr. 7:S408- S415.

[28] Pulinilkunnil T, Abrahani A, Varghese J, Chan N, Tang I, Ghosh S, Kulpa J, Allard M, Brownsey R, Rodrigues B (2003) Evidence for rapid "metabolic switching" through lipoprotein lipase occupation of endothelial-binding sites. J. Mol. Cell Cardiol. 35: 1093-1103.

[29] Basford JE, Wancata L, Hofmann SM, Silva RA, Davidson WS, Howles PN, Hui DY (2011) Hepatic deficiency of low density lipoprotein receptor-related protein-1 reduces high density lipoprotein secretion and plasma levels in mice. J. Biol. Chem. 286: 13079-13087.

[30] Kanda T, Brown JD, Orasanu G, Vogel S, Gonzalez FJ (2009) PPARgamma in the endothelium regulates metabolic responses to high-fat diet in mice. J Clin Invest.;119:110–124.

[31] Belaïd-Nouira Y, Bakhta H, Bouaziz M, Flehi-Slim I, Haouas Z, Ben Cheikh H (2012) Study of lipid profile and parieto-temporal lipid peroxidation in AlCl3 mediated neurotoxicity. modulatory effect of fenugreek seeds. Lipids Health Dis. 11: 16-26.

[32] Dubé JB, Boffa MB, Hegele RA, Koschinsky ML (2012) Lipoprotein(a): more interesting than ever after 50 years. Curr. Opin. Lipidol. 23:133-140.

[33] Tsimikas S, Miller YI (2011) Oxidative modification of lipoproteins: mechanisms, role in inflammation and potential clinical applications in cardiovascular disease. Curr. Pharm. Des. 17: 27-37.

[34] Tani S, Saito Y, Anazawa T, Kawamata H, Furuya S, Takahashi H, Iida K, Matsumoto M, Washio T, Kumabe N, Nagao K, Hirayama A (2011) Low-density lipoprotein cholesterol/apolipoprotein B ratio may be a useful index that differs in statin-treated patients with and without coronary artery disease: a case control study. Int Heart J. 52: 343-347.

[35] Perona JS, Avella M, Botham KM, Ruiz-Gutierrez V (2008) Differential modulation of hepatic very low-density lipoprotein secretion by triacylglycerol-rich lipoproteins derived from different oleic-acid rich dietary oils. Br. J. Nutr. 99: 29-36.

[36] Subramanian S, Chait A (2012) Hypertriglyceridemia secondary to obesity and diabetes. Biochim. Biophys. Acta. 1821: 819-825

[37] Dallinga-Thie GM, Franssen R, Mooij HL, Visser ME, Hassing HC, Peelman F, Kastelein JJ, Péterfy M, Nieuwdorp M (2010) The metabolism of triglyceride-rich lipoproteins revisited: new players, new insight. Atherosclerosis. 211: 1-8.

[38] Chang BH, Li L, Saha P, Chan L (2010) Absence of adipose differentiation related protein upregulates hepatic VLDL secretion, relieves hepatosteatosis, and improves whole body insulin resistance in leptin-deficient mice. J. Lipid Res. 51: 2132-2142.

[39] Annema W, Tietge U (2011) Role of hepatic lipase and endothelial lipase in high-density lipoprotein-mediated reverse cholesterol transport. J.Curr Atheroscler. Rep. 13: 257-265.

[40] Hansen MK, McVey MJ, White RF, Legos JJ, Brusq JM, Grillot DA, Issandou M, Barone FC (2010) Selective CETP inhibition and PPAR alpha agonism increase HDL cholesterol and reduce LDL cholesterol in human ApoB100/human CETP transgenic mice. J. Cardiovasc. Pharmacol. Ther. 15: 196-202.

[41] Salerno AG, Patrício PR, Berti JA, Oliveira HC (2009) Cholesteryl ester transfer protein (CETP) increases postprandial triglyceridaemia and delays triacylglycerol plasma clearance in transgenic mice. Biochem J. 419: 629-634.

[42] Swarbrick MM, Stanhope KL, Elliott SS, Graham JL, Krauss RM, Christiansen MP, Griffen SC, Keim NL, Havel PJ (2008) Consumption of fructose-sweetened beverages for 10 weeks increases postprandial triacylglycerol and apolipoprotein-B concentrations in overweight and obese women. Br. J. Nutr. 100: 947-952.

[43] Benn M, Stene MC, Nordestgaard BG, Jensen GB, Steffensen R, Tybjaerg-Hansen A (2008) Common and rare alleles in apolipoprotein B contribute to plasma levels of low-density lipoprotein cholesterol in the general population. J. Clin. Endocrinol. Metab. 93:1038-1045.

[44] Lewis RM, Hanson MA, Burdge GC (2011) Umbilical venous-arterial plasma composition differences suggest differential incorporation of fatty acids in NEFA and cholesteryl ester pools. Br. J. Nutr. 106: 463-467.

[45] Calabresi L, Franceschini G (2010) Lecithin:cholesterolacyltransferase, high-density lipoproteins, and atheroprotection in humans. Trends Cardiovasc. Med. 20: 50-53.

[46] Kolovou GD, Anagnostopoulou KK, Kostakou PM, Mikhailidis DP (2009) Cholesterol ester transfer protein (CETP), postprandial lipemia and hypolipidemic drugs..Curr Med Chem. 16: 4345-4360.

[47] Eu CH, Lim WY, Ton SH, bin Abdul Kadir K (2010) Glycyrrhizic acid improved lipoprotein lipase expression, insulin sensitivity, serum lipid and lipid deposition in high-fat diet-induced obese rats. Lipids Health. Dis. 29: 79-81.

[48] Storey SM, Atshaves BP, McIntosh AL, Landrock KK, Martin GG, Huang H, Ross Payne H, Johnson JD, Macfarlane RD, Kier AB, Schroeder F (2010) Effect of sterol carrier protein-2 gene ablation on HDL-mediated cholesterol efflux from cultured primary mouse hepatocytes. Am J Physiol Gastrointest Liver Physiol. 299: G244- G254.

[49] Sekiya M, Osuga J, Yahagi N, Okazaki H, Tamura Y, Igarashi M, Takase S, Harada K, Okazaki S, Iizuka Y, Ohashi K, Yagyu H, Okazaki M, Gotoda T, Nagai R, Kadowaki T, Shimano H, Yamada N, Ishibashi S (2008) Hormone-sensitive lipase is involved in hepatic cholesteryl ester hydrolysis. J. Lipid Res. 49: 1829-1838.

[50] Lindi V, Schwab U, Louheranta A, Vessby B, Hermansen K, Tapsell L, Riccardi G, Rivellese AA, Laakso M, Uusitupa MI (2008) The G-250A polymorphism in the hepaticlipase gene promoter is associated with changes in hepatic lipase activity and LDL cholesterol: The KANWU Study. KANWU Study Group. Nutr. Metab. Cardiovasc. Dis. 18: 88-95.

[51] Kolovou GD, Anagnostopoulou KK, Kostakou PM, Mikhailidis DP (2009) Cholesterol ester transfer protein (CETP), postprandial lipemia and hypolipidemic drugs. Curr. Med. Chem. 16: 4345-4360.

[52] Behre HM, Simoni M, Nieschlag E (1997) Strong association between serum levels of leptin and testosterone in men. Clin. Endocrinol. 47: 237–240.

[53] Huang Y, Liu XQ, Rall SC, Taylor JM, von Eckardstein A, Assmann G, Mahley RW (1998) Overexpression and accumulation of apolipoprotein E as a cause of hypertriglyceridemia. J. Biol. Chem. 273: 26388-26393.

[54] Xie X, Zhu Y (2007) Insights into cholesterol efflux in vascular endothelial cells. Cardiovasc Hematol Disord Drug Targets. 7: 127-134.

[55] Miida T, Seino U, Miyazaki O, Hanyu O, Hirayama S, Saito T, Ishikawa Y, Akamatsu S, Nakano T, Nakajima K, Okazaki M, Okada M (2008) Probucol markedly reduces HDL phospholipids and elevated prebeta1-HDL without delayed conversion into alpha-migrating HDL: putative role of angiopoietin-like protein 3 in probucol-induced HDL remodeling. Atherosclerosis. 200: 329-335.

Chapter 4

The Anti-Atherogenic Effects of Lycopene

Amany M. M. Basuny

Additional information is available at the end of the chapter

1. Introduction

Cardiovascular diseases (CVD) are one of the leading causes of death in world. Many epidemiological studies have concluded that a diet rich in fruits and vegetables reduces the incidence of heart disease in humans (Khachik *et al.*, 2002). Carotenoids are important photochemical those are considered to be responsible for the health protective effects of fruits and vegetables (Omoni & Aluko, 2005). The carotenoids are a group of over 600 fat soluble pigments that are responsible for the natural yellow, orange, and red colors of fruits and vegetables (Giovannucci, 2002). Lycopene is one of such carotenoids, and is the pigment principally responsible for the distinctive red color of ripe tomato (*Lycopersicon esculentum*) and tomato products (Shi, 2000). Several epidemiological studies have suggested that a high consumption of tomatoes and tomato products containing lycopene may protect against CVD (Wu *et al.*, 2003). These epidemiological leads have stimulated a number of animal model studies designed to test this hypothesis and to establish the beneficial effects of lycopene. Evidence from these studies suggests that lycopene has anti-atherogenic effects both in vitro and in vivo. The focus of this chapter is the anti-atherogenic effects of lycopene. This chapter will also highlight the chemical composition of lycopene, its sources and function, as well as potential impact an human health.

2. Sources and function of lycopene

Animals and humans do to not synthesize lycopene, and thus depend on dietary sources. Tomatoes and tomato products are the major dietary sources of lycopene. Other sources include watermelon, pink grapefruit, apricots, pink guava and papaya (Willis & Wians, 2003). Lycopene is the most abundant carotenoid in ripe tomatoes, comprising approximately 80-90% of the pigments present. The amount of lycopene in fresh tomatoes depends on the variety, maturity, and environmental conditions in which the fruit matures (Shi, 2000).

Source	Lycopene content (mg/100g wet basis)
Tomatoes fresh	0.72 – 20
Tomato juice	5.00 – 11.60
Tomato sauce	6.20
Tomato paste	5.40 – 15.00
Tomato soup	7.99
Ketchup	9.90 – 13.44
Pizza sauce	12.71
Watermelon	2.30 – 7.20
Pink guava	5.23 – 5.50
Pink grapefruit	0.35 – 3.36
Papaya	0.11 – 5.30
Carrot	0.65 – 0.78
Pumpkin	0.38 – 0.46
Sweet potato	0.02 – 0.11
Apricot	0.01- 0..05

Table 1. shows the lycopene content of tomatoes, some commonly consumed tomato products and other lycopene containing fruits and vegetables.

Lycopene is also widely distributed in the human body. It is one of the major carotenoids found in the human serum (between 21 and 43% of total carotenoids) with plasma levels ranging from 0.22 to 1.06 nmol/ml (Cohen, 2002). It is also found in various tissues throughout the body such as the liver, kidney, adrenal glands, tests, ovaries and the prostate gland (Basu & Imrhan, 2006). Unlike other carotenoids like α-and β-carotene, lycopene lacks the β.:onone rang structure common to other carotenooids (Agarwal & Rao, 2000). Although it lacks provitamine an activity, lycopene is known to be a potent antioxidant (Livny *et al.*, 2002). Reactive oxygen (ROS) species have been implicated in playing a major role in the causation and progression of several chronic diseases. These ROS are highly reactive oxidant molecules that are generated endogenously through regular metabolic activity. They react with cellular components, causing oxidative damage to such critical cellular biomolecules as lipids, proteins and DNA. Antioxidants are protective agents that inactive ROS and therefore, significantly delay or prevent oxidative damage associated with chronic disease risk. Lycopene is one of the most potent antioxidants among the dietary carotenoids and may help lower the risk of chronic diseases including cancer and heart disease.

3. Chemical composition of lycopene

Lycopene is a lipophelic, 40-carbon atom highly unsaturated, straight chain hydrocarbon containing 11 conjugated and 2 non-conjugated double bonds. The all-trans isomer of lycopene is the most predominant isomer in fresh tomatoes and is the most thermodynamically stable from (figure 1). The many conjugated double bonds of lycopene make it a potentially powerful antioxidant, a characteristic believed to be responsible for its beneficial effects. The antioxidant activity of lycopene is high light by its singlet oxygen-quenching property and its ability to trap peroxyl radicals. This singlet quenching ability of lycopene is twice as high as that of β-carotene and 10 times higher than that of α-tocopherol and butylated hydroxyl toluene.

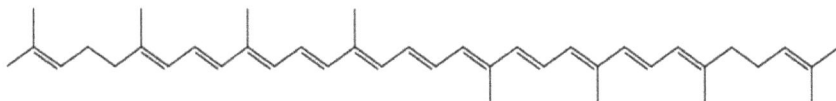

Figure 1. All-*trans* Lycopene.

As a result of the 11 conjugated carbon-carbon double bonds in its backbone, lycopene can theoretically assume 211 or 2048 geometrical configurations (Omani & Aluko, 2005).

However, it is now known that the biosynthesis in plants leads to the all-*trans*-form, and this is independent of its thermodynamic stability. In human plasma, lycopene is an isomeric mixture, containing at least 60% of the total lycopene as cis- isomers (Kim *et al.*, 2012).

All-*trans*, 5-*cis*, 9-*cis*, 13-*cis*, and 15-*cis* are the most commonly identified isomeric forms of lycopene with the stability sequence being 5-*cis*>all-*trans*>9-*cis*>13-*cis*>15- *cis*>7-*cis*>11-*cis*, (Agarwal & Rao, 2000) so that the 5-*cis*-form is thermodynamically more stable than the all-*trans*-isomer. Whereas a large number of geometrical isomers are theoretically possible for all-*trans* lycopene, according to only certain ethylenic groups of a lycopene molecule can participate in *cis-trans* isomerization because of steric hindrance. In fact, only about 72 lycopene *cis* isomers are structurally favorable. Figure 2 illustrates the structural distinctions of the predominant lycopene geometrical isomers.

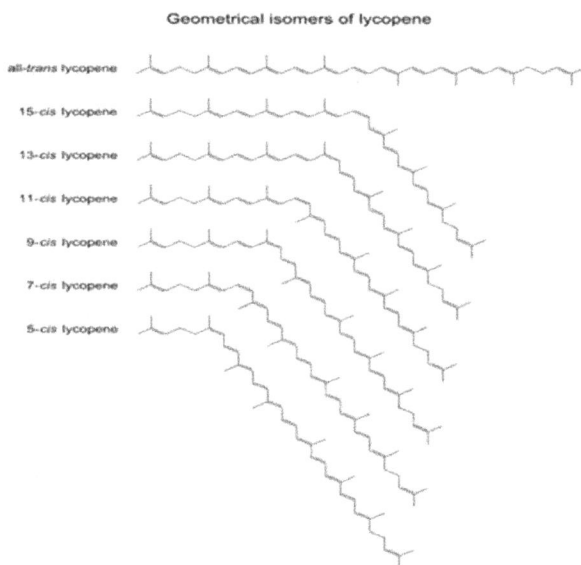

Figure 2. Geometrical isomers of lycopene

4. Mechanisms action of lycopene

A cellular and molecular study have shown lycopene to be one of the most potent antioxidants and has been suggested to prevent atherogenesis by protecting critical bimolecules such as DNA, proteins, lipids and low density lipoproteins (Pool-zobel et al., 1997). Lycopene, because of its high number of conjugated double bonds, exhibits higher singlet oxygen quenching ability compared to β-carotene or α-tocopherol (Di-Mascio et al., 1989). Cis lycopene has been shown to predominate in both benign and malignant prostate tissues, suggesting a possible beneficial effect of high cis-isomer concentrations, and also the involvement of tissue isomerases in vivo isomerization from all trans to cis form (Clinton et al., 1996). Where as Levin et al., (1997) have shown that 9- cis- β-carotene is a better antioxidant than its all-trans counterpart, no such mechanistic data have been reported in case of individual lycopene isomers. Handley et al., (2003) reported a significant increase in 5-cis lycopene concentrations following a 1- week lycopene-restricted diet, and a subsequent reduction in 5-cis, and a concomitant increase in cis-β, cis-D and cis-E lycopene isomers during the 15-day dietary intervention with tomato products in healthy individuals. Although this study reported a decrease in LDL oxidizability due to the intervention with tomato lycopene, the individual antioxidant role of lycopene isomers and their inter conversions remain unclear. At a physiological concentration of 0.3 μmol/1, lycopene has been shown to inhibit growth of non-neoplastic human prostate epithelial cells in vitro, through cell cycle arrest which may be of significant implications in preventing benign prostate hyperplasia, a risk factor for prostate cancer (Obermuller-Jevic et al., 2003). Lycopene has also been shown to significantly reduce LNCaP human prostate cancer cell survival in a dose-dependent manner, and this anti-neoplastic action may be explained by increased DNA damage at high lycopene concentrations (> 5μm), whereas lower levels of lycopene reduced malondialdehyde formation, with no effects on DNA (Hwang & Bowen, 2005). Physiologically attainable concentrations of lycopene have been shown to induce mitochondrial apoptosis in LNCaP human prostate cancer cells, although no effects were observed on cellular proliferation or necrosis (Hantz et al., 2005). Lycopene has also been shown to interfere in lipid metabolism, lipid oxidation and corresponding development of atherosclerosis. Lycopene treatment has been shown to cause a 37% suppression of cellular cholesterol synthesis in J-774A.1 macrophage cell line, and augment the activity of macrophage LDL receptors (Fuhrman et al., 1997). Oxidized LDLs are highly atherogenic as they stimulate cholesterol accumulation and foam cell formation, initiating the fatty streaks of atherosclerosis (Libby, 2006). LDL susceptibility to oxidative modifications is decrease by an acyl analog of platelet-activating (PAF), acyl-PAF, which experts its beneficial role during the initiation and progression of atherosclerosis. Purified lycopene in association with α-tocopherol or tomato lipophillic extracts has been shown to enhance acyl-PAF biosynthesis in endothelial cells during oxidative stress (Balestrieri et al., 2004). Fuhrman et al., (2000) further reported comparative data in which tomato oleoresin exhibited superior capacity to inhibit in vitro LDL oxidation in comparison with pure lycopene by up to fivefold. A combination of purified lycopene (5μmol/I) with α-toopherol in the concentration range of 1-10μmol/I resulted in a significant greater inhibition of in vitro LDL oxidation, than the expected additive individual inhibitions. In this study, purified lycopene was also shown to act synergistically with other natural antioxidants like the flavonoid glabridin, the phenolics rosmarinic acid and carnosic acid, and garlic acid in inhibiting LDL oxidation in vitro. These observations suggested a superior antiatherogeneic characteristic of tomato oleoresin over pure lycopene. The combination of lycopene with other natural antioxidants, as in tomatoes, may be more potent in inhibiting lipid peroxidation, than lycopene per se. The antiatherogenic effects of lycopene are generally believed to be due to its antioxidant properties. Dietary lycopene increases blood and tissue lycopene levels and acting as an antioxidant, lycopene traps reactive oxygen species and reduce the oxidative damage to lipids (lipoproteins and membrane lipids), proteins including important enzymes, and DNA, therapy lowering oxidative stress. This reduced oxidative stress then leads to a reduced risk for chronic diseases associated with oxidative stress such as cardiovascular disease (Omani & Aluko 2005). Alternatively, some non-oxidative mechanisms may be responsible for the beneficial effects of lycopene. The increased lycopene status in the body may regulate gene functions, improve intercellular communication, modulate hormone and immune response, or regulate metabolism, thus lowering the risk for chronic disease (Agarwl & Rao, 2000). A possible mechanism speculated for the protective role of lycopene in heart disease is via the inhibition of cellular HMGCoA reducate, the rate-limiting enzyme in cholesterol synthesis (Fuhrman *et al.*, 1997).

5. Lycopene stability

Being acyclic, lycopene possesses symmetrical planarity and has no vitamin A activity, and as a highly conjugated polyene, it is particularly susceptible to oxidative degradation. Physical and chemical factors known to degrade other carotenoids, including elevated temperature, exposure to light, oxygen, extremes in pH, and molecules with active surfaces that can destabilize the double bonds, apply to lycopene as well (Rao *et al.*, 2003).

In a study to determine the photoprotective potential of dietary antioxidants including lycopene carried out by Handley *et al.*, (2003) carotenoids were prepared in special nanoparticle formulations together with vitamin C and/or vitamin E. The presence of vitamin E in the formulation further increased the stability and cellular uptake of lycopene, which suggests that vitamin E in the nanoparticle, protects lycopene against oxidative transformation. Their findings suggest that lycopene stability may be improved by nanoparticle formulation and incorporation of vitamin E in the lycopene formulation.

Badimon *et al.*, 2010 studied the stability of lycopene during heating and illumination. They carried out various pretreatment steps to the all-trans lycopene standard, which included; dissolving the lycopene standard into hexane and evaporating to dryness under nitrogen in vials, after which a thin film formed at the bottom surface. The resulting lycopene was heated at 50, 100, and 150°C or illuminated at a distance of 30 cm with illumination intensity in the range of 2000–3000 lux (25°C) for varied lengths of time (up to100 hours for heating and 5 days for illumination). After analysis, the degradation of total lycopene (all-*trans* plus *cis* forms) during heating or illumination was found to fit a firstorder model. At 50°C, the isomerization dominated in the first 9 hours; however, degradation was favored afterwards. At 100 and 150°C, the degradation proceeded faster than the isomerization, whereas, during illumination, isomerization was the main reaction. The degradation rate constant (min–1) of lycopene was found to rise with increasing temperature with an activation energy calculated as 61.0 kJ/mol.

The stability of crystalline lycopene was determined under various temperature conditions (5, 25, and 35°C) while stored in airtight containers, sealed under inert gas, and protected from light. After 30 months of storage, crystalline lycopene remained stable when stored under the recommended conditions (Barros *et al.*, 2011).

Lycopene (synthetically prepared by the Wittig reaction) 5% TG (Tablet Grade) and lycopene 10% WS (Water Soluble) beadlet formulations tested for over 24 months of storage, and Lycopene 10% FS (Fluid Suspension) liquid formulation tested for over 12 months of storage under various temperature conditions (5 and 25°C), were all found to be stable.(25) For the 10% WS lycopene beadlet formulations, an important market application form, stability with respect to oxidation under ambient light conditions and room temperature for 12 months in beverages was found to be 93% of the initial content of the beverage lycopene (Pool-zobel *et al.*, 1997).

6. Dietary intake of lycopene

The human body is unable to synthesize carotenoids, which qualifies diet as the only source of these components in blood and tissues. At least 85% of our dietary lycopene comes from tomato fruit and tomato-based products, the remainder being obtained from other fruits such as watermelon, pink grapefruit, guava, and papaya, Tomatoes are an integral part of the human diet and are commonly consumed in fresh form or in processed form such as tomato juice, paste, puree, ketchup, soup, and sauce. Kim *et al.*, (2012) used a tomato products consumption frequency questionnaire to estimate the average daily consumption of different tomato products in the Canadian population.

Di-Mascio *et al.*, (1989) estimated that 50% of the dietary lycopene was obtained from fresh tomatoes, while the average daily intake of lycopene was estimated to be 25 mg in the Canadian population. In a British study conducted with elderly females, the daily consumption of lycopene-rich food, such as tomatoes and baked beans in tomato sauce (measured by weight of foods eaten), was equivalent to a daily lycopene intake of 1.03 mg per person (Omani & Aluko, 2005) developed a database from which the carotenoid intake of the German population, stratified by sex and age, was evaluated on the basis of the German National Food Consumption Survey (NVS). The mean total carotenoid intake amounted to 5.33 mg/day. The average intake of lycopene was 1.28 mg/day with tomatoes and tomato products providing most of the lycopene.

A study presenting data on dietary intake of specific carotenoids in The Netherlands, based on a food composition database for carotenoids, was done by Furhman *et al.*, (1997). Regularly eaten vegetables, the main dietary source of carotenoids, were sampled comprehensively and analyzed with modern analytic methods. The database was complemented with data from literature and information from food manufacturers. Intake of carotenoids was calculated for participants of the Dutch Cohort Study on diet and cancer, aged 55 to 69 in 1986, and the mean intake of lycopene was 1.0 mg/day for men and 1.3 mg/day for women.

6.1. Bioavailability of lycopene

Although 90% of the lycopene in dietary sources is found in the linear, all-trans conformation, human tissues (Particularly liver, adrenal, adipose tissue, testes and prostate) contain mainly cis-isomers. Hollowy *et al.*, (2002) reported that a dietary supplementation of tomato pure for 2 weeks in healthy volunteers led to a completely different isomer pattern of plasma lycopene in these volunteers, versus those present in tomato pure. 5-cis, 13-cis and 9-cis lycopene isomers, not detected in tomato puree, were predominant in the serum (Hollowary *et al.*, 2000).Analysis of plasma lycopene in male participants in the health professionals follow-up study revealed 12 distinct cis-isomers and the total cis-lycopene contributed about 60-80% of total lycopene concentrations (Wu *et al.*, 2003). Studies conducted with lymph cannulated ferrts have shown better absorption of cis-isomers and their subsequent enrichment in tissues (Boileau *et al.*, 1999). Physiochemical studies also suggest that cis-isomer geometry accounts for more efficient incorporation of lycopene into mixed micelles in the lumen of the intestine and into chylomicrons by the enterocyte. Cis-isomers are also preferentially incorporated by the liver into very low-density lipoprotein (VLDL) and get secreted into the blood (Britton, 1995). Research has shown convincing evidence regarding the isomerization of all trans-lycopene to cis-isomers, under acidic conditions of the gastric juice. Incubation of lycopene derived from capsules with simulated gastric juice for 1-min shown a 40% cis-lycopene content, whereas the levels did not exceed 20% even after 3h incubation with water as a control. However, when tomato puree was incubated for 3h with simulated gastric juice, the cis-lycopene content was only 18% versus 10% on incubation with water. Thus, gastric pH and food matrix influence isomerization and subsequent absorption and increased bioavailability of cis-lycopene (Re *et al.*, 2001).

The process of cooking which releases lycopene from the matrix into the lipid phase of the meal increases its bioavailability, and tomato paste and tomato puree are more bioavailable sources of lycopene than raw tomatoes (Gartner *et al.*, 1997 & Porrini *et al.*, 1998). Factors such as certain fibers, fat substituents, plant sterols and cholesterol-lowering drugs can interfere with the incorporation of lycopene into micelles, thus lowering its absorption (Boileau *et al.*, 2002). Several clinical trials have also shown the bioavailability of lycopene from processed tomato products (Table 2). Agarwal and Rao (1998), reported a significant increase in serum lycopene levels following a 1-week daily, consumption of spaghetti sauce (39mg of lycopene), tomato juice (50mg of lycopene) or tomato oleoresin (75 or 150 mg of lycopene), in comparison with the placebo, in healthy human volunteers. There was also indication that the lycopene levels increased in a dose-dependent manner in the case of tomato sauce and tomato oleoresin. Reboul *et al.*, (2005) further demonstrated that enrichment of tomato paste with 6% tomato peel increases lycopene bioavailability in men, thereby suggesting the beneficial effects of peel enrichment, which are usually eliminated during tomato processing. Richelle *et al.*, (2002) compared the bioavailability of lycopene from tomato paste and from lactolycopene formulation (Lycopene from tomato oleoresin embedded in a whey protein matrix), and reported similar bioavailability of lycopene from the two sources in healthy subjects. Dietary fat has been shown to promote lycopene absorption, principally via stimulating bile production for the formation of bile acid micelles. Consumption of tomato products with olive oil or sunflower oil has been shown to produce an identical bioavailability of lycopene, although plasma antioxidant activity improved with olive oil consumption, suggesting a favorable impact of monounsaturated fatty acids on lycopene absorption and its antioxidant mechanism (Lee *et al.*, 2000). In an interesting study Unlu *et al.*, (2005) reported the role of avocado lipids in enhancing lycopene absorption. In this study, in healthy, nonpregnant, nonsmoking adults, the addition of avocado oil (12 or 24g) to salsa (300g) enhanced lycopene absorption, resulting in 4.4 times the mean area under the concentration-versus-time curve after intake of avocado-free salsa. This study demonstrates the favorable impact of avocado consumption on lycopene absorption and has been attributed to the fatty acid distribution of avocados (66.00% oleic acid), which may facilitate the formation of chylomicrons. In a comparative study by Hoppe et al., (2003), both synthetic and tomato –based lycopene supplementation showed similar significant increases of serum total lycopene above baseline whereas no significant changes were found in the placebo group. In an attempt to study lycopene metabolism, Diwadkar-Navsariwala *et al.*, (2003) developed a physiological pharmcokinetic model to describe the disposition of lycopene, administered as a tomato beverage formulation at five graded does (10, 30, 60, 90, or 120 mg) in healthy men. Blood was collected before dose administration and at scheduled study intervals until 672h. The overall results of this study showed that independent of dose, 80% of the subjects absorbed less than 6mg of lycopene, suggesting a possible saturation of absorptive mechanisms. This may have important implications for planning clinical trials with pharmacological doses of lycopene in the control and prevention of chronic disease, if absorption saturation occurs at normally consumed levels of dietary lycopene.

6.2. The anti-atherogenic effects of lycopene

In a previous study (Basuny *et al.*, 2006 and 2009) was to study the effect of tomato lycopene on hypercholesterolemia. Lycopene of tomato wastes was extracted and determination. The level of tomato lycopene was 145.50ppm. An aliquots of the concentrated tomato lycopene, represent 100, 200, 400 and 800ppm; grade lycopene (200ppm) and butylated hydroxyl toluene (BHT, 200ppm) were investigated by 1,1-diphenyl-2-picrylhydrazyl (DPPH) free radical scavenging method. These compounds were administered to rats fed on hypercholestrolemic diet daily from 10 weeks by stomach tube. Serum lipid contents (total lipids, total cholesterol, high density lipoprotein cholesterol and low density lipoprotein cholesterol), oxidative biomarkers (glutathione peroxidase and malonaldhyde), the liver (aspartate aminotransferase, alanine aminotranseferase and alkaline phosphatase activities) and kidney (uric acid, urea and creatinine) function testes were measured to assess the safety limits of the lycopene in tomato wastes. The data of the aforementioned measurements indicated that the administration of tomato lycopene did not cause any changes in liver and kidney functions. On the contrary, rats fed on hypercholesterolemic diet induced significant increases in the enzymes activities and the serum levels of total lipids, total cholesterol and low and high density lipoproteins cholesterol and decreased levels of the glutathione peroxidase and malonaldhyde. In conclusion, presently available data from epidemiological and a number of animal studies have provided evidence to suggest that lycopene, the naturally present carotenoid in tomatoes and other fruits and vegetables, possesses anti-atherogenic effects. However, there is a need for more human dietary intervention studies in order to better understand the role of lycopene in human health.

Scientific evidence indicates that oxidation of low density lipoprotein (LDL), which carry cholesterol in the blood stream plays an important role in the development of atherosclerosis, the underlying disorder leading to heart attacks and ischemic strokes (Rao, 2002). Several studies indicate that consuming the antioxidant lycopene that is contained in tomatoes and tomato lycopene products can reduce the risk of cardiovascular diseases (CVD). Available evidence from the Kuopio Ischaemic Heart Disease Risk Factor (KIHD) study suggests that the thickness of the innermost wall of blood vessels and the risk of myocardial infarction reduced in persons with higher serum and adipose tissue concentrations of lycopene (Rissanen *et al.,* 2003). This finding suggests that the serum lycopene concentration may play a role in the early stages of atherosclerosis. A thick artery wall is a sign of early atherosclerosis, and increased thickness of the intima media has been shown to predict coronary events. Similarly, the relationship between plasma lycopene concentration and intima-media thickness of the common carotid artery wall (CCA-IMT) was investigation in 520 middle-aged men and women 45-69 years as parts of the Antioxidant Supplementation in Atherosclerosis Prevention (ASAP) study (Rissanen *et al.,* 2000). Low levels of plasma lycopene were associated with a 17.80% increment in CCA-IMT in men, while there was no significant difference among women. These findings also suggest that low plasma lycopene concentrations are associated with early atherosclerosis, evidenced by increased CCA-IMT in middle-aged men.

Findings from the Rotterdam Study (Klipstein-Grobusch *et al.,* 2000) showed modest inverse associations between levels of serum lycopene and atherosclerosis, assessed by the presence of calcified plaques in the abdominal aorta. Study population comprised of 108 cases of aortic atherosclerosis and 109 controls aged 55 years and over. The association between serum lycopene levels and atherosclerosis was most pronounced among subjects who were current and former smokers. No association with risk of aortic calcification for the serum carotenoids α-carotene, β-carotene, lutein and zeaxanthin was observed. These results suggest that lycopene may play a protective role in the development of atherosclerosis. Results from the European Study of Antioxidant, Myocardial Infarction, and Cancer of the breast (the EURAMIC study) also show that men with the highest concentration of lycopene in their adipose tissue biopsy had a 48% reduction in risk of myocardial information compared with men with the lowest adipose lycopene concentrations (Kohlmeir *et al.,* 1997). An increase in LDL oxidation is known to be associated with an increased risk of atherosclerosis and coronary heart disease (Parthasarathy, 1998). Agarwal and Rao (1998) investigated the effect of dietary supplementation of lycopene on LDL oxidation in 19 healthy human subjects. Dietary lycopene was provided using tomato juice, spaghetti sauce and tomato oleoresin for a period of 1 week each. Blood samples were collected at the end of each treatment, and TBARS and conjugated dienes were measured to estimate LDL oxidation. In addition to significantly increasing serum lycopene levels by a least twofold, lycopene supplementation significantly reduced serum lipid peroxidation and LDL oxidation. The average decrease of LDL –TBARS and LDL-conjugated diene for the tomato products treatment over placebo was 25 and 13%, respectively. These results suggest significance for lycopene in decreasing risk for coronary heart disease. Results from the ongoing Women's Health Study (WHS) showed that women with the highest intake of tomato-based foods rich in lycopene had a reduced risk for CVD compared to women with a low intake of those foods (Sesso et al., 2003). Results showed that women who consumed seven servings or more of tomato based foods like tomato sauce and pizza each week had a nearly 30% risk reduction in total CVD compared to the group with intakes of less than one serving per week. The researchers also found out that women who ate more than 10 servings per week had an even more pronounced reduction in risk (65%) for specific CVD outcomes such as heart attack or stroke. Though not statistically significant, the strongest association of dietary lycopene with CVD protection was seen among women with a median dietary lycopene intake of 20.20 mg/day, who had a 33% reduction in risk of the disease when compared with women with the lowest dietary lycopene intake (3.3 mg/day).

Lycopene has also been shown to have a hypercholesterolemic effect both in vivo and in vitro. In a small dietary supplementation study, six healthy male subjects were fed 60 mg/day lycopene for 3 months. At the end of the treatment period, a significant 14% reduction in plasma LDL cholesterol levels was observed in vivo with no effect on HDL cholesterol concentration (Fuhrman et al., 1997) & Lorenz et al., 2012).

6.3. Safety of lycopene

The safety issue for carotenoids attracted much attention after the publication of the β-carotene supplementation trials, which yielded negative results. It is interesting that in thus studies an increased risk for lung cancer was related to a 12- and 16 fold increase in β-carotene plasma levels due to supplementation. β-carotene plasma levels increased from 0.32μml before supplementation up to 3.90 and 5.90 μm, respectively. Rao et al., (2003), which showed no effect for β-carotene supplementation, only a 5-fold increase in the carotenoid serum level was achieved. Interestingly, the only study with positive results after supplementation with β-carotene was achieved in linxian, a chinese community with very low carotenoid levels (0.11μm) before the intervention (Jonker et al., 2003). Although supplementation caused an 11-fold increase in β-carotene level, the final concentration of β-carotene reached was a relatively low 1.5μm. Interestingly, reviewing many studies which measured serum levels of β-carotene and lycopene after supplementation suggests that β-carotene serum levels are significantly higher than those found for lycopene. Serum levels reached for β-carotene are around 3μm and may exceed 5μm after supplementation; on the other hand lycopene levels above 1.2μm are rarely seen even after long-term application. Moreover, the serum level achieved for lycopene was not directly correlated to the amount of the supplementation carotenoid (Nahum et el., 2001). For example, supplemented as high as 75 mg/day did not increase lycopene serum levels more than 1µm (Agarwl & Rao 1998). In conclusion, by some unknown mechanism, lycopene plasma levels after supplementation remain relatively low, which may provide a safety value.

6.4. Lycopene relationship with other micronutrients

When reviewing data related to the chemoprevention of various diseases, it become evident that the use of a single carotenoid, or any other micronutrient which has been successful in vitro and animal models, does not prove as favorable in human intervention studies. That is, there is no magic bullet. In fact, accumulating evidence suggests that a concerted, synergistic action of various micronutrients is, more likely to be the basis of the disease-prevention activity of a diet rich in vegetables and fruits. Indeed, the sources of lycopene used in most of the human studies reviewed there were either prepared tomato products or tomato extracts containing lycopene and other tomato micronutrients and carotenoids in various proportions. Pure lycopene has not been tested as a single in human prevention studies. On the other hand, many studies showing the beneficial effect of lycopene in alleviating chronic conditions have been conducted in which the subjects were provided with tomato-based foods, or tomato extracts, but not with the pure compound. For example, the oleoresin preparation used in many of these studies also contained other tomato carotenoids such as phytoene, phytofluene and β-carotene (Amir et al., 1999; Pastori et al., 1998 & Stahl et al., 1998). In a recent study (Bioleau et al., 2003) that compared the potency of freeze-dried whole tomatoes (tomato powder) or pure lycopene in a rat model of prostate cancer. Rats were treated with the carcinogen (N-methyl1-N-nitrosourea) combined with androgens to stimulate prostate carcinogenesis, and the ability of these two preparations containing lycopene to enhance survival was compared. Mortality with prostate cancer was lower by 25 % (p- 0.09) for rats fed the tomato powder diet than for rats fed control feed. Prostate cancer morality of rats fed our lycopene was similar to that of the control group. The authors concluded that consumption of tomato powder but not pure lycopene inhibited prostate carcinogenesis, suggesting that tomato products contain other compounds, besides lycopene, that modify prostate carcinogenesis.

6.5. Epidemiologic studies: lycopene and cardiovascular diseases

Epidemiological observations also report an inverse association between plasma of tissue lycopene levels and the incidence of cardiovascular diseases. In the Kuopio Ischemic Heart Disease Risk Factor Study, lower levels of plasma lycopene were seen in men who had a coronary event compared with men who did not. In addition, a higher concentration of serum lycopene was inversely correlated with a decrease in the mean and maximal intima-mediated thickness of the common carotid artery (CCA-IMT) with lo lycopene, resulting in an 18% increase in CCA-IMT (Rissanen et al., 2003). The European Multiccnter Case-Control Study on antioxidants, Myocardial Infarction and Breast Cancer Study (EURAMIC Study) reported that a higher lycopene concentration was independently protective against cardiovascular diseases (Basu & Imrhan 2006). The Women's Health Study further revealed that a decreased risk for developing cardiovascular diseases was more strongly associated with higher tomato intake than with lycopene intake (Sesso et al., 2003). Processed tomato products definitely provide a bioavailability source of lycopene and have a positive correlation with plasma and tissue lycopene levels. However, these studies do not suggest a role of lycopene perse, in reducing the risks for cardiovascular diseases, as plasma level of lycopene, in epidemiologic studies, only reflects the consumption of tomato and tomato products.

7. Conclusion

Thus, it can be concluded that moderate amounts of whole food-based supplementation (2–4 servings) of tomato soup, tomato puree, tomato paste, tomato juice or other tomato beverages, consumed with dietary fats, such as olive oil or avocados, leads to increases in plasma carotenoids, particu- larly lycopene. The recommended daily intake of lycopene has been set at 35 mg that can be obtained by consuming two glasses of tomato juice or through a combination of tomato products (Rao and Agarwal, 2000). These foods may have both chemopreventive as well as chemotherapeutic values as outlined in Figure 3. In the light of recent clinical trials, a combination of naturally occurring carotenoids, including lycopene, in food sources and supplements, is a better approach to disease prevention and therapy, versus a single nutrient. Lycopene has shown distinct antioxidant and anticarcinogenic effects at cellular levels, and definitely contributes to the health benefits of consumption of tomato products. However, until further research establishes sig- nificant health benefits of lycopene supplementation per se, in humans, the conclusion may be drawn that consumption of naturally occurring carotenoid-rich fruits and vegetables, particularly processed tomato products containing lycopene, should be encouraged, with positive implications in health and disease.

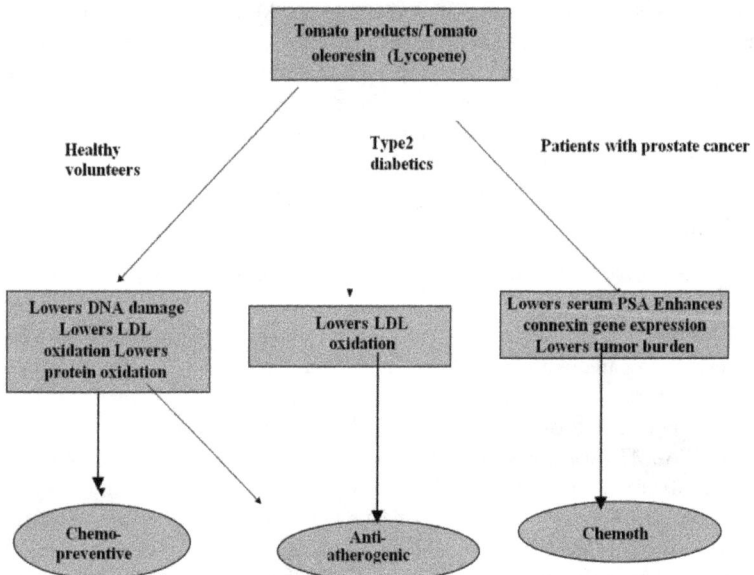

Figure 3. Summary of mechanisms of action of tomato products or tomato oleoresin supplementation, containing lycopene, in health and disease.

		Type and duration of lycopene supplementation	Effects on biomarkers of oxidative stress/ carcinogenesis	
Agarwal and Rao (1998)	19 healthy subjects (mean age 29 years, BMI 2472.8 kg/m²)	0 mg lycopene (placebo), 39 mg lycopene (spaghetti sauce), 50 mg lycopene (tomato juice), or 75 mg lycopene (tomato oleoresin) per day for 1 week	25% decrease in LDL-TBARS 13% decrease in LDL-CD for all groups versus placebo (P<0.05)	Increase at 7 days in all groups versus placebo (P<0.05)
Riso et al. (1999)	10 healthy subjects (mean age 23.171.1 years, BMI 20.571.5 kg/m²)	16.5 mg lycopene (60 g tomato puree), per day for 21 days	38% decrease in DNA damage in lymphocytes (P<0.05)	Increase at 21 days versus baseline (P<0.001)
Bub et al. (2000)	23 healthy volunteers (mean age 3474 years, BMI 2372 kg/m²)	40 mg lycopene (330 ml tomato juice) for 2 weeks	12% decrease in plasma TBARS 18% increase in LDL lag time (P<0.05) no effects on water-soluble antioxidants, FRAP, glutathione peroxidase and reductase activities (P<0.05)	Increase at 2 weeks versus baseline (P<0.05)
Chopra et al. (2000)	34 healthy females (mean age 37.578.5 years, BMI 2473.5 kg/m²)	440 mg lycopene (200 g tomato puree þ 100 g watermelon) per day for 7 days	Significant decrease in LDL oxidizability in nonsmokers (P<0.05); no effects in smokers (P<0.05)	Increase at 7 days versus baseline (P<0.05)
Porrini and Riso (2000)	9 healthy subjects (mean age 25.472.2years, BMI 20.371.5 kg/m²)	7 mg lycopene (25 g tomato puree), per day for 14 days	50% decrease in DNA damage in lymphocytes (P<0.05)	Increase at 14 days versus baseline (P<0.001)
Upritchard et al. (2000)	15 well-controlled type II diabetics (mean age 6378years, BMI30.977 kg/m²)	Tomato juice (500 ml) per day or placebo for 4 weeks	Decreased LDL oxidizability versus baseline (P<0.001)	Increase at 4 weeks versus baseline (P<0.001)
Hininger et al. (2001)	175 healthy volunteers (mean age 33.571 years, BMI-24.370.5 kg/m²)	15 mg lycopene (natural tomato extract) or placebo per day for 12 weeks	No effects on LDL oxidation, reduced glutathione, protein SH groups and antioxidant metalloenzyme activities (P<0.05)	Increase at 12 weeks versus baseline (P<0.05)
Chen et al. (2001)	32 patients with localized prostate adenocarcinoma (mean age 63.776.1 years, BMI 28.074.9 kg/m²)	30 mg lycopene (200 g spaghetti sauce) per day for 3 weeks before surgery or a reference group with no supplementation	Decreased leukocyte and prostate tissue oxidative DNA damage; decreased serum PSA levels (P<0.05)	Increase at 3 weeks versus baseline (P<0.001)
Kucuk et al. (2001)	26 patients with newly diagnosed, clinically localized prostate cancer (mean age 62.1571.85 years, BMI not reported)	15 mg lycopene (Lyc-O-Mato capsules) twice daily or no supplementation for 3 weeks before surgery	Decreased tumor growth in the intervention group versus control(P<0.05); decreased plasma PSA levels and increased expression of connexin43 in prostate tissue in the intervention group versus control (P<0.05);decreased plasma IGF-1 levels in intervention and control groups(P<0.05)	No effects at 3 weeks versus baseline (P<0.05)

		Type and duration of lycopene supplementation	Effects on biomarkers of oxidative stress/ carcinogenesis	
Porrini et al. (2002)	9 healthy subjects (mean age 25.272.2 years, BMI 20.271.6 kg/m²)	7 mg lycopene (25 g tomato puree) with 150 g of spinach and 10 g of olive oil per day for 3 weeks	Decreased DNA oxidative damage (P<0.05)	Not reported

Table 2. Summary of clinical trials investigating the effects of supplementation of tomato products, tomato oleoresin or purified lycopene on biomarkers of oxidative stress and Carcinogenesis

Author details

Amany M. M. Basuny
Department of Fats & Oils, Food Technology Research Institute,
Agriculture Research Centre, Giza, Egypt

8. References

Agarwal, A. & Rao, A. (1998): Tomato lycopene and low density lipoprotein oxidation: a human dietary intervention study. Lipids, 33: 981-984.

Agrawal, S. & Rao, V. (2000): Tomato lycopene and its role in human health and chronic diseases. Canadian Medical Association Journal, 163: 739-744.

Amir, H.; Karas, M. & Giat, J. (1999): Lycopene 1, 25 di-hydroxy vitamin-D3 cooperate in the inhibition of cell cycle progression and induction of differentiation in Hl-60 leukemic cells. Nutrition Cancer, 33: 105-112.

Badiman, L.; Vilahur, G. & Padro, T. (2010): Nutraceuticals and atherosclerosis: Human trials. Cardiovascular Therabeutics, 28: 202-215.

Barros, L.; Carbrita, L.; Boas, M.; Carvaiho, A. & Ferreira, I. (2011): Chemical, biochemical and electrochemical assays to evaluate phytochemicals and antioxidant activity of wild plants. Food Chemistry, 127: 1600-1608.

Basu, A. & Imrhan, V. (2006): Tomato versus lycopene in oxidative stress and carcinogenesis: conclusions from clinical trials. European Journal OF Clinical Nurition, 1-9.

Basuny, A. M.; Mostafat, D. M. & Azouz, A. (20060: Supplementation of polyunsaturated oils with lycopene as natural antioxidant and antipolymerization during heating process. Minia Journal of Agricultural Research and Development, 26: 449-469.

Basuny, A. M.; Gaafar, A. M. & Arafat, S. M. (2009): Tomato lycopene is a natural antioxidant and cn alleviate hypercholesterolemia. African Journal of Biotechnology, 23: 6627-6633.

Boileau, T. W.; Liao, Z.; Kim, S.; Lemeshow, S.; Erdman, J. & Clinton, S. (2003): Prostate carcinogensis in N-methyl-N-nitrosourea (NMW-testosterone-treated rats fed tomato powder, lycopene, or energy-restricted diets. J Natl. Cancer Inst. 95: 1578-1586.

Boileau AC, Merchen NR, Wasson K, Atkinson CA, Erdman JW (1999). cis-Lycopene is more bioavailable than trans-lycopene in vitro and in vivo in lymph-cannulated ferrets. J Nutr 129, 1176–1181.

Boileau TWM, Boileau AC, Erdman JW (2002). Bioavailability of alltrans and cis-isomers of lycopene. Exp Biol Med 227, 914–919.

Britton, G. (1995): Structure and properties of carotenoids in relation to function. FASEB J 9, 1551–1558.

Briviba, K.; Schnabele, K.; Rechkemmer, G.; Bub, A. (2004): Supplementa- tion of a diet low in carotenoids with tomato or carrot juice does not affect lipid peroxidation in plasma and feces of healthy men. J Nutr 134, 1081–1083.

Bub, A.; Watzl, B.; Abrahamse, L.; Delincee, H.; Adam, S. & Wever, J. (2000): Moderate intervention with carotenoid-rich vegetable products reduces lipid peroxidation in men. J Nutr 130, 2200–2206.

Bub, A.; Barth, S. W.; Watzl, B.; Briviba, K. & Rechkemmer, G. (2005): Araoxonase 1 Q192R (PON1-192) polymorphism is associated with reduced lipid peroxidation in healthy young men on a low- carotenoid diet supplemented with tomato juice. Br J Nutr 93, 291–297.

Chen, L.; Stacewicz-Sapuntzakis, M.; Duncan, C.; Sharifi, R.; Ghosh, L. & Van Breemen, R. (2001): Oxidative DNA damage in prostate cancer patients consuming tomato sauce- based entrees as a whole-food intervention. J Natl Cancer Inst 93, 1872–1879.

Chopra, M.; O'Neill, M. E.; Keogh, N.; Wortley, G.; Southon, S. & Thurnham, D. I. (2000): Influence of increased fruit and vegetable intake on plasma and lipoprotein carotenoids and LDL oxidation in smokers and nonsmokers. Clin Chem 46, 1818–1829.

Clinton, S. K.; Emenhiser, C.; Schwartz, S. J.; Bostwick, D. G.; Williams, A. W. & Moore, B. J. (1996): Cis–trans lycopene isomers, carotenoids, and retinol in the human prostate. Cancer Epidemiol Biomarkers Prev 5, 823–833.

Cohen, L. (2002): A review of animal model studies of tomato carotenoids, lycopene and cancer chemoprevention. Experimental Biology and Medicine, 277: 864–868.

Di Mascio, P.; Kaiser, S. & Sies, H. (1989): Lycopene as the most efficient biological carotenoid singlet oxygen quencher. Arch Biochem Biophys 274, 532–538.

Diwadkar-Navsariwala, V.; Novotny, J. A.; Gustin, D. M.; Sosman, J. A.; Rodvold, K. A. & Crowell, J. A. (2003): A physiological pharmacokinetic model describing the disposition of lycopene in healthy men. J Lipid Res 44, 1927–1939.

Fuhrman, B.; Elis, A. & Aviram, M. (1997): Hydpocholesterolemic effect of lycopene and β-carotene is related to suppression of cholesterol synthesis and augmentation of LDL receptor activity in macrophages-Biochemical and Biophysical Research Communications, 233: 658-662.

Gartner, C.; Stahl, W. & Sies, H. (1997): Lycopene is more bioavailable from tomato paste than from fresh tomatoes. Am J Clin Nutr 66, 116–122.

Giovannucci, E. (2002): A review of epidemiologic studies of tomatoes, lycopene and prostate cancer. Experimental Biology and Medicine, 227: 852-859.

Hadley, C. W.; Clinton, S. K. & Schwartz, S. J. (2003): The consumption of processed tomato products enhances plasma lycopene concentrations in association with reduced lipoprotein sensitivity to oxidative damage. J Nutr 133, 727–732.

Hantz, H. L.; Young, L. F.; Martin, K. R. (2005): Physiologically attainable concentrations of lycopene induce mitochondrial apoptosis in LNCaP human prostate cancer cells. Exp Biol Med 230, 171–179.

Hininger, I. A.; Meyer-Wenger, A.; Moser, U.; Wright, A.; Southon, S. & Thurnham, D. (2001): No significant effects of lutein, lycopene or b-carotene supplementation on biological markers of oxidative stress and LDL oxidizability in healthy adult subjects. J Am Coll Nutr 20, 232–238.

Holloway, D. E.; Yang, M.; Paganga, G.; Rice-Evans, C. A. & Bramley, P. M. (2000): Isomerization of dietary lycopene during assimilation and transport in plasma. Free Radical Res 32, 93–102.

Hoppe, P. P.; Kramer, K.; Van den Berg, H.; Steenge, G. & Vliet, T. (2003): Synthetic and tomato-based lycopene have identical bioavailability in humans. Eur J Nutr 42, 272–278.

Hwang, E. S. & Bowen, P. E. (2005): Effects of lycopene and tomato paste extracts on DNA and lipid oxidation in LNCaP human prostate cancer cells. Biofactors 23, 97–105.

Jonker, D.; Kuper, C.; Fraile, N.; Estrella, A. & Otero, C. (2003): Ninety-day oral toxicity study of lycopene from Blakeslea trispora in rats. Regul Toxicol Pharmacology, 37: 396–406.

Kim, Y.; Park, Y.; Lee, K.; Jeon, S.; Gregor, R. & Choi, S. (2012): Dose dependent effects of lycopene enriched tomato wino on liver and adipose tissue in high fat diet fed rats. Food Chemistry, 130: 42-48.

Kiokias, S. & Gordon, M. H. (2003): Dietary supplementation with a natural carotenoid mixture decreases oxidative stress. Eur J Clin. Nutr 57, 1135–1140.

Klipstein-Grobusch, K.; Launer, L.; Geleijnse, J.; Boeing, H.; Hofman, A. & Wtteman, J. (2000): Serum caroteniods and atherosclerosis. The Rotterdam study. Atherosclerosis, 148: 49-56.

Khachik, F.; Carvalho, L.; Bernstein, P.S; Muir, G.; Zhao, D. & Katz, N. (2002): Chemistry, distribution and metabolism of tomato carotenoids and their impact on human health. Experimental Biology and Medicine, 227: 845-851.

Kohlmeir, L.; Kark, J.; Gomez-Garcia, E.; Martin, B.; Steck, S. & Kardinaal, A. (1997): Lycopene and myocardial infraction risk in the EURAMIC study. American Journal of Epidemiology, 146: 618-626.

Kucuk, O.; Sarkar, F. H.; Sakr, W.; Djurie, Z.; Pollak, M. N. & Khachik, F. (2001): Phase II randomized clinical trial of lycopene supplemen- tation before radical prostatectomy. Cancer Epidemiol Biomarkers Prev 10, 861–868.

Lee, A.; Thurnham, D. & Chopra, M. (2000): Consumption of tomato products with olive oil but not sunflower oil increases the antioxidant activity of plasma. Free Radical Biol Med 29, 1051–1055.

Levin, G.; Yeshurun, M. & Mokady, S. (1997): In vivo antiperoxidative effect of 9-cis b-carotene compared with that of the all-trans isomer. Nutr Cancer 27, 293–297.

Libby, P. (2006): Inflammation and cardiovascular disease mechanisms. Am J Clin Nutr 83, 456S–460S.

Liu, C.; Russell, R. M. & Wang, X. D. (2006): Lycopene supplementation prevents smoke-induced changes in p53, p53 phosphorylation, cell proliferation, and apoptosis in the gastric mucosa of ferrets. J Nutr 136, 106–111.

Lorenz, M.; Fechner, M.; Kalkowski, J.; Frohlich, K.; Trautman, A.; Bohm, V.; Liebisch, G.; Lehneis, S.; Schmitz, G.; Ludwing, A.; Baumann, G.; Stangl, K. & Stangle, V. (2012): Effects of lycopene on the initial state of atherosclerosis in New Zealand white rabbits. PLoS one 7: 1-8.

Livny, O.; Kaplan, I.; Reifen, R.; Polak, S.; Madar, Z. & Schwartz, B. (2002): Lycopene inhibits proliferation and enhances gap-junctional communication of KB-1 human oral tumor cells. Journal of Nutrition, 132: 3754-3759.

Nahum, A.; Hirsch, K. & Danilenko, M. (2000): Lycopene inhibition of cell cycle progression in breast and endometrial cancer cells in associated with reduction in cyclin D levels and retention of P 27 in the cyclin E- cdk 2 complexes. Oncogene, 26: 3428-3436.

Omoni, O. & Aluko, R. (2005): The anticarcinogenic and antiatherogenic effects of lycopene: a review. Trends in Food Science & Technology, 16: 344-350.

Parthasarathy, S. (1998): Mechanisms by which dietary antioxidants may prevent cardiovascular diseases. Journal of Medicinal Food, 1: 45-51.

Paster, M.; Fander, H.; Boscoboinik, D. & Azzi, A. (1998): Lycopene in association with α-tocopherol inhibits at physiological concentrations proliferation of prostate carcinoma cells. Biochemistry Biophysics Research communication, 35: 582-585.

Obermuller-Jevic, U. C.; Olano-Martin, E.; Corbacho, A. M.; Eiserich, J. P. Van der Vliet. A. & Valacchi, G. (2003): Lycopene inhibits the growth of normal human prostate epithelial cells in vitro. J Nutr 133, 3356–3360.

Pool-Zobel, B. L.; Bub, A.; Muller, H.; Wollowski, I. & Rechkemmer, G. (1997): Consumption of vegetables reduces genetic damage in humans: first result of a human intervention trial with carotenoid-rich foods. Carcinogenesis 18, 1847–1850.

Porrini, M.; Riso, P. & Testolin, G. (1998): Absorption of lycopene from single or daily portions of raw and processed tomato. Br J Nutr 80, 353–361.

Porrini, M. & Riso, P. (2000): Lymphocyte lycopene concentration and DNA protection from oxidative damage is increased in women after a short period of tomato consumption 130, 189–192.

Porrini, M.; Riso, P. & Oriani, G. (2002): Spinach and tomato consumption increases lymphocyte DNA resistance to oxidative stress but this is not related to cell carotenoid concentrations. Eur J Nutr 41, 95-100.

Porrini, M.; Riso, P.; Brusamolino, A.; Berti, C.; Guarnieri, S. & Visioli, F. (2005): Daily intake of a formulated tomato drink affects carotenoid plasma and lymphocyte concentrations and improves cellular antioxidant protection. Br J Nutr 93, 93–99.

Rao, A. V. & Shen, H. (2002): Effect of low dose lycopene intake on lycopene bioavailability and oxidative stress. Nutr Res 22, 1125-1131.

Rao, G.; Guns, E. & Rao, A. (2003): Lycopene: Its role in human health and disease. Agro Food Industry In Tech, 8: 25-30.

Rao, A. V. (2004): Processed tomato products as a source of dietary lycopene: bioavailability and antioxidant properties. Can J Diet Pract Res 65, 161–165.

Reboul, E. Borel, P.; Mikail, C.; Abou, L.; Charbonnier, M. & Caris-Veyrat. C. (2005): Enrichment of tomato paste with 6% tomato peel increases lycopene and b-carotene bioavailability in men. J Nutr 135, 790–794.

Re, R.; Fraser P. D.; Long M, Bramley P. M. & Rice-Evans C. (2001): Isomerization of lycopene in the gastric milieu. Biochem Biophys Res Commun 281, 576–581.

Richelle, M.; Bortlik, K.; Liardet, S.; Hager, C.; Lambelet, P. & Baur, M. (2002): A food-based formulation provides lycopene with the same bioavailability to humans as that from tomato paste. J Nutr 132, 404–408.

Riso, P.; Pinder, A.; Santangelo, A. & Porrini, M. (1999): Does tomato consumption effectively increase the resistance of lymphocyte DNA to oxidative damage? Am J Clin Nutr 69, 712–718.

Riso, P.; Visioli, F.; Erba, D.; Testolin, G. & Porrini, M. (2004): Lycopene and vitamin C concentrations increased in plasma and lymphocytes after tomato intake. Effects on cellular antioxidant protection. Eur J Clin Nutr 58, 1350–1358.

Rissanen, T.; Voutilainen, S.; Nyyssonen, K.; Salonen. & Salonen J. T. (2000): Low plasma lycopene concentrations is associated with increased intima-media thickness of the carotid artery wall. Arteisclerosis, Thrombosis and Vascular Biology, 20: 677-2681.

Rissanen, T.; Voutilainen, S.; Nyyssonen, K.; Salonon, J. Kaplan, G. & Salonen, J. (2003): Serum lycopene concentration and carotid atherosclerosis: the Kuopio Ischemic Heart Disease Risk Factor Study. Am J Clin Nutr 77, 133–138.

Sesson, H. D.; Liu, S.; Gaziano, M. & Buring, J. (2003): Dietary lycopene, tomato-based food products and cardiovascular disease in women. Journal of Nutrition, 133: 2336-341.

Shi, J. (2000): Lycopene in tomatoes: Chemical and physical properties affected by food processing. Critical Reviews In Food Science and Nutrition, 40: 1-42.

Stahl, W.; Junghans, A.; Boer, B.; Driomina, E.; Briviba, K. & Sies, H. (1998): Carotenoid mixtures protect multilamellar liposomes against oxidative damage: synergistic effects of lycopene and lutein. FEBS Lett. 42: 305-308.

Unlu, N. Z.; Bohn, T.; Clinton, S. K. & Schwartz, S. J. (2005): Carotenoid absorption from salad and salsa by humans is enhanced by the addition of avocado or avocado oil. J Nutr 135, 431–436.

Upritchard, J. E.; Sutherland, W. H. F.; Mann, J, I. (2000): Effect of supplementation with tomato juice, vitamin E, and vitamin C on LDL oxidation and products of inflammatory activity in Type 2 diabetes. Diabetes Care 23, 733–738.

Visioli, F.; Riso, P.; Grande, S.; Gall, C. & Porrini, M. (2003): Protective activity of tomato products on in vivo markers of lipid oxidation. Eur J Nutr 42, 201–206.

Willis, M. S. & Wiams, F. H. (2003): The role of nutrition in preventing prostate cancer: a review of the proposed mechanisms of action of various dietary substances. Clinica Cimica Acta, 330:57-83.

Wu, K.; Schwaz, S. J.; Platz, A.; Clinton, S.; Erdman, J. & Ferruzzi, M. (2003): Variations in plasma lycopene and specific isomers over time in a cohort of US men. Journal of Nutrition, 133; 1930-1936.

Zhao, X.; Aldini, G.; Johnson, E. J.; Rasmussen, H. Kraemer, K. & Woolf, H. (2006): Modification of lymphocyte DNA damage by carotenoid supplementation in postmenopausal women. Am J Clin Nutr 83, 163–169.

Chapter 5

HDL-Associated Paraoxonase 1 Gene Polymorphisms as a Genetic Markers for Wide Spread Diseases

Ivana Pejin-Grubiša

Additional information is available at the end of the chapter

1. Introduction

The story of paraoxonase 1 (PON1) begins in 1946, when Abraham Mazur reported the presence of an enzyme in human and rabbit tissues which was able to hydrolyse organophosphate compounds [1]. In 1950s, enzyme was named "paraoxonase" according to its ability to hydrolyse paraoxon, the toxic metabolite of the organophosphate insecticide parathion [2,3]. Later it was discovered that it exhibits a broad spectrum of activities and has diverse substrates. Mackness and colleagues linked PON1 to cardiovascular diseases in 1991 and demonstrated that PON1 could prevent the accumulation of oxidized lipids in low-density lipoprotein (LDL) [4]. However, despite intensive research over sixty years the exact physiological function of PON1 is still unclear.

2. Body

2.1. Paraoxonase 1

The paraoxonase (PON) family of the enzymes consists of three members, PON1, PON2 and PON3 that share approximately 65% similarity at the amino acid level. These were named in order of their discovery, but according to the structural homology and predicted evolutionary distance between them it seems that PON2 is the oldest and PON1 is the youngest family member [5].

PON1 and PON3 enzymes are secreted from liver cells and associate with HDL in the circulation [6]. Low levels of PON1 may be expressed in a number of tissues, primarily in epithelia. PON2 in humans is more widely expressed and is found in nearly every human tissue including heart, kidney, liver, lung, placenta, small intestine, spleen stomach, testis [6-

8]. Also, human PON2 mRNA is detected in the cells of the artery wall, including endothelial cells, smooth muscle cells and macrophages and is undetectable in HDL, LDL or the media of cultured cells [6, 7]. Of the three PON proteins, PON3 is the most recently identified and the least characterized.

PON1 is the most studied and best understood. This calcium-dependent esterase is consisting of 354 amino acids with a molecular mass od approximately 45 kDa [9, 10] and requires calcium ions for structural stability and enzymatic activities [11]. It is capable of hydrolyzing organophosphates such as oxon metabolites of a insecticides parathion, diazinon and chlorpyrihos and nerve agents sarin and soman, aromatic esters such is phenyl acetate (arylesterase activity) and a variety of aromatic and aliphatic lactones (lactonase activity) [12-18]. Beside its protective role against dietary and environmental lactones, PON1 also catalyzes the reaction of lactonization of γ- and δ-hydroxycarboxylic acids [18]. It is capable of hydrolyse the oxidised lipid derivates 5-hydroxy-eicosatetraenoic acid lactone acid (5-HETEL) and 4-hydroxy-docosahexaeonic acid (4-HDoHE) which are potent triggers of an inflammatory response and therefore determinants of atherosclerotic disease [16,19].

2.2. Paraoxonase 1 gene polymorphisms

Genes that code for three PON proteins (*pon1, pon2 and pon3*) are located to each other on the long arm of chromosome 7 in humans (7q21.3-22.1) and share approximately 70% similarity at the nucleotide level [8]. Earlier studies on different human populations showed that the hydrolytic activity of serum PON1 was polymorphically distributed [20-22] and a number of research demonstrated that the molecular basis of these differences were Q192R, L55M and C(-107)T polymorphisms in *pon1* gene.

The *pon1* gene contains functional polymorphisms in both the coding and promoter regions. In the coding region, two common polymorphism are a glutamine (Q) to arginine (R) substitution at codon 192 (Q192R) and a leucine (L) to methionine (M) substitution at position 55 (L55M). In Q192R polymorphism the exchange of codon CAA to CGA in exon 6 of *pon1* gene determines isoforms of the enzyme which differ greatly in the rate of hydrolysis a number of substrates. Paraoxon is hydrolysed at a far greater rate by the R192 isoform compared to the Q192, but some organophospates and lactones are hydrolyzed faster by Q192 [12,13]. Recent study showed that R isoform of the enzyme has higher lactonase activity and increased antiatherogenic potential [23].

L55M polymorphism (exchange of codon TTG to ATG in exon 3) is correlated with blood enzyme level with isoform L55 associated with higher serum enzymatic activity. Still, it is not clear whether this is because of a decreased stability of the M55 alloenzyme [24] and/or because of the linkage disequilibrium with -107/-108 T allele [25, 26]. Isoform M55 showed lower stability and loses activity more rapidly and to a greater extent than the L isoform [24]. This is due to the key role of L55 in packing in the propeller's central tunnel, and of its neighboring residues which ligate calcium iones [27].

At least five polymorphisms have been detected in the human *pon1* gene promoter region: C(-107/-108)T, G(-126)C, G(-162)A, G(-832)A and G(-909)C, but only C(-107T) apear to affects expression level of PON1 enzyme [25,26]. This single nucleotide polymorphism (SNP) is within stimulating protein-1 (Sp1) binding site with allele T that disrupts the recognition sequence for Sp1 and results in decreased affinity for it [28].

The frequencies of alleles of Q192R, L55M and C(-107)T polymorphisms are different among populations worldwide (Table 1). Data for European population showed predominance of Q192 and -107C alleles over R192 and -107T alleles. Spanish and Serbian populations showed higher frequency of the -107T allele. For codon 55 polymorphism, populations worldwide show predominance of L55 over M55 allele. In Asia allele Q192 is more frequent only in Indian Punjabis and Iranians and allele -107C is predominant among examined populations. Afro-Americans and Amerindian tribes showed higher frequency of allele R compared with allele Q and predominance of allele-107C. Only Mexicans showed higher frequency of -107T allele. There is a very little data from „black" continent and it concerns only Q192R polymorphism frequency with higher frequency of allele R only in Beninese (Table 1).

More than 200 single nucleotide polymorphisms (SNPs) have been identified in the human *pon1* gene but only these three have been associated with a number of pathophysiological conditions.

2.3. Pon1 variants and oxidative stress-related disorders

The central role of HDL is in the process of reverse cholesterol transport (RHC). Also it has antioxidative, antiinflammatory and antifibrinolytic functions that contribute to its antiatherosclerotic effects. Mackness and coworkers were the first that showed that HDL acted at a specific point in the oxidation cascade: it metabolises oxidized phospholipids on LDL [29]. Although several other HDL-associated proteins such as apo AI, lecithin:cholesterol acyltransferase (LCAT) and platelet-activating factor acetyltransferase (PAFAH) also have antioxidant properties, PON1 seems to be the predominant antioxidant enzyme [4, 29-31]. HDL isolated from the blood of PON1 knock-out mice or from avian species which naturally lack PON1, has at best, no effect on LDL-oxidation and at worst promotes LDL-oxidation [32,33]. Conversely, HDL isolated from mice overexpressing human PON1 completely abolishes LDL-oxidation [34]. Several human studies have shown an inverse linear relationship between the concentration of oxidised-LDL in the circulation and PON1 activity, strongly implicating PON1 in the metabolism of oxidised-LDL *in vivo* [35,36].

Enzymatic and nonenzymatic systems of antioxidative protection are included in scavenging free radicals and their metabolic products and in maintaining normal cellular physiology. Increased level of free radicals and impairment of antioxidant status are processes underlying pathophysiologic mechanisms in a variety of diseases including atherosclerosis, diabetes mellitus, cancer, chronic liver impairment, several neurological diseases, many infectious diseases and association studies have identified links between *pon1* gene polymorphisms and susceptibility and outcome of these diseases.

pon1 polymorphisms	Q192R		L55M		C(-107)T		References
	Q	R	L	M	C	T	
Populations of Europe							
Finnish	0.69	0.31	0.67	0.33	-	-	76
Dutch	0.68	0.32	0.63	0.37	-	-	77
Spanish	0.7	0.3	0.63	0.37	0.46	0.54	78
Italians	0.65	0.35	0.66	0.34	0.57	0.43	79
English	0.78	0.22	0.7	0.3	0.52	0.48	80
Turkish	0.69	0.31	0.7	0.3	-	-	81
Croatian	0.77	0.23	0.66	0.34	0.54	0.46	82
Czecs	0.54	0.46	0.69	0.31	0.59	0.41	83
Serbian	0.77	0.23	0.68	0.32	-	-	84
Populations of Asia							
Asian Indians Punjabis	0.74	0.26	0.81	0.19	0.52	0.48	85
Japanese	0.4	0.6	0.94	0.06	0.48	0.52	86
Koreans	0.38	0.620	0.94	0.06	-	-	87
Chinese	0.42	0.58	0.95	0.05	0.57	0.43	88, 89
Iranian	0.69	0.31	0.59	0.41	-	-	90
Populations of America							
Caucasian-Americans	0.73	0.27	0.64	0.36	0.5	0.5	26
Canadians	0.73	0.27	0.64	0.36	0.48	0.52	91, 28
African-Americans	0.37	0.63	0.79	0.21	0.85	0.15	92
Amazonian Amerindian tribes	0.27	0.730	0.967	0.033	-	-	93
Caribean-Hispanics	0.540	0.460	0.71	0.29	0.65	0.35	92
Mexicans	0.510	0.490	0.84	0.16	0.45	0.55	94
Peruvians	0.539	0.461	-	-	0.61	0.39	95
Populations of Africa							
Beninese	0.388	0.612	-	-	-	-	96
Ethiopians	0.592	0.408	-	-	-	-	96
Egyptians	0.67	0.33	-	-	-	-	97

Table 1. The allele frequencies of *pon1* gene polymorphisms Q192R, L55M and C(-107)T in populations worldwide

According to World Health Organization (WHO data for 2010), 95% of mortality in Serbia is caused by chronic noncontagious diseases, wherefrom 58% of it is caused by cardiovascular diseases (CVD) [37]. Although patients with CVD commonly have at least one identifiable risk factor, many ischemic events occur in the absence of any of it [38]. Atherogenesis, one of the main risk factors for CVD, is initiated by oxidation of the low-density lipoprotein (LDL) and by impairment in oxidative stress-antioxidant balance.

Enhanced oxidative stress such as in diabetes, leads to the development of accelerated atherosclerosis. Atherosclerosis in patients with diabetes tends to occur earlier and be more aggressive. People with type 2 diabetes have a 3–4 fold increased risk of developing atherosclerosis compared to people without type 2 diabetes. Serbia falls into the group of European countries with the highest diabetes mortality rates where diabetes is the fifth leading cause of death and the fifth cause of the burden of disease [39]. At least a half of the persons with non-insulin dependent diabetes mellitus (NIDDM) have not been diagnosed and are not aware of their disease [40,41].

Due to the abovementioned, there has been a marked interest in discovering additional markers of oxidative stress, including gene variants, which may have a role in predicting wide spread diseases risk. Because controversial results have been reported so far, the aim of studies performed in our laboratory was to evaluate possible interactions between *pon1* gene polymorphisms and clinical manifestations of atherosclerosis and diabetes mellitus type 2 in our population.

Allele and genotype frequencies for Q192R, L55M and C(-107)T did not show significant difference between cases with clinical manifestations of atherosclerosis (60 subjects) and controls (100 subjects) (P>0.05). Although the M allele (L55M) has shown a somewhat higher risk (OR=1.23) and the T allele (-107C/T) has shown a 1.49 times lower risk of occurence of the disease (OR=0.67) the difference did not reach statistical significance, most likely due to low number of subjects (Grubisa et al., unpublished data).

Also, we investigated the association between these polymorphisms and atherosclerosis in patients with type 2 diabetes mellitus (140 subjects). Our results have shown that R allele is a risk factor for atherosclerosis in these patients (OR=2.22, P<0.0001). Although M allele has shown a little higher risk (OR=1.26) and allele T has shown a slightly lower risk (OR=0.85) the results obtained do not support an association between these *pon1* gene variants and atherosclerosis in NIDDM patients (Grubisa et al., unpublished data).

Lactones are hydrolyzed preferentially by either PON1 Q or R isoformes, depending of their structure. R192 is more efficient at hydrolyzing homocysteine thiolactone, while δ-valerolactone and 2-coumaranone are more rapidly hydrolyzed by PON1Q192 [12]. In 1990's the results obtained indicated that the Q192R polymorphism may play the role in coronay heart disease (CHD) etiology because this genotype is associated with LDL oxidation; the PON1-192 R isoform is less effective at hydrolysing lipid peroxides than the Q isoform [42,43]. It have been shown that position 192 is involved in HDL binding as a part of amphipathic helix H2 of active site [27]. Gaidukov and coworkers reported from *in vitro* and sera tests that the PON1-192Q izoform binds HDL with a 3-fold lower affinity than the R isozyme and consequently exhibits significantly reduced stability, lipolactonase activity, and macrophage cholesterol efflux [27]. The higher lactonase activity is manifested by increased antiatherogenic potency: the observed rate of HDL-mediated cholesterol efflux from macrophages is 2.2-fold higher for the 192R [27]. Also it was shown that the affinity and stability of the PON1 on HDL was lower in sera of individuals with the Q192 variant than in individuals with the 192R variant [27]. Low levels of HDL particles is one of the strongest risk factors for coronary heart disease and one of the characteristic features of diabetic dyslipidemia and it seems that proteins on HDL play a major role in the protection against atherosclerois-based cardiovascular diseases. HDL carryng apolipoprotein A-I binds PON1 with high affinity, stabilizes the enzyme and stimulates its lipolactonase activity [44].

PON1 is also an extracellular homocysteine-thiolactonase (Hcy-thiolactonase). Hcy-thiolactone is a toxic metabolite linked to immune activation and thrombogenesis in human cardiovascular diseases and is elevated under conditions predisposing atherosclerosis [45-47]. A small fraction of Hcy, a sulfur-containing amino acid, is metabolized to a Hcy-thiolactone in an error-editing reaction in protein biosynthesis when Hcy is mistakenly selected instead of dietary methonine (Met) [48]. Hcy-thiolactone is neutral at physiological pH and can diffuse out of the cell and accumulate in the extracellular fluids where is hydrolyzed to Hcy by extracellular Hcy-thiolactonase-paraoxonase 1 [49]

Hcy-thiolactonase activity is strongly associated with *pon1* genotype in diverse human populations [15]. High Hcy-thiolactonase activity is associated with L55 and R192 alleles, more frequent in blacks than in whites and low activity is associated with M55 and Q192 alleles, more frequent in whites than in blacks [15]. Despite the impact of *pon1* genotype on Hcy-thiolactonase activity, these genetic variations are not asociated with atherosclerosis-based cardiovascular diseases. It seems that PON1 phenotype is better predictor [16,50].

Human clinical studies suggest that PON1 phenotype, i.e., paraoxonase activity is a much stronger predictor of cardiovascular disease status than PON1 genetic polymorphisms [51-55] a finding that has been confirmed in other studies [52,55]. Bhattacharyya and colleagues demonstrated that both the *pon1* Q192R polymorphism and serum PON1 activity are associated with prevalent coronary artery disease and incident adverse cardiovascular events [56]. This study complemented the study of Gaidukov, demonstrating that individuals with the arginine (R) at position 192 have higher serum levels of PON1 activity, lower systemic indices of systemic oxidative stress and corresponding reductions in both prevalent coronary artery disease and prospective cardiac events [56]. Plasma PON1 activity can vary up to 40 to 50-fold, and differences in PON1 protein levels up to 13–15-fold are also present within a single PON1 Q192R genotype in adults [57,58]. A number of studies indicated that measurement of an individual's PON1 function (serum activity) takes into account all polymorphism and other factors that might affect PON1 activity or expression. However, modulation of PON1 by alcohol, smoking, drugs, diet, certain physiological and pathological conditions should also be considered. These factors can increase or decrease PON1 activity [59] as well as HDL status.

However, PON1 activity is partially inactivated during the detoxification of lipid hydroperoxides [60]. This effect can be possibly related to displacement of calcium ions or inhibition through free radicals directly. It has been suggested that other antioxidant enzymes might prevent this inhibition of PON1 activity. Antioxidant enzymes, all show co-activity and might work in a collaboration against oxidative stress and elevation in oxidative stress might inhibit these enzymes [61].

Paraoxonases are important detoxifying and anti-oxidative enzymes, which establishes their role in organophosphate poisoning, diabetes, obesity, cardiovascular diseases, and innate immunity [62, 63] Consequently, PON2 has been the focus of a great deal of research in recent years. Both PON1 and PON2 protect against atherosclerosis development and share ability to hydrolyze lactones with both overlaping and distinct substrate specificities [19]. Although PON1 is associated with circulating serum HDL and reduces oxidative stress in lipoproteins, macrophages in arterial walls and in atherosclerotic lesion by its ability to hydrolyze specific oxidized lipids, PON2 acts as an intracelullar antioxidant [7,64-68] associated with plasma membrane [6-8]. The mechanism how PON2 modulates oxidative stress is still unknown, although Altenhöfer demonstrated that PON2 prevents superoxide generation, but was ineffective against existing radicals [69]. Oxidative stress affects PON2 expression too, but additional studies are needed to highlight the PON2 expression level under oxidative stress since controversial results both from *in vivo* and *in vitro* experiments have been reported [6,7,66,70-74].

3. Conclusion

Paraoxonase 1 is found to be associated with HDL particles within circulation and therefore promotes some of HDL's functions. There is no consistent evidence for involvement of *pon1* genotypes in atherosclerosis and diabetes mellitus type 2. Studies analyzed the role of *pon1* polymorphisms in oxidative stress-based diseases showed a great variation in ethnics, environmental background, age and gender of case and control groups. Allele frequencies appeared to be dependent on geographic locations, perhaps also due to genetic drift. Probably the effect of each polymorphism alone of he so called oxidative stress-associated genes is not strong enough to affect initiation and progression of atherosclerosis as well as PON1 enzyme status (activity levels and catalytic efficiency specified by the Q192R polymorphism) [75].

Author details

Ivana Pejin-Grubiša
Department of Human Genetics and Prenatal Diagnostics, Zvezdara University Medical Center, Belgrade, Serbia

4. References

[1]   Mazur A (1946) An Enzyme in Animal Tissue Capable of Hydrolyzing the Phosphorus-fluorine Bond of Alkyl Fluorophosphates. J. Biol. Chem. 164:271–289.

[2] Aldridge WN (1953a) Serum Esterases I. Two Types of Esterase (A and B) Hydrolysing p-nitrophenyl Acetate, Propionate and Butyrate and a Method for Their Determination. Biochem. J. 53:110–117.

[3] Aldridge WN (1953b) Serum Esterases II. An Enzyme Hydrolysing Diethyl p-nitrophenyl Acetate (E600) and Its Identity With the A-esterase of Mammalian Sera. Biochem. J. 53:117–124.

[4] Mackness M, Arrol S, Durrington PN (1991) Paraoxonase Prevents Accumulation of Lipoperoxides in Low-Density Lipoprotein. FEBS Lett. 286:152–154.

[5] Draganov DI, La Du BN (2004) Pharmacogenetics of Paraoxonses, a Brief Review. Naunyn-Schmiedeberg's Arch. Pharmacol. 369:78–88.

[6] Ng CJ, Shih DM, Hama SY, Villa B, Navab M, Reddy ST (2005) The Paraoxonase Gene Family and Atherosclerosis. Free Radic. Biol. Med. 38:153–163.

[7] Ng CJ, Wadleigh DJ, Gangopadhyay A, Hama S, Grijalva VR, Navab M, Fogelman AM, Reddy ST (2001) Paraoxonase-2 is a Ubiquitously Expressed Protein With Antioxidant Properties and is Capable of Preventing Cell-mediated Oxidative Modification of Low Density Lipoprotein. J. Biol. Chem. 276(48): 44444–44449.

[8] Primo-Parmo SL, Sorenson RC, Teiber J, La Du BN, (1996) The Human Serum Paraoxonase/Arylesterase Gene (PON1) is One Member of a Multigene Family. Genomics 33:498–507.

[9] Furlong CE, Richter RJ, Seidel SL, Costa LG, Motulsky AG (1989) Spectrophotometric Assays for the Enzymatic Hydrolysis of the Active Metabolites of Chlorpyrifos and Parathion by Plasma Paraoxonase/Arylesterase. Anal. Biochem. 180: 242–247.

[10] Hassett C, Richter RJ, Humbert R, Chapline C, Crabb JW, Omiecinski CJ, Furlong CE (1991) Characterization of cDNA Clones Encoding Rabbit and Human Serum Paraoxonase: the Mature Protein Retains its Signal Sequence. Biochemistry 30:10141–10149.

[11] Kuo CL, La Du BN (1998) Calcium Binding by Human and Rabbit Serum Paraoxonases. Structural Stability and Enzymatic Activity. Drug Metab. Dispos. 26:653–660.

[12] Billecke S, Draganov D, Counsell R, Stetson P, Watson C, Hsu C, La Du BN (2000) Human Serum Paraoxonase (PON1) Isozymes Q and R Hydrolyze Lactones and Cyclic Carbonate Esters. Drug Metab. Dispos. 28:1335–1342.

[13] Davies HG, Richter RJ, Keifer M, Broomfield CA, Sowalla J, and Furlong CE (1996) The Effect of the Human Serum Paraoxonase Polymorphism is Reversed With Diazoxon, Soman and Sarin. Nat. Genet. 14:334–336.

[14] Furlong CE, Li WF, Brophy VH, Jarvik GP, Richter RJ, Shih DM, Lusis AJ, Costa LG (2000) The PON1 Gene and Detoxication. Neurotoxicology 21:581–587.

[15] Jakubowski H, Ambrosius WT, Pratt JH (2001) Genetic Determinants of Homocysteine Thiolactonase Activity in Humans: Implications for Atherosclerosis. FEBS Lett. 491:35–39.

[16] Khersonsky O, Tawfik DS (2005) Structure-Reactivity Studies of Serum Paraoxonase PON1 Suggest That its Native Activity is Lactonase. Biochemistry 44:6371–6382.

[17] Rodrigo L, Mackness B, Durrington PN, Hernandez A, Mackness MI (2001) Hydrolysis of Platelet-Activating Factor by Human Serum Paraoxonase. Biochem. J. 354:1–7.

[18] Teiber JF, Draganov DI, La Du BN (2003) Lactonase and Lactonizing Activities of Human Serum Paraoxonase (PON1) and Rabbit Serum PON3. Biochem. Pharmacol. 66:887–896.

[19] Draganov DI, Teiber JF, Speelman A, Osawa Y, Sunahara R, La Du BN (2005) Human Paraoxonases (PON1, PON2, and PON3) are Lactonases With Overlapping and Distinct Substrate Specificities. J. Lipid Res. 46:1239–1247.

[20] Eckerson HW, Wyte CM, LaDu BN (1983) The Human Serum Paraoxonase/Arylesterase Polymorphism. Am. J. Hum. Genet. 35:1126–1138.

[21] Mueller RF, Hornung S, Furlong CE, Anderson J, Giblett ER, Motulsky AG (1983) Plasma Paraoxonase Polymorphism: a New Enzyme Assay, Population, Family Biochemical and Linkage Studies. Am. J. Hum. Genet. 35:393–408.

[22] Playfer JR, Eze LC, Bullen MF, Evans DA (1976) Genetic Polymorphism and Interethnic Variability of Plasma Paraoxonase Activity. J. Med. Genet. 13:337–342.

[23] Gaidukov L, Rosenblat M, Aviram M, Tawfik DS (2006) The 192R/Q Polymorphs of Serum Paraoxonase PON1 Differ in HDL Binding, Lipolactonase Stimulation, and Cholesterol Efflux: J. Lipid Res. 47:2492–2502.

[24] Leviev I, Deakin S, James RW (2001) Decreased Stability of the M54 Isoform of Paraoxonase as a Contributory Factor to Variations in Human Serum Paraoxonase Concentrations. J. Lipid Res. 42: 528–535.

[25] Leviev I, James RW (2000) Promoter Polymorphisms of Human Paraoxonase PON1 Gene and Serum Paraoxonase Activities and Concentrations. Arterioscler. Thromb.Vasc. Biol. 20:516–521.

[26] Brophy VH, Jampsa RL, Clendenning JB, McKinstry LA, Jarvik GP, Furlong CE (2001) Effects of 5′ Regulatory-Region Polymorphisms on Paraoxonase-Gene (PON1) Expression. Am. J. Hum. Genet. 68:1428–1436.

[27] Harel M, Aharoni A, Gaidukov L, Brumshtein B, Khersonsky O, Meged R, Dvir H, Ravelli RBG, McCarthy A, Toker L, Silman I, Sussman JL, Tawfik DS (2004) Structure and Evolution of the Serum Paraoxonase Family of Detoxifying and Anti-Atherosclerotic Enzymes. Nat. Struct. Mol. Biol. 11:412–419.

[28] Deakin S, Leviev I, Brulhart-Meynet MC, James RW (2003) Paraoxonase-1 Promoter Haplotypes and Serum Paraoxonase: a Predominant Role for Polymorphic Position -107, Implicating the Sp1 Transcription Factor. Biochem. J. 372:643–649.

[29] Mackness MI, Arrol S, Abbott CA, Durrington PN (1993) Protection of Low-Density Lipoprotein Against Oxidative Modification by High-Density Lipoprotein Associated Paraoxonase. Atherosclerosis. 104:129–135.

[30] Ahmed Z, Ravandi A, Maguire GF, Emili A, Draganov D, La Du BN, Kuksis A, Connelly PW (2001) Apolipoprotein AI Promotes the Formation of Phosphatidylcholine Core Aldehydes That are Hydrolysed by Paraoxonase (PON1) During High Density Lipoprotein Oxidation With a Peroxynitrite Donor. J. Biol. Chem. 276:24473–24481.

[31] Watson AD, Berliner JA, Hama SY, La Du BN, Fault KF, Fogelman AM, Navab M (1995) Protective Effect of High Density Lipoprotein Associated Paraoxonase-Inhibition of the Biological Activity of Minimally Oxidised Low-Density Lipoprotein. J. Clin. Invest. 96: 2882–2891.

[32] Shih DM, Gu L, Xia YR, Navab M, Li WF, Hama S, Castellani LW, Furlong CE, Costa LG, Fogelman AM, Lusis AJ (1998) Mice Lacking Serum Paraoxonase are Susceptible to Organophosphate Toxicity and Atherosclerosis. Nature. 394:284–287.

[33] Mackness B, Mackness M.I, Arrol S, Turkie W, Durrington PN (1998). Effect of the Human Serum Paraoxonase 55 and 192 Genetic Polymorphisms on the Protection by High Density Lipoprotein Against Low Density Lipoprotein Oxidative Modification. FEBS Letts. 423:57–60.

[34] Tward A, Xia YR, Wang XP, Shi YS, Park C, Castellani LW, Lusis A, Shih DH (2002) Decreased Atherosclerotic Lesion Formation in Human Serum Paraoxonase Transgenic Mice. Circulation 106:484–490.

[35] Sampson MJ, Braschi S, Willis G, Astley SB (2005) Paraoxonase-1 (PON1) Genotype and Activity and in vivo Oxidised, Plasma Low-Density Lipoprotein in Type II Diabetes. Clin. Sci 109:189–197.

[36] Tsuzura S, Ikeda Y, Suehiro T, Ota K, Osaki F, Arii K, Kumon Y, Hashimoto K (2004) Correlation of Plasma Oxidized Low-Density Lipoprotein Levels to Vascular Complications and Human Serum Paraoxonase in Patients With Type 2 Diabetes. Metabolism 53:297–302.

[37] World Health Organization (WHO) (2011) Noncommunicable Diseases Country Profiles 2011. Available: http://www.who.int/nmh/publications/ncd_profiles2011/en/

[38] Futterman LG, Lemberg L (1998) Fifty Percent of Patients With Coronary Artery Disease Do Not Have Any of the Conventional Risk Factors. Am J Crit Care. 7:240-4.

[39] Incidence and Mortality of Diabetes in Serbia 2010. Serbian Diabetes Registry. Report No5. Institute of Public Health of Serbia „Dr Milan Jovanović Batut" 2011.

[40] Ford ES (2005) Risks for All-Cause Mortality, Cardiovascular Disease, and Diabetes Associated With the Metabolic Syndrome: a Summary of the Evidence. Diabet. Care 28:1769–1778.

[41] McEwan P, Williams JE, Griffiths Bagust A, Peters JR, Hopkinson P, Currie CJ (2004) Evaluating the Performance of the Framingham Risk Equations in a Population With Diabetes. Diabet. Med. 21:318–323.

[42] Aviram M, Billecke S, Sorenson R, Bisgaier C, Newton R, Rosenblat M, Erogul J, Hsu C, Dunlop C, La Du BN (1998) Paraoxonase Active Site Required for Protection Against LDL Oxidation Involves its Free Sulphydryl Group and is Different From That Required for its Arylesterase/Paraoxonase Activities: Selective Action of Human Paraoxonase Alloenzymes Q and R. Arterioscl. Thromb. Vasc. Biol. 10:1617–1624.

[43] Mackness B, Durrington PN, Mackness MI (1999) Polymorphisms of Paraoxonase Genes and Low-Density Lipoprotein Lipid Peroxidation: Lancet 353:468–469.

[44] Gaidukov L, Tawfik DS (2005) High Affinity, Stability, and Lactonase Activity of Serum Paraoxonase PON1 Anchored on HDL With ApoA-I. Biochemistry 44:11843–11854.

[45] Jakubowski H (1997) Metabolism of Homocysteine Thiolactone in Human Cell Cultures. Possible Mechanism for Pathological Consequences of Elevated Homocysteine Levels. J. Biol. Chem. 272:1935–1942.

[46] Jakubowski H, Zhang L, Bardeguez A, Aviv A (2000) Homocysteine Thiolactone and Protein Homocysteinylation in Human Endothelial Cells: Implications for Atherosclerosis. Circ. Res. 87: 45–51.

[47] Jakubowski H (2006) Pathophysiological Consequences of Homocysteine Excess. J. Nutr. 136:1741S–1749S.

[48] Jakubowski H (2004) Molecular Basis of Homocysteine Toxicity in Humans. Cell Mol. Life Sci. 61:470–487.

[49] Chwatko G, Jakubowski H (2005) The Determination of Homocysteine-Thiolactone in Human Plasma. Anal. Biochem. 337:271–277.

[50] Domagała TB, Łacinski M, Trzeciak WH, Mackness B, Mackness MI, Jakubowski H (2006) The Correlation of Homocysteine-Thiolactonase Activity of the Paraoxonase (PON1) Protein With Coronary Heart Disease Status. Cell Mol. Biol. (Noisy-le-grand) 52:3–9.

[51] Jarvik GP, Rozek LS, Brophy VH, Hatsukami TS, Richter RJ, Schellenberg GD, Furlong CE (2000) Paraoxonase (PON1) Phenotype is a Better Predictor of Vascular Disease Than is PON1192 or PON155 Genotype. Arterioscler. Thromb. Vasc. Biol. 20:2441-2447.

[52] Jarvik GP, Jampsa R, Richter RJ, Carlson CS, Rieder MJ, Nickerson DA, Furlong CE (2003) Novel Paraoxonase (PON1) Nonsense and Missense Mutations Predicted by Functional Genomic Assay of PON1 Status. Pharmacogenetics. 13:291-295.

[53] Mackness B, Davies GK, Turkie W, Lee E, Roberts DH, Hill E, Roberts C, Durrington PN, Mackness MI (2001) Paraoxonase Status in Coronary Heart Disease: Are Activity and Concentration More Important Than Genotype? Arterioscler. Thromb. Vasc. Biol. 21:1451–1457.

[54] Mackness B, Durrington P, McElduff P, Yarnell J, Azam N, Watt M, Mackness M (2003) Low Paraoxonase Activity Predicts Coronary Events in the Caerphilly Prospective Study. Circulation. 107:2775–2779.

[55] Mackness MI, Durrington PN, Mackness B (2004) The Role of Paraoxonase 1 Activity in Cardiovascular Disease: Potential for Therapeutic Intervention. Am. J. Cardiovasc. Drugs. 4:211–217.

[56] Bhattacharyya T, Nicholls SJ, Topol EJ, Zhang R, Yang X, Schmitt D, Fu X, Shao M, Brennan DM, Ellis SG, Allayee H, Lusis AJ, Hazen SL (2008) Relationship of Paraoxonase 1 (PON1) Gene Polymorphisms and Functional Activity With Systemic Oxidative Stress and Cardiovascular Risk. JAMA 299:1265–1276.

[57] Costa LG, Cole TB, Jarvik GP, Furlong CE (2003) Functional Genomics of the Paraoxonase (PON1) Polymorphisms: Effect on Pesticide Sensitivity, Cardiovascular Disease, and Drug Metabolism. Annu Rev. Med. 54:371–392.

[58] Richter RJ, Jarvik GP, Furlong CE (1999) Determination of Paraoxonase 1 (PON1) Status Without the Use of Toxic Organophosphate Substrates. Circ Cardiovasc Genet. 1:147–152.

[59] Costa LG, Cole TB, Furlong CE. (2005) Paraoxonase (PON1): From Toxicology to Cardiovascular Medicine. Acta Biomed. Suppl 2:50-57.

[60] Karabina SA, Lehner AN, Frank E, Parthasarathy S, Santanam N(2005) Oxidative Inactivation of Paraoxonase-Implications in Diabetes Mellitus and Atherosclerosis. Biochim. Biophys. Acta. 1725:213–221.

[61] Sozmen EY, Sagin FG, Kayikcioglu M, Sozmen B (2008) Oxidative Stress & Antioxidants and PON1 in Health and Disease. In: Mackness B, Mackness M, Aviram M, Paragh G, editors. The Paraoxonases: Their Role in Disease Development and Xenobiotic Metabolism. Dordrecht:Springer.pp 61-73.

[62] Camps J, Marsillach J, Joven J (2009) The Paraoxonases: Role in Human Diseases and Methodological Difficulties in Measurement.Crit. Rev. Clin. Lab. Sci. 46:83–106.

[63] Shih DM, Lusis AJ (2009) The Roles of PON1 and PON2 in Cardiovascular Disease and Innate Immunity. Curr. Opin. Lipidol. 20: 288–292.

[64] Ng CJ, N. Bourquard N, Grijalva V, Hama S, Shih DM, Navab M, Fogelman AM, Lusis AJ, Young S, Reddy ST (2006) Paraoxonase-2 Deficiency Aggravates Atherosclerosis in Mice Despite Lower Apolipoprotein-B-containing Lipoproteins: Anti-atherogenic Role for Paraoxonase-2. J. Biol. Chem. 281:29491–29500.

[65] Ng CJ, Hama SY, Bourquard N, Navab M, Reddy ST (2006) Adenovirus Mediated Expression of Human Paraoxonase 2 Protects Against the Development of Atherosclerosis in Apolipoprotein E-deficient mice. Mol. Genet. Metab. 89:368–373.

[66] Fortunato G, Di Taranto MD, Bracale UM, Del Guercio L, Carbone F, Mazzaccara C, Morgante A, D'Armiento FP, D'Armiento M, Porcellini M, sacchetti L, Bracale G, Salvatore F (2008) Decreased Paraoxonase-2 Expression in Human Carotids During the Progression of Atherosclerosis. Arterioscler. Thromb. Vasc. Biol. 28:594–600.

[67] Horke S, Witte I, Wilgenbus P, Kruger M, Strand D, Forstermann U (2007) Paraoxonase-2 Reduces Oxidative Stress in Vascular Cells and Decreases Endoplasmic Reticulum Stress-induced Caspase Activation. Circulation. 115:2055–2064.

[68] Devarajan A, Bourquard N, Hama S, Navab M, grijalva VR, Morvardi S, Clarke CF, Vergnes L, Reue K, Teiber JF, Reddy ST (2011) Paraoxonase 2 Deficiency Alters Mitochondrial Function and Exacerbates the Development of Atherosclerosis. Antioxid. Redox Signal. 14:341-351.

[69] Altenhöfer S, Witte I, Teiber JF, Wilgenbus P, Pautz A, Li H, Daiber A, Witan H, Clement AM, Förstermann U, Horke S (2010) One Enzyme, Two Functions: PON2 Prevents Mitochondrial Superoxide Formation and Apoptosis Indenpendent From its Lactonase Activity. J. Biol. Chem. 285:24398-24403.

[70] Rosenblat M, Draganov D, Watson CE, Bisgaier CL, La Du BN, Aviram M (2003) Mouse Macrophage Paraoxonase 2 Activity is Increased Whereas Cellular Paraoxonase 3 Activity is Decreased Under Oxidative Stress. Arterioscler. Thromb. Vasc. Biol. 23:468–474.

[71] Shiner M., B. Fuhrman B, Aviram M (2004) Paraoxonase 2 (PON2) Expression is Upregulated via a Reduced-nicotinamide-adenine-dinucleotide-phosphate (NADPH)-oxidase-dependent Mechanism During Monocytes Differentiation Into Macrophages. Free Radic. Biol. Med. 37: 2052–2063.

[72] Shiner M, Fuhrman B, Aviram M (2006) A Biphasic U-shape Effect of Cellular Oxidative Stress on the Macrophage Anti-oxidant Paraoxonase 2 (PON2) Enzymatic Activity. Biochem. Biophys. Res. Commun. 349:1094–1099.

[73] Rosenblat M, Hayek T, Hussein K, Aviram M (2004) Decreased Macrophage Paraoxonase 2 Expression in Patients with Hypercholesterolemia is the Result of Their Increased Cellular Cholesterol Content: Effect of Atorvastatin Therapy. Arterioscler. Thromb. Vasc. Biol. 24:175–180.

[74] Levy E, Trudel K, Bendayan M, Seidman EG, Delvin E, Lavoie JC, Precourt LP, Amre D, Sinnett D (2007) Biological Role, Protein Expression, Subcellular Localization and Oxidative Stress Response of Paraoxonase 2 in the Intestine of Humans and Rats. Am. J. Physiol. Gastrointest. Liver Physiol. 293:G1252–G1261.

[75] Li WF., Costa LG, Furlong CE (1993) Serum Paraoxonase Status: a Major Factor in Determining Resistance to Organophosphates. J. Toxicol. Environ. Health. 40:337–346.

[76] Clarimon J, Eerola J, Hellströrm O, Tienari PJ, Singleton A (2004) Paraoxonase 1 (PON1) Gene Polymorphisms and Parkinson's Disease in a Finnish Population. Neurosci. Lett. 367:168–170.

[77] Leus FR, Zwart M, Kastelein JJP, Voorbij HAM (2001) PON2 Gene Variants are Associated With Clinical Manifestations of Cardiovascular Disease in Familial Hypercholesterolemia Patients. Atherosclerosis. 154:641–649.

[78] Parra S, Alonso-Villaverde C, Coll B, Ferré N, Marsillach J, Aragonès G, Mackness M, Mackness B, Masana L, Joven J, Camps J (2007) Serum Paraoxonase-1 Activity and Concentration are Influenced by Human Immunodeficiency Virus Infection. Atherosclerosis. 194:175-181.

[79] Sardo MA, Campo S, Bonaiuto M, Bonaiuto A, Saitta C, Trimarchi G, Castaldo M, Bitto A, Cinquegrani M, Saitta A (2005) Antioxidant Effect of Atorvastatin is Independent of PON1 Gene T(–107)C, Q192R and L55M Polymorphisms in Hypercholesterolaemic Patients. Curr. Med. Res. Opin. 21:777–784.

[80] O'Leary KA, Edwards RJ, Town MM, Boobis AR (2005) Genetic and Other Sources of Variation in the Activity of Serum Paraoxonase/Diazoxonase in Humans: Consequences for Risk From Exposure to Diazinon. Pharmacogenet. Genomics 15:51-60.

[81] Aynacioglu AS, Cascorbi I, Mrozikiewich PM, Nacak M, Tapanyigit EE, Roots I (1999) Paraoxonase 1 Mutations in a Turkish Population. Toxico.L. Appl. Pharmacol. 157:174–177.

[82] Grdić M, Barišić K, Rumora L, Salamunić I, Tadijanović M, Žanić-Grubišić T, Pšikalova R, Fleger-Meštrić Z, Juretić D (2008) Genetic Frequencies of Paraoxonase 1 Gene Polymorphisms in Croatian Population. Croat. Chem. Acta 81:105-111.

[83] Flekač M, Škrha J, Zídková K, Lacinová Z, Hilgertová J (2008) Paraoxonase 1 Gene Polymorphisms and Enzyme Activities in Diabetes Mellitus. Physiol Res. 57:717-726.

[84] Pejin-Grubiša I, Buzadžic I, Jankovic-Oreščanin B, Barjaktarović-Vučinić N (2010) Distribution of Paraoxonase 1 Coding Region Polymorphisms in Serbian Population. Genetika 42:235-247.

[85] Gupta N, Singh S, Maturu N, Sharma YP, Gill KD (2011) Paraoxonase 1 Polymorphisms, Haplotypes and Activity in Predicting CAD Risk in North-West Indian Punjabis. PloS One 6(5). Available:http://www.plosone.org/article/info%3Adoi%2F10.1371%2Fjournal.pone.0017 805. Accessed 2011 May 24

[86] Suehiro T, Nakamura T, Inoue M, Shiinoki T, Ikeda Y, Kumon Y, Shindo M, Tanaka H, Hashimoto K (2000) A Polymorphism Upstream From the Human Paraoxonase (PON1) Gene and its Asociation With PON1 Expression. Atherosclerosis. 150:295-298.

[87] Hong SH, Song J, Min WK, Kim JQ (2001) Genetic Variations of the Paraoxonase Gene in Patients With Coronary Artery Disease. Clin. Biochem. 34:475-481.

[88] Mohamed Ali S, Chia SE (2008) Interethnic Variability of Plasma Paraoxonase (PON1) Activity Towards Organophosphates and PON1 Polymorphism Among Asian Populations-a Short Review. Ind. Health 46:309-317.

[89] Zhang F, Liu HW, Fan P, Bai H, Song Q (2011) The -108 C/T Polymorphism in Paraoxonase 1 Gene in Chinese Patients With Polycystic Ovary Syndrome. Sichuan Da Xue Xue Bao Yi Xue Ban. 42:24-28.

[90] Sepahvand F, Rahimi-Moghaddam P, Shafiei M, Ghaffari SM, Rostam-Shirazi M, Mahmoudian M (2007) Frequency of Paraoxonase 192/55 Polymorphism in an Iranian Population. J. Toxicol. Environ. Health 70:1125–1129.

[91] McKeown-Eyssen C, Baines C, Cole DEC, Riley N, Tyndale RF, Marshall L, Jazmaji V (2004) Case-Control Study of Genotypes in Multiple Chemical Sensitivity: CYP2D6, NAT1, NAT2, PON1, PON2 and MTHFR. Int. J. Epidem. 33:1–8.

[92] Chen J, Kumar M, Chen W, Berkowitz G, wetmur JG (2003) Increased Influence of Genetic Variation on PON1 Activity in Neonates. Environ. Health Perspect. 111:1403–1409.

[93] Santos NPC, Santos AKCR, Santos SEB (2005) Frequency of the Q192R and L55M Polymorphisms of the Human Serum Paraoxonase Gene (PON1) in Ten Amazonian Amerindian Tribes. Genet. Mol. Biol. 28:36-39.

[94] Rojas-Garcia AE, Solis-Heredia MJ, Pina-Guzman B, Vega L, LopezCarrillo L, Quintanilla-Vega B (2005) Genetic Polymorphisms and Activity of PON1 in a Mexican Population. Toxicol. Appl. Pharmacol. 205:282–289.

[95] Cataño HC, Cueva JL, Cardenas AM, Izaguirre V, Zavaleta AI, Carranca E, Hernández AF (2006) Distribution of Paraoxonase-1 Gene Polymorphisms and Enzyme Activity in a Peruvian Population. Environ. Molecul. Mutagen. 47:699-706.

[96] Scacchi R, Corbo RM, Rickards O, De Stefano GF (2003) New Data on the World Distribution of Paraoxonase (PON1Gln192→Arg) Gene Frequencies. Hum. Biol. 75:365-373.

[97] El-Fasakhany FM, El-Segeaya O, Alahwal L, Abu Al-Nooman S (2007) Paraoxonase 1 Activity and Paraoxonase 192 Gene Polymorphism in Non Insulin Dependent Diabetes Mellitus Patients Among Egyptian Population. Tanta. Med. Scien. J. 2:68-77.

Animal Models for Lipoprotein Research

Animal Models as Tools for Translational Research: Focus on Atherosclerosis, Metabolic Syndrome and Type-II Diabetes Mellitus

Isaac Karimi

Additional information is available at the end of the chapter

1. Introduction

Close to one century ago, Joslin, an American diabetologist, proposed the link between diabetes and obesity [1]. He concluded that "diabetes is largely a penalty of obesity, and the greater the obesity, the more likely is Nature to enforce it". In the 1950s, Vague [2] described that central obesity predisposes not only to diabetes but also to atherosclerosis. In the 1970s, for the first time, Haller [3] used the term "metabolic syndrome" (MetS) for associations of obesity, diabetes mellitus (DM), hyperlipoproteinemia, hyperuricemia, and hepatic steatosis when describing the additive effects of risk factors on atherosclerosis. Phillips developed the concept of metabolic risk factors for myocardial infarction and described a cluster of abnormalities including glucose intolerance, hyperinsulinemia, hyperlipidemia, and hypertension [4,5]. In 1988, Reaven, an American endocrinologist, propounded that insulin resistance (IR) was the cause of glucose intolerance, hyperinsulinaemia, increased very-low-density lipoprotein cholesterol (VLDL-C), decreased high-density lipoprotein cholesterol (HDL-C) and hypertension and named the constellation of abnormalities "syndrome X" [6]. Reaven did not include abdominal obesity, which has also been hypothesized as the underlying factor, as part of the condition. In the late 1990s and the early 21st century, MetS was widely recognized as a leading risk factor for cardiovascular morbidity and mortality and variously defined by World Health Organization [7], International Diabetes Federation (IDF [8]), the European Group for the Study of Insulin Resistance [9] and the National Cholesterol Education Program Adult Treatment Panel III [10] based on the reference intervals of its components. Accordingly to these definitions, MetS is thought to represent a combination of cardiometabolic risk determinants, including obesity, glucose intolerance and IR, dyslipidemia (including hypertriglyceridaemia, increased free fatty acids (FFAs) and decreased HDL-C) and hypertension and more recently a growing list of clinical manifestations like polycystic ovarian syndrome (PCOS), atherosclerosis, proinflammatory state, oxidative stress and non-alcoholic fatty liver disease (NAFLD) has been associated to it.

The MetS is increasingly recognized as a strong predictor of patient risk for developing coronary artery disease. It is associated with an atherogenic dyslipidemia characterized by elevated levels of triglycerides (TGs), reduced levels of HDL-C and a preponderance of small dense low-density lipoprotein (LDL) particles [11]. An atherogenic dyslipidemia is an integral component of MetS, and a major contributor to the cardiovascular risks in patients. These alarming situations increase the priority for developing new methods and technologies to investigate and to fight the MetS and its related comorbidities. Translational physiology offers us specific animal models for investigating these conditions to help support biomedical research efforts towards finding the necessary cures. This chapter summarizes various types of animal models that used as a tool in lipoprotein clinical researches and critically evaluates the physiological fidelity of these animal models to the human condition. The animal models are used to investigate biological or pathobiological phenomena or employed to find therapeutic and/or toxic effects of a xenobiotic or food ingredients. The laboratory animal models are developed and used to study the cause, nature, and cure of human lipoprotein disorders. They may conveniently be categorized in one of the following two groups:

1.   Experimental animal models of lipoprotein disorders
2.   Spontaneous animal models of lipoprotein disorders

2. Experimental animal models of lipoprotein disorders

Experimental (induced) models are healthy animals in which the condition (usually disease) to be investigated is experimentally induced, for instance, the induction of DM with encephalomyocarditis virus or alloxan. Although homologous animal models that completely show symptoms and the course of the lipoprotein metabolic disorders are very rare, the most induced models are exploratory, helping to understand mechanisms operative in fundamental normal biology or mechanisms associated with an abnormal biological function. Generally, induced models of metabolic disorders are prepared by genetic manipulation, dietary intervention, surgery, applying xenobiotics (drugs or toxins), and a combination of mentioned methods (see review [12]). This chapter will focus mainly on diet-induced and spontaneous animal models commonly used to investigate lipoprotein metabolic disorders. Readers referred to chapter 22 to study transgenic models of lipoprotein disorders.

Nowadays, obesity, particularly visceral (or central) obesity, is accepted as network backbone of the other MetS components and their manifestations. It has been reported that the incidence of MetS and type 2 diabetes (T2D) increases with the severity of obesity [13]. In this context, increasing body mass index is positively associated with prevalence of both impaired glucose tolerance and T2D and also correlated with dyslipidemia component of MetS that characterized by (a) increased flux of free fatty acids (FFA), (b) raised TGs values,

(c) low HDL-C values, (d) increased small, dense low density lipoprotein particles (e) increased TC and LDL-C and (f) raised apolipoprotein (apo) B levels

The intake of high energy diet and sedentary behavior in developed countries has an accrual effect on the incidence of obesity. Although the association between visceral fat and MetS is strong, the mechanism is not fully elucidated. The adipose tissue is not an inert tissue and constitutively produces adipocytokines that involve in pathogenesis of MetS and IR (see review [14]). Diets play a fundamental role in inducing obesity-related diseases in human, and most animal models do use diet as a way to precipitate the obesity-related diseases. Today, most diet-driven animal disease models are generated using open source, purified ingredient diets. The "open source" nature of purified ingredient diets allows researchers to compare data from different studies, since the diet formulas are generally freely available to the public, while the "close source", chow diets are differently formulated. Purified ingredients, on the other hand, are highly refined and contain just a single nutrient (ie. fructose). These ingredients have little variability and therefore provide consistency between batches, and so help to minimize data variability. There are numerous differences between chows and purified diets, creating countless variables, thus making it difficult to interpret the results when these diets are used together in a study. Chow is a nonpurified diet composed of a mixture of intact feed. In contrast, purified diets provide macronutrients as purified ingredients. For example, carbohydrate in chow diets is derived from complex mixtures of corn and wheat flakes, wheat middlings, ground corn, and dried whey. In addition to carbohydrate, these ingredients provide variable amounts of protein, fat, vitamins, minerals, and various phytochemicals and other (anti)nutrients. Some of these compounds, in particular the phytoestrogens, may act as endocrine disruptors that alter endocrine milieu and disease progression and so are usually unwanted variables. Finally, purified ingredient diet formulas can be easily modified so that researchers can intentionally and specifically change one ingredient at a time, allowing them to study the effects of large or small changes in the nutritional quantity and quality of the diet. Because of these advantages, most metabolic disease animal research uses and requires purified ingredient diets. In addition to purified and chow diets, some scientists used what is known as the cafeteria diet (CAF) to induce obesity. In this model, animals are allowed free access to standard chow and water while concurrently offered highly pleasant, energy dense, unhealthy human junk foods including cookies, candy, cheese, and processed meats *ad libitum*. These foods contain a substantial amount of salt, sugar, and fat and are meant to simulate the human "Western diet". However, the nutritive and nonnutritive components of these foods are not well defined. In addition, the animal may choose a different selection of foods each day. In this section, I discuss how high diets influence the phenotypes of the obesity and/or MetS in translated animals.

2.1. The mouse models

The advantages of mouse models that made them suitable for translating human conditions include a well-known genome, relative ease of genetic manipulation, a short breeding span, access to physiological and invasive testing, short reproductive cycle, large litter size, much lower cost and possibility of conducting longitudinal studies using larger numbers of animals, rapid development of atherosclerotic plaques, only partial resemblance to humans, very high levels of blood lipids, useful for noninvasive imaging and large experience.

Normal mice have traditionally not been ideal models of cardiovascular disease research since they typically have very low levels of TC and LDL-C but high levels of HDL-C. This is in contrast to humans in whom the reverse is true because unlike humans and several other animals, mice do not possess plasma cholesteryl ester transfer protein (CETP) and, therefore, about 70% of the plasma TC is found in HDL particles. Mouse models have proved to be useful to study development and progression of atherosclerotic lesion, and several reviews have extensively discussed the different available models (see review [15]). The ability of mice to maintain their cholesterol profile even in the face of high-cholesterol diets means that very little actual atherosclerosis develops [16]. As wild-type mice are resistant to lesion development, the current mouse models for atherosclerosis are based on genetic modifications of lipoprotein metabolism with additional dietary changes. In order to 'force' the atherosclerosis phenotype on normal mice, it is usually necessary to combine high concentrations of dietary cholesterol with 0.25%-0.5% cholic acid which promotes fat and cholesterol absorption from the intestine [17]. However, cholic acid can also promote liver inflammation, decrease bile acid production, and alter circulating TG and HDL-C, it may independently affect the development of atherosclerosis [15]. Atherosclerosis is a complex multifactorial disease with different etiologies that synergistically promote lesion development. High-fat diets (HFDs) are used to model obesity, dyslipidaemia, atherosclerosis, IR and MetS in rodents (see reviews [18,19]. High-fat diet (HFD) feeding in mice increased systolic blood pressure and induced endothelial dysfunction [20] and some kind of nephropathy [21]. Different types of HFDs have been used with fat fractions ranging between 20% and 60% energy as fat as either animal-derived fats, such as lard or beef tallow, or plant oils such as olive or coconut oil [22]. Long-term feeding of rats (60% of energy) and mice (35% fat wt/wt) with HFD increased body weight compared to standard chow-fed controls [23]. Although the increase in body weight was significant after as little as 2 weeks, the diet-induced phenotype became apparent after more than 4 weeks of HFD feeding [23]. Long-term feeding with both animal and plant fat-enriched diets eventually led to moderate hyperglycaemia and impaired glucose tolerance in most rat and mouse strains [24]. Lard, coconut oil and olive oil (42% of energy content) increased body weight, deposition of liver TGs, plasma TGs and FFAs concentrations and plasma insulin concentrations [22]. Lard, coconut oil and olive oil caused hepatic steatosis with no signs of inflammation and fibrosis [22]. Although HFD induces most of the symptoms of human MetS in rodents, it does not resemble the diet causing MetS and associated complications, as the human diet is more complex than a HFD. Other major components of modern diets are refined carbohydrates and fructose. The epidemiologic data has proved that a significant correlation in the prevalence of diabetes with fat, carbohydrate, corn syrup (source of fructose), and total energy intakes. The striking features of these studies are the fact that intake of corn syrup was positively associated with T2D, while protein and fat were not (see review [25]). Most studies have utilized mice as animal models to define the role of carbohydrate enriched diets in formation of different aspects of MetS. The interested readers will have to go to the current literature in order to understand more fully the fidelity of mice for translation of similar conditions in humans. In this context, high-fat high-carbohydrate (HFHC) diet contains 55% fructose and 45% sucrose (wt/vol) in drinking water has been fed to nongenetically modified adult male C57Bl/6 mice for 16 weeks led to obesity and nonalcoholic steatohepatitis (NASH) [26]. HFHC has been used to induce hyperglycemia, glucose intolerance, IR, increased fat pad weight and adipocyte hypertrophy and commonly HFHC-fed mouse models used to screen therapeutic effects of various drugs and diets against MetS and its comorbidities (e.g., [27]). Charlton and colleagues recently proposed an obese mouse model of NASH that induced by feeding fast food (high SFs, cholesterol, and fructose) diet [28]. C57BL/6J mouse is T2D model by simply feeding HFD to nonobese, nondiabetic C57BL/6J mouse strain. It is characterized by marked obesity, hyperinsulinaemia, IR and glucose intolerance [29]. Diets contribute to T2D in mouse identical to human, and a HFD and sucrose administration appears to speed up the development of the disease in mice. In this context, impaired insulin secretion and/or impaired insulin action also contribute to the diabetic phenotype for these mice [30]. However, the mouse models are observed to develop diabetes in relation to profound obesity and do not display the same islet pathology as humans with T2D [31]. A large number of investigations also have been carried out within recent years, concerning the therapeutic action of various (bio)pharmaceutical and nutraceutical compounds on mouse models of lipoprotein disorders (e.g., [32]).

2.2. The rat models

Rats, like humans, showed different vulnerability to diet-induced obesity. At first, an animal model of diet-induced obesity is one introduced by Levin and coworkers [33] and developed into a purified diet model [34]. In this model, Sprague-Dawley (SD) rats fed a purified moderately high-fat (MHF) diet exhibit a bimodal pattern in body weight gain similar to that observed in humans. Approximately half of the rats gain weight rapidly compared with chow-fed rats (obesity prone [OP] or diet-induced obese [DIO]), whereas the other half gain BW at a rate similar to or lower than that of the chow-fed animals (obesity resistant [OR] or diet-resistant [DR]) [34-36]. Most rodents tend to become obese on HFD and very high-fat diet (VHFD), but there can be variable responses in glucose tolerance, IR, TGs, and other parameters depending on the strain and gender, and source of dietary fat [22,37]. When outbred SD and Wistar rats were placed on HFD (32 or 45 kcal% fat), there was a wide distribution in body weight gain and a subset of animals became obese, whereas others remained as lean as the animals fed with a low-fat diet (LFD) have shown that the rat model of diet-induced obesity develops mild hypertension accompanied by vascular and renal changes similar to those observed in obese hypertensive humans [35,38]. The MHF diet that they used contains 32% kcal fat, a value similar to the average Western diet, as opposed to many other models that have very high levels of fat [38]. All rats fed the MHF diet did not become obese and their body weight displayed a bimodal distribution. The increased body weight reflects an increase in the adipose mass in the OP rats versus the chow-fed rats [38].

Elevation of plasma TGs and FFAs was commonly observed in patients with diabetic dyslipidemia or obesity [39]. Evidence showed that hyperglycemia and hypertriglyceridemia had direct effects on arterial wall and induced endothelial dysfunction [40]. The elevation of TGs and fasting plasma glucose was noted in HFD studies [41]. However, the levels of TGs and TC in high-fat fed DR rats were no more than chow-fed control rats. The HDL-cholesterol level decreases in hypercholesterolemic and MHF-fed rats [38]. The TGs content of plasma, LDL, and VLDL has been increased in OP rats fed MHF diet after 3 weeks [38]. As opposed to cholesterol content, this difference is even greater after 10 weeks of the MHF diet therefore authors concluded that factors other than diet like reduced growth hormone secretion are also responsible for the high levels of TGs in OP rats [38,42]. The underlying mechanism is not known. Insight into the differences in endocrine and lipoprotein metabolism may provide further evidences. For example, Yang et al's study showed that DR rats had higher levels of plasma peptide YY, a gut-derived anorexigen, than DIO and the control groups. This indicates that a difference in appetite control is responsible for the lower caloric intake and weight gain in DR rats [43]. One of the common features of obesity in humans is dyslipidemia which occurs in rat model of diet-induced obesity and is frequently associated with hypertension [38]. I have decided to ignore molecular mechanisms of hypertension in OP rat because of limited space. However, hypertension developed in OP, but not OR, rats, is a multifactorial disorder and diet is not the major factor that causes the high blood pressure in this model.

According to Barker hypothesis, adult metabolic diseases are programmed during fetal life [44]. To investigate the mechanisms by which altered intrauterine *milieu* predisposes to later development of MetS, different animal models have been developed (see review [45]). Interestingly, offspring of rats fed high saturated fats during pregnancy have fetal IR [46], abnormal cholesterol metabolism [47] and raised adult blood pressure [48]. Furthermore, the outbred Sprague-Dawley DIO and DR rats have been selectively bred over time such that their future body weight response to a HFD is known *in utero*, allowing the researcher to look early in life (prior to the onset of obesity) for genetic traits that may later predispose them to their DIO or DR phenotypes [37,49].

The inbred obese Zucker diabetic fatty (ZDF) rat is high-fidelity model with close resemblance to human case in obesity and T2D. The males become obese and diabetic on a LFD, but HFD feeding promotes more robust disease. The female ZDF rat is unique in that while they are obese, they do not develop diabetes unless fed a diet (in this case, chow-based) containing 48 kcal% fat [50]. The female ZDF rat is also suitable model mimics pre-diabetic state in humans because she shows a prolonged period of insulin sensitivity prior to the onset of diet-induced diabetes [51]. The ZDF rats show profound dyslipoproteinaemia with increased TC and TGs levels and lower chylomicra disposal rates that mimics conditions occurred in human case of obesity [52]. Although normal rats are not ideal model of cardiovascular disease research since they typically have very low levels of TC and LDL-C but high levels of HDL-C, they are mild diet-responsive. The ability of rats to sustain their cholesterol profile even in the face of high-cholesterol diets means that very little actual atherosclerosis develops [16]. However, feeding Wistar rats a high calorie "Western diet"

(45% fat) for up to 48 weeks induces obesity and cardiac dysfunction, while a high fat diet (60% fat) induces obesity only [53]. The "Western diet" composed of a purified ingredient SF-rich HFD, and cholesterol (~0.2% by weight) can elevate TC and LDL-C and in turn cause atherosclerosis in certain rodent models and humans [54]. A mixture of high levels of dietary cholesterol with 0.25%-0.5% cholic acid has been used to induce atherosclerosis phenotype on normal rats and mice for many years ago [55]. More recently, Zaragosa and colleagues introduced various animal models of cardiovascular diseases (see review [56]). Surprisingly rat does not develop atheroma in the process of atherosclerosis (see review [56]). Generally rats are highly resistant to the development of atherosclerosis because they lack physiological resemblance on many aspects with humans that are pathophysiologically important [57]. For example, HDL is dominating lipoprotein in these animals and rat platelets are generally resistant in hyperlipidemic condition (see review [58]). Rats are potentially practical model for studying hypercholesterolemia along with hypertension (see review [58]). They exhibit augmented thrombotic response and develop coronary atherosclerotic lesions under hypertensive and hyperlipidemic conditions (e.g., [59]). Triglyceride-rich diets containing various amounts of cholesterol, with or without cholic acid have been used to induce hypercholesterolemia in rats. The fat sources vary from lard to soybean, canola or sunflower oils. Nevertheless, the question of the caloric value of the employed diets has not yet been considered properly since their high fat content, which is the strategy used in order to induce hypercholesterolemia, leads to lower ingestion by the animals and induces malnutrition. To overcome this shortcoming, Matos and colleagues [60] proposed a diet containing 25% soybean oil, 1.0% cholesterol, 13% fiber (cellulose) and 4,538.4 Kcal/Kg that led to an increase in LDL-C, a decrease in the HDL-C fraction and affected less the hepatic function of the rats during eight weeks. Roberts and colleagues presented a rat model of diet-induced syndrome X and they explored potential mechanisms of hypercholesterolemia in diet-induced syndrome X [61]. To induce syndrome X, female Fischer rats were fed a high-fat (primarily from lard plus a small amount of corn oil), refined-carbohydrate (sucrose) diet for 20 months [61]. Sampey and colleagues [62] have demonstrated that the CAF is a more robust model of MetS than lard-based HFD and that the rapid-onset of weight gain, obesity, multiorgan dysfunctions and pathologies observed in the CAF model more closely reflect the modern human condition of early onset obesity. However, they did not repot possible lipid-lipoprotein disorders that may be occurred in their model. Recently, Manting and colleagues [63] have shown that a combination of chronic stress and HFD (83.25% basal feed, 10% lard, 1.5% cholesterol, 0.2% sodium taurodeoxycholate, 5% sugar and 0.05% propylthiouracil) can induce lipid metabolism disorder in Wistar rats and they claimed that their multiple factor model better mimics the disease characteristics of human beings.

2.3. The hamster models

Hamsters are another animal model can be used to assess some aspects of MetS. Like rats and mice, HDL-C is predominant plasma cholesterol-rich lipoprotein in these animal, but in contrast, dietary cholesterol (~ 0.1%) can significantly elevate LDL-C and like humans, SF can increase these levels further [64]. The combination of high dietary SF and cholesterol is commonly used to promote atherosclerosis in these animals and atherosclerotic lesions similar to those found in humans can be found after prolonged feeding periods [65]. Actually, cholesterol itself may not always be necessary for this phenotype, since a purified diet with no cholesterol but high concentrations of SF can promote more aortic cholesterol accumulation compared to a diet with both cocoa butter and 0.15% cholesterol [66]. Cholesterol-fed hamsters have been used to screen therapeutic anti-atherosclerotic and hypolipidemic properties of (phyto)medicines (e.g., [67,68]). Hamsters have been proposed as an animal model to evaluate diet-induced atherosclerosis since the 1980s [69]. Relative to other normal rodent models, hamsters have a low rate of endogenous cholesterol synthesis, cholesteryl ester transfer protein (CETP) activity and tissue specific editing of apolipoprotein (apo) B mRNA and secretion of apo B-100 from the liver and apo B-48 from the small intestine. Hamsters, like humans, take up approximately 80% of LDL-C *via* the LDL receptor pathway. The morphology of aortic foam cells and lesions in hamsters fed atherogenic diets was reported to be similar to human lesions [70]. Recently, in a systematic review Dillard and colleagues concluded that the Golden-Syrian hamster does not appear to be a constructive model to determine the mechanism(s) of diet-induced development of atherosclerotic lesions (see review [71]) however the authors only concentrated on atherogenecity of cholesterol- and fat-rich diets in hamster models of atherosclerosis.

Leung and colleagues investigated intestinal lipoprotein production and the response to insulin sensitization in the high fat-fed Syrian Golden hamster for 5 weeks [72]. They concluded that Syrian Golden Hamsters were fed 60% fat is a good model of nutritionally-induced IR that intestinal overproduction of lipoproteins appear to contribute to the hypertriglyceridemia of IR in this animal model and insulin sensitization with rosiglitazone (an insulin sensitizer) ameliorates intestinal apoB48 particle overproduction in this model. An appropriate dyslipidemic animal model that has diabetes would provide an important tool for research on the treatment of diabetic dyslipidemia. Ten days of high fat feeding in golden Syrian hamsters resulted in a significant increase in IR and baseline serum lipid levels accompanied by a prominent dyslipidemia. Thirteen days of treatment with fenofibrate, a peroxisome proliferator-activated receptor alpha (PPAR alpha) selective agonist, produced a dose-dependent improvement in serum lipid levels characterized by lowered VLDL-C and LDL-C and raised HDL-C in a fashion similar to that seen in man [73]. Various diet formula, fat resources and time tables have been found to induce some aspects of MetS in the literature. For example, a diet consisted of 80 g of anhydrous butterfat, 100 g of corn oil, 20 g of Menhaden fish oil and 1.5 g of cholesterol has been used to encourage hypercholesterolemia in male golden Syrian hamsters for 21 days [68]. Male golden hamsters were given 15% HFD contained 100 g of lard and 50 g of soybean oil and 100 g of sucrose showed diabetic dyslipidemia for eight weeks [74]. F1B hamster is a genetically-defined hamster, derived from two highly inbred lines, namely by crossbreeding between Bio 87.20 female with a Bio 1.5 male. F1B hamster is an exciting animal model for hyperlipidemic-related applications. The F1B strain is very responsive to SF and cholesterol by increasing the non-HDL fraction to a greater extent than the HDL fraction [75]. Dietary fatty acid chain length, degree of saturation and *cis-trans* conformation have been shown to alter several metabolic pathways involving cholesterol throughout the body, the combined effect of which is reflected in plasma lipid and lipoprotein profiles (see review [76]). Interestingly, intake of *trans*-fatty acids in shortenings and margarines has been linked to increased risk of cardiovascular disease through effects on lipoprotein metabolism and substituting *trans*-fatty acids for either saturated or polyunsaturated fatty acids results in more deleterious lipid-lipoprotein profiles [77]. Hamsters are candidate model to investigate cardiometabolic risks of different fat resource and fat-rich diets [78]. Similarity with the human LDL receptor gene, makes hamster ideal to study LDL receptor antagonists and also useful for drugs which interfere with CETP activities and reverse cholesterol transport (RCT) from peripheral tissues to the liver for biliary and fecal excretion [79]. A considerable amount of experimental attention is currently directed at understanding the *in vivo* mechanisms of RCT. Although not established *in vivo*, RCT is thought to be impaired in patients with MetS, in which liver steatosis prevalence is relatively high. In this sense, Briand and coworkers [80] introduced a hamster model of MetS to study RCT. These scientists with the help of HFT diet containing 27% fat, 0.5% cholesterol, and 0.25% deoxycholate as well as 10% fructose in drinking water for 4 weeks induced promoted IR, dyslipidemia with significantly higher plasma non-HDL-C concentrations and CETP activity, and hepatic steatosis. *In vivo* RCT was assessed by intraperitoneally injecting (3)H-cholesterol labeled macrophages. Finally their results indicate a significant increase in macrophage-derived cholesterol fecal excretion, which may not compensate for the diet-induced dyslipidemia and liver steatosis [80]. One of the main target organs of MetS is liver, in which it manifests itself as NAFLD [81]. Bhathena and colleagues currently developed a triumphant BioF1B Golden Syrian hamster model of MetS that successfully manifested hyperlipidemia, IR and NAFLD [81]. They induced this model by feeding hamsters a high-fat, high-cholesterol, inadequate methionine- and choline-containing diet. In addition to F1B hamster strain from Biobreeders (Watertown, MA) that commonly used to study diet-induced metabolic disorders other three outbred strains are Charles River (CR), Sasco and Harlan (see review [71]). All these strains are derived from inbred or outbred Golden-Syrian hamster.

Similar to rats, hamsters fed high fructose diets (~60% of energy) may develop IR and hypertriglyceridemia in TG after only two weeks compared to diets low in fructose [72,73]. Interestingly, hamsters fed high-sucrose diets did not have elevated TG levels and developed only mild IR relative to those fed diets high in fructose [72]. Avramoglu and colleagues reviewed mechanisms of metabolic dyslipidemia in insulin resistant states (see review [82]). They developed an explanatory fructose-fed hamster model of insulin resistance to study hepatic lipid metabolism as its lipoprotein metabolism as described previously [83,84]. Hamsters exhibit obesity, hypertriglyceridemia, increase plasma FFAs concentration and IR if fed fructose-rich diet for a two week period. Fructose feeding induced a noteworthy increase in synthesis and secretion of total TGs as well as VLDL-TG by primary hamster hepatocytes [73]. The microsomal triglyceride transfer protein plays a pivotal role in VLDL assembly and its activity showed a striking 2.1-fold elevation in hepatocytes derived from fructose-fed versus control hamsters [73]. The apoB production also has been increased in the fructose-fed hamsters [73]. Fructose-fed hamster also has been introduced as an exploratory animal model to excavate role of intestinal lipoprotein overproduction in the dyslipidemia of insulin-resistant states [74]. The authors have shown that fructose feeding for 3 weeks increases secretion of apoB48-containing lipoproteins in the fasting state and during steady state fat feeding. Wang and coworkers [75] investigated the composition of plasma lipoproteins in hamsters fed high-carbohydrate diets of varying complexity (60% carbohydrate as chow, cornstarch, or fructose) for 2 weeks. They showed that hamsters fed the high-fructose diet showed significantly increased VLDL–triglyceride (92.3%), free cholesterol (68.6%), and phospholipid (95%), whereas apolipoprotein B levels remained unchanged. Fructose feeding induced a 42.5% increase LDL–triglyceride concurrent with a 20% reduction in LDL–cholesteryl ester. Compositional changes were associated with reduced LDL diameter. In contrast, fructose feeding caused elevations in all HDL fractions.

2.4. The guinea pig models

A number of seminal reviews on the details of the criteria that make guinea pigs suitable animal models for studying lipoprotein metabolism are available (see reviews [85-87]) and a summary will be presented here. Guinea pigs in contrast to other rodents have higher levels of plasma LDL-C compared to HDL-C. As humans, guinea pigs have higher concentrations of free compared to esterified cholesterol found in the liver and they show evidence of moderate rates of hepatic cholesterol synthesis and catabolism. Similar to humans, the binding domain for the LDL receptor of guinea pigs discriminates normal and familial binding defective apo B-100 and apo B mRNA editing in liver is scarce (< 1%) compared to 18 to 70% in other species [88]. The three important proteins involved in lipoprotein remodeling and RCT (CETP, lecithin:cholesterol acyltransferase (LCAT), and lipoprotein lipase (LPL) have been reported in guinea pigs.

Guinea pigs have been used as models to dissect the mechanisms by which various dietary fat resources influences plasma lipid-lipoprotein profiles. In contrast to hamsters they do not possess a fore-stomach fermentation which modifies dietary macronutrients before reaching the small intestine [89]. Guinea pigs are not only superior models for studying the mechanisms by which statins [90], cholestyramine [91], apical sodium bile acid transport inhibitors [92] and microsomal transfer protein inhibitors [93] lower plasma LDL-C but also are selected to investigate the mechanisms by which certain drugs or toxins affect lipid-lipoprotein metabolism (e.g.,[94]). Guinea pigs respond to dietary fat saturation, dietary cholesterol and dietary fiber by alterations in LDL-C (see review [87]). For example, the SF-rich diet will increase TC and LDL-C levels much more than polyunsaturated fat (PF)-rich diet in guinea pigs and cholesterol-rich diets can further increase TC and LDL-C levels [95].

The suitability of guinea pigs as models of atherosclerosis is augmented by an array of review and assessment features (see review [59]). However, guinea pigs do not develop advanced atherosclerotic lesions, and are not an entrenched model for atherosclerosis progression [96].

High plasma level of lipoprotein (a) (also called Lp(a)) is associated with coronary heart disease and other forms of atherosclerosis in humans (see review [97]), and as primates, hedgehogs [98] and guinea pigs possess Lp(a) among normal animal models [99]. Guinea pigs are practical model to study the role of oxidized LDL (oxLDL) in progression of atherosclerosis [100]. Initial atherosclerosis induced by various formula of HFD in guinea-pigs. Intake of HFD (guinea-pig pellet diet + 0.2% w/w cholesterol) can induce the onset of early atherosclerotic changes in coronary artery, aorta and major organs at least for one month [101]. High SF diet supplemented with high cholesterol (0.25%) will advance an atherosclerotic process for twelve weeks in guinea pigs [102]. Yang and colleagues [103] introduced a hyperlipidemic guinea pig model in a comparative investigation. They concluded that different response of TG metabolism to a HFD (0.1% cholesterol and 10% lard) in guinea pigs and rats suggests that Hartley guinea pigs could be a better hypertriglyceridemia animal model than rats for research on lipid metabolism disorders and hypolipidemic drugs. It seems that chronic dyslipidemia associated with hypertriglyceridemia may reduce auditory function. In this context guinea pigs fed a HFD used as an animal model to find the relationship between of sensorineural hearing loss and dyslipidemia [104].

As rats, guinea pigs are accepted models for studying fetal programming of cardiovascular diseases. Interestingly, adipogenesis in the guinea pig is very active during early postnatal life and was altered by a maternal HFD; thus, it is an adequate model for intrauterine fat deposition [105]. More studies are requested to explore lipid-lipoprotein profiles of guinea pigs that received a maternal HFD. Caillier and colleagues [106] currently generated a guinea pig model of MetS by 150-day exposure to diabetogenic high fat high sucrose or the high fat high fructose diets. To my knowledge, it would be early to consider guinea pig as an animal model of MetS since the literatures are scarce.

2.5. The rabbit models

A century ago, rabbits were used as translated animal models of atherosclerosis [107]. Since then, a number of animal models have been used to explain the relationship between disorders of lipid metabolism and atherogenesis (see reviews [108,109]). In this sense, dietary lipid manipulation and use of naturally defective animals, such as Watanabe heritable hyperlipidemia (WHHL) rabbits, have been the focus of most experimental settings (see chapter 22). Rabbits are appropriate animal models for studying lipoprotein metabolism and its disorders because they share with humans several aspects of lipoprotein metabolism, such as similarities in composition of apolipoprotein B containing lipoproteins, hepatic production of apo B 100-containing VLDL, plasma CETP activity, human-like apo B, low hepatic lipase activity and high absorption rate of dietary cholesterol. Unlike humans, rabbits are hepatic lipase–deficient and do not have an analogue of human apo A-II. Rabbits do not form spontaneous atherosclerotic lesions and therefore require very high cholesterol levels to induce more advanced disease (see review [110]). Rabbits also have significant differences in their lipid metabolism from humans, which can result in their development of "cholesterol storage syndrome" while on high-cholesterol diets (0.5–3%), with cholesterol deposited in their liver, adrenal cortex, and reticuloendothelial and genitourinary systems

[108]. We found that a high-cholesterol diet contained in 0.47% cholesterol would be tolerable for adult male rabbits for 4 weeks but our high-cholesterol diet was mildly atherogenic [94,111]. The atherosclerotic lesions of rabbits do not completely resemble those in humans [108] and the formed lesions are more fatty and macrophage rich than human [112]. Atherogenic diets are usually associated with hypercholesterolemia and the development of atherosclerotic lesions in the aortic arch and thoracic aorta rather than in the abdominal aorta that is almost always affected in humans. New Zealand White rabbits are the strain commonly used in atherosclerosis research. Although they have low plasma TC concentrations and HDL as dominant lipoprotein [113], βVLDL becomes the major class of plasma lipoproteins when exposed to cholesterol-rich diet (see review [59]). In conjunction with chylomicron remnants, βVLDL becomes highly atherogenic. Long-term experiments using diets high in cholesterol are discouraging in rabbits, because they cannot increase the excretion of sterols and resulting hepatotoxicity does not allow the animal to survive (see review [114]).

Various HFD and intervention period have been used to induce MetS in rabbits or its components like IR, visceral obesity, hypertension, dyslipidemia (e.g., [115,116]). Rabbits are suitable animals to investigate MetS-associated multiorgan dysfunctions. Helfenstein and colleagues recently proposed an experimental model of impaired glucose tolerance combined with hypercholesterolaemia induced by diets (high-fat/high-sucrose (10/40%) and cholesterol-enriched diet for 24 weeks) that gained weight, increased blood glucose, TC, LDL-C, TGs, and decreased HDL-C in New Zealand male rabbits [117]. Their cheap model reproduced several metabolic characteristics of human DM and promoted early signs of retinopathy. Corona and colleagues [118] reviewed relationships between hypogonadism and MetS emphasizing their possible interaction in the pathogenesis of cardiovascular diseases. However they concluded that the clinical significance of the MetS-associated hypogonadism needs further clarifications. Vignozzi and colleagues [119] described an animal model of MetS obtained by feeding male rabbits for 12 weeks. In their experiment, HFD-animals develop hypogonadism and all the MetS features like hyperglycemia, glucose intolerance, dyslipidemia, hypertension, and visceral obesity. A recently established rabbit model of HFD-induced MetS showed hypogonadism and the presence of prostate gland alterations, including inflammation, hypoxia and fibrosis [120]. Rabbits fed a cholesterol-rich diet (1% cholesterol) for 8 weeks and 12 weeks share several physiopathological aspects of NAFLD [121]. Because this model is not insulin resistant and obese, it may be useful for elucidating the mechanism of NAFLD related mainly to hyperlipidemia.

3. Spontaneous animal models of lipoprotein disorders

The pathophysiology of disorders of lipoprotein metabolism of humans cannot highly translated to wild-type rodents since they are very resistant to atherogenesis and have no similarity to human lipid and lipoprotein metabolism; further, they do not develop cardiovascular diseases identical to humans. Therefore search for more reliable model is still continuing. In this regard, pig, is a considered a very good model of human atherosclerosis, because it is similar to humans in terms of body size and other physiological features (see

review [122]). Pigs spontaneously develop atherosclerosis even on a normal porcine diet and
dietary modification lead to sever atherosclerosis [123]. As humans, pigs transport most
cholesterol in LDL-C and dietary modification alters their plasma lipoproteins closely
resemble those occurring in humans. In contrast to rodents, swine atherosclerosis, like the
human illness, progresses to advanced stages (see review [122]). For example, mini-pigs fed
fat-enriched food showed fatty streaks in their abdominal and thoracic aorta and coronary
arteries during 18 months [124]. Cholesterol contents of diets also affects the extension and
exacerbation of atherosclerotic changes in pigs (see review [122]). Johansen and colleagues
[125] suggested an obese Göttingen minipig model of MetS that was highly responsive to a
high fat high energy diet. Several swine models of T2D and IR have been proposed (see
review [18]). However, spontaneity in development of MetS and IR, is not common in this
species [58]. Dogs do not naturally show atherosclerosis and cholesterol- and SF-rich diets
combined with thyroid suppression is required for atherosclerosis development [126].
Beagles and miniature Schnauzer dogs show useful similarities with human in cholesterol
synthesis, and lipoproteins level [127,128]. Feline DM, in both spontaneous and inducible
forms, therefore provides a reliable animal model of human T2D and may provide
additional insights into the clinical, physiological, and pathological features of this disease
(see review [129]). Considerably more studies must be forthcoming to establish firmly how
lipoprotein profile participates in pathogenesis of atherosclerosis, DM and possibly MetS in
pet animal since companion animal obesity would be a serious veterinary medical concern
in near future [130]. The subject of suitability of other domesticated animal species such as
pigs and sheep, as well as feral, migrating and hibernating species for studying lipid and
lipoprotein metabolism has been concisely reviewed (see review [131]). For thorough
coverage of this aspect of animal models, this work is recommended.

A number of other interesting wild rodents that are explanatory or exploratory animal
models of different disorders of lipid and/or carbohydrate metabolism have been
introduced. Recently, *Octodon degus* (degu) has been proposed as an animal model of diet-
induced development of atherosclerosis [132]. To induce atherosclerosis, degus were fed for
16 weeks chow containing 0.25% cholesterol and 6% palm oil. Cholesterol-fed degus
exhibited 4- to 5-fold increases in TC, principally in the VLDL-C and LDL-C fractions and
developed cholesteryl ester-rich atherosclerotic lesions throughout the aorta [132].
Hedgehogs are homologous animal models for studying roles of Lp(a) in atherosclerosis
[98]. Sand rat, Tuco-Tuco and spiny mouse are unusual models of diet-induced obesity and
T2D (see review [12]. In laboratory condition, sand rat (*Psammomys obesus*) develops obesity
and diabetes when fed on standard chow (high energy diet) instead of its usual energy-
diluted vegetable diet composed mainly of saltbush *Atriplex* [133]. Surprisingly, sand rats
are studied extensively and serve as more statable polygenic model for the study of
diabesity syndrome [133,134]. Spiny mouse (*Acomys calirinus*) is another small rodent living
in semiarid areas of eastern Mediterranean. They gain weight and exhibit marked impaired
pancreatic beta cell when they are placed in captivity on high energy rodent lab chow [135].
Ctenomis talarum (Tuco-tuco) is another feral species which exhibit similar characteristic
features of sand rat and spiny mice when fed on high energy rodent diet [136]. Brandt's vole

(*Lasiopodomys brandtii*) is another rodent model used in diet-induced obesity [137]. The Nile grass rat (NGR), *Arvicanthis niloticus*, is a herbivorous rodent inhabiting dry savanna, woodlands, and grasslands in Africa. Noda and colleagues [138] recently showed that the NGR is a precious, spontaneous model for exploring the etiology and pathophysiology of MetS as well as its various complications.

Avian models of human atherosclerosis include pigeon, chicken, Japanese quail, turkey and parrots. Although these avian models are not frequently used in studying atherosclerosis, it is worthy to note, that spontaneous atherosclerosis in the chicken was first described in 1914 [139]. Use of pigeon models of atherosclerosis has been extensively reviewed (see review [140]). Briefly, the most important key points supporting the use of pigeons as models for human atherosclerosis include: 1. Pigeons are hypercholesterolemic compared to humans. 2. Pigeons are primarily HDL-C carriers but βVLDL-C and LDL-C become major lipid carriers when these animals are fed cholesterol-rich diet. 3. Their lipid metabolism and lesion progression are similar to humans. 4. Pigeons also resemble humans in cellular and vascular dysfunctions involving in atherogenesis. 5. Pigeons are negative animal models to study relevance of apoE, apo B48, chylomicra or LDL receptor in atherosclerosis pathology. 6. Pigeons are susceptible to both spontaneous and diet-induced atherosclerosis. Parrots are exceptional animal models to assess the impacts of various risk factors include elevated cholesterol level, diet composition, social stress and inactivity (similar to sedentary behavior in humans) on occurrence and progression of atherosclerosis (see review [59]).

Phylogenetically, nonhuman primates are more similar to humans than other models in terms of lipid-lipoprotein profiles, pathophysiology of atherosclerosis, feeding habits, and genotype. It is demonstrated that, along with aging, some rhesus monkeys spontaneously develop diabesity (e.g. [141]). In this context, *Macaca nigra* is very valuable in studies focused on the interactions between atherosclerosis and diabetes [142]. Spontaneous diabetes has been documented in nonhuman primates include cynomolgus, rhesus, bonnet, Formosan rock, pig-tailed, celebes macaques, African green monkeys, and baboons (see reviews [18,143]). Diabetic nonhuman primates have detrimental changes in plasma lipid and lipoprotein concentrations and lipoprotein composition which may contribute to progression of atherosclerosis. As both the prediabetic condition (similar to MetS in humans) and overt diabetes become better translated in monkeys, their use in pharmacological studies is increasing. Monkeys can be categorized into hyperresponders and hyporesponders based on initiation, progression and severity of atherosclerotic lesions. Several nonhuman primates, such as squirrel monkeys, baboons, and wooly and spider monkeys, may develop spontaneous early stage (fatty streaks) atherosclerosis at different anatomical locations (see reviews [59,108]). Rare cases of LDL receptor deficiency in a rhesus monkey family associated with increased levels of LDL-C, Lp(a), and advanced atherosclerotic lesions in the aorta, and to a lesser extent in coronary arteries, were reported [144]. Nonhuman primates are more reliable model to study cardiovascular disease plus MetS rather than rodents, since they develop MetS and cardiovascular diseases as they age. They develop spontaneous (in some species) and high fat high cholesterol diet-induced atherosclerotic lesions [145]. Nonhuman primates are good model of hypertension and its harmful effect on atherosclerosis development. The close similarity of plasma lipoprotein-lipid level, plaque development and its calcification and mineralization with humans makes nonhuman primates practical model to explore the correlation between plasma lipids and plaque development (see review [59]). Kaufman and colleagues measured some anthropometric indices and metabolic parameters in 250 laboratory-born bonnet macaques living in social groups and maintained on commercial monkey chow [146]. Finally they concluded socially reared and housed bonnet macaques may provide a useful model for studying the pathogenesis, prevention, and treatment of the MetS. Recently, in an excellent investigation, Zhang and colleagues established a rhesus monkey model of spontaneous MetS using population screening approaches suitable to explore the pathogenesis of MetS in relation to cardiovascular disease and DM [147]. To sum, the inconsistency in anatomic location of atherosclerotic lesions, high cost of husbandry and veterinary services, limited animal availability, difficult handling, together with ethical queries are major obstacles in the use of monkeys as common animal models in studying MetS and its comorbidities.

4. Conclusion

The incidence of metabolic syndrome is increasing on a pandemic level. One of the major underlying cause and/or outcome of metabolic syndrome is dyslipidemia, which contribute greatly to the cardiovascular problems associated with the syndrome. The animal models have a vital role to play in extending our understanding of metabolic syndrome and its related comorbidities. Conventional laboratory animals such as mice, rats, hamsters, guinea pigs and rabbits have been examined to gain a better perceptive of the relationship between disorders of lipid metabolism and their clinical correlations. High-fat diets frequently used to induce different aspects of metabolic syndrome in rodent models. However, nonconventional animal models like pig, pigeon, and feral animals (e.g., spiny mice, sand rat, hedgehogs) can consider as spontaneous animal models suitable for studying both the pathogenesis and potential therapeutic agents in lipoprotein disorders. The attempts to find animal models relevant to the study of metabolic syndrome are continuing.

Author details

Isaac Karimi

Division of Biochemistry, Physiology and Pharmacology, Department of Basic Veterinary Sciences, School of Veterinary Medicine, Razi University, Kermanshah, Iran

5. References

[1] Joslin EP (1922) The prevention of diabetes mellitus. JAMA. 76(2):79-84.

[2] Vague J (1956) The degree of masculine differentiation of obesities: a factor determining predisposition to diabetes, atherosclerosis, gout, and uric calculous disease. Am J Clin Nutr 4: 20–34.

[3] Haller H (1977) Epidermiology and associated risk factors of hyperlipoproteinemia. Z Gesamte Inn Med. 15;32(8):124-8. [Article in German]

[4] Phillips GB (1978) Sex hormones, risk factors and cardiovascular disease. Am J Med. 65:7-11.

[5] Phillips GB (1977) Relationship between serum sex hormones and glucose, insulin, and lipid abnormalities in men with myocardial infarction. Proc Natl Acad Sci USA.74: 1729-1733.

[6] Reaven GM (1988) Banting lecture 1988. Role of insulin resistance in human disease. Diabetes 37: 1595–1607.

[7] World Health Organization (1999) WHO consultation: definition, diagnosis and clasification of diabetes mellitus and its complications. Geneva: WHO.

[8] Alberti KG, Zimmet P, Shaw J (2005) The metabolic syndrome: a new worldwide definition. Lancet. 24; 366:1059–62.

[9] Balkau B, Charles MA (1999) Comment on the provisional report from the WHO consultation. European Group for the Study of Insulin Resistance (EGIR). Diabet Med.16:442–3.

[10] Executive Summary of The Third Report of The National Cholesterol Education Program (NCEP) (2001) Expert Panel on Detection, Evaluation, And Treatment of High Blood Cholesterol In Adults (Adult Treatment Panel III). JAMA.285:2486–97.

[11] Cziraky MJ (2004) Management of dyslipidemia in patients with metabolic syndrome. J Am Pharm Assoc. 44: 478-488.

[12] Srinivasan K, Ramarao P (2007) Animal models in type 2 diabetes research: An overview. Indian J Med Res. 125: 451-472

[13] Weiss R, Dziura J, Burgert TS, Tamborlane WV, Taksali SE, Yeckel CW, Allen K, Lopes M, Savoye M, Morrison J, Sherwin RS, Caprio S (2004) Obesity and the metabolic syndrome in children and adolescents. N Engl J Med.350:2362–74.

[14] Bruce KD, Byrne CD (2009) The metabolic syndrome: common origins of a multifactorial disorder. Postgrad Med J. 85:614–621.

[15] Getz GS, Reardon CA (2006) Diet and murine atherosclerosis. Arterioscler Thromb Vasc Biol. 26:242-249.

[16] Maxwell KN, Soccio RE, Duncan EM, Sehayek E, Breslow JL (2003) Novel putative SREBP and LXR target genes identified by microarray analysis in liver of cholesterol-fed mice. J Lipid Res. 44:2109-2119.

[17] Nishina PM, Lowe S, Verstuyft J, Naggert JK, Kuypers FA, Paigen B (1993) Effects of dietary fats from animal and plant sources on diet-induced fatty streak lesions in C57BL/6J mice. J Lipid Res. 34:1413-1422.

[18] Cefalu WT (2006) Animal models of type 2 diabetes: clinical presentation and pathophysiological relevance to the human condition. ILAR J. 47(3).

[19] Panchal SK, Brown L (2011) Rodent models for metabolic syndrome research. J Biomed Biotechnol. 2011:351982.

[20] Kobayasi R, Akamine EH, Davel AP, Rodrigues MAM, Carvalho CRO and Rossoni LV (2010) "Oxidative stress and inflammatory mediators contribute to endothelial dysfunction in high-fat diet-induced obesity in mice," J Hypertension. 28: 2111–2119.

[21] Deji N, Kume S, Araki SI et al (2009) "Structural and functional changes in the kidneys of high-fat diet-induced obese mice,". Am J Physio. 296: F118–F126.

[22] Buettner R, Parhofer KG, Woenckhaus M, Wrede CE, Kunz-Schughart LA, Scholmerich J, Bollheimer LC (2006) Defining high-fat-diet rat models: metabolic and molecular effects of different fat types. J Mol. Endocrinol. 36:485-501.

[23] Sutherland LN, Capozzi LC, Turchinsky NJ, Bell RC, Wright DC (2008) "Time course of high-fat diet-induced reductions in adipose tissue mitochondrial proteins: potential mechanisms and the relationship to glucose intolerance," Am J Physio. 295: E1076–E1083.

[24] Sweazea KL, Lekic M, Walker BR (2010) "Comparison of mechanisms involved in impaired vascular reactivity between high sucrose and high fat diets in rats," Nutr Metabol. 7:48.

[25] Basciano H, Federico L, Adeli K (2005) Fructose, insulin resistance, and metabolic dyslipidemia. Nutr Metabol. 2:5 doi:10.1186/1743-7075-2-5.

[26] Kohli R, Kirby M, Xanthakos SA, Softic S, Feldstein AE, Saxena V, Tang PH, Miles L, Miles MV, Balistreri WF, Woods SC, Seeley RJ (2010) High-fructose, medium chain trans fat diet induces liver fibrosis and elevates plasma coenzyme Q9 in a novel murine model of obesity and nonalcoholic steatohepatitis. Hepatology. 52(3):934-44.

[27] Nascimento FA, Barbosa-da-Silva S, Fernandes-Santos C, Mandarim-de-Lacerda CA, Aguila MB (2010) Adipose tissue, liver and pancreas structural alterations in C57BL/6 mice fed high-fat-high-sucrose diet supplemented with fish oil (n-3 fatty acid rich oil). Exp Toxicol Pathol. 62(1):17-25.

[28] Charlton M, Krishnan A, Viker K, Sanderson S, Cazanave S, McConico A, Masuoko H, Gores G (2011) Fast food diet mouse: novel small animal model of NASH with ballooning, progressive fibrosis, and high physiological fidelity to the human condition. Am J Physiol Gastrointest Liver Physiol. 301(5):G825-34.

[29] Surwit RS, Kuhn CM, Cochrane C, McCubbin JA, Feinglos MN (1988) Diet-induced type II diabetes in C57BL/6J mice. Diabetes. 37: 1163-7.

[30] Ikegami H, Fujisawa T, Ogihara T (2004) Mouse models of type 1 and type 2 diabetes derived from the same closed colony: Genetic susceptibility shared between two types of diabetes. ILAR J. 45:268-277.

[31] Harmon JS, Gleason CE, Tanaka Y, Poitout V, Robertson RP (2001) Antecedent hyperglycemia, not hyperlipidemia, is associated with increased islet triacylglycerol content and decreased insulin gene mRNA level in Zucker diabetic fatty rats. Diabetes. 50:2481-2486.

[32] Hu Y, Davies GE (2010) Berberine inhibits adipogenesis in high-fat diet-induced obesity mice. Fitoterapia. 81: 358-366.

[33] Levin BE, Triscari J, Sullivan AC (1983) Relationship between sympathetic activity and diet-induced obesity in two rat strains. Am J Physiol. 245: R364-R371.

[34] Lauterio TJ, Bond JP, Ulman EA (1994) Development and characterization of a purified diet to identify obesity-susceptible and -resistant rat populations. J Nutr. 124:2172-2178.

[35] Chang S, Graham B, Yakubu F, Lin D, Peters JC, Hill JO (1990) Metabolic differences between obesity-prone and obesity-resistant rats. Am J Physiol 259:R1103-R1110.

[36] Levin BE, Dunn-Meynell AA (2006) Differential effects of exercise on body weight gain and adiposity in obesity-prone and –resistant rats. Int J Obes(Lond). 30:722-727.

[37] Levin BE, Keesey RE (1998) Defense of differing body weight set points in diet-induced obese and resistant rats. Am J Physiol. 274:R412-R419.

[38] Dobrian AD, Davies MJ, Prewitt RL, Lauterio TJ (2000) Development of Hypertension in a Rat Model of Diet-Induced Obesity. Hypertension.35:1009-1015.

[39] Krauss RM (2004) Lipids and lipoproteins in patients with type 2 diabetes. Diabetes Care.27:04–1496.

[40] Monti LD, Landoni C, Setola E, Galluccio E, Lucotti P, Sandoli EP, Origgi A, Lucignani G,Piatti P, Fazio F (2004) Myocardial insulin resistance associated with chronic hypertriglyceridemia and increased FFA levels in Type 2 diabetic patients. Am J Physiol Heart Circ Physiol. 287:H1225–31.

[41] Akiyama T, Tachibana I, Shirohara H, Watanabe N, Otsuki M (1996) High-fat hypercaloric diet induces obesity glucose intolerance and hyperlipidemia in normal adult male Wistar rat. Diabetes Res Clin Pract. 31: 27–35.

[42] Lauterio TJ, Barkan A, DeAngelo M, DeMott-Friberg R, Ramirez R (1998) Plasma growth hormone secretion is impaired in obesity-prone rats before onset of diet-induced obesity. Am J Physiol. 275:E6-E11.

[43] Yang N, Wang C, Xu M, Mao L, Liu L, Sun X (2005) Interaction of dietary composition and PYY gene expression in diet-induced obesity in rats. J Huazhong Univ Sci Technolog Med Sci.25:243–6.

[44] Barker DJP, Gluckman PD, Godfrey KM, Harding JE, Owens JA, Robinson JS (1993) Fetal nutrition and cardiovascular disease in adult life. Lancet 341: 938–941.

[45] Bertram CE, Hanson MA (2001) Animal models and programming of the metabolic syndrome. Brit Med Bull. 60:103–121

[46] Guo F, Jen KLC (1995) High fat feeding during pregnancy and lactation affects offspring metabolism in rats. Physiol Behav.57: 681–6

[47] Brown SA, Rogers LK, Dunn JK, Gotto AM, Jr, Patsch W (1990) Development of cholesterol homeostatic memory in the rat is influenced by maternal diets. Metab Clin Exp. 39: 468–73.

[48] Langley-Evans SC (1996) Intrauterine programming of hypertension in the rat: nutrient interactions. Comp Biochem Physiol. 114: 327–31

[49] Ricci MR, Levin BE (2003) Ontogeny of diet-induced obesity in selectively bred Sprague-Dawley rats. Am J Physiol Regul. Integr. Comp Physiol. 285:R610-R618.

[50] Corsetti JP, Sparks JD, Peterson RG, Smith RL, Sparks CE (2000) Effect of dietary fat on the development of non-insulin dependent diabetes mellitus in obese Zucker diabetic fatty male and female rats. Atherosclerosis. 148:231-241.

[51] Owens D (2006) Spontaneous, surgically and chemically induced models of disease. In The Laboratory Rat. Suckow MA, Weisbroth SH, Franklin CL, Eds. Elsevier Academic Press. 711-732 p.

[52] Blay M, Peinado-Onsurbe J, Julve J, Rodríguez V, Fernández-López JA, Remesar X, Alemany M (2001) Anomalous lipoproteins in obese Zucker rats. Diabetes Obes Metab. 3(4):259-70.

[53] Ballal K, Wilson CR, Harmancey R, Taegtmeyer H (2010) Obesogenic high fat western diet induces oxidative stress and apoptosis in rat heart. Mol Cell Biochem. 344(1-2):221-30.

[54] Hegsted DM, Ausman LM, Johnson JA, Dallal GE (1993) Dietary fat and serum lipids: an evaluation of the experimental data. Am J Clin Nutr. 57:875-883.

[55] Fillios LC, Andrus SB, Mann GV, Stare FJ (1956) Experimental production of gross atherosclerosis in the rat. J Exp Med. 104: 539.

[56] Zaragoza C, Gomez-Guerrero C, Martin-Ventura JL, Blanco-Colio L, Lavin B, Mallavia B, Tarin C, Mas S, Ortiz A, Egido J (2011) Animal models of cardiovascular diseases. J Biomed Biotechnol. doi:10.1155/2011/497841

[57] Russell JC, Proctor SD (2006) Small animal models of cardiovascular disease: tools for the study of the roles of metabolic syndrome, dyslipidemia, and atherosclerosis. Cardiovasc Pathol. 15: 318-30.

[58] Singh V, Tiwari RL, Dikshit M, Barthwal MK (2009) Models to study atherosclerosis: a mechanistic insight. Curr Vasc Pharmacol. 7: 75-109.

[59] Gomibuchi H, Okazaki M, Iwai S, Kumai T, Kobayashi S, Oguchi K (2007) Development of hyperfibrinogenemia in spontaneously hypertensive and hyperlipidemic rats: a potentially useful animal model as a complication of hypertension and hyperlipidemia. Exp Anim.56: 1-10.

[60] Matos SL, Paula Hd, Pedrosa ML, dos Santos RC, de Oliveira EL, Júnior DAC, Silva ME (2005) Dietary models for inducing hypercholesterolemia in rats. Braz Arch Biol Techn. 48:203-209.

[61] Roberts CK, Liang K, Barnard RJ, Kim CH,Vaziri ND (2004) HMG-CoA reductase, cholesterol 7a-hydroxylase, LDL receptor, SR-B1, and ACAT in diet-induced syndrome X. Kidney Int. 66:1503-1511.

[62] Sampey BP, Vanhoose AM, Winfield HM, Freemerman AJ, Muehlbauer MJ, Fueger PT, Newgard CB, Makowski L (2011) Cafeteria diet is a robust model of human metabolic syndrome with liver and adipose inflammation; comparison to high-fat diet. Obesity.19: 1109-1117.

[63] Manting L, Haihong Z, Jing L, Shaodong C, Yihua L (2011) The model of rat lipid metabolism disorder induced by chronic stress accompanying high-fat-diet. Lipids Health Dis. 10:153.

[64] Spady DK, Woollett LA, Dietschy JM (1993) Regulation of plasma LDL- cholesterol levels by dietary cholesterol and fatty acids. Annu Rev Nutr. 13:355-381.

[65] Pien CS, Davis WP, Marone AJ, Foxall TL (2006) Characterization of diet induced aortic atherosclerosis in Syrian F1B Hamsters. J Exp Anim Sci. 42:65-83, 2006

[66] Alexaki A, Wilson TA, Atallah MT, Handelman G, Nicolosi RJ (2004) Hamsters fed diets high in saturated fat have increased cholesterol accumulation and cytokine production in the aortic arch compared with cholesterol-fed hamsters with moderately elevated plasma non-HDL cholesterol concentrations. J Nutr. 134:410-415.

[67] Vinson JA, Mandarano M, Hirst M, Trevithick JR, Bose P (2003) Phenol antioxidant quantity and quality in foods: beers and the effect of two types of beer on an animal model of atherosclerosis. J Agric Food Chem. 51(18):5528-33.

[68] Rimando AM, Nagmani R, Feller DF, Yokoyama W (2005) Pterostilbene, a new agonist for the peroxisome proliferator-activated receptor r-isoform, lowers plasma lipoproteins and cholesterol in hypercholesterolemic hamsters. J Agric Food Chem. 53: 3403–3407

[69] Nistor A, Bulla A, Filip DA, Radu A (1987) The hyperlipidemic hamster as a model of experimental atherosclerosis. Atherosclerosis. 68:159-173.

[70] Kahlon T, Chow F, Irving D, Sayre R (1996) Cholesterol response and foam cell formation in hamsters fed two levels of saturated fat and various levels of cholesterol. Nutr Res. 16:1353-1368.

[71] Dillard A, Matthan NR, Lichtenstein AH (2010) Use of hamster as a model to study diet-induced Atherosclerosis. Nutr Metab. 7:89

[72] Leung N, Naples M, Uffelman K, Szeto L, Adeli K, Lewis GF (2004) Rosiglitazone improves intestinal lipoprotein overproduction in the fat-fed Syrian Golden hamster, an animal model of nutritionally-induced insulin resistance. Atherosclerosis. 174(2):235-41.

[73] Wang PR, Guo Q, Ippolito M, Wu M, Milot D, Ventre J, Doebber T, Wright SD, Chao YS (2001) High fat fed hamster, a unique animal model for treatment of diabetic dyslipidemia with peroxisome proliferator activated receptor alpha selective agonists. Eur J Pharmacol. 427(3):285-93.

[74] Li S-Y, Chang C-Q, Ma F-Y, Yu C-L (2009) Modulating effects of chlorogenic acid on lipids and glucose metabolism and expression of hepatic peroxisome proliferator-activated receptor-α in golden hamsters fed on high fat diet. Biomed Environ Sci. 22:122-129.

[75] Tyburczy C, Major C, Lock AL, Destaillats F, Lawrence P, Brenna JT, Salter AM, Bauman DE (2009) Individual trans octadecenoic acids and partially hyrdogenated vegetable oil differentially affect hepatic lipid and lipoprotein metabolism in golden Syrian hamsters. J Nutr.139:257–263.

[76] Kritchevsky D (2001) Diet and atherosclerosis. J Nutr Health Aging. 5:155–159.

[77] Dorfman SE, Laurent D, Gounarides JS, Li X, Mullarkey TL, Rocheford EC, Sari-Sarraf F, Hirsch EA, Hughes TE, Commerford SR (2009) Metabolic implications of dietary trans-fatty acids. Obesity.17:1200–1207.

[78] Costa RRS, Villela NR, Souza MdGS, Boa BCS, Cyrino FZGA, Silva SV, Lisboa PC, Moura EG, Barja-Fidalgo TC, Bouskela E (2011) High fat diet induces central obesity, insulin resistance and microvascular dysfunction in hamsters. Microvascular Res. 82:416-422.

[79] Kothari HV, Poirier KJ, Lee WH, Satoh Y (1997) Inhibition of cholesterol ester transfer protein CGS 25159 and changes in lipoproteins in hamsters. Atherosclerosis. 128: 59-66.

[80] Briand F, Thiéblemont Q, Muzotte E, Sulpice T (2012). High-Fat and Fructose Intake Induces Insulin Resistance, Dyslipidemia, and Liver Steatosis and Alters In Vivo Macrophage-to-Feces Reverse Cholesterol Transport in Hamsters. J Nutr. 142(4):704-9.

[81] Bhathena J, Kulamarva A, Martoni C, Malgorzata A, Urbanska, Malhotra M, Paul A, Prakash S (2011) Diet-induced metabolic hamster model of nonalcoholic fatty liver disease. Diabetes, Metabolic Syndrome and Obesity: Targets and Therapy.4:195–203

[82] Avramoglu RK, Qiu W, Adeli K (2003) Mechanisms of metabolic dyslipidemia in insulin resistant states: deregulation of hepatic and intestinal lipoprotein secretion. Frontiers in Bioscience. 8: 464-476.

[83] Sullivan MP, Cerda JJ, Robbins FL, Burgin CW, Beatty RJ (1993) The gerbil, hamster, and guinea pig as rodent models for hyperlipidemia. Lab Anim Sci. 43,575-578.

[84] Hoang VQ, Botham KM, Benson GM, Eldredge EE, Jackson B, Pearce N, Suckling KE (1993) Bile acid synthesis in hamster hepatocytes in primary culture: sources of cholesterol and comparison with other species. Biochim Biophys Acta. 1210:73-80.

[85] Fernandez ML(2001) Guinea pigs as models for cholesterol and lipoprotein metabolism. J Nutr.131:10-20.

[86] West KL, Fernandez ML (2004) Guinea pigs as models to study the hypocholesesterolemic effects of drugs. Cardiovasc Rev. 22:7-22.

[87] Fernandez ML, Volek JS (2006) Guinea pigs: a suitable animal model to study lipoprotein metabolism, atherosclerosis and inflammation. Nutr Metab (Lond). 3: 17.

[88] Greeve J, Altkemper I, Dietrich J-H, Greten H, Windler E (1993) Apolipoprotein mRNA editing in 12 different mammalian species: hepatic expression is reflected in low concentrations of apoB-containing plasma lipoproteins. J Lipid Res. 34:1367-1383.

[89] Fernandez ML, Yount NY, McNamara DJ (1990) Whole body cholesterol synthesis in the guinea pig. Effects of dietary fat quality. Biochim Biophys Acta. 1044:340-348.

[90] Madsen CS, Janovitz E, Zhang R, Nguyen-Tran V, Ryan CS, Yin X, Monshizadegan H, Chang M, D'Arienzo C, Scheer S, Setters R, Search D, Chen X, Zhuang S, Kunselman L, Peters A, Harrity T, Apedo A, Huang C, Cuff CA, Kowala MC, Blanar MA, Sun CQ, Robl JA, Stein PD (2008). The Guinea pig as a preclinical model for demonstrating the efficacy and safety of statins. J Pharmacol Exp Ther. 324(2):576-86.

[91] Fernandez ML, Roy S, Vergara-Jimenez M (2000) Resistant starch and cholestyramine have distinct effects on hepatic cholesterol metabolism in guinea pigs fed a hypercholesterolemic diet. Nutr Res.20:837-850.

[92] West K, Ramjiganesh T, Roy S, Keller BT, Fernandez ML (2002) SC-435, an ileal, apical sodium-dependent bile acid transporter inhibitor (ASBT) alters hepatic cholesterol metabolism and lowers plasma low-density-lipoprotein-cholesterol concentrations in guinea pigs. J Pharmacol Exp Theraup.303:291-299.

[93] Aggarwal D, West KL, Zern TL, Shrestha S, Vergara-Jimenez M, Fernandez ML (2005) JTT-130, a microsomal triglyceride transfer protein (MTP) inhibitor lowers plasma triglycerides and LDL cholesterol concentrations without increasing hepatic triglycerides in guinea pigs. BMC Cardiovasc Disord. 27:5:30.

[94] Karimi I, Hayatgheybi H, Shamspur T, Kamalak A, Pooyanmehr M, Marandi Y (2011) Chemical composition and effect of an essential oil of Salix aegyptiaca L. (Musk willow) in hypercholesterolemic rabbit model. Braz J Pharmacog. 21(3): 407-414.

[95] Lin ECK, Fernandez ML, Tosca MA, McNamara DJ (1994) Regulation of hepatic LDL metabolism in the guinea pig by dietary fat and cholesterol. J Lipid Res.35:446-57.

[96] Lynch SM, Gaziano JM, Frei B (1996) Ascorbic acid and atherosclerotic cardiovascular disease. Subcell Biochem. 25: 331-67.

[97] McCormick SPA (2004) Lipoprotein(a): biology and clinical importance. Clin Biochem Rev. 25(1): 69–80.

[98] Hrzenjak A, Frank S, Wo X, Zhou Y, Van Berkel T, Kostner GM (2003) Galactose-specific asialoglycoprotein receptor is involved in lipoprotein (a) catabolism. Biochem J. 376:765-71.

[99] Rath M, Pauling L (1990) Hypothesis: lipoprotein(a) is a surrogate for ascorbate. Proc Natl Acad Sci USA.87: 6204-7.

[100] Leite JO, DeOgburn R, Ratliff J, Su R, Smyth JA, Volek JS, McGrane MM, Dardik A, Fernandez ML (2010) Low-carbohydrate diets reduce lipid accumulation and arterial inflammation in guinea pigs fed a high-cholesterol diet. Atherosclerosis. 209(2):442-8.

[101] Mangathayaru K, Kuruvilla S, Balakrishna K, Venkhatesh J (2009) Modulatory effect of *Inula racemosa* Hook. f. (Asteraceae) on experimental atherosclerosis in guinea-pigs. J Pharm Pharmacol. 61(8):1111-8.

[102] Leite JO, Vaishnav U, Puglisi M, Fraser H, Trias J, Fernandez ML (2009) A-002 (Varespladib), a phospholipase A2 inhibitor, reduces atherosclerosis in guinea pigs. BMC Cardiovasc Disord. 17:7.

[103] Yang R, Guo P, Song X, Liu F, Gao N (2011) Hyperlipidemic guinea pig model: mechanisms of triglyceride metabolism disorder and comparison to rat. Biol. Pharm. Bull. 34(7): 1046-1051.

[104] Evans MB, Tonini R, Shope CD, Oghalai JS, Jerger JF, Insull W Jr, Brownell WE (2006) Dyslipidemia and auditory function. Otol Neurotol. 27(5):609-14.

[105] Castañeda-Gutiérrez E, Pouteau E, Pescia G, Moulin J, Aprikian O, Macé K (2011) The guinea pig as a model for metabolic programming of adiposity. Am J Clin Nutr. 94:1838S-1845S

[106] Caillier B, Pilote S, Patoine D, Levac X, Couture C, Daleau P, Simard C, Drolet B (2012) Metabolic syndrome potentiates the cardiac action potential-prolonging action of drugs: a possible 'anti-proarrhythmic' role for amlodipine. Pharmacol Res. 65(3):320-7.

[107] Ignatowski AC (1908) Influence of animal food on the organism of rabbits. Izv Imp Voyenno-Med Akad Peter. 16:154–173.

[108] Moghadasian MH, Frohlich JJ, McManus BM (2001) Advances in experimental dyslipidemia and atherosclerosis. Lab Invest. 81:1173–1183.

[109] Xiangdong L, Yuanwu L, Hua Z, Liming R, Qiuyan L, Ning L (2011) Animal models for the atherosclerosis research: a review. Protein Cell. 2(3):189-201.

[110] Sider KL, Blaser MC, Simmons CA (2011) Animal models of calcific aortic valve disease. International Journal of Inflammation.ID 364310, 18 pages doi:10.4061/2011/364310.

[111] Karimi I, Hayatgheybi H, Razmjo M, Yousefi M, Dadyan A, Hadipour MM (2010) Anti-hyperlipidaemic effects of an essential oil of *Melissa officinalis*. L in cholesterol-fed rabbits. J Appl Biologl Sci. 4(1): 23-28.

[112] Badimon L (2001) Atherosclerosis and thrombosis: lessons from animal models. Thromb Haemost. 86: 356-65.

[113] Taylor JM, Fan J (1997) Transgenic rabbit models for the study of atherosclerosis. Front Biosci. 2: d298-308.

[114] Yanni AE (2004) The laboratory rabbit: an animal model of atherosclerosis research. Lab Anim. 38: 246–256.

[115] Morelli A, Vignozzi L, Maggi M, Adorini L (2011) Farnesoid X receptor activation improves erectile dysfunction in models of metabolic syndrome and diabetes. Biochim Biophys Acta. 1812(8):859-66.

[116] Mallidis C, Czerwiec A, Filippi S, O'Neill J, Maggi M, McClure N (2011)
Spermatogenic and sperm quality differences in an experimental model of metabolic
syndrome and hypogonadal hypogonadism. Reproduction. 142(1):63-71.

[117] Helfenstein T, Fonseca FA, Ihara SS, Bottós JM, Moreira FT, Pott H Jr, Farah ME,
Martins MC, Izar MC (2011) Impaired glucose tolerance plus hyperlipidaemia induced
by diet promotes retina microaneurysms in New Zeala6nd rabbits. Int J Exp Pathol.
92(1):40-9.

[118] Corona G, Rastrelli G, Morelli A, Vignozzi L, Mannucci E, Maggi M (2011)
Hypogonadism and metabolic syndrome. J Endocrinol Invest. 34(7):557-67.

[119] Vignozzi L, Morelli A, Sarchielli E, Comeglio P, Filippi S, Cellai I, Maneschi E, Serni S,
Gacci M, Carini M, Piccinni MP, Saad F, Adorini L, Vannelli GB, Maggi M (2012)
Testosterone protects from metabolic syndrome-associated prostate inflammation: an
experimental study in rabbit. J Endocrinol. 212(1):71-84.

[120] Morelli A, Comeglio P, Filippi S, Sarchielli E, Cellai I, Vignozzi L, Yehiely-Cohen R,
Maneschi E, Gacci M, Carini M, Adorini L, Vannelli GB, Maggi M (2012) Testosterone
and farnesoid X receptor agonist INT-747 counteract high fat diet-induced bladder
alterations in a rabbit model of metabolic syndrome. J Steroid Biochem Mol Biol. DOI:
10.1016/j.jsbmb.2012.02.007

[121] Kainuma M, Fujimoto M, Sekiya N, Tsuneyama K, Cheng C, Takano Y, Terasawa K,
Shimada Y (2006) Cholesterol-fed rabbit as a unique model of nonalcoholic, nonobese,
non-insulin-resistant fatty liver disease with characteristic fibrosis. J Gastroenterol.
41:971-80.

[122] Pasławski R, Pasławska U, Szuba A, Nicpoń J (2011) Swine as a Model of Experimental
Atherosclerosis. Adv Clin Exp Med. 20: 211–215.

[123] Royo T, Alfón J, Berrozpe M, Badimon L (2000) Effect of gemfibrozil on peripheral
atherosclerosis and platelet activation in a pig model of hyperlipidemia. Eur J Clin
Invest. 30(10): 843–852.

[124] Jacobsson L (1998) Experimental hyperlipidemia and atherosclerosis in mini-pigs;
influence of certain drugs. Scand J Lab Anim Sci. 25: 85–91.

[125] Johansen T, Hansen HS, Richelsen B, Malmlof R (2001) The obese Göttingen minipig as
a model of the metabolic syndrome: Dietary effects on obesity, insulin sensitivity, and
growth hormone profile. Comp Med. 51:150-155.

[126] Narayanaswamy M, Wright KC, Kandarpa K (2000) Animal models for atherosclerosis,
restenosis, and endovascular graft research. J Vasc Interv Radiol.11: 5-17.

[127] Berkhout TA, Simon HM, Jackson B, Yates J, Pearce N, Groot PH, Bentzen C, Niesor E,
Kerns WD, Suckling KE (1997) SR-12813 lowers plasma cholesterol in beagle dogs by
decreasing cholesterol biosynthesis. Atherosclerosis. 133: 203-12.

[128] Xenoulis PG, Suchodolski JS, Levinski MD, Steiner JM (2007) Investigation of
hypertriglyceridemia in healthy Miniature Schnauzers. J Vet Intern Med. 21: 1224-30.

[129] Henson MS, O'Brien TD (2006) Feline models of type 2 diabetes mellitus. ILAR J. 47:
234-242.

[130] German AJ (2006) The growing problem of obesity in dogs and cats. J Nutr. 136:1940S-
1946S

[131] Clarke I (2008) Models of 'obesity' in large animals and birds. Obesity and Metabolism. 36: 107-117.

[132] Homan R, Hanselman JC, Bak-Mueller S, Washburn M, Lester P, Jensen HE, Pinkosky SL, Castle C, Taylor B (2010) Atherosclerosis in *Octodon degus* (degu) as a model for human disease. Atherosclerosis. 212(1):48-54.

[133] Shafrir E, Ziv E (1998) Cellular mechanism of nutrionally induced insulin resistance: the desert rodent *Psammomys obesus* and other animals in which insulin resistance leads to detrimental outcome. J Basic Clin Physiol Pharmacol. 9: 347-85.

[134] Kaiser N, Nesher R, Donath MY, Fraenkel M, Behar V, Magnan C (2005) *Psammomys obesus*, a model for environment-gene interactions in type 2 diabetes. Diabetes. 54: S137-44.

[135] Velasquez MT, Kimmel PL, Michaelis OE (1990) Animal models of spontaneous diabetic kidney disease. FASEB J. 4: 2850-9.

[136] Weir BJ (1974) The development of diabetes in the tuco-tuco (*Ctenomys talarum*). Proc R Soc Med. 67(9):843-6.

[137] Zhao Z-J, Chen J-F, Wang D-H (2010) Diet-induced obesity in the short-day-lean Brandt's vole. Physiol Behav. 99(1):47-53.

[138] Noda K, Melhorn MI, Zandi S, Frimmel S, Tayyari F, Hisatomi T, Almulki L, Pronczuk A, Hayes KC, Hafezi-Moghadam A (2010) An animal model of spontaneous metabolic syndrome: Nile grass rat. FASEB J. 24: 2443–2453.

[139] Roberts JC, Jr, Straus R (Eds) (1965) Comparative atherosclerosis; the morphology of spontaneous and induced atherosclerotic lesions in animals and its relation to human disease, Harper & Row, New York

[140] Anderson JL, Smith SC, Taylor Jr RL (2012) Spontaneous Atherosclerosis in Pigeons: A good model of human disease, atherogenesis, Prof. Sampath Parthasarathy (Ed.), ISBN: 978-953-307-992-9, InTech.

[141] Tigno XT, Gerzanich G, Hansen BC (2004) Age-related changes in metabolic parameters of nonhuman primates. J Gerontol A Biol Sci Med Sci. 59:1081–1088.

[142] Kojic ZZ (2003) Animal models in the study of atherosclerosis. Srp Arh Celok Lek. 131: 266-70.

[143] Wagner JD, Kavanagh K, Ward GM, Auerbach BJ, Harwood Jr HJ, Kaplan JR (2006) Old World Nonhuman Primate Models of Type 2 Diabetes Mellitus. ILAR J. 47: 259-271.

[144] Kusumi Y, Scanu AM, McGill HC, and Wissler RW (1993) Atherosclerosis in a rhesus monkey with genetic hypercholesterolemia and elevated plasma Lp(a). Atherosclerosis. 99:165–174.

[145] Pick R, Johnson PJ, Glick G (1974) Deleterious effects of hypertension on the development of aortic and coronary atherosclerosis in stumptail macaques (*Macaca speciosa*) on an atherogenic diet. Circ Res. 35: 472-82.

[146] Kaufman D, Smith ELP, Gohil BC, Banerji M-A, Coplan JD, Kral JG, Rosenblum LA (2005) Early Appearance of the metabolic syndrome in socially reared bonnet macaques. J Clin Endocrinol Meta. 90:404–408.

[147] Zhang X, Zhang R, Raab S, Zheng W, Wang J, Liu N, Zhu T, Xue L, Song Z, Mao J, Li K, Zhang H, Zhang Y, Han C, Ding Y, Wang H, Hou N, Liu Y, Shang S, Li C, Sebokova E, Cheng H, Huang PL (2011) Rhesus macaques develop metabolic syndrome with reversible vascular dysfunction responsive to pioglitazone. Circulation. 124(1):77-86.

Chapter 7

Genetically Modified Animal Models for Lipoprotein Research

Masashi Shiomi, Tomonari Koike
and Tatsuro Ishida

Additional information is available at the end of the chapter

1. Introduction

Coronary artery disease (CAD) is a leading cause of death in the world and one of the major risk factors for CAD is dyslipidemia. In understanding dyslipidemia and developing therapeutics, animal models, especially genetically modified animals, have played important roles and contributed greatly to progress in this field. Before the development of genetically modified animals, the Watanabe heritable hyperlipidemic (WHHL) rabbit, the first animal model for familial hypercholesterolemia, developed by Yoshio Watanabe in 1980 (Watanabe, 1980), helped to verify a low-density lipoprotein (LDL) receptor-pathway in vivo and to clarify lipoprotein metabolism in humans (Goldstein, 1983), in addition to the process by which atherosclerosis develops (Shiomi, 2009). Furthermore, WHHL rabbits have contributed to the development of hypocholesterolemic agents, statins, (Watanabe, 1981; Tsujita, 1986) and to clarifying anti-atherosclerotic effects (Watanabe, 1988; Shiomi, 1995; 2009). In the present, WHHL rabbits were improved by selective breeding to produce the WHHLMI strain, which suffers from severe and vulnerable coronary atheromatous plaques and myocardial infarction due to coronary occlusion with progression of atherosclerotic plaques (Shiomi, 2003). However, WHHL or WHHLMI rabbits were not suitable for studying the role of genes in lipid metabolism, because it is difficult to apply genetic modification techniques to rabbits.

The first transgenic mice were developed in 1982 (Gordon, 1982) and the first knockout mice in 1984 (Bradley, 1984). Genetically modified mice are commonly used to study lipoprotein metabolism and atherosclerosis. The first transgenic mice for lipoprotein metabolism were LDLR-overexpressing mice, developed in 1988 (Hofmann, 1988), and the first knockout (KO) mice for lipoprotein metabolism were apolipoprotein (apo) E-KO mice, developed in 1992 (Zhang, 1992). Thereafter, numerous genetically modified mouse models were produced and these mice have contributed to a better understanding of lipoprotein metabolism. However, we should recognize that the lipoprotein metabolism of genetically modified mice is not entirely the same as that of humans, despite hyperlipidemia or hypercholesterolemia. In addition, recent studies have demonstrated different phenotypes manifested in mice and rabbits after the same gene transfer (Fan, 2003). The first transgenic rabbit was developed in 1985 (Hammer, 1985) and the first transgenic rabbit for lipoprotein metabolism, the hepatic lipase-overexpressing rabbit, was developed in 1994 (Fan, 1994). The differences in phenotype following gene transfer between mice and rabbits may be due to species differences in lipoprotein metabolism. When using animal models in experimental research, one has to be careful interpreting the results. Since other chapters explain in detail the functions of enzymes, apolipoproteins, and receptors relating to lipoprotein metabolism, this chapter concentrates on introducing various genetically modified animals and species differences in phenotype expression after gene modification for researchers wishing to study lipoprotein metabolism.

2. Species differences in lipoprotein metabolism

2.1. Species differences in lipoprotein profiles

Fig. 1 shows lipoprotein profiles of mice, rabbits, and humans analysed with high performance liquid chromatography (HPLC). The main lipoprotein is LDL in human normal subjects but high-density lipoprotein (HDL) in wild-type mice. LDL is increased markedly in the plasma of patients with hypercholesterolemia (Yin, 2012) and WHHLMI rabbits, but lipoproteins that elute at the position of very low-density lipoprotein (VLDL) are increased in apoE-deficient mice, one of the most commonly used genetically modified mice (Piedrahita, 1992). Although serum lipid levels and the lipoprotein profile of apoE-KO mice vary depending on the colony, the lipoprotein profile of apoE-KO mice is similar to that of cholesterol-fed rabbits (Yin, 2012). In LDL receptor (LDLR)-deficient individuals, LDL increased markedly in patients (familial hypercholesterolemia) and rabbits (WHHLMI rabbits) despite a normal diet or chow, while plasma LDL levels were not so high in homozygous LDLR-KO mice fed standard chow as described by Ishibashi et al. (1993), although serum lipid levels and lipoprotein profiles of LDLR-KO mice vary in each colony similar to those of apoE-KO mice. After consumption of a cholesterol-enriched diet, plasma cholesterol levels increased markedly in LDLR-KO mice, similar to FH patients and WHHL rabbits, but the increased lipoprotein fraction eluted at the position of VLDL (Ishibashi, 1994) and the lipoprotein profile was similar to that of cholesterol-fed rabbits (Yin, 2012). These differences in lipoprotein profiles between mice and humans or rabbits are probably due to the species differences in lipoprotein metabolism.

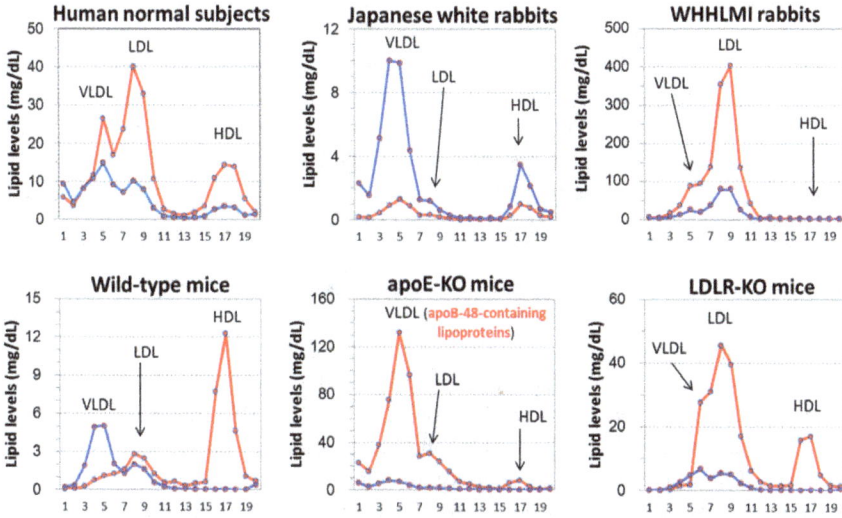

Figure 1. Lipoprotein profiles of a healthy human, rabbits, and mice. Lipoprotein profiles were analyzed with high performance liquid chromatography. Animals were fed standard chow. Red lines indicate cholesterol and blue lines indicate triglyceride

2.2. Lipoprotein metabolism in mice and rats

Fig. 2 shows a schematic diagram of lipoprotein metabolism in wild-type mice and rats. Dietary cholesterol is absorbed at the intestine mediated by Niemann-Pick C1-like 1 protein (NPC1L1) (Altmann, 2004) and ATP-binding cassette transporters (ABC) (Berge, 2000). Dietary fat is hydrolyzed to form monoglycerides and fatty acids in the intestine. These lipolytic products are translocated to the enterocyte membrane, and migrate to the endoplasmic reticulum. Overexpression of the tis7 gene in mice increases triglyceride absorption (Wang, 2009), and the transport of saturated fatty acids and cholesterol is decreased in fatty acid-binding protein (FABP)-KO mice (Newberry, 2009). Monoglycerides and fatty acids are re-esterified into triglycerides at the cytoplasmic surface of the endoplasmic reticulum. Synthesized triglycerides, cholesterol, and apolipoproteins are assembled in chylomicron particles. In the process of chylomicron assembly, MTP assists in binding lipids to apolipoproteins (Hussain, 2012). Since the apoB mRNA editing enzyme (apobec-1) functions at the intestinal wall, chylomicron particles contain apoB-48 as a major apolipoprotein but do not contain apoB-100. Chylomicron particles are released into the lacteal vessel. ApoB-48 is a marker of exogenous lipoprotein in humans and rabbits. At the capillary of peripheral tissue, chylomicron particles release free fatty acids (FFA) to adipose tissues mediated by lipoprotein lipase (LPL) and are converted to chylomicron remnants (Goldstein, 1983). Chylomicron remnants rapidly disappear from the circulation through apoE-receptors (apoER, remnant receptors) expressed in the liver. The metabolism of exogenous lipoproteins is preserved across species. In endogenous lipoprotein metabolism, cholesterol and other lipids synthesized in the liver are assembled in VLDL particles. Since apobec-1 is not expressed in the liver in humans and rabbits, VLDL particles contain apoB-100 but not apoB-48. However, VLDL particles of mice and rats contain both apoB-100 and apoB-48, because apobec-1 is expressed in the liver (Greeve, 1993). At the capillary in peripheral tissue, VLDL particles release FFA similar to chylomicron particles, which is then transformed into intermediate-density lipoprotein (IDL) (Goldstein, 1983). Part of the remaining VLDL particles and/or partially catalyzed VLDL particles bind to VLDL receptors expressed in peripheral tissue (Takahashi, 2004). Finally, apoB-48-containing VLDL and IDL in mice and rats disappear from the circulation through apoER expressed in the liver similar to chylomicron remnants. Some IDL particles containing apoB-100 bind to LDLRs in the liver, the rest release fatty acids via hepatic lipase (HL) and are transformed into LDL. LDL particles bind to LDLRs expressed at the surface of somatic cells. Therefore, a marker of endogenous lipoproteins is apoB-100 in humans and rabbits but endogenous lipoproteins of mice and rats contain both apoB-100 and apoB-48. Since the fractional catabolic rate for apoB-48-containing lipoproteins is very high compared to that for apoB-100-containing lipoproteins (Li, 1996), concentrations of VLDL and LDL are very low in wild-type mice and rats compared to humans. In reverse cholesterol transport from peripheral tissue to the liver, high-density lipoproteins (HDLs) receive free cholesterol from macrophages through scavenger receptor type B-I (SR-BI) and ABCs such as ABCA1 (Rohrer, 2009). The free cholesterol transported from macrophages is esterified by lecithin:cholesterol acyltransferase (LCAT) in plasma. The esterified cholesterol in HDL particles is transferred to VLDL, IDL, and LDL in plasma by cholesterol ester transfer protein (CETP) in humans and rabbits (Son, 1986), while mice and rats do not have CETP activity in plasma (Agellon, 1991). Therefore, the cholesterol ester in HDL particles is not transferred to apoB-containing lipoproteins in mice and rats. This is one of the major reasons why HDL is the predominant lipoprotein in these two species. Circulating HDL particles bind to SR-BI in the liver. Therefore, the large difference in lipoprotein metabolism between mice / rats and humans / rabbits is characterized by both the expression of apobec-1 in the liver and absence of CETP in the plasma of mice and rats. Dyslipidemia develops when lipoprotein metabolism is impaired. For example, LDLR deficiency causes familial hypercholesterolemia, and impaired LPL function causes hypertriglyceridemia. A number of animal models including transgenic animals, knockout animals, and spontaneous mutant animals can develop hypercholesterolemia, hypertriglyceridemia, and postprandial hypertriglyceridemia. However, the lipoprotein profile of these mice is greatly different from that of human patients.

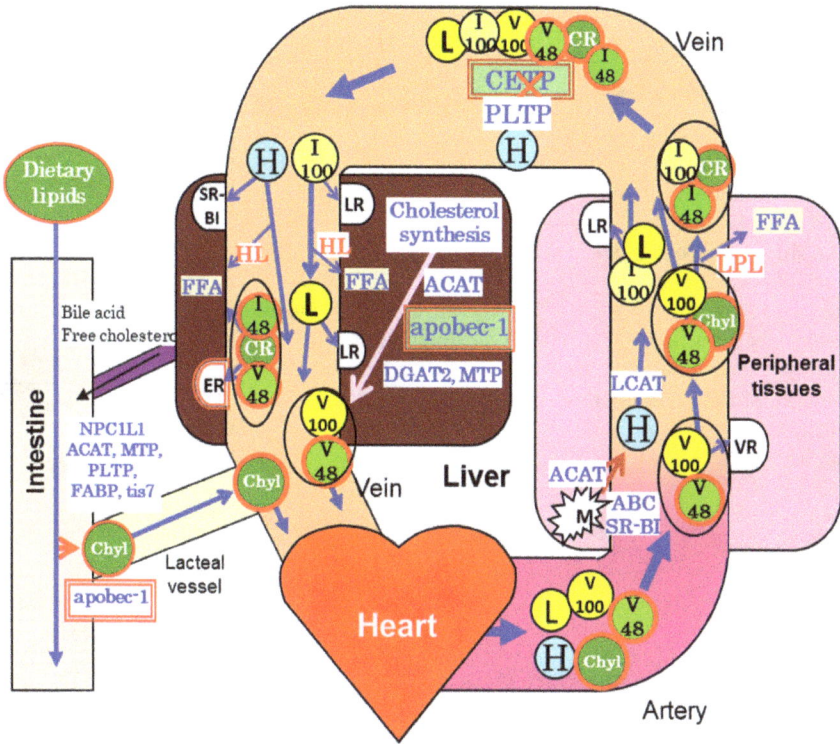

Figure 2. Lipoprotein metabolism in mice and rats (wild-type). Abbreviations: ACAT, acyl-CoA:cholesterol acyltransferase; apoB, apolipoprotein B; CETP, cholesterol ester transfer protein; Chyl, chylomicron; CR, chylomicron remnant; DGAT2, acyl-Co-A:diacylglycerol acyltransferase 2; ER, apoE receptor; FABP, fatty acids-binding protein; FFA, free fatty acids; H, high-density lipoprotein; HL, hepatic lipase; I48, intermediate-density lipoprotein (IDL) with apoB-48; I100, IDL with apoB-100; L, low-density lipoprotein (LDL); LCAT, lecithin:cholesterol acyltransferase; LPL, lipoprotein lipase; LR, LDL receptor; M, macrophage; MTP, microsomal triglyceride transfer protein; NPC1L1, Niemann-Pick C1-like 1; PLTP, phospholipid transfer protein; SR-BI, scavenger receptor type B-I; V48, very low-density lipoprotein (VLDL) with apoB-48; V100, VLDL with apoB-100; VR, VLDL receptor

3. Genetically modified animal models as tools for studying lipoprotein metabolism

Genetically modified animals for studying lipoprotein metabolism are summarized in Table 1. The list includes genetically modified mice, rats, rabbits, and chicken, but not double- or triple-modified animals. Several double- or triple-modified mice were developed by cross-breeding, for example, the apoE/LDLR-dKO mouse, the apoE/SRBI-dKO mouse, and others. Genetically modified animals can be good tools for clarifying the role of genes in lipoprotein metabolism and atherosclerosis if researchers take into consideration species differences.

	Transgenic animals			Knockout animals		
	mouse	rat	rabbit	mouse	rat	chick
Apolipoprotein						
apoA-I	Walsh, 1989	Swanson, 1992	Duverger, 1996	Plump, 1997		
apoA-II	Marzal-Casacub, 1996		Koike, 2009	Weng, 1996		
apoB100	Farese, 1995		Fan, 1995	Young, 1995		
apoC-I	Simonet, 1991			Gautier, 2002		
apoC-II	Shachter, 1994					
apoC-III	Aalto-Setala, 1992			Ding, 2011		
apoC-IV	Allan, 1996					
apoE	Shimano, 1992		Fan, 1998	Piedrahita, 1992		
apo(a)	Chiesa, 1992		Rouy, 1998			
apoM	Christoffersen, 2008		Christoffersen, 2008			
Cholesterol absorption in intestine						
NPC1L1				Altmann, 2004		
ABCA1				McNeish, 2003	Mulligan, 2003	
ACAT1				Buhman, 2000		
Apobec 1				Kendrick, 2001		
MTP				Xie, 2006		
PLTP				Liu, 2007		
SR-BI				Mardones, 2001		
FABP				Newberry, 2009		
tis7	Wang, 2005					
VLDL secretion from liver						
DGAT2				Liu, 2008		
MTP				Raabe, 1998		
Apobec-1				Morrison, 1996		
Lipolytic enzyme						
LPL	Shimada, 1993		Fan, 2001	Coleman, 1995		
HL	Braschi, 1998		Fan, 1994	Gonzalez-Navarro, 2004		
EL	Ishida, 2003;			Ishida, 2003; Ma, 2003		
Lipoprotein metabolism						
LDLR	Hofmann, 1998			Ishibashi, 1993	Asahina, 2012	
PCSK9	Herbert, 2010			Rashid, 2005		
VLDLR				Frykman, 1995		
SR-type A				Suzuki, 1997		
Reverse cholesterol transport						
ABCA1	Vaisman, 2001			McNeish, 2000		
ABCG1	Kennedy, 2005			Kennedy, 2005		
SR-BI	Wang, 1998			Rigotti, 1997		
LCAT	Vaisman, 1995		Hoeg, 1996			
CETP	Aggellon, 1991	Herrera, 1999				
PLTP	Jiang, 1996		Masson, 2011	Jiang, 1999		

Table 1. List of genetically modified animals regarding lipoprotein metabolism.

3.1. Cholesterol absorption in the intestine

Recent studies using genetically modified animals have help to clarify the mechanism of cholesterol absorption in the jejunum. Dietary cholesterol forms micelles with bail acids in the lumen of the jejunum, which are then transported through NPC1L1. Thereafter, the free cholesterol is esterified by acyl coenzyme A:cholesterol acyltransferase 2 (ACAT2) and chylomicron particles are formed by the packaging of esterified cholesterol, triglyceride, and apolipoprotein B by MTP (Hussain, 2012). NPC1L1 was found by Altman et al (2002). NPC1L1, highly expressed in the jejunum and located on the surface of absorptive enterocytes, is critical for the intestinal absorption of dietary and biliary cholesterol (Altmann, 2004). NPC1L1 mediates cholesterol uptake through vesicular endocytosis. Davis et al (2004) produced NPC1L1 KO mice, which had substantially reduced intestinal uptake of cholesterol and sitosterol. NPC1L1-deficiency resulted in the up-regulation of intestinal hydroxymethylglutaryl-CoA synthase mRNA expression and an increase in intestinal cholesterol synthesis, the down-regulation of ABCA1 mRNA expression, and no change in ABCG5 and ABCG8 mRNA levels. Therefore, NPC1L1 is required for intestinal uptake of both cholesterol and phytosterols and plays a major role in cholesterol homeostasis. These findings in NPC1L1-KO mice were similar to results obtained with an inhibitor of cholesterol absorption, ezetimibe (Garcia-Calvo, 2005). Phospholipid transfer protein (PLTP) is also involved in cholesterol absorption in the intestine (Liu R, 2007). PLTP-KO mice absorb significantly less cholesterol than wild-type mice. In addition, mRNA levels of NPC1L1 and ABCA1 and MTP activity levels were significantly decreased in the small intestine of PLTP-KO mice. The free cholesterol taken up through NPC1L1 and PLTP is esterified by ACAT2. Experiments with ACAT2-KO mice demonstrated that a deficiency of ACAT2 activity inhibits cholesterol absorption in the intestine (Buhman, 2000). In the intestine, lipids absorbed are packaged with apolipoproteins and form chylomicron particles. The major structural apolipoprotein in chylomicron particles is apoB-48. ApoB-48 is produced by apobec 1, which inserts a stop codon. A deficiency of apobec-1 in the intestine resulted in reduction in the secretion and assembly of chylomicron particles (Kendrick, 2001). These results from apobec-1-KO mice suggested that apoB-48 is involved in the assembly of chylomicron particles (Lo, 2008). Finally, absorbed lipids and synthesized apolipoproteins are assembled by MTP. MTP-KO mice demonstrated a decrease in cholesterol absorption and chylomicron secretion, in addition to manifestations of steatorrhea (Xie, 2006). Although ABC and SR-BI were considered important to cholesterol absorption until the year 2000, SR-BI is not essential for intestinal cholesterol absorption (Mardones, 2001). Cholesterol absorption was independent of ABCA1 in KO mice (McNeish, 2000) and ABCA1-mutant chickens (Mulligan, 2003). Thus, studies with genetically modified animals have verified the mechanisms of dietary lipid absorption revealed by experiments in vitro.

3.2. Formation and secretion of VLDL particles from liver

The liver is the main organ in lipoprotein metabolism. Endogenous lipoprotein (VLDL) particles are produced in liver. The production and secretion of VLDL consist of the synthesis of cholesterol, triglyceride, phospholipids, and apolipoproteins, and assembly of these components. As described, apobec-1 is expressed in the liver in mice and rats, but not in humans and rabbits. Therefore, apoB-48-containing VLDL particles are secreted from the mouse and rat liver. Compared to those containing apoB-100, VLDL particles containing apoB-48 are rapidly cleared from circulation through apoER expressed on hepatocytes, similar to chylomicron remnants (Fig 2). To better approximate human lipoprotein metabolism, apobec-1-deficient mice were developed by gene targeting (Morrison, 1996). The LDL levels increased and HDL levels decreased in the circulation. However, overexpression of human apoB-100 showed different results between mice and rabbits. Plasma cholesterol levels decreased in apoB-100-overexpresing mice (Farese, 1996), although plasma cholesterol, triglyceride, and LDL levels increased and HDL levels decreased in apoB-100-overexpressing rabbits (Fan, 1995). This difference may be due to differences in CETP activity in the circulation between mice and rabbits. In addition, suppression of acyl-CoA: diacylglycerol acyltransferase 2 (DGAT2) expression by antisense treatment decreased VLDL secretion in mice (Liu, 2008). This suggests that suppression of triglyceride synthesis decreases VLDL secretion. Studies using inhibitors of MTP suggested that MTP plays a key role in the production of VLDL particles and inhibition of MTP activity decreases VLDL secretion in WHHL rabbits (Shiomi, 2001). Indeed, MTP+/- mice fed a high-fat diet demonstrated decreased levels of apoB-containing lipoproteins in plasma (Raabe, 1998). Furthermore, hepatocytes synthesize and/or secrete a lot of apolipoproteins, such as apoA, apoB, apoC, apoE, apoM, and apo(a). The function of these apolipoproteins has been clarified using genetically modified animals. However, influences of the overexpression of apoB, apoE, and apo(a) differ between mice and rabbits. ApoE overexpression resulted in a marked decrease in non-HDL cholesterol in mice (Shimano, 1992), while in rabbits, cholesterol of LDL and HDL increased and the fractional catabolic rate of chylomicron also increased (Fan, 1998). These differences between mice and rabbits may be due to CETP activity and the expression of abobec-1 in mouse liver. Lipoprotein (a), an atherogenic lipoprotein, is formed by the binding of apo(a) to LDL particles and is detected in plasma of only humans and monkeys. In human-apo(a) transgenic mice (Chiesa, 1992), apo(a) does not bind to mouse LDL particles, while human apo(a) binds to rabbit LDL particles and lipoprotein (a) is also atherogenic in rabbits (Rouy, 1998; Fan, 1999). Therefore, the role of endogenous apoB-containing lipoproteins (VLDL, IDL, and LDL) in the regulation of plasma lipid levels differs between genetically modified mice and rabbits or humans.

3.3. Lipolysis of apoB-containing lipoproteins

Lipoproteins are transporters in circulation that provide cholesterol as a material for steroid hormones and the cytoskeleton, and triglycerides (fatty acids) for energy to peripheral tissue. In the transportation of fatty acids, lipoprotein lipase (LPL), hepatic lipase (HL), and endothelial lipase (EL) function at capillaries and apoC affects lipolytic activities. LPL mediates the lipolysis of VLDL and chylomicrons, and these lipoprotein particles are transformed into IDL and chylomicron remnants, respectively. Although LPL-/- mice die within a day after birth because of dramatic hypertriglyceridemia, impaired fat tolerance, and hypoglycemia (Weinstock, 1995), these LPL-KO mice could be rescued by transient LPL expression induced by adenoviral-mediated gene transfer (Strauss, 2001). Rescued adult LPL-KO mice exhibit severe hypertriglyceridemia as patients with homozygous LPL-deficiency. LPL+/- mice showed increases in plasma triglyceride levels due to increases in the fraction of VLDL and chylomicrons in the circulation. Overexpression of LPL in mice (Shimada, 1993) and rabbits (Fan, 2001) caused decreases in plasma triglyceride, VLDL, and LDL levels, in addition to the suppression of atherosclerotic lesions (Shimada, 1996). ApoC-I modulates this metabolism. Although knockout of apoC-I gene did not affect serum lipid levels, expression of human CETP markedly increased levels of cholesterol ester in plasma, VLDL, and LDL in apoC-I-KO mice (Gautier, 2002). In contrast, in transgenic mice overexpressing human apoC-I, plasma triglyceride and total cholesterol levels were increased compared to those in wild-type mice. In addition, overexpression of apoC-I, apoC-II, apoC-III, and apoC-IV also increased plasma triglyceride and total cholesterol levels, and suppressed LPL activity in mice (Simonet, 1991; Shachter, 1994; Aalto-Setala, 1992; Allan, 1996) and rabbits (Ding, 2011). HL modulates the metabolism of both apoB-containing and apoA-containing lipoproteins. In apoE-KO mice, deficiency of HL showed a decrease in the fractional catabolic rate of apoB-48-containing VLDL, IDL and LDL despite no effects on apoB-100-containing LDL, in addition to increases in total cholesterol and triglyceride levels in apoB-containing and apoA-containing lipoproteins (Mezdour, 1997 & Gonzalez-Navarro H, 2004). However, development of atherosclerotic lesions was reduced in HL-KO mice. In HL-transgenic mice (Brashci, 1998) and rabbits (Fan, 1994), catabolism of both HDL and apoB-48-containing lipoproteins is enhanced, and plasma total cholesterol and triglyceride levels are decreased. Therefore, HL may be associated with catabolism of not only apoB-containing lipoproteins but HDL. EL is located in arterial endothelial cells and has phospholipase activity against phospholipids in HDL particles (Broedl, 2003; Ishida, 2003). EL hydrolyzes phospholipids on HDL particles and promotes catabolism of HDL. Overexpression of EL decreases in HDL cholesterol and apoA-I levels decreased in mice (Ishida, 2003; Jaye, 1999). By contrast, a deficiency of EL increases HDL levels (Ishida, 2003; Ma, 2003), in addition to atherogenic action (Ishida, 2004) and allergic asthma (Otera, 2009). Another study confirmed the high HDL-C levels in EL-/- mice but did not document an association with atherosclerosis (Ko, 2005). Thus, the role of EL in reverse cholesterol transport and atherosclerosis has not been fully elucidated. Further studies are required to clarify the function of EL in the metabolism of HDL and atherosclerosis. In studies about lipolysis, genetically modified animals have demonstrated no species differences, and are useful in this field.

3.4. Receptor-mediated catabolism of apoB-containing lipoproteins

3.4.1. LDL receptor

Lipoprotein receptors, such as LDLRs, VLDL receptors (VLDLRs), apoE receptors (remnant receptors, apoERs), and scavenger receptors (SRs), take up lipoprotein particles into parenchymal cells and/or phagocytes. LDLRs are expressed on the surface of parenchymal cells and bind to circulating LDL. The ligands are apoB-100 and apoE. A deficiency or the suppression of LDLRs results in the accumulation of LDL in the circulation, a condition known as human familial hypercholesterolemia. Several animal models for LDLR-deficiency have been developed. One of the better known models is the WHHL (Watanabe, 1980) or WHHLMI (Shiomi, 2003; 2009) rabbit. WHHL or WHHLMI rabbits show hypercholesterolemia due to LDLR deficiency even when fed standard chow. However, LDLR-KO mice (Ishibashi, 1993 & 1994) and LDLR-KO rats (Asahina, 2012) showed mildly increased serum cholesterol levels. Sanan et al. (1998) reported that LDLR-KO mice expressing human apoB-100 showed hypercholesterolemia due to the accumulation of LDL in the plasma even in chow feeding. In addition, Teng et al (1997) demonstrated that adenovirus-mediated gene delivery of apobec-1 reduced plasma apoB-100 levels, leading to the almost complete elimination of LDL particles and a reduction in LDL cholesterol in LDLR-KO mice. These studies suggests that the absence of any increase in plasma cholesterol levels in LDLR-KO mice is due to the expression of apobec-1 in the liver and apoB-100-containing LDL is a key player in LDLR deficiency to increased plasma cholesterol levels. Expression of LDLRs on the cell surface is regulated by proprotein convertase subtilisin/kexin type 9 (PCSK9). Recently, Huijgen et al. (2012) reported that plasma levels of PCSK9 were associated with LDL cholesterol levels in patients with familial hypercholesterolemia. In addition, overexpression of PCSK9 induced negative modulation of LDLR expression and decreased plasma LDL clearance, also promoting atherosclerosis (Herbert, 2010). In contrast, knockout of PCSK9 resulted in an increase in the LDL receptor protein (Mbikay, 2010) and a decrease in plasma cholesterol levels (Rashid, 2005). Furthermore, PCSK9 regulates the expression of not only LDLRs but VLDLRs and apoERs (Poirier, 2008). Therefore, PCSK9 can be considered a new target in the treatment of hypercholesterolemia.

3.4.2. VLDL receptor

VLDL particles are incorporated through VLDLRs. VLDLRs are expressed in heart, muscle, adipose tissues, and macrophages but not in liver in humans and rabbits. In mice, however, VLDLRs are not expressed in macrophages (Takahashi, 2011), suggesting the process of atherogenesis to be somewhat different between mice and humans or rabbits. Knockout of VLDLRs does not affect lipoprotein metabolism but decreases body weight, BMI, and epididymal fat in mice (Frykman, 1995). In addition, LPL activity is decreased by VLDLR-deficiency. These observations suggest VLDLRs to be associated with metabolic syndrome. Furthermore, a recent study suggests that the expression of VLDLRs is affected by PCSK9 (Roubtsova, 2011). Surprisingly, adipose tissues of apoE-KO mice did not express LDLR, VLDLR, and LDLR-related proteins (Huang, 2009), although wild-type mice developed these receptors in adipose tissue. Since the VLDLR has various functions, genetically modified animals may contribute to further studies.

3.4.3. Remnant receptor

Remnant receptors are mainly expressed in liver and contribute to the metabolism of exogenous lipoproteins, which contain apoB-48, in humans and rabbits. Therefore, down-regulation of remnant receptor function causes the accumulation of chylomicron remnants in plasma. As their ligand is apolipoprotein E, remnant receptors are also called apoE receptors (apoERs). In mice and rats, VLDL, IDL, and LDL contain apoB-48 due to the expression of apobec-1 in the liver. These apoB-48-containing lipoproteins bind to apoERs through interaction with the apoE ligand and disappear from the circulation rapidly. The fractional catabolic rate for apoB-48-containing VLDL is remarkably high compared to that for apoB-100-containing VLDL (Gonzalez-Navarro, 2004). This is one of the reasons why concentrations of VLDL and LDL are very low in plasma of mice and rats. In contrast, apoE-KO mice have very high VLDL concentration and the VLDL fraction consists of apoB-48 (Gonzalez-Navarro, 2004). Since apoE is a ligand of apoER, lipoproteins containing only apoB-48 cannot bind to apoERs in apoE-KO mice. This is the reason for the hypercholesterolemia in apoE-KO mice. Consequently, the hypercholesterolemia of apoE-KO mice due to the accumulation of apoB-48-containing lipoproteins is different from human hypercholesterolemia due to the accumulation of apoB-100-containing lipoproteins. This difference affects the development of hypocholesterolemic agents.

3.4.4. Scavenger receptor type A

Scavenger receptor type A (SR-A) is expressed on phagocytes and plays an important role in the removal of modified lipoproteins, such as oxidized-LDL, acetyl-LDL, and glycated-LDL. Therefore, SR-A plays an important role in atherogenesis. Knockout of SR-A decreases the uptake of modified LDL, but does not affect plasma lipid levels (Suzuki, 1997).

3.5. Reverse cholesterol transport

The plasma concentration of HDL is inversely related to the risk of atherosclerotic vascular diseases. HDL plays a key role in the reverse transport of cholesterol from peripheral tissue to liver. Recent studies suggest that HDL is also associated with anti-inflammation, anti-thrombosis, anti-oxidation, and the enhancement of endothelial function. Newly synthesized apoA-I binds to ABCs (particularly ABC-A1) or SR-BI of macrophages and takes up free cholesterol from macrophages. The complex of apoA-I and free cholesterol is transformed into discoidal nascent HDL (pre-β HDL). These nascent HDLs become HDL particles (HDL3) after esterification of the free cholesterol by LCAT in plasma. In the process of the transformation from discoidal HDL to HDL3, HDL takes up apoE and free cholesterol from macrophages mainly by ABCG1. Therefore, as the HDL matures, its cholesterol content increases. In humans and rabbits, CETP in plasma exchanges the cholesterol ester of HDL particles with triglyceride in apoB-containing lipoproteins. Therefore, peripheral cholesterol is transported by two pathways; an LDLR pathway mediated by CETP function and a SR-BI pathway. However, mice and rats do not have CETP activity in plasma.

Therefore, the pathway of reverse cholesterol transport is markedly different between mice / rats and humans / rabbits.

3.5.1. Apolipoproteins of HDL particles

HDL particles contain apoA, apoC, apoE, and apoM. ApoA, the main structural apolipoprotein of HDL particles, is mainly classified as apoA-I, apoA-II, and apoA-IV. ApoA and apoE play an important role in the efflux of cholesterol from macrophages to discoidal and small HDL, respectively. Recent studies suggested that apoM is related to the anti-oxidative function of HDL (Elsoe, 2012).

ApoA-I is a major structural apolipoprotein of HDL particles. ApoA is synthesized mainly in the liver and intestine, and from HDL particles hydrolyzed by HL. Humans and mice have two types of HDLs. One contains only apoA-I and the other, both apoA-I and apoA-II. However, rabbit HDLs are apoA-I particles (Chapman, 1980 & Koike, 2009). ApoA-I plays an important role in the reverse transport of cholesterol. Overexpression of apoA-I increases HDL-cholesterol levels in mice (Walsh, 1989), rats (Swanson, 1992), and rabbits (Duverger, 1996). In contrast, knockout of the apoA-I gene in mice decreases cholesterol levels in HDL, VLDL, and whole plasma (Plump, 1997). In addition, apoA-I-deficient HDL is a poor substrate for HL and LCAT. These studies using genetically modified animals indicate that apoA-I plays an important role in cholesterol reverse transport. Another major apolipoprotein of HDL is apoA-II. Overexpression of human apoA-II in mouse liver resulted in a decrease in plasma cholesterol levels due to a decrease in HDL cholesterol but an increase in plasma triglyceride levels (Marzal-Casacub, 1996). In addition, LCAT activity and mouse apoA-II levels in plasma were decreased in human apoA-II transgenic mice. Consequently, the changes in plasma lipid levels in human apoA-II transgenic mice may be due to a reduction in levels of mouse apoA-II. These results suggest species differences in apoA-II. In addition, apoA-II is dimer in human but monomer in mice. Knockout of apoA-II in mice resulted in a decrease in not only HDL-cholesterol but non-HDL cholesterol. In addition, the fractional catabolic rate for apoA-I was increased by a deficiency of apoA-II (Weng, 1996). Furthermore, the deficiency was associated with lower free fatty acid, glucose, and insulin levels, suggesting insulin hypersensitivity, while apoA-II does not relate to insulin sensitivity in humans. Therefore, the function of apoA-II is very confusing in mouse models. Conversely, rabbits overexpressing human apoA-II, which do not have apoA-II, lipid levels in plasma and non-HDL lipoproteins were increased but HDL-cholesterol levels and activities of LPL and HL were decreased (Koike, 2009). Therefore, effects of human apoA-II overexpression may be different between mice and rabbits. To clarify the function of apoA-II, more studies are required.

3.5.2. Transfer of cholesterol from macrophages to HDL

The start of the reverse cholesterol transport process is the transfer of cholesterol from macrophages to apoA-I, in which ABCs play important roles. Several strains of mice with genetically modified ABCA1 and ABCG1 have been produced. Overexpression of ABCA1 in mice increases cholesterol efflux from macrophages, in addition to increases in levels of cholesterol, apoA-I, and apoA-II in HDL (Vaisman, 2001). In contrast, ABCA1-KO mice showed a marked decrease in HDL-cholesterol, LDL-cholesterol, and plasma apoB levels, and an absence of apoA-I in plasma, but an increase in cholesterol absorption and the accumulation of lipid-laden macrophages (McNeish, 2000). Furthermore, in ABCG1-KO mice, cholesterol efflux from macrophages to HDL is decreased (Kennedy, 2005). After its efflux from macrophages to apoA-I and HDL, free cholesterol is esterified by LCAT in the plasma. Overexpression of LCAT in mice increases levels of cholesterol, apoA-I, apoA-II, and apoE in plasma, in addition to HDL cholesterol (Vaisman, 1995). LCAT-overexpressing rabbits showed an increase in HDL-cholesterol on a chow diet but non-HDL cholesterol was not increased on a cholesterol diet (Hoeg, 1996). Therefore, LCAT plays an important role in the esterification of free cholesterol in HDL. Studies using ABCA1-KO mice, ABCG1-KO mice, and ABCA1/ABCG1-dKO mice (Out, 2008) have elucidated how cholesterol is transported from macrophages to HDL. ApoE also promotes reverse cholesterol transport by enhancing the efflux of free cholesterol from peripheral macrophages to maturing HDL particles (Hayek, 1994). The efflux of free cholesterol to apoE-binding HDL is mediated by ABCG1/4 (Matsuura, 2006). The efflux from macrophages to apoE-containing HDL3 (small HDL particles) depends on apoE. HDL-cholesterol levels are markedly low in apoE-KO mice, despite being high HDL in wild-type mice (Zhang, 1992). Therefore, the low HDL-cholesterol levels in apoE-KO mice are due to a decrease in the efflux of cholesterol from macrophages to HDL. By contrast, the overexpression of apoE induced a marked decrease in apoB-containing lipoproteins in mice (Shimano, 1992), but human apoE3 overexpression increased cholesterol levels in not only HDL but LDL in rabbits (Fan, 1998). These differences between mice and rabbits are due to the activity of CETP in plasma, since no CETP activity is detected in mice but strong activity is found in rabbits.

3.5.3. Transfer of lipids from HDL to apoB-containing lipoproteins

One of the main courses of reverse cholesterol transport depends on CETP in plasma in human and rabbits (Son, 1986). However, mice and rats do not have CETP activity in plasma (Agellon, 1991). Therefore, in mice, the HDL cholesterol level is high but cholesterol levels of apoB-containing lipoproteins are markedly low. Overexpression of human CETP in mice induces a decrease in HDL-cholesterol but no changes in cholesterol levels in VLDL and LDL (Agellon, 1991). Similar findings were made in Dahl rats (Herrera, 1999). One of the reasons for no changes in cholesterol levels of non-HDL fraction is due to a rapid clearance of apoB-48-containing lipoproteins through apoER. These results suggest that CETP transfers cholesterol ester from HDL to apoB-containing lipoproteins. Furthermore, CETP expression led to atherosclerosis in Dahl rats fed a cholesterol diet. Therefore, CETP-overexpression can cause complex responses depending on diet. PLTP is a plasma protein, which transfers phospholipids from apoB-containing lipoproteins to HDL. Knockout of PLTP resulted in a decrease in cholesterol (Jiang, 1999; 2001). In PLTP transgenic mice, PLTP did not affect lipid levels of apoB-containing lipoproteins but increased phospholipid and cholesterol levels in HDL (Jiang, 1996). However, overexpression of PLTP in rabbits increased cholesterol levels in apoB-containing lipoproteins but had no effect on HDL lipids in a high-cholesterol deiet feeding (Masson, 2011). These differences in the function of PLTP between mice and rabbits may be due to fundamental differences in lipoprotein metabolism, such as CETP activity in the plasma and apobec-1 expression in the liver. Therefore, one has to be deliberate in interpreting results from gene modification studies.

3.5.4. HDL receptors

HDL particles are incorporated by SR-BI, a HDL receptor, expressed in liver. In humans and rabbits, cholesterol is transferred from peripheral macrophages to liver through two pathways, via CETP-LDLR and SR-BI, while in mice and rats, cholesterol is transported to liver via SR-BI expressed in liver. Overexpression of SR-BI in mice induces a decrease in plasma lipids and an increase in the fractional catabolic rate for HDL (Wang, 1998). In contrast, SR-BI-KO mice show increases in plasma cholesterol levels, HDL particle size, and levels of apoE and apoA-I in HDL particles (Rigotti, 1997). These results demonstrate the function of SR-BI in reverse cholesterol transport. However, these changes in plasma lipid levels reflect HDL lipid levels, because mice do not have CETP activity in the plasma.

4. Species differences in phenotypes between genetically modified animals

Table 2 shows species differences in phenotypes of lipoprotein metabolism between genetically modified animals when the same genes were modified. Overexpression of apoB-100 increased HDL-cholesterol but had no effect on non-HDL-cholesterol in mice fed a chow diet (Farese, 1996) but increased in cholesterol levels in plasma and LDL in rabbits (Fan, 1995). Overexpression of apoE decreased levels of apoB-containing lipoproteins in mice (Shimano, 1992) but increased cholesterol levels in plasma and LDL in rabbits (Fan, 1998). In addition, overexpression of PLTP increased HDL-cholesterol and apoA-I levels but did not affect LDL-cholesterol in mice (Jiang, 1996), while it increased LDL-cholesterol and did not affect HDL in rabbits (Masson, 2011). These differences may be due to high CETP activity in rabbits and no CETP activity in mice. Lipoprotein(a) is detected in humans and monkeys, and is an atherogenic lipoprotein. In mice overexpressing human apo(a) (Chiesa, 1992), apo(a) did not bind to mouse LDL particles, while in rabbits, human apo(a) can bound to rabbit LDL particles and formed Lp(a) (Rouy, 1998; Fan, 1999). Finally, LDLR-KO increased cholesterol levels mildly in plasma and LDL in mice (Ishibashi, 1993) and rats (Asahina, 2012), while spontaneous LDLR-deficient rabbits (WHHL or WHHLMI rabbits) show severe hypercholesterolemia due to the accumulation of LDL in plasma even on a normal diet (Goldstein, 1993; Shiomi, 2003; 2009). In humans, LDLR-deficiency produces severe hypercholesterolemia due to the accumulation of LDL in plasma. These differences in effects of LDLR between humans / rabbits and mice / rats may be due to the expression of apobec-1 in the liver and CETP activity in the plasma. Therefore, one has to consider species differences when using animal models. LDLR-KO mice on a high-fat diet showed dramatic hypercholesterolemia (Ishibashi, 1994), and the plasma cholesterol level is comparable with or higher than that of apoE-KO mice. In contrast, the degree of atherosclerosis is greater in apoE-KO mice than LDLR-KO mice. In addition, LDLR-KO mice on a chow diet did not show massive atherosclerotic lesions at the age of 12 months (Ishibashi, 1994), while apoE-KO mice on a chow diet for the same period showed massive atherosclerotic lesions (Zhang, 1992 and Reddick, 1994). These differences in plasma lipid profiles and atherosclerosis are not fully understood but are likely to be attributable to the quality or subtype of the circulating lipoproteins.

Genes modification	Mice and/or rats	Rabbits
Apo-B100 overexpression	Increase in HDL-cholesterol and no effect on non-HDL-cholesterol on chow feeding	Increase in plasma and LDL cholesterol
Apo-E overexpression	Decrease in non-HDL cholesterol	Increase in plasma and LDL cholesterol
Apo (a) overexpression	Not bind to mouse LDL	Binds to rabbit LDL and forms Lp(a)
PLTP overexpression	Increase in HDL-cholesterol and apoA-I but no effects on LDL-cholesterol	Increase in LDL-cholesterol but no effect on HDL
LDLR-deficiency	Mild increase in plasma cholesterol	Marked increase in plasma cholesterol

Table 2. Species differences in phenotype on overexpression of the same genes.

5. Genetically modified animal models for human dyslipidemia

Table 3 summarizes plasma lipid profiles of genetically modified animal models for human dyslipidemia. Plasma lipid and/or lipoprotein profiles of genetically modified mice resemble those for human diseases involving apoC-II deficiency, LPL deficiency, and CETP deficiency. However, plasma lipid levels and lipoprotein profiles of genetically deficient mice are markedly different from those of humans with a deficiency of ABCA1, apoE or LDLR. In ABCA1-KO mice, cholesterol levels markedly decreased in not only HDL but whole plasma, while patients with Tangier disease, who do not have ABCA1 and show very low levels of HDL-cholesterol, exhibit a mild decrease in plasma cholesterol levels. This difference between ABCA1-KO mice and patients with Tangier disease may be due to CETP activity in plasma. ApoE-KO mice fed normal chow show hypercholesterolemia and the increased lipoprotein fraction is VLDL, which contains apoB-48, although plasma triglyceride levels are almost normal. In addition, HDL-cholesterol levels are markedly low. However, patients with apoE deficiency show type III hyperlipidemia by the WHO classification. The increased lipoprotein is VLDL and IDL, and both cholesterol and triglyceride levels are increased. HDL cholesterol is almost normal (Mabuchi, 1989). These differences in lipoprotein metabolism between mice and patients may be due to the expression of apobec-1 in mouse liver and the triglyceride content of the VLDL fraction. Considering these observations, the hypercholesterolemia in apoE-KO mice may not reflect human hypercholesterolemia. In LDLR deficiency, although humans and rabbit models show marked hypercholesterolemia due to the accumulation of LDL in plasma despite a normal diet (Watanabe, 1980; Shiomi, 2009), the accumulation of LDL in plasma in homozygous LDLR-KO mice is mild (Ishibashi, 1993; 1994). These differences in plasma lipid levels and lipoprotein profiles between LDLR-KO mice and familial hypercholesterolemia or WHHL rabbits are due to the rapid clearance of apoB-48-containing VLDL, IDL, and LDL through apoER in mouse liver. Recently, LDLR-KO rats were developed (Asahina, 2012). These animals have a similar lipoprotein profile to LDLR-KO mice. Studies demonstrate that both the expression of apobec-1 in liver and a deficiency of CETP in plasma greatly affect lipoprotein metabolism and plasma lipid levels in mice and rats. To solve these problems with LDLR-KO mice and apoE-KO mice, cross breeding with apobec-1-KO/CETP-expressing animals may be required in studies of lipoprotein metabolism. In the development of statins, potent anti-hyperlipidemic agents used by more than 40 million patients world-wide, no cholesterol-lowering effect was observed in mice and rats, although strong cholesterol-lowering effects were found in rabbits, dogs, monkeys and chickens (Tsujita, 1986). In addition, simvastatin, a statin, did not decrease serum cholesterol levels in LDLR-KO mice and CETP(+/-)LDLR(-/-)mice, and increased serum cholesterol levels in apoE-KO mice at a dose of 30 mg/kg/day (Yin, 2012), although an extremely high dose of statins (0.168% in diet, 200-300 mg/kg/day) decreased serum cholesterol levels in LDLR-KO mice (Krause, 1998). In contrast, WHHL rabbits, an animal model of familial hypercholesterolemia, have played important roles in studies of the hypocholesterolemic effects and anti-atherosclerotic effects of statins (Shiomi, 1995; 2009). The cholesterol-lowering effect of statins is mainly mediated by an increase in LDLR in liver. Therefore, the effect is weak when the contribution of LDLR to the regulation of plasma cholesterol levels is small, as in mice and rats. These studies suggest the need to select animal models based on study purposes.

Gene modification	Mice and/or rats	Human
ABCA1 deficiency	Marked decrease in cholesterol levels in both plasma and HDL	Tangier disease, Marked decrease in HDL cholesterol but mild decrease in plasma cholesterol
apoC-II deficiency	Hypertriglyceridemia	Hypertriglyceridemia
apoE-deficiency	Hypercholesterolemia Increase in VLDL and decrease in HDL	Combined hyperlipidemia Increase in IDL and no changes in HDL
LDLR-deficiency	Mild increase in plasma cholesterol Mild increase in LDL	Hypercholesterolemia Marked increase in LDL
LPL deficiency	Hypertriglyceridemia Lethal right after birth in homozygotes	Hypertriglyceridemia
CETP deficiency	High HDL cholesterol (wild-type mice)	High HDL cholesterol

Table 3. Plasma lipid profiles of genetically modified animal models for human dyslipidemia

6. Conclusion

In this chapter, the authors summarized achievements of studies using genetically modified animal models in lipoprotein research. The cross-breeding of genetically modified animals, such as double KO mice, triple KO mice, and others, has contributed to studies of lipoprotein metabolism. Studies using genetically modified mice have elucidated the mechanisms of cholesterol absorption in the intestine, lipolysis of apoB-containing lipoproteins, lipoprotein receptor function, and cholesterol efflux from macrophages to HDL. Although genetically modified animals are useful to elucidate the function of genes related to lipoprotein metabolism, we have to carefully select animal species to know the effect of these genes on lipid levels of whole plasma and lipoprotein profiles in humans. In addition, genetically modified mice have limitations in studies about the development of hypocholesterolemic agents, because of the expression of apobec-1 in liver and a deficiency of CETP activity in plasma. Consequently, these fundamental differences in lipoprotein metabolism between mice and humans affect the interpretation of results of gene modification about lipoprotein metabolism in mice. To solve these problems, genetically modified mice should be produced using CETP-transgenic/apobec-1-KO mice or animals having a background of no expression of apobec-1 in the liver and expression of CETP in plasma. We have to be careful in the interpretation of results obtained using genetically modified animals, and to select animal models in response to study purposes to extrapolate the results to humans. Recently, techniques of X-linked severe combined immunodeficiency (X-SCID) using zinc-finger nucleases, transcriptional activator-like effector nucleases (TALEN), and mutation using an N-ethyl-N-nitrosourea mutagenesis have become available for knockout gene expression. These techniques will be able to produce KO-animals other than mice. These animals will contribute further to studies of lipoprotein metabolism and lipid disorders in humans.

Author details

Masashi Shiomi, Tomonari Koike and Tatsuro Ishida
Kobe University Graduate School of Medicine, Japan

Acknowledgement

This work was supported in part by grants-in-aid for scientific research from the Ministry of Education, Culture, Sports and Technology, Japan (23300157).

7. References

Aalto-Setala, K. (1992). Transgenic animals in lipoprotein research. *Ann Med*, Vol.24, No.5, pp.405-409

Agellon, L.B.; Walsh, A.; Hayek, T.; Moulin, P.; Jiang, X.C.; Shelanski, S.A.; Breslow, J.L. & Tall, A.R. (1991). Reduced high density lipoprotein cholesterol in human cholesteryl ester transfer protein transgenic mice. *J Biol Chem*, Vol.266, No.17, pp.10796-10801

Allan, C.M. & Taylor, J.M. (1996). Expression of a novel human apolipoprotein (apoC-IV) causes hypertriglyceridemia in transgenic mice. *J Lipid Res*, Vol 37, No. 7, pp. 1510-1518

Altmann, S.W.; Davis, H.R.; Yao, X.; Laverty, M.; Compton, D.S.; Zhu, L.J.; Crona, J.H.; Caplen, M.A.; Hoos, L.M.; Tetzloff, G.; Priestley, T.; Burnett, D.A.; Strader, C.D. & Graziano, M.P. (2002). The identification of intestinal scavenger receptor class B, type I (SR-BI) by expression cloning and its role in cholesterol absorption. *Biochim Biophys Acta*, Vol.1580, No.1, pp. 77-93

Altmann, S. W.; Davis, H. R.; Zhu, L.; Yao, X.; Hoos, L. M.; Tetzloff, G.; Iyer S. N.; Maguire, M.; Golovko, A.; Zeng, M.; Wang, L.; Murgolo, N. & Graziano, M. P. (2004). Niemann-Pick C1 like 1 protein is critical for intestinal cholesterol absorption. *Science*, Vol.303, No.5661, pp.1201-1204

Asahina, M.; Mashimo, T.; Takeyama, M.; Tozawa, R.; Hashimoto, T.; Takizawa, A.; Ueda, M.; Aoto, T.; Kuramoto, K. & Serikawa, T. (2012). Hypercholesterolemia and atherosclerosis in low density lipoprotein receptor mutant rats. *Biochem Biophys Res Commun*, Vol. 418, No.3, pp. 553-558

Berge, K.E.; Tian, H.; Graf, G.A.; Yu, L.; Grishin, N.V.; Schultz, J.: Kwiterovich, P.; Shan, B.; Barnes, R. &, Hobbs, H.H. (2000). Accumulation of dietary cholesterol in sitosterolemia caused by mutations in adjacent ABC transporters. *Science*, Vol.290, No.5497, pp.1771-1775.

Bradley A, Evans M, Kaufman MH, & Robertson E. (1984). Formation of germ-line chimaeras from embryo-derived teratocarcinoma cell lines. *Nature*, Vol.309, No.5965, pp. 255-256

Braschi, S., Couture, N.; Gambarotta, A.; Gauthier, B.R.; Coffill, C.R.; Sparks, D.L.; Maeda, N. & Schultz, J.R. (1998). Hepatic lipase affects both HDL and ApoB-containing lipoprotein levels in the mouse. *Biochim Biophys Acta*, Vol.1392, No.2-3, pp.276-290

Broedl, U.C; Maugeais, C.; Marchadier, D.; Glick, J.M. & Rader, D.L. (2003). Effects of nonlipolytic ligand function of endothelial lipase on high density lipoprotein metabolism in vivo. J Biol Chem, Vol.278, No.42, pp.40688-40693

Buhman, K.K.; Accad, M.; Novak, S; Choi, R.S.; Wong, J.S.; Hamilton, R.L.; Turley, S. &, Farese, R.V. Jr. (2000). Resistance to diet-induced hypercholesterolemia and gallstone formation in ACAT2-deficient mice. *Nat Med*, Vol.6, No.12, pp.1341-1347

Chapman, M.J. (1980). Animal lipoproteins: Chemistry, structure, and comparative aspects. *J Lipid Res*. Vol.21, No., pp.789–853

Chiesa, G.; Hobbs, H.H.; Koschinsky, M.L.; Lawn R.M.; Maika, S.D. & Hammer, R.E. (1992). Reconstitution of lipoprotein(a) by infusion of human low density lipoprotein into transgenic mice expressing human apolipoprotein(a). *J Biol Chem*, Vol.267, No.34, pp. 24369-24374

Christoffersen, C.; Ahnström, J.; Axler, O.; Christensen, E.I.; Dahlbäck, B. & Nielsen, L.B. (2008). The signal peptide anchors apolipoprotein M in plasma lipoproteins and prevents rapid clearance of apolipoprotein M from plasma. J Biol Chem, Vol.283, No.27, pp.18765-18772

Christoffersen, C,; Jauhiainen, M.; Moser, M.; Porse, B.; Ehnholm, C.; Boesl, M.; Dahlbäck, B. & Nielsen, L.B. (2008). Effect of apolipoprotein M on high density lipoprotein metabolism and atherosclerosis in low density lipoprotein receptor knock-out mice. *J Biol Chem*, Vol.283, No.4, pp.1839-1847

Coleman, T.; Seip, R.L.; Gimble, J.M.; Kee, D.; Meda, N. & Semenkovich, C.F. (1995). COOH-terminal disruption of lipoprotein lipase in mice is lethal in homozygotes, but heterozygotes have, elevated triglycerides and impaired enzyme activity. *J Biol Chem*, Vol.270, No. 21, pp. 12518-12525

Davis, H.R. Jr.; Zhu, L.J.; Hoos, L.M.; Tetzloff, G.; Maguire, M.; Liu, J.; Yao, X.; Iyer, S.P.N.; Lam, M.H.; Lund, E.G.; Detmers, P.A.; Graziano, M.P. & Altman, S.W. (2004). Niemann-Pick C1 like 1 (NPC1L1) is the intestinal phytosterol and cholesterol transporter and a key modulator of whole-body cholesterol homeostasis. *J Biol Chem*, Vol. 279, No.32, pp. 33586-33592

Ding, Y.; Wang, Y.; Zhu, H.; Fan, J.; Liu, G. & Liu. E. (2011). Hypertriglyceridemia and delayed clearance if fat load in transgenic rabbits expressing human apolipoprotein CIII. *Transgenic Res*, Vol 20, No.4, pp.867-875

Duverger, N.; Kruth, H.; Emmanuel, F.; Caillaud, J.M.; Viglietta, C.; Castro, G.; Tailleux, A.; Fievet, C.; Fruchart, J.C.; Houdebine, L.M. & Denefle, P. (1996). Inhibition of atherosclerosis development in cholesterol-fed human apolipoprotein A-I-transgenic rabbits. *Circulation*. Vol.94, No.4, pp.713-717

Elsøe, S.; Ahnström, J.; Christoffersen, C.; Hoofnagle, A.N.; Plomgaard, P.; Heinecke, J.W.; Binder, C.J.; Björkbacka, H.; Dahlbäck, B. & Nielsen, L.B. (20129. Apolipoprotein M binds oxidized phospholipids and increases the antioxidant effect of HDL. *Atherosclerosis*, Vol.221, No.1, pp.91-97

Fan, J.; Araki, M.; Wu, L.; Challah, M.; Shimoyamada, H.; Lawn, R.M.; Kakuta, H.; Shikama, H. & Watanabe, T. (1999). Assembly of lipoprotein (a) in transgenic rabbits expressing human apolipoprotein (a). *Biochem Biophys Res Commun*, Vol.255, No.3, pp.639-644.

Fan, J.; Ji, Z.S.; Huang, Y.; de Silva, H.; Sanan, D.; Mahley, R.W.; Innerarity, T.L. & Taylor, J.M. (1998). Increased expression of apolipoprotein E in transgenic rabbits results in reduced levels of very low density lipoproteins and an accumulation of low density lipoproteins in plasma. *J Clin Invest*, Vol.101, No.10, pp.2151-2164

Fan, J.; McCormick, S.P.; Krauss, R.M.; Taylor, S.; Quan, R.; Taylor, J.M. & Young, S.G. (1995). Overexpression of human apolipoprotein B-100 in transgenic rabbits results in increased levels of LDL and decreased levels of HDL. *Arterioscler Thromb Vasc Biol*, Vol.15, No.11, pp.1889-1899

Fan, J.; Unoki, H.; Kojima, N.; Sun, H.; Shimoyamada, H.; Deng, H.; Okazaki, M.; Shikama, H.; Yamada, N. & Watanabe, T. (2001). Overexpression of lipoprotein lipase in transgenic rabbits inhibits diet-induced hypercholesterolemia and atherosclerosis. *J Biol Chem*, Vol.276, No.43, pp.40071-40079

Fan, J., Wang. J. Bensadoun, A., Lauer, S.J., Dang, Q., Mahley, R.W. & Taylor, J.M. (1994). Overexpression of hepatic lipase in transgenic rabbits leads to a marked reduction of plasma high density lipoprotein and intermediate density lipoproteins. *Proc Natl Acd Sci USA*,Vol. 91, No. 18, pp. 8724-8728

Fan, J. & Watanabe, T. (2003). Transgenic rabbits as therapeutic protein bioreactors and human disease models. *Pharmacology and Therapeutics*, Vol.99, No.3, pp. 261-282

Farese, R.V.Jr.; Cases, S.; Ruland, S.L.; Kayden, H.J.; Wong, J.S.; Young, S.G. & Hamilton, R.L. (1996). A novel function for apolipoprotein B: lipoprotein synthesis in the yolk sac is critical for maternal-fetal lipid transport in mice. *J Lipid Res*, Vol.37, No.2, pp.347-360

Farese, R.V.Jr.; Ruland, S.L.; Flynn, L.M.; Stokowski, R.P. & Young, S.G. (1995). Knockout of the mouse apolipoprotein B gene results in embryonic lethality in homozygotes and protection against diet-induced hypercholesterolemia in heterozygotes. *Proc Natl Acad Sci USA*, Vol .92, No.5, pp.1774-1778

Frykman, P.K.; Brown, M.S.; Yamamoto, T.; Goldstein, J.L. & Herz, J. (1995). Normal plasma lipoproteins and fertility in gene-targeted mice homozygous for a disruption in the gene encoding very low density lipoprotein receptor. *Proc Natl Acad Sci U S A*, Vol.92, Mo.18, pp.8453-8457

Garcia-Calvo, M.; Lisnock, J.; Bull, H.G.; Hawes, B.E.; Burnett, D.A.; Braun, M.P.; Crona, J.H.; Davis, H.R. Jr.; Dean. D.C.; Detmers, P.A.; Graziano, M.P.; Hughes, M.; Macintyre, D.E.; Ogawa, A.; O'neill, K.A.; Iyer, S.P.; Shevell, D.E.; Smith, M.M.; Tang, Y.S.; Makarewicz, A.M.; Ujjainwalla, F.; Altmann, S.W.; Chapman, K.T. & Thornberry, N.A. (2005). The target of ezetimibe is Niemann-Pick C1 like 1 (NPC1L1). *Proc Natl Acd Sci U S A*, Vol. 102, No. 23, pp. 8132-8137

Gautier, T.; Masson, D.; Jong, M.C.; Duverneuil, L.; Guern, N.L.; Deckert, V.; de Barros, J.P.; Dumont, L.; Bataille, A.; Zak, Z.; Jiang, X.; Tall, A.R.; Havekes, L.M. & Lagrost, L. (2002). Apolipoprotein CI deficiency markedly augments plasma lipoprotein changes mediated by human cholesterol ester transfer protein (CETP) in CETP transgenic/apoCI-knocked out mice. *J Biol Chem*, Vol 277, No.35, pp. 31354-31363

Goldstein, J.L.; Kita, T. & Brown, M.S. (1983). Defective lipoprotein receptors and atherosclerosis: Lessons from an animal counterpart of familial hypercholesterolemia. *N Engl J Med*, Vol. 309, No.5, pp. 288-296

Gonzalez-Navarro, H.; Nong, Z.; Amar, M.J.A.; Shamburek, R.D.; Najib-Fruchart, J.; Paigen, B.J.; Brewer, B. Jr. & Santamarina-Fojo, S. (2004). The ligand-binding function of hepatic lipase modulates the development of atherosclerosis in transgenic mice. *J Biol Chem*, Vol.279, No.44 , pp. 45312-45321

Gordon, J. W. & Ruddle, F.H. (1982). Germ line transmission in transgenic mice. *Progress in Clinical and Biological Research*, Vol.85, pp. 111-124.

Greeve, J.; Altkemper, I.; Dieterich, J.H.; Greten, H. & Windler, E. (1993). Apolipoprotein B mRNA editing in 12 different mammalian species: hepatic expression is reflected in low concentrations of apoB-containing plasma 1ipoproteins. *J Lipid Res*, Vol.34, No. 8, pp. 1367-1383.

Hammer, R.E., Pursel, V.G., Rexroad, C.E. Jr., Wall, R.J., Bolt, D.J., Ebert, K.M., Palmiter, R.D. & Brinster, R.L. (1985). Production of transgenic rabbits, sheep and pigs by microinjection. *Nature*, Vol.315, No.6021, pp. 680-683

Hayek, T.; Oiknine, J.; Brook, J.G. & Aviram, M. (1994). Role of HDL apolipoprotein E in cellular cholesterol efflux: studies in apo E knockout transgenic mice. *Biochem Biophys Res Commun*, Vol.205, No.2, pp.1072-11078

Herbert B, Patel D, Waddington SN, Eden ER, McAleenan A, Sun XM, Soutar AK. (2010). Increased secretion of lipoproteins in transgenic mice expressing human D374Y PCSK9 under physiological genetic control. *Arterioscler Thromb Vasc Biol*,Vol.30, No.7, pp.1333-1339

Herrera, V.L.; Makrides, S.C.; Xie, H.X.; Adari, H.; Krauss, R.M.; Ryan, U.S. & Ruiz-Opazo, N. (1999). Spontaneous combined hyperlipidemia, coronary heart disease and decreased survival in Dahl salt-sensitive hypertensive rats transgenic for human cholesteryl ester transfer protein. *Nat Med*, Vol.5, No.12, pp.1383-1389

Hoeg, J.M.; Santamarina-Fojo, S.; Bérard, A.M.; Cornhill, J.F.; Herderick, E.E.; Feldman, S.H.; Haudenschild, C.C.; Vaisman, B.L.; Hoyt, R.F.Jr.; Demosky, S.J.Jr.; Kauffman, R.D.; Hazel, C.M.; Marcovina, S.M. & Brewer, H.B. Jr. (1996). Overexpression of lecithin:cholesterol acyltransferase in transgenic rabbits prevents diet-induced atherosclerosis. *Proc Natl Acad Sci U S A*, Vol.93, No.21, pp.11448-11453

Hoeg, J. M.; Vaisman, B. L.; Demosky, S.J. Jr.; Meyn, S.M.; Talley, G.D.; Hoyt, R.F. Jr.; Feldman, S.; Berard, A.M.; Sakai, N.; Wood, D.; Brousseau, M.E.; Marcovina, S.; Brewer, H.B.Jr. & Santamarina-Hojo, S. (1996). "Lecithin:cholesterol acyltransferase overexpression generates hyperalpha-lipoproteinemia and a nonatherogenic lipoprotein pattern in transgenic rabbits. *J Biol Chem*, Vol.271, No.8, pp. 4396-4402

Hofmann, S.L.; Russell, D.W.; Brown, M.S.; Goldstein, J.L. & Hammer, R.E. (1988). Overexpression of low density lipoprotein (LDL) receptor eliminates LDL from plasma in transgenic mice. *Science*, Vol.239, No.4845, pp.1277-1281

Huang, Z.H.; Minshall, R.D. & Mazzone, T. (2009). Mechanism for endogenously expressed apoE modulation of adipocyte very low density lipoprotein metabolism. *J Biol Chem*, Vol.284, No.46, pp.31512-13522

Huijgen, R.; Fouchier, S.W.; Denoun, M.; Hutten, B.A.; Vissers, M.N.; Lambert, G. & Kastelein, J.J.P. (2012). Plasma levels of proprotein convertase subtilisin Kexin type 9 (PCSK9) and phenotypic variability in familial hypercholesterolemia. *J Lipid Res*, (in press)

Hussain, M.M.; Rava, P.; Walsh, M.; Rana, M. & Iqbal, J. (2012). Multiple functions of microsmal triglyceride transfer protein. *Nutr Metab*, Vol. 9, No. , pp. 14-

Ishibashi, S.; Brown, M.S.; Goldstein, J.L.; Gerard, R.D.; Hammer, R.E. & Herz, J. (1993). Hypercholesterolemia in low density lipoprotein receptor knockout mice and its reversal by adenovirus-mediated gene delivery. *J Clin Invest*, Vol.92, No.2, pp.883-893

Ishibashi, S., Goldstein, J.L., Brown, M.S., Herz, J. & Burns, D.K. (1994). Massive xanthomatosis and atherosclerosis in cholesterol-fed low density lipoprotein receptor-negative mice. *J Clin Invest*, Vol.93, No.5, pp.1885-1893

Ishida, T.; Choi, S.; Kundu, R.K.; Hirata, K.; Rubin, E.M.; Cooper ,A.D. & Quertermous, T. (2003). Endothelial lipase is a major determinant of HDL level. *J Clin Invest*, Vol.111, No.3, pp.347-355.

Ishida T, Choi SY, Kundu RK, Spin J, Yamashita T, Hirata K, Kojima Y, Yokoyama M, Cooper AD, Quertermous T. (2004). Endothelial lipase modulates susceptibility to atherosclerosis in apolipoprotein-E-deficient mice. J Biol Chem. Vol.279, No.43, pp45085-92.

Jiang, X.C.; Bruce, C.; Mar, J.; Lin, M.; Ji, Y.; Francone, O.L; & Tall, A.R. (1999). Targeted mutation of plasma phospholipid transfer protein gene markedly reduces high-density lipoprotein levels. *J Clin Invest*, Vol.103, No.6, pp.907-914

Jiang, X.; Francone, O.L.; Bruce, C.; Milne, R.; Mar, J.; Walsh, A.; Breslow, J.L. & Tall, A.R. (1996). Increased prebeta-high density lipoprotein, apolipoprotein AI, and phospholipid in mice expressing the human phospholipid transfer protein and human apolipoprotein AI transgenes. *J Clin Invest*, Vol.98, No.10, pp.2373-2380

Jiang, X.C.; Qin, S.; Qiao, C.; Kawano, K.; Lin, M.; Skold, A.; Xiao, X. & Tall, A.R. (2001). Apolipoprotein B secretion and atherosclerosis are decreased in mice with phospholipid-transfer protein deficiency. *Nat Med*, Vol.7, No.7, pp.847-852

Kennedy, M.A.; Barrera, G.C.; Nakamura, K.; Baldán, A.; Tarr, P.; Fishbein, M.C.; Frank, J.; Francone, O.L. & Edwards, P.A. (2005). ABCG1 has a critical role in mediating cholesterol efflux to HDL and preventing cellular lipid accumulation. *Cell Metab*, Vol.1, No.2, pp.121-131

Kendrick, J.S.; Chan, L. & Higgins, J.A. (2001). Superior role of apolipoprotein B48 over apolipoprotein B100 in chylomicron assembly and fat absorption: an investigation of apobec-1 knock-out and wild-type mice. *Biochem J*, Vol.356, No.3, pp. 821-827

Ko KW, Paul A, Ma K, Li L, Chan L. (2005). Endothelial lipase modulates HDL but has no effect on atherosclerosis development in apoE-/- and LDLR-/- mice. J Lipid Res. Vol.46, No.12. pp2586-94.

Koike, T.; Kitajima, S.; Yu, Y.; Li, Y.; Nishijima, K.; Liu, E.; Sun, H.; Wagar, A.B.; Shibata, N.; Inoue, T.; Wang, Y.; Zhang, B.; Kobayashi, J.; Morimoto, M.; Saku, K.; Watanabe, T. & Fan, J. (2009). Expression of human apoAII in transgenic rabbits leads to dyslipidemia: a new model for combined hyperlipidemia. *Arterioscler Thromb Vasc Biology*, Vol.29, No.12, pp. 2047-2053.

Koo, C.; Innerarity, T.L. & Mahley, R.W. (1985). Obligatory role of cholesterol and apolipoprotein E in the formation of large cholesterol-enriched and receptor-active high density lipoproteins. *J Biol Chem*, Vol.260, No.22, pp. 11934-11943

Krause, B.R. & Princen, H.M.G. (1998). Lack of predictablity of classical animal models for hypolipidemic activity: a ggod time for mice? Atherosclerosis, Vol 140, No.1 , pp.15-24

Li, X.; Catalina, F.; Grundy, S.M. & Patel, S. (1996). Method to measure apolipoprotein B-48 and B-100 secretion rates in an individual mouse: evidence for a very rapid turnover of VLDL and preferential removal of B-48- relative to B-100-containing lipoproteins. *J. Lipid Res*. Vol. 37, No.1, pp. 210-220.

Liu, R.; Iqbal, J.; Yeang, C.; Wang, D.Q.; Hussain, M.M. & Jiang, X.C. (2007). Phospholipid transfer protein-deficient protein-deficient mice absorb less cholesterol. *Arterioscler Thromb Vasc Biol*, Vol. 27, No.9, pp. 2014-2021

Liu, Y.; Millar, J.S.; Cronley, D.A,; Graham, M.; Crooke, R, Bilheimer, JT. & Rader, D.J. (2008). Knockout of acyl-Coa:diacylglycerol acyltransferase 2 with antisense oligonucleotide reduces VLDL TG and apoB secretion in mice. *Biochim Biophys Acta*, Vol.1781, No. 3, pp.97-104

Lo, C.M.; Nordskog, B.K.; Nauli, A.M.; Zheng, S.; Vonlehmden, S.B.; Yang, Q.; Lee, D.; Swift, L.L.; Davidson, N.O. & Tso, P. (2008). Why does the gut choose apolipoprotein B48 but not B100 for chylomicron formation? *Am J Physiol Gastrointest Liver Physio,* Vol.294, No.1, pp.G344-G352.

Ma K, Cilingiroglu M, Otvos JD, Ballantyne CM, Marian AJ, Chan L. (2003). Endothelial lipase is a major genetic determinant for high-density lipoprotein concentration, structure, and metabolism. Proc Natl Acad Sci U S A. Vol.100, No.2, pp2748-53.

Mabuchi, H.; Itoh, H.; Takeda, M.; Kajinami, K.; Wakasugi, T.; Koizumi, J.; Takeda, R. & Asagami, C. (1989). A young type III hyperlipoproteinemic patient associated with apolipoprotein E deficiency. *Metabolism,* Vol.38, No.2, pp.115-119.

Mardones, P.; Quiñones, V.; Amigo, L.; Moreno, M.; Miquel, J.F.; Schwarz, M.; Miettinen, H.E.; Trigatti, B.; Krieger, M.; VanPatten, S.; Cohen, D.E. & Rigotti, A. (2001). Hepatic cholesterol and bile acid metabolism and intestinal cholesterol absorption in scavenger receptor class B type I-deficient mice. *J Lipid Res,* Vol.42, No.2, pp.170-180

Marzal-Casacuberta, A.; Blanco-Vaca, F.; Ishida, B.Y.; Julve-Gil, J.; Shen, J.; Calvet-Márquez, S.; González-Sastre, F. & Chan, L. (1996). Functional lecithin:cholesterol acyltransferase deficiency and high density lipoprotein deficiency in transgenic mice overexpressing human apolipoprotein A-II. *J Biol Chem,* Vol.271, No.12, pp.6720-6728

Masson, D.; Deckert, V.; Gautier, T.; Klein, A.; Desrumaux, C.; Viglietta, C.; Pais de Barros, J.P.; Le Guern, N.; Grober, J.; Labbé, J.; Ménétrier, F.; Ripoll, P.J.; Leroux-Coyau, M.; Jolivet, G.; Houdebine, & Lagrost, L. (2011). Worsening of diet-induced atherosclerosis in a new model of transgenic rabbit expressing the human plasma phospholipid transfer protein. *Arterioscler Thromb Vasc Biol,* Vol.31,No.4, pp.766-774

Matsuura, F.; Wang, N.; Chen, W.; Jiang, X.C. & Tall, A.R. (2006). HDL from CETP-deficient subjects shows enhanced ability to promote cholesterol efflux from macrophages in an apoE- and ABCG1-dependent pathway. *J Clin Invest,* Vol.116, No.5, pp.1435-1442

Mbikay M, Sirois F, Mayne J, Wang GS, Chen A, Dewpura T, Prat A, Seidah NG, Chretien M, Scott FW. (2010). PCSK9-deficient mice exhibit impaired glucose tolerance and pancreatic islet abnormalities. *FEBS Lett,* Vol.584, No.4, pp.701-706

McNeish, J.; Aiello, R.J.; Guyot, D,; Turi, T.; Gabel, C.; Aldinger, C.; Hoppe, K.L.; Roach, M.L.; Royer, L.J.; de Wet, J.; Broccardo, C.; Chimini, G. & Francone, O.L. (2000). High density lipoprotein deficiency and foam cell accumulation in mice with targeted disruption of ATP-binding cassette transporter-1. *Proc Natl Acad Sci U S A,* Vol.97, No.8, pp.4245-4250

Mezdour, H.R.; Jones, R.; Dengremont, C.; Castro, G. & Maeda, N. (1997). Hepatic lipase deficiency increases plasma cholesterol but reduces susceptibility to atherosclerosis in apolipoprotein E-deficient mice. J. Biol. Chem., Vol.272, No.21, pp. 13570–13575Morrison, J.R.; Paszty, C.; Stevens, M.E.; Hughes, S.D.; Forte, T.; Scott, J. & Rubin, E.M. (1996). Apolipoprotein B RNA editing enzyme-deficient mice are viable despite alterations in lipoprotein metabolism. *Proc Natl Acad Sci USA,* Vol. 93, No.14, pp.7154-7159

Mulligan, J.D.; Flowers, M.T.; Tebon, A.; Bitgood, J.J.; Wellington, C.; Hayden, M.R. & Attie, A.D. (2003). ABCA1 is essential for efficient basolateral cholesterol efflux during the absorption of dietary cholesterol in chickens. *J Biol Chem,* Vol.278, No.15, pp.13356-3366

Newberry, E.P.; Kennedy, S.M.; Xie, Y.; Luo, J. & Davidson, N.O. (2009). Diet-induced alterations in intestinal and extrahepatic lipid metabolism in liver fatty acid binding protein knockout mice. *Mol Cell Biochem*, Vol 326, No. 1-2, pp. 79-86

Otera, H.; Ishida, T.; Nishiuma, T.; Kobayashi, K.; Kotani, Y.; Yasuda, T.; Kundu, R.K.; Quertermous, T.; Hirata, K. & Nishiuma, Y. (2009). Targeting inactivation of endothelial lipase attenuates lung allergic inflammation through raising plasma HDL level and inhibiting eosinophil infiltration. Am J Physiol Lung Cell Mol Physiol, Vol 296, No. 4, pp. L594-602

Out, R.; Hoekstra, M.; Habets, K.; Meurs, I.; de Waard, V.; Hildebrand, R.B.; Wang, Y.; Chimini, G.; Kuiper, J.; Van Berkel, T.J. & Van Eck, M. (2008). Combined deletion of macrophage ABCA1 and ABCG1 leads to massive lipid accumulation in tissue macrophages and distinct atherosclerosis at relatively low plasma cholesterol levels. *Arterioscler Thromb Vasc Biol*, Vol.28, No.2, pp.258-264

Piedrahita, J.A.; Zhang, S.H.; Hagaman, J.R.; Oliver, P.M. & Maeda N. (1992). Generation of mice carrying a mutant apolipoprotein E gene inactivated by gene targeting in embryonic stem cell. *Proc Natl Acd Sci USA*, Vol.89, No.10, pp. 4471-4475

Poirier, S.; Mayer, G.; Benjannet, S.; Bergeron, E.; Marcinkiewicz, J.; Nassoury, N.; Mayer, H.; Nimpf, J.; Prat, A. & Seidah, N.G. (2008). The proprotein convertase PCSK9 induces the degradation of low density lipoprotein receptor (LDLR) and its closest family members VLDLR and ApoER2. *J Biol Chem*, Vol.283, No.4, pp.2363-2372

Plump, A.S.; Azrolan, N.; Odaka, H.; Wu, L.; Jiang, X.; Tall, A.; Eisenberg, S. & Breslow, J.L. (1997). ApoA-I knockout mice: characterization of HDL metabolism in homozygotes and identification of a post-RNA mechanism of apoA-I up-regulation in heterozygotes. *J Lipid Res*, Vol.38, No.5, pp.1033-1047

Raabe, M.; Flynn, L.M.; Zlot, C.H.; Wong, J.S.; Veniant, M.M.; Hamilton, R.L. & Young, S.G. (1998). Knockout of the abetalipoproteinemia gene in mice: reduced lipoprotein secretion in heterozygotes and embryonic lethality in homozygotes. *Proc Natl Acd Sci USA*, Vol.95, No.15, pp.8686-8691

Rashid, S.; Curtis, D.E.; Garuti, R.; Anderson, N.N.; Bashmakov, Y. ; Ho, Y.K.; Hammer, R.E.; Moon, Y.A. & Horton, J.D. (2005). Decreased plasma cholesterol and hypersensitivity to statins in mice lacking Pcsk9. *Proc Natl Acad Sci USA*, Vol.102, No.15, pp.5374-5379

Reddick RL, Zhang SH, Maeda N. Atherosclerosis in mice lacking apo E. Evaluation of lesional development and progression. Arterioscler Thromb. 1994;14(1):141-7.

Rigotti, A.; Trigatti, B.L.; Penman, M.; Rayburn, H.; Herz, J. & Krieger, M. (1997). A targeted mutation in the murine gene encoding the high density lipoprotein (HDL) receptor scavenger receptor class B type I reveals its key role in HDL metabolism. *Proc Natl Acad Sci U S A*, Vol.94, No.23, pp.12610-12615

Rohrer, L.; Ohnsorg, R.M.; Landolt, F.; Rinninger, F. & von Eckardstein, A. (2009). High0density lipoprotein transport through aortic endothelial cells involves scavenger receptor BI and ATP-binding cassette transporter G1. *Cir Res*, Vol.104, No.10, pp.1142-1150

Rouy, D.; Duverger, N.; Lin, S.D.; Emmanuel, F.; Houdebine, L.M.; Denefle, P.; Viglietta, C.; Gong, E.; Rubin, E.M. & Hughes, S.D. (1998). Apolipoprotein(a) yeast artificial chromosome transgenic rabbits. Lipoprotein(a) assembly with human and rabbit apolipoprotein B. *J Biol Chem*, Vol. 273, No. 2, pp. 1247-1251

Sanan, D.A.; Newland, D.L.; Tao, R.; Marcovina, S.; Wang, J.; Mooser, V.; Hammer, R,E, & Hobbs, H.H. (1998). Low density lipoprotein receptor-negative mice expressing human apolipoprotein B-100 develop complex atherosclerotic lesions on a chow diet: No accentuation by apolipoprotein(a). *Proc Natl Acad Sci USA*; Vol.95, No. 8, pp. 4544-4549

Shachter, N.S.; Hayek, T.; Leff, T.; Smith, J.D.; Rosenberg, D.W.; Walsh, A.; Ramakrishnan, R.; Goldberg, I.J.; Ginsberg, H.N. & Breslow, J.L. (1994). Overexpression of apolipoprotein CII causes hypertriglyceridemia in transgenic mice. *J Clin Invest*, Vol.93, No.4, pp.1683-1690

Shimada, M.; Ishibashi, S.; Inaba, T.; Yagyu, H.; Harada, K.; Osuga, J.; Ohashi, K.; Yazaki, Y. & Yamada, N. (1996). Suppression of diet-induced atherosclerosis in low density lipoprotein receptor knockout mice overexpressing lipoprotein lipase. Proc Natl Acad Sci U S A, Vol.93, No.14, pp.7242-7246

Shimada, M.; Shimano, H.; Gotoda, T.; Yamamoto, K.; Kawamura, M.; Inaba, T.; Yazaki, Y. & Yamada, N. (1993). Overexpression of human lipoprotein lipase in transgenic mice. Resistance to diet-induced hypertriglyceridemia and hypercholesterolemia. *J Biol Chem*, Vol.268, No.24, pp.17924-17929

Shimano, H.; Yamada, N.; Katsuki, M.; Shimada, M.; Gotoda, T.; Harada, K.; Murase, T.; Fukazawa, C.; Takaku, F. & Yazaki, Y. (1992). Overexpression of apolipoprotein E in transgenic mice: marked reduction in plasma lipoproteins except high density lipoprotein and resistance against diet-induced hypercholesterolemia. *Proc Natl Acad Sci U S A*, Vol.89, No.5, pp.1750-1754

Shiomi, M. & Ito, T. (2001). MTP inhibitor decreases plasma cholesterol levels in LDL receptor-deficient WHHL rabbits by lowering the VLDL secretion. *Eur J Pharmacol* Vol.431, No.1, pp 127-131

Shiomi, M. & Ito, T. (2009). The Watanabe heritable hyperlipidemic (WHHL) rabbit, its characteristics and history of development: A tribute to the late Dr. Yoshio Watanabe. *Atherosclerosis*, Vol.207, No.1, pp. 1-7.

Shiomi, M.; Ito, T.; Tsukada, T.; Yata, T.; Watanabe, Y.; Tsujita, Y.; Fukami, M.; Fukushige, J.; Hosohawa, T. & Tamura, A. (1995). Reduction of serum cholesterol levels alters lesional composition of atherosclerotic plaques: Effect of pravastatin sodium on atherosclerosis in mature WHHL rabbits. *Arterioscler Theromb Vasc Biol*, Vol.15, No.11, pp.1938-1944

Shiomi, M.; Ito, T.; Yamada, S.; Kawashima, S. & Fan, J. (2003). Development of an animal model for spontaneous myocardial infarction (WHHLMI rabbit). *Arterioscler Thromb Vasc Biol*, Vol.23, No.7, pp. 1239-1244

Simonet, W.S.; Bucay, N.; Pitas, R.E.; Lauer, S.J. & Taylor, J.M. (1991). Mulyiple tissue-specific elements control the apolipoprotein E-C-I gene locus in transgenic mice. *J Biol Chem*, Vol.266, No. 14, pp. 8651-8654

Son, Y.S. & Zilversmit, D.B. (1989). Increased lipid transfer activities in hyperlipidemic rabbit plasma. *Arteriosclerosis*, Vol. 6, No.3, pp. 345-351

Strauss, J.G.; Frank, S.; Kratky, D.; Hammerle, G.; Hrzenjak, A.; Knipping, G.; von Eckardstein, A.; Kostner, G.M. & Zechner, R. (2001). Adenovirus-mediated rescueof lipoprotein lipase-deficient mice. Lipolysis of triglyceride-rich lipoproteins is essential for high density lipoprotein maturation in mice. *J Biol Chem*, Vol. 276, No.39, pp.36083–36090.

Suzuki, H.; Kurihara, Y.; Takeya, M.; Kamada, N.; Kataoka, M.; Jishage, K.; Ueda, O.; Sakaguchi, H.; Higashi, T.; Suzuki, T.; Takashima, Y.; Kawabe, Y.; Cynshi, O.; Wada, Y.; Honda, M.; Kurihara, H.; Aburatani, H.; Doi, T.; Matsumoto, A.; Azuma, S.; Noda, T.; Toyoda, Y.; Itakura, H.; Yazaki, Y.; Horiuchi, S.; Takahashi, K.; Kruijt, J.K.; van Berkel, T.J.C.; Steinbrecher, U.P.; Ishibashi, S.; Maeda, N.; Gordon, S. & Kodama, T. (1997). A role for macrophage scavenger receptors in atherosclerosis and susceptibility to infection. *Nature*, Vol.386, No.6622, pp. 292-296

Swanson, M. E.; Hughes, T. E.; Denny, I.S.; France, D.S.; Paterniti, J.R.Jr.; Tapparelli, C.; Gfeller, P. & Burki, K. (1992). High level expression of human apolipoprotein A-I in transgenic rats raises total serum high density lipoprotein cholesterol and lowers rat apolipoprotein A-I. *Transgenic Res*, Vol.1, No.3, pp. 142-147

Takahashi, S.; Ito, T.; Zenimaru, Y.; Suzuki, J.; Miyamori, I.; Takahashi, M.; Ishida, T.; Hirata, K.; Yamamoto, T.; Iwasaki. T.; Hattori, H. & Shiomi, M. (2011). Species differences of macrophage very low-density-lipoprotein (VLDL) receptor protein expression. *Biochem Biophys Res Commun*. Vol.407, No.4, pp.656-662

Takahashi, S.; Sakai, J.; Hattori, H.; Zenimaru, Y.; Suzuki, J.; Miyamori, I. & Yamamoto T. (2004). The very low-density lipoprotein (VLDL) receptor: characterization and function as a peripheral lipoprotein receptor. *J Atheroscler Thromb*, Vol.11, No.4, pp. 200-208

Teng, B.; Ishida, B.; Forte, T.M.; Blumenthal, S.; Song, L.Z.; Fotto, A.M. Jr, & Cham, L. (1997). Effective lowering of plasma, LDL, and esterified cholesterol in LDL receptor-knockout mice by adenovirus-mediated gene delivery of apoB mRNA editing enzyme (Apobec-1). *Arterioscler Thromb Vasc Biol*, Vol. 17, No. 5, pp. 889-897

Tsujita, Y.; Kuroda, M.; Shimada, Y.; Tanzawa, K.; Arai, M.; Kaneko, I.; Tanaka, M.; Masuda, H.; Tarumi, C.; Watanabe, Y. & Fujii, S. (1986). CS-514, a competitive inhibitor of 3-hydroxy-3-methylglutaryl coenzyme A reductase: Tissue-selective inhibition of sterol synthesis and hypolipidemic effect on various animal species. *Biochim Biophys Acta*, Vol.877, No.1, pp.50-60

Vaisman, B.L.; Klein, H.G.; Rouis, M.; Bérard, A.M.; Kindt, M.R.; Talley, G.D.; Meyn, S.M.; Hoyt, R.F.Jr.; Marcovina, S.M. ; Albers, J.J.; Hoeg, J.M.; Brewer, H.B.Jr. & Santamarin-Fojo, S. (1995). Overexpression of human lecithin cholesterol acyltransferase leads to hyperalphalipoproteinemia in transgenic mice. *J Biol Chem*, Vol.270, No.20, pp.12269-12275

Vaisman, B.L.; Lambert, G.; Amar, M.; Joyce, C.; Ito, T.; Shamburek, R.D.; Cain, W.J.; Fruchart-Najib, J.; Neufeld, E.B.; Remaley, A.T.; Brewer, H.B.J. & Santamarina-Fojo, S. (2001). ABCA1 overexpression leads to hyperalphalipoproteinemia and increased biliary cholesterol excretion in transgenic mice. *J Clin Invest*. Vol.108, No.2, pp.303–309

Walsh, A.; Ito, Y. & Breslow, J.L. (1989). High levels of human apolipoprotein A-I in transgenic mice result in increased plasma levels of small high density lipoprotein (HDL) particles comparable to human HDL3. *J Biol Chem*, Vol. 264, No.11, pp.6488-6494

Wang, N.; Arai, T.; Ji, Y.; Rinninger, F. & Tall. A.R. (1998). Liver-specific overexpression of scavenger receptor BI decreases levels of very low density lipoprotein ApoB, low density lipoprotein ApoB, and high density lipoprotein in transgenic mice. *J Biol Chem*, Vol.273, No.49, pp.32920-3296

Wang, Y.; Iordanov, H.; Swietlicki, E.A.; Wang, L.; Fritsch, C.; Coleman, T.; Semenkovich, C.F.; Levin, M.S. & Rubin, D.C. (2005). Targeted intestinal overexpression of the immediate early gene tis7 in transgenic mice increases triglyceride absorption and adiposity. *J Biol Chem*, Vol. 280, No. 41, pp. 34764-34775

Watanabe, Y. (1980). Serial inbreeding of rabbits with hereditary hyperlipidemia (WHHL-rabbit): Incidence and development of atherosclerosis and xanthoma. *Atherosclerosis*, Vol.36, No.2, pp.261-268

Watanabe, Y.; Ito. T.; Saeki, M.; Kuroda, M.; Tanzawa, K.; Mochizuki M.; Tsujita Y. & Arai, M. (1981). Hypolipidemic effects of CS-500 (ML-236B) in WHHL-rabbit, a heritable animal model for hyperlipidemia. *Atherosclerosis*, Vol. 38, No.1-2, pp.27-31

Watanabe, Y.; Ito, T.; Shiomi, M.; Tsujita, Y.; Kuroda, M.; Arai, M.; Fukami, M. & Tamura, A. (1988). Preventive effect of pravastatin sodium, a potent inhibitor of 3-hydroxy-3-methylglutaryl coenzyme A reductase, on coronary atherosclerosis and xanthoma in WHHL rabbits. *Biochim Biophys Acta*, Vol. 960, No.3, pp.294-302

Weinstock PH, Bisgaier CL, Aalto-Setälä K, Radner H, Ramakrishnan R, Levak-Frank S, Essenburg AD, Zechner R, Breslow JL. (1995). Severe hypertriglyceridemia, reduced high density lipoprotein, and neonatal death in lipoprotein lipase knockout mice. Mild hypertriglyceridemia with impaired very low density lipoprotein clearance in heterozygotes. J Clin Invest. Vol.96, No.6, pp2555-68.

Weng, W. & Breslow, J.L. (1996). Dramatically decreased high density lipoprotein cholesterol, increased remnant clearance, and insulin hypersensitivity in apolipoprotein A-II knockout mice suggest a complex role for apolipoprotein A-II in atherosclerosis susceptibility. *Proc Natl Acad Sci USA*, Vol.93, No.25, pp. 14788-14794

Xie, Y.; Newberry, E.P.; Young, S.G.; Robine, S.; Hamilton, R.L.; Wong, J.S.; Luo, J.; Kennedy, S. & Davidson, N.O. (2006). Compensatory increase in hepatic lipogenesis in mice with conditional intestine-specific Mttp deficiency. *J Biol Chem*, Vol.281, No.7, pp.4075-86

Yagyu, H.; Lutz, E.P.; Kako, Y.; Marks, S.; Hu, Y.; Choi, S.Y.; Bensadoun, A. & Goldberg, I.J. (2002). Very low density lipoprotein (VLDL) receptor-deficient mice have reduced lipoprotein lipase activity. Possible causes of hypertriglyceridemia and reduced body mass with VLDL receptor deficiency. *J Biol Chem*, Vol.277, No.12, pp.10037-10043

Yin, W.; Carballo-Jane, E.; McLaren, D.G.; Mendoza, V.H.; Gagen, K.; Geoghagen, N.S.; McNamara, L.A.; Gorski, J.N.; Eiermann, G.J.; Petrov, A.; Wolf, M.; Tong, X.; Wilsie, L.C.; Akiyama, T.E.; Chen, J.; Thankappan, A.; Xue, J.; Ping, X.; Andrews, G.; Wickham, L.A.; Gai, C.L.; Trinh, T. Kulick, A.A.; Donnelly, M.J.; Voronin, G.O.; Rosa, R.; Cumiskey, A.M.; Bekkari, K.; Mitnaul, L.J.; Puig, O.; Chen, F.; Raubertas, R.; Wong, P.H.; Hansen, B.C.; Koblan, K.S.; Roddy, T.P.; Hubbard, B.K. & Strack, A.M. (2012). Plasma lipid profiling across species for the identification of optimal animal models of human dyslipidemia. *J Lipid Res*, Vol 53, No.1, pp. 51-65.

Young, S. G.; Cham, C.M.; Pittas , R.E.; Burri, B.J.; Connolly, A.; Flynn, L.; Pappu, A.S.; Wong, J.S.; Hamilton, R.L. & Farese, R.V.Jr. (1995). A genetic model for absent chylomicron formation: mice apolipoprotein B in the liver, but not in the intestine. *J Clin Invest*, Vol.96, No. 6, pp.2932-2946

Zhang, S.H.; Reddick, R.L.; Piedrahita, J.A. & Maeda, N. (1992). Spontaneous hypercholesterolemia and arterial lesions in mice lacking apolipoprotein E. *Science*, Vol.258, No.5081, pp.468-471

Role of Lipoproteins in Neurodegenerative Diseases

Chapter 8

Lipoproteins and Apolipoproteins in Alzheimer's Disease

Etsuro Matsubara

Additional information is available at the end of the chapter

1. Introduction

Alzheimer's disease (AD) represents the so-called "storage disorder" of amyloid β (Aβ). The AD brain contains soluble and insoluble Aβ, both of which have been hypothesized to underlie the development of cognitive deficits or dementia (1-3). The steady-state level of Aβ is controlled by the generation of Aβ from its precursor, the degradation of Aβ within the brain, and transport of Aβ out of the brain. The imbalance among three metabolic pathways results in excessive accumulation and deposition of Aβ in the brain, which may trigger a complex downstream cascade (e.g., primary amyloid plaque formation or secondary tauopathy and neurodegeneration) leading to memory loss or dementia in AD. Accumulated lines of evidence indicate that such a memory loss represents a synaptic failure caused directly by soluble Aβ oligomers (AβOs) (4-6), whereas amyloid fibrils may cause neuronal injury indirectly via microglial activation (7). Many attentions are paid to understand the mechanism underlying the neurotoxic action of AβOs so far. However, the exact metabolic conditions controlling the *in vivo* generation of soluble AβOs has been out of attention.

Several lines of evidence indicated that lipidic environments in the central nervous system (CNS) represent one of the prevailing metabolic conditions. We then hypothesized that an alteration of the lipoprotein-soluble Aβ interaction in the CNS is capable of initiating and/or accelerating the cascade favoring Aβ assembly (8). We found that dissociation of Aβ42 from lipoprotein in the cerebrospinal fluid from AD accelerates Aβ42 assembly (9). Thus, lipoprotein is a key molecule to maintain monomeric soluble Aβ42 in CNS.

In this chapter, we review the issue regarding how lipoprotein and apolipoproteins contribute to physiological metabolic conditions. Then, we focus on how they constitute the

AD-related metabolic conditions in the CNS. We are certain that these points of view introduce a novel approach to find a therapeutic intervention for AD.

2. Lipoproteins, apolipoproteins, and Aβ metabolism in the CNS

In the CNS, we need to be aware that cholesterol metabolism is quite different from that in systemic circulation. Lipidic environments in the CNS were regulated by HDL-like lipoproteins, mainly lipidated apolipoprotein E (apoE), which is in charge of cholesterol transport to and from neurons (10, 11). This is also the case in lipidated apolipoprotein J (apoJ) (12). In addition to lipid trafficking, apoE or apoJ as a form of HDL-like lipoprotein plays a major role in Aβ metabolism in the CNS. Both apolipoproteins are well known as major carrier proteins for Aβ (13-17). Interestingly, transgenic mouse models of AD (apoE$^{-/-}$/apoJ$^{-/-}$) revealed that both apolipoproteins regulate in a cooperative manner the clearance and the deposition of Aβ in brain (18). The hypothetical pathways involved in the clearance of CNS Aβ are efflux of Aβ into the plasma via blood-brain barrier (BBB). Two lipoprotein-receptors, LRP-1 and LRP-2, seem to be responsible for efflux of lipoprotein-free or lipoprotein-associated (apoJ-associated) Aβ from the brain to blood, respectively (19). *In vivo* relevance of LRP-1-mediated Aβ transport has been confirmed in transgenic mice expressing low LRP-1-receptor and APP, which develops extensive Aβ accumulation much faster than transgenic mice expressing high level of APP (20). Reduced expression of brain endotherial LRP-1 was also observed in AD patients, which was associated with impaired Aβ clearance and cerebrovascular accumulation. LPR-2 appeared to function bi-directionally (influx vs efflux) at BBB. In contrast to LRP2-mediated influx (21), LPR2-mediated efflux of brain Aβ was actively operated under physiological concentration of either Aβ or apoJ (19). Interestingly, a recent study shows that apoE4 binding to Aβ redirects its clearance from LRP-1 to VLDLR, which resulted in slower efflux of brain Aβ than LRP-1 (22). In contrast, apoE2-Aβ and apoE3-Aβ complexes are cleared at BBB via both LPR-1 and VLDLR at a substantially faster rate than apoE4-Aβ complexes(22). Impairment of the above-mentioned receptor-mediated clearance at BBB could contribute to the pathogenesis of AD. Alternatively, ApoE4-HDL shows less cholesterol exchange between lipid particles and the neuronal membrane as compared with apoE3-HDL (23), leading to altered membrane functions, e.g., signal transduction, enzyme activities, ion channel properties, and conformation of sAβ peptides, which contribute to the disease-related metabolic conditions. Furthermore, when the generation of HDL-like lipoproteins in the AD mouse model is suppressed or overexpressed via the specific regulation of ATP-binding cassette A1 (ABCA1), Aβ deposition exhibits augmentation or reduction, respectively, which depends on the degree of ABCA1-mediated lipidation of apoE in the CNS (24, 25). From these points of view, lipidic environments in the CNS represent one of the prevailing metabolic conditions. We hypothesized that an alteration of the lipoprotein-sAβ interaction in the CNS is capable of initiating and/or accelerating the cascade favoring Aβ assembly. Thus, we postulate that lipoproteins or apolipoproteins may regulate the metabolic conditions controlling the *in vivo* generation of soluble AβOs.

3. Aß is present in either lipoprotein-free or lipoprotein-associated form in brain parenchyma

To assess the above-mentioned issue, we examined whether the dissociation of sAß from lipoprotein-particles occurs in the brain. The combination of size exclusion chromatography (SEC) and enzyme-linked immunosorbent assay (ELISA) revealed that the dissociation of sAß from lipoprotein-particles occurs in brain parenchyma and the presence of soluble dimeric lipoprotein-free Aß in AD brains (8). These findings may support the hypothesis that functionally declined lipoproteins may be major determinants in the production of metabolic conditions leading to higher levels of soluble dimeric SDS-resistant form of Aβ in AD brains (8, 26). At this moment, it remains undetermined whether dissociation of Aβ from lipoprotein or less association of Aβ to lipoproteins accounts for such a metabolic conditions. To further verify this hypothesis, we focused on the entorhinal cortex (EC), followed by biochemical analyses using an anti-oligomer specific antibody, namely 2C3 (9, 27). Fifty brains obtained from healthy elderly are composed of three Braak NFT stages; Braak NFT stages I-II (n=35, normal control); Braak NFT stages III-IV (n=13, MCI stage); Braak NFT stages IV-V (n=2, AD stages). Immunoblot analysis of the delipidated EC employing monoclonal 2C3 revealed that the accumulation of soluble 12-mers precedes the appearance of neuronal loss or cognitive impairment, and is enhanced as the Braak neurofibrially tangle (NFT) stages progress, indicating that the ECs of AD patients indeed bear metabolic conditions that accelerate Aβ assembly.

4. Aß is present in either lipoprotein-free or lipoprotein-associated form in cerebrospinal fluid (CSF)[9]

The presence of lipoprotein-free sAβOs in CSF was also assessed in age-matched normal controls (NCs) and patients with Alzheimer's disease (AD) by SEC and ELISA specific for either AβOs or AβMs. The SEC experiment using pooled CSF revealed that the dissociation of sAβMs from lipoprotein particles indeed occurs in CSF, which was lower in AD than in NCs. Furthermore, the SEC experiment using lipoprotein-depleted pooled CSF (LPD-CSF) confirmed the presence of oligomeric 2C3 conformers (4- to 35-mers), which appeared to be higher in AD patients than in NCs. To address the issue on the presence of any metabolic conditions favoring Aβ assembly, we compared the levels of lipoprotein-free sAβMs and sAβOs in LPD-CSF from the 12 sporadic AD patients and 13 NCs to evaluate the AβOs/AβMs ratio (the O/M index). The levels of 2C3 oligomeric conformers composed of Aβ42 are significantly higher in AD patients than in NCs. The O/M index for either Aβ42 or Aβ40 is also significantly higher in AD patients than in NCs. Of note, the relative amounts of total lipoprotein-associated sAβMs (~70%) versus lipoprotein-free sAβMs (~30%) remained

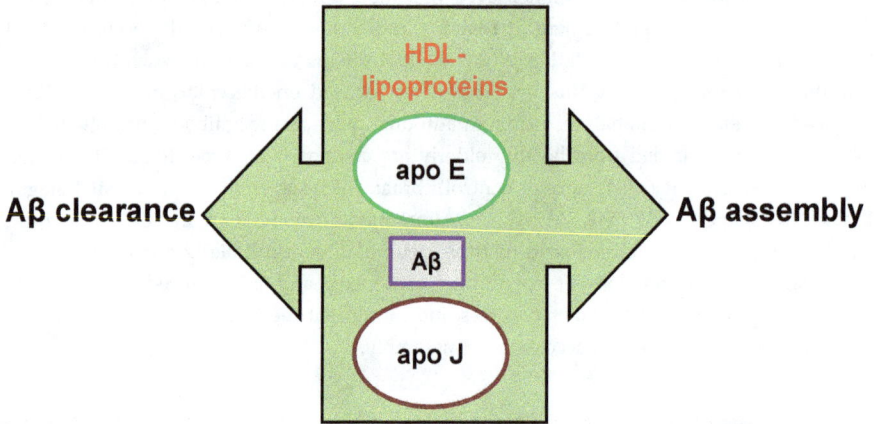

Figure 1. Hypothetical metabolic conditions favoring Aβ assembly. Functionally declined lipoproteins may accelerate the generation of metabolic conditions leading to higher levels of soluble Aβassembly in the CNS.

essentially unchanged in sporadic AD patients as compared with NCs. However, the relative amounts of lipoprotein-free Aβ42 was significantly lower in the sporadic AD patients (9.3 ± 3.9 %) than in NCs (13.2 ± 4.5 %), which is in accordance with our above-mentioned finding that the level of oligomeric 2C3 conformers composed of Aβ42 was significantly elevated in AD patients. Thus, it is likely that the conversion of lipoprotein-free monomeric soluble Aβ42 into oligomeric assembly preferentially occurs in AD CSF, mirroring the disease-related metabolic conditions in the brain parenchyma.

5. Summary

We previously reported that ~90% of sAβMs that circulate in normal plasma is associated with lipoprotein particles (27). From the above data, it is plausible to assume that about 70% of CSF sAβMs is normally associated with lipoprotein particles, indicating that CNS constitutes a risky environment where the lipoproteins-sAβMs interaction is impaired, leading to Aβ assembly. From this point of view, a key molecule to maintain monomeric sAβ42 metabolism in CNS appears to be HDL-like lipoprotein particles. In this sense, the dissociation of sAβ42 from or the lack of association with HDL-like lipoprotein particles not only constitutes a potential mechanism to initiate and/or accelerate the cascade favoring Aβ42 assembly in the brain, but also results in a reduced clearance of physiological lipoprotein-associated sAβ42 peptides in the brain. Thus, above-mentioned CNS environments may strongly affect conformation of sAβ peptides, resulting in the conversion of sAβ42 monomers into sAβ42 assembly. The findings suggest that functionally declined lipoproteins may accelerate the generation of metabolic conditions leading to higher levels of sAβ42 assembly in the CNS.

Author details

Etsuro Matsubara
Department of Neurology, Institute of Brain Science, Hirosaki Graduate School of Medicine, Japan

6. References

[1] Hardy J, Allsop D: Amyloid deposition as the central event in the aetiology of Alzheimer's disease. *Trends Pharmacol Sci* 1991, 12: 383-388.

[2] Lue LF, Kuo YM, Roher AE, Brachova L, Shen Y, Sue L, Beach T, Kurth JH, Rydel RE, Rogers J: Soluble amyloid β peptide concentration as a predictor of synaptic change in Alzheimer's disease. *Am J Pathol* 1999, 155: 853-862.

[3] McLean CA, Cherny RA, Fraser FW, Fuller SJ, Smith MJ, Beyreuther K, Bush AI, Masters CL: Soluble pool of Abeta amyloid as a determinant of severity of neurodegeneration in Alzheimer's disease. *Ann Neurol* 1999, 46: 860-866.

[4] Klein WL, Krafft GA, Finch CE: Targeting small Abeta oligomers: the solution to an Alzheimer's disease conundrum? *Trends Neurosci* 2001, 24: 219-224.

[5] Selkoe DJ: Alzheimer's disease is a synaptic failure. *Science* 2002, 298: 789-791.

[6] Hass C, SelkoeDJ: Soluble protein oligomers in neurodegeneration: lessons from the Alzheimer's amyloid β-peptide. *Nat Rev Mol Cell Biol* 2007, 8: 101-112.

[7] Akiyama H, Barger S, Barnum S, Bradt B, Bauer J, Cole GM, Cooper NR, Eikelenboom P, Emmerling M, Fiebich BL, Finch CE, Frautschy S, Griffin WS, Hampel H, Hull M, Landreth G, Lue L, Mrak R, Mackenzie IR, McGeer PL, O'Banion MK, Pachter J, Pasinetti G, Plata-Salaman C, Rogers J, Rydel R, Shen Y, Streit W, Strohmeyer R, Tooyoma I, Van Muiswinkel FL, Veerhuis R, Walker D, Webster S, Wegrzyniak B, Wenk G, Wyss-Coray T: Inflammation and Alzheimer's disease. *Neurobiol Aging* 2000, 21: 383-421.

[8] Matsubara E, Sekijima Y, Tokuda T, Urakami K, Amari M, Shizuka-Ikeda M, Tomidokoro Y, Ikeda M, Kawarabayashi T, Harigaya Y, Ikeda S, Murakami T, Abe K, Otomo E, Hirai S, Frangione B, Ghiso J, Shoji M. 2004. Soluble Abeta homeostasis in AD and DS: impairment of anti-amyloidogenic protection by lipoproteins. *Neurobiol Aging* 25:833-841.

[9] Takamura A, Kawarabayashi T, Yokoseki T, Shibata M, Morishima-Kawashima M, Saito Y, Murayama S, Ihara Y, Abe K, Shoji M, Michikawa M, Matsubara E. The Dissociation of Aβ from Lipoprotein in Cerebrospinal Fluid from Alzheimer's Disease accelerates Aβ42 assembly. J Neurosci Res. 2011;89(6):815-821.

[10] Michikawa M, Gong JS, Fan QW, Sawamura N, Yanagisawa K. 2001. A novel action of alzheimer's amyloid beta-protein (Abeta): oligomeric Abeta promotes lipid release. *J Neurosci* 21:7226-7235.

[11] Gong JS, Sawamura N, Zou K, Sakai J, Yanagisawa K, Michikawa M. 2002. Amyloid beta-protein affects cholesterol metabolism in cultured neurons: implications for pivotal role of cholesterol in the amyloid cascade. *J Neurosci Res* 70:438-46.

[12] DeMattos RB, Brendza RP, Heuser JE, Kierson M, Cirrito JR, Fryer J, Sullivan PM, Fagan AM, Han X, Holtzman DM. Purification and characterization of astrocyte-secreted apolipoprotein E and J-containing lipoproteins from wild-type and human apoE transgenic mice. Neurochem Int. 2001, 39(5-6):415-25.

[13] Ghiso J, Matsubara E, Koudinov A, Choi-Miura NH, Tomita M, Wisniewski T, Frangione B. The cerebrospinal-fluid soluble form of Alzheimer's amyloid ß is complexed to SP-40,40 (apolipoprotein J), an inhbitor of the complement membrane-attack complex. Biochem J.,1993;293:27-30.

[14] Wisniewski T, Golabek A, Matsubara E, Ghiso J, Frangione B. Apolipoprotein E: binding to soluble Alzheimer's ß –amyloid. Biochem Biophys Res Commun., 1993;192:359-365.

[15] Koudinov A, Matsubara E, Frangione B, Ghiso J. The soluble form of Alzheimer's amyloid ß protein is complexed to high density lipoprotein 3 and very high density

lipoprotein in normal human plasma. Biochem Biophys Res Commun., 1994;205:1164-1171.

[16] Matsubara E, Frangione B, Ghiso J. Characterization of apolipoprotein J-Alzheimer's A beta interaction. *J Biol Chem.* 1995, 270:7563-7567.

[17] Matsubara E, Soto C, Governale S, Frangione B, Ghisom J. Apolipoprotein J and Alzheimer's amyloid ß solubility. Biochem J, 1996;316:671-679.

[18] DeMattos RB, Bales KR, Cummins DJ, Paul SM, Holtzman DM. Brain to plasma amyloid-beta efflux: a measure of brain amyloid burden in a mouse model of Alzheimer's disease. Science 295: 2264-2267, 2002

[19] Bell RD, Sagare AP, Friedman AE, Bedi GS, Holtzman DM, Deane R, Zlokovic BV. Transport pathways for clearance of human Alzheimer's amyloid beta-peptide and apolipoproteins E and J in the mouse central nervous system. J Cereb Blood Flow Metab. 27:909-918, 2007.

[20] Shibata M, Yamada S, Kumar SR, Calero M, Bading J, Frangione B, Holtzman DM, Miller CA, Strickland DK, Ghiso J, Zlokovic BV. Clearance of Alzheimer's amyloid-ss(1-40) peptide from brain by LDL receptor-related protein-1 at the blood-brain barrier. J Clin Invest. 2000 Dec;106(12):1489-99.

[21] Shayo M, McLay RN, Kastin AJ, Banks WA. The putative blood-brain barrier transporter for the beta-amyloid binding protein apolipoprotein J is saturated at physiological concentration. Life Sci 60: 115-118, 1997.

[22] Deane R, Sagare A, Hamm K, Parisi M, Lane S, Finn MB, Holtzman DM, Zlokovic BV. apoE isoform-specific disruption of amyloid beta peptide clearance from mouse brain. J Clin Invest. 2008 Dec;118(12):4002-13.

[23] Zou K, Gong JS, Yanagisawa K, Michikawa M. 2002. A novel function of monomeric amyloid beta-protein serving as an antioxidant molecule against metal-induced oxidative damage. J Neurosci 22:4833-4841.

[24] Wahrle SE, Jiang H, Parsadanian M, Hartman RE, Bales KR, Paul SM, Holtzman DM. 2005. Deletion of Abca1 increases Abeta deposition in the PDAPP transgenic mouse model of Alzheimer disease. J Biol Chem 280:43236-43242.

[25] Wahrle SE, Jiang H, Parsadanian M, Kim J, Li A, Knoten A, Jain S, Hirsch-Reinshagen V, Wellington CL, Bales KR, Paul SM, Holtzman DM. 2008. Overexpression of ABCA1 reduces amyloid deposition in the PDAPP mouse model of Alzheimer disease. J Clin Invest 118:671-682.

[26] Matsubara E, Ghiso J, Frangione B, Amari M, Tomidokoro Y, Ikeda Y, Harigaya Y, Okamoto K, Shoji M. 1999. Lipoprotein-free amyloidogenic peptides in plasma are elevated in patients with sporadic Alzheimer's disease and Down's syndrome.Ann Neurol 45:537-541.

[27] Takamura A, Okamoto Y, Kawarabayashi T, Yokoseki T, Shibata M, Mouri A, Nabeshima T, Sun H, Abe K, Shoji M, Yanagisawa K, Michikawa M, Matsubara E.

Extracellular and Intraneuronal HMW-AbetaOs Represent a Molecular Basis of Memory Loss in Alzheimer's Disease Model Mouse. Mol Neurodegener. 2011;20: 6.

Chapter 9

Genetics of Ischemic Stroke: Emphasis on Candidate-Gene Association Studies

Sanja Stankovic, Milika Asanin and Nada Majkic-Singh

Additional information is available at the end of the chapter

:

1. Introduction

Stroke is the leading cause of neurological disability and among the leading causes of death worldwide. It is a focal neurological deficit that results from events that decrease or stop cerebral blood flow. As the consequence neurons cease functioning and irreversible neuronal ischemia and injury occur.

Broadly, strokes are classified into two main types-ischemic and hemorrhagic. Ischemic stroke (IS) is characterized by blockage in blood flow to a focal area of the brain, until hemorrhagic stroke is caused by bleeding into the brain. Acute IS is more common than hemorrhagic stroke. Although according the previous literature data about 80% of strokes were ischemic, the retrospective review from a stroke center found that about 60% were ischemic [1]. Except their causes and pathophysiology ischemic and hemorrhagic types differ in their treatments and outcomes [2].

Based on the system of categorizing stroke developed in multicenter Trial of Org 10172 in Acute Stroke Treatment (TOAST), IS may be divided into the following major subtypes: large artery infarction, small-vessel (lacunar) infarction, and cardioembolic infarction. This classification on the basis of inferred origin of cerebrovascular occlusion [3] is the most frequently used. Other studies used systems based on clinical presentation or location and size of the lesion within the brain (such as the Oxfordshire Community Stroke Project system) [4]. It classifies patients in five infarct types: cerebral infarction, lacunar infarct, total anterior circulation infarct, partial anterior circulation infarct, and posterior circulation infarcts. Many other classifications have been proposed, such as those from the Lausanne Stroke Registry and the Étude du profil Génétique de l'Infarctus Cérébral (GÉNIC) study [5,6]. The first one included atherosclerosis with stenosis, atherosclerosis without stenosis, emboligenic heart disease, hypertensive arteriopathy, cerebrall hemorrhage, mixed causes and undetermined causes. The former included atherothrombotic stroke, cardioembolic stroke, lacunar stroke, arterial dissection, unknown causes stroke. Although stroke is often considered a disease of elderly persons, one third of strokes occur in persons younger than 65 years.

Risk factors for IS includes modifiable and non-modifiable etiologies. Non-modifiable risk factors include: age, sex, race, ethnicity, heredity, etc. Modifiable risk factors include the followings: hypertension, diabetes mellitus, hypercholesterolemia, atrial fibrillation, lifestyle factors, etc. Unfortunately, modifiable risk factors accounts for only approximately 60% of the population-attributable risk for stroke [7].

2. Genetic risk factors in stroke

Evidence continues to accumulate to suggest important roles for genetic factors in stroke. Genetic risk factors are particularly interesting, because they can offer a direct clue to the biological pathways involved. Genetic factors might affect stroke risk at various levels. They could act through conventional risk factors, interact with conventional and environmental risk factors, or contribute directly to an established stroke mechanism. They could further affect the latency of stroke or infarct size, and stroke outcome [8]. Stroke may be the outcome of a number of monogenic disorders or, more commonly, a polygenic multifactorial disease.

Evidence shows that genetic factors are more important in small- and large-vessel stroke than in cardioembolic stroke [9,10]. Some intermediate phenotypes also exhibit high heritability, such as carotid intima-medial wall thickness and white-matter lesions [8].

Genetic predisposition to stroke has been proven in animal models and in humans (twins, affected sibling pair, families). Several studies demonstrated higher rates of stroke among relatives of patients who died from stroke than among relatives of healthy control subjects. In a large study of stroke patients and age and sex matched controls, the odds ratios (ORs) of having a family history of stroke were 2.24 for large vessel-disease and 1.93 for small vessel disease [9]. Twin studies have confirmed a significant genetic component to stroke, with the stroke prevalence fivefold higher in monozygotic than in dizygotic twins [11]. Touze and Rothwell [12] in a meta–analysis based on 18 studies confirmed sex differences in heritability of IS; women with stroke were more likely than men to have a parental history of stroke, which is accounted for by an excess maternal history of stroke. Also, genetic predisposition could differ depending on age and IS subtype.

The initial expectancy to find only one or a few common mutations that substantially contribute to the risk of IS shifted toward the hypothesis of a large number of small-effect genetic variants with complex gene-gene and gene-environment interactions. The first approach used in identification of genetic variants contributing to stroke was linkage studies. Linkage analysis relies on the cosegregation of known polymorphic DNA marker with nearby, unknown disease-causing alleles in families. This approach was successful in monogenic diseases, but was less successful in the identification of genetic loci that contribute to the occurrence of polygenic stroke. The second approach was candidate gene approach.

2.1. Candidate-gene association studies of ischemic stroke

Until recently, candidate gene approach was the most common in genetic investigation of IS. A gene identified as a "candidate" is hypothesized to be involved in IS risk, and then, genetic variants, usually single nucleotide polymorphisms (SNPs), are identified within that gene. The SNPs are selected on the basis of their localization in genes which encode proteins with a known function in a biological pathway implicated in the pathophysiology of the disease. Then, the frequency of the SNPs is determined in a series of cases and controls and the obtained results are compared. They use a case-control study design. A gene variant that is more common in patients than in controls may cause stroke or be located close to the true causal variant.

Genes encoding products involved in lipid metabolism, thrombosis, and inflammation are believed to be potential genetic factors for IS [13-15]. Although a large group of candidate genes have been studied, most of the epidemiological results are conflicting. Especially great interest is shown in exploring potential links between polymorphisms in genes encoding proteins involved in lipid metabolism and the risk of IS.

This chapter summarize the results of meta-analyses and case-control studies assessing the linkage of specific candidate genes with the risk of IS and specific subtypes. Electronic databases (Medline (http://www.ncbi.nlm.nih. gov/pubmed/), Embase (http://www. embase.com/), Google Scholar (http://scholar.google.com/), Yahoo (http://www.yahoo.com/), Kobson (http://www.kobson. nb.rs/) were searched until March 2012 and the obtained results were included in the text.

It is very well known that individuals with higher levels of plasma cholesterol, decreased high-density lipoprotein (HDL) and increased low-density lipoprotein (LDL) have a higher risk of premature atherosclerosis. The phenotype may arise not only from single gene disorders, but also from a number of genetic and environmental factors, including polymorphic variants of genes encoding the apolipoproteins, lipoprotein receptors and the key enzymes of plasma lipoprotein metabolism.

Apolipoprotein E. One of the most intensively investigated candidate genes for IS that received widespread attention is the apolipoprotein (apo) E gene. It forms a cluster with certain apoC genes on the long arm of chromosome 19 (19q13.2). The human apoE gene is polymorphic, with three common alleles (ε2, ε3, ε4) coding for three isoforms (E2, E3, E4). The association studies of apoE gene polymorphisms with IS gave conflicting results based on 9 meta-analyses [16-24] and 77 case-control studies [25-101]. In small case-control or cross-sectional studies, both the ε2ε3 genotype and the ε4 allele have been over-represented in patients with IS. Other groups have examined the role of the apoE genotype in modulating the outcome of cerebral infarction as this lipoprotein appears to be an important regulator of lipid turnover within the brain and of neuronal membrane maintenance and repair. McCarron et al. [102] found a favorable effect of the ε4 allele on stroke outcome. Stankovic and colleagues [85] reviewed the conflicting results on the importance of the apoε alleles in predisposition to IS.

Seven meta–analyses [17-19,21-24] gave a positive association between the ε4 allele and IS. The first one [22], published in 1999, revealed a significantly higher apoε4 allele frequency in affected patients compared with controls (OR 1.68, 95% CI 1.36–2.09, P<0.001). In the next decade, five meta–analyses [17-19, 21,23] confirmed that ε4 allele carriers have a higher risk of IS compared with pooled ε2 and ε3 allele carriers in European populations, persons of non-European descent, Asians, Han Chinese and persons with early-onset IS. Performing large-scale meta-analysis (10674 cases/33430 controls) consisted of four meta-analysis [19,21-23] and 9 case-control studies [33,35,36,54,59,65,66,84,88], Hamzi et al. [24] calculated OR for the apoε4 allele to be 0.95 (95%CI 0.77-1.14, P=0.002).

Approximately half of all case–control studies [26,27,29,33,38,41,45,47,49,51,53,54,57,58, 60,64,67,69,71,73-76,78-80,82,84,85,89-91,93-96,99,100] showed an increased frequency of the ε4 allele in stroke patients, making it a highly probable risk factor for IS; in four, significant association with large-vessel IS was observed. Three groups described the ε2 allele as a risk factor for IS [76,85,94]. The status of the E2/3 genotype as a protective or risk factor is controversial. One report [100] demonstrated a protective role of the ε4 allele for small-vessel disease, and another [93] concluded that the E3/4 genotype could be a risk factor for lacunar stroke compared with the E3/3 genotype.

Several SNPs have been described in the 5' regulatory region (c.491A>T, c.427T>C, c.219G>T, and c.113G>C), but current information is very preliminary. A higher risk of IS was associated with the G allele of the tightly linked c.219G>T and c113G>C promoter polymorphisms [96], and with the T allele of c.427T>C polymorphism [94]. One paper [94] reported the C allele of c.427T>C polymorphism as protective for IS.

Other apolipoproteins. Except apoE gene polymorphism that was frequently investigated polymorphism in patients with IS, another apolipoprotein genes have undergone intense investigation (apo AI/CIII, apoAIV, apoAV, apoB, apoH). The most published studies investigating the relationship between these polymorphisms and IS are small in sample size and inconclusive in their results.

Some authors have studied the association between IS and DNA polymorphisms in apoAI gene (*SstI* (rs5128), *MspI*, c.75G>A, c.84T>C), apoCIII gene (c.641C>A, c.482 C>T, c.455C>T, c.1100C>T, c3175C>G, c3206T>G), apoAIV (p.Thr347Ser, p.Gln360His), and apoH (c.1025G>C, c.341G>A), mainly with negative results [28,30,31,34,52,103,104].

The apoB gene is located on chromosome 2q23, spanning approximately 43 kb and has 29 exons and 28 introns. ApoB polymorphisms (T71I (c>t; rs17246849), A591V (c>t; rs17240681), *Bfa*I (P2712L; c>t; rs17240903), *Msp*I (R3611Q; g>a; rs17247291), *Eco*RI (E4154K; g>a; rs1042031), and *Eco*57I (N4311S; a>g; rs17240958), p.Arg3500Gln, c.4311A>G) were examined in patients with IS. Only two studies found that apoB polymorphisms [105,106] were associated with IS risk. Zhang et al. [107] found that C7673T polymorphism in apoB gene is associated with risk of ischemic cerebral infarction with family history in 47 Han Chinese patients. In Danish prospective study (the Copenhagen City Heart Study) [108] with 23-yr follow-up the E4154K KK homozygosity was associated with an 80% reduction in risk of IS (0.2 (0.1-0.7)) compared with non-carriers. The other SNPs or haplotypes examined in this study were not associated with risk of IS.

The most promising results in IS studies are connected with apoA5 and apo(a) gene polymorphisms. It is well konown that apoAV is a member of apoAI/CIII/AIV gene cluster. apoAV gene consists of 4 exons and codes 369 amino acids protein. The common variants within the apoAV gene are associated with plasma tryglicerides (TG) levels, by enhancing the intravascular triglyceride hydrolysis by activating lipoprotein lipase (LPL), or can decrease the serum concentration of triglycerides through the inhibition of the hepatic very low density lipoprotein (VLDL) production. Literature data suggest significant association between apoAV gene polymorphisms (c.1131T>C, c.12238T>C, c.553G>T) and IS risk [34,109-112]. The association of apoAV 56G allele was observed in the large-vessel associated stroke group compared to the healthy controls [113]. The same group of authors [114] examined three polymorphism in apoAV gene in small-vessel, large-vessel and mixed subgroups of 378 patients with stroke and healthy controls. They found that patients carriers of -1131C and IVS3+476A alleles confer risk for all IS types, In this study the T1259C variant was not associated with IS that is in agreement with previous study of Jeromi et al.[112]. Recently published study on Han Chinese population confirmed the previously found association between c.1131T>C polymorphism in apoAV gene and IS risk [115].

There is growing and convincing evidence that elevated lipoprotein (a) levels have a significant role in stroke. Genetic studies demonstrated that Lp(a) is an inherited trait determined almost entirely by the apo(a) gene locus. Variations at the apo(a) gene locus beyond the kringle IV-2 domain seem to influence Lp(a) concentrations [116]. The pentanucleotide TTTTA repeat (PNTR) polymorphism located at the 5' untranslated region of the apo(a) gene accounts for 10% to 14% of the variation in plasma Lp(a) concentrations [117], and was reported to be inversely correlated with Lp(a) levels. Low numbers of apo(a) TTTTA VNTR were associated with IS in three studies [118-120] that were included with the only meta-analysis [19] that evaluated the association of apo(a) TTTTA VNTR polymorphism and IS.

The Precocious Coronary Artery Disease (PROCARDIS) study identified 2 single-nucleotide polymorphisms (SNPs) at the Lp(a) locus (LPA) on chromosome 6q26–27 (rs3798220 (T/C) and rs10455872(A/G)) that each was strongly and independently related to Lp(a) levels and risk of coronary disease [121]. Wang et al. [122] in meta-analysis of 3550 IS cases and 6560 controls showed no significant association of LPA variants previously associated with Lp(a) levels with IS (OR per allele 0.96, 95% CI 0.88-1.04, for rs1853021 and 0.95, 95% CI 0.88-1.03, for rs1800769). Also, there was the lack of evidence of an association of LPA score and prevalent or incident stroke in Heart Protection Study (1326 prevalent and 507 incident IS cases) [123]. It does not exclude the possibility that lowering Lp(a) could have beneficial effects on the risk of stroke or stroke subtypes. On the contrary, theWomen's Health Study (123 IS cases) suggested a positive association of rs3798220 with stroke [124].

Future studies are warranted to assess whether the analysis of previously mentioned polymorphisms may be useful for the clinical approach to evaluate risk factors for IS.

Cholestryl ester transfer protein (CETP). CETP participates in HDL metabolism by facilitating the transfer of cholesteryl esters from HDL to apoB-containing lipoproteins in exchange for triglycerides being transferred to HDL This glycoprotein is secreted mainly from the liver and circulates in plasma, bound mainly to HDL. A deficiency of CETP is connected with anti-atherogenic profile, with increased HDL and decreased LDL levels. The CETP gene is located on chromosome 16q21 and consists of 16 exons. Several polymorphisms have been described, including (Taq1 B in intron-1(rs708272), 405V and A373P (rs5880) in exon 12, R451Q (rs 1800777) in exon 15, and -629A/C (rs 1800775). Of these, the most widely studied is the *Taq*I B polymorphism which results from a nucleotide substitution at position 277 of the first intron (rs708272). CETP Taq1 B2B2 genotype is associated with decreased CETP activity, higher HDL-cholesterol concentrations [125,126], decreased risk of coronary artery disease [126,127], lower carotid intimal medial thickness and stenosis [128], lower incidence of microangiopathy in patients with type 2 diabetes [129], and atrial fibrillation [130].

The relationship between CETP polymorphisms and the risk of IS has been the subject of eight reports [28,30,34,131-135]. An association with CETP *Taq*1 B polymorphism was found in one study [133] but not in another [132]. Some isolated reports of a significant association relate to the rs12720922 and rs9939244 [134] and the rs5883 [135] polymorphisms. Clearly, more extensive investigations in this area are warranted.

ATP-binding cassette transporter I(ABCAI). ABCA1 is a transmembrane protein present on peripheral tissue cells, crucial in the initial step of HDL formation. It mediates the transfer of cellular phospholipids and cholesterol to acceptor apolipoproteins such as apolipoprotein A-I [136]. The ABCA1 locus is located on chromosome 9q22-q31, and is composed of 50 exons ranging in size from 33bp to 249bp. More than 100 common and rare variants have been described [137]. Several polymorphisms of the ABCA1 gene have been investigated for their association with IS.

The first published study in IS on 244 Hungarian patients [138] suggests a protective role for the *ABCA1*-R219K and V771M polymorphisms. Pasdar et al. [139] studied four common polymorphisms in ABCA1 gene: G/A-L158L, G/A-R219K, G/A-G316G and G/A-R1587K in 400 Caucasian IS patients. There was no significant difference in allele frequencies of all polymorphisms, as the haplotypes arrangement. This study did not support a major role for the ABCA1 gene as a risk factor for IS. Following a report of an association of -14C/T polymorphism in the promoter region of the ABCA1 gene with IS [140], extensive studies to confirm this association in different populations are essential.

Lipoprotein lipase (LPL). Lipoprotein lipase (LPL) is a member of the lipase gene family [141] that may play a central role in lipid metabolism. The major sources of LPL synthesis are skeletal and heart muscle as well as adipose tissue, from which the mature enzyme is then secreted and transported to the vascular endothelium, the physiological site of the enzyme's action [142]. The physiological action of LPL consists of the hydrolysis of the triacylglycerol component of triglycerides and VLDL, resulting in the production of chylomicron remnants, and in the case of VLDL, resulting in the production of smaller, intermediate-density lipoproteins [143]. LPL is also synthesized by macrophages and macrophage-derived foam cells in atherosclerotic lesions [144-146], and this fraction of the enzyme has been linked to LPL-related proatherogenic effects. LPL possess a noncatalytic activity on lipoproteins such as molecular bridging [147] and retention of LDL-C by proteoglycans of the subendothelial matrix occurs, thereby proposing LPL activity in the arterial wall to promote atherosclerosis.

The human LPL gene is localized to chromosome 8p22, spanning 35kb. It contains 10 exons. The gene locus is highly polymorphic and contains many single nucleotide polymorphisms (SNPs) in both coding and non-coding regions. Some cause loss of enzymatic activity and others have only mild detrimental effects on LPL function, or serve more as markers for genetic variation elsewhere in the genome [148].

Epidemiological evidence on the potential role of LPL in IS remains scarce and controversial. Two SNPs in the coding DNA (cSNPs) that have been studied extensively cause point mutations in exons 2 and 6, with substitution of an aspartic acid to an asparagine residue at position 9 (D9N, p.Asp9Asn), and an asparagine to a serine residue at position 291 (N291S, p.Asn291Ser), respectively. These mutations occur at high frequencies in the general population (up to 5%) and are associated with elevated TG, decreased HDL-cholesterol levels, and concomitantly with a higher incidence of cardiovascular disease compared with non-carriers [149]. Polymorphism Ser447Ter is a consequence of a C to G transversion at nucleotide 1595 in exon 9, which converts the serine 447 codon (TCA) to a premature termination codon (TGA). This polymorphism is associated with increased lipolytic function and beneficial effects on lipid homeostasis and atheroprotection [148]. HindIII polymorphisms of the LPL gene in intron 8, which identifies a two-allele polymorphism with restriction fragments of 6 kb (H1) and 11 kb (H2), is associated with elevated TG levels [150], low HDL-cholesterol levels [151], and was considered as a possible IS-associated polymorphism [152] Also, Pvu II polymorphism in intron 6 has been associated with high TG levels and coronary artery disease.

Four meta–analysis [16,153,154], and 17 case–control studies have been reported [28,30,34,72,88,94,132,152,155-163] about the association of LPL gene polymorphisms and IS. In a meta–analysis of six studies [153] the inverse association between LPL Ser447Ter polymorphism and IS risk was of borderline significance (OR=0.88, 95%CI 0.79–0.99, P=0.033). In recently published meta-analysis [154] of 4681 IS patients and 8516 controls from 13 studies LPL Ter447 variant was associated with a significantly reduced risk for IS (OR 0.79, 95%CI 0.68-0.93, P=0.005) in Causcasian and East-Asian population. According the data of four studies (387 cases/589 controls), this association was of great importance in atherosclerotic stroke (OR 0.44, 95%CI 0.32-0.62, P<0.00001). In the meta-analysis of same authors [154] that included 7 studies (3669 cases and 6693 controls) no significant association between Ser291 variant and IS stroke risk was found. This is in accordance with the conclusion of previously published meta-analysis of LPL Asn291Ser polymorphism and IS [16]. A positive association between S447X variant and stroke has been reported in specific subtypes, as in the study of Shimo-Nakanishi et al. [152], Zhao et al. [160], Guan et al. [161], and Xu et al. [163] which reported a relationship with atherosclerotic stroke, and in the prospective cohort study of Morrison et al. [72] who described a positive association between S447X and asymptomatic stroke lesions, and in the study of Kostulas et al. [162] where the protective role of G-allele of LPL S447X polymorphism had a lower frequency in males. Shimo-Nakanishi et al. [152] observed a protective role of H- H- and H-H+ genotypes vs. H+H+ (*Hind*III polymorphism), and Xu et al. [163] noted a protective role of the P allele (*Pvu*II polymorphism) for IS. In conclusion, there is evidence to support an association between LPL gene polymorphism and IS, but this notion needs to be strengthened by further investigations.

Hepatic lipase. Despite the numerous association studies of LPL gene polymorphisms and IS, and these have generated consistently negative results [28,30,164].

Paraoxonase(PON). Paraoxonase is a glycoprotein, HDL-associated esterase, that hydrolyzes products of lipid peroxidation and prevents the oxidation of LDL. It has antioxidant and anti-atherogenic properties [165,166]. The paraoxonase gene maps to chromosome 7q21.3. It codes three isoforms, PON1, PON2, and PON3, that share 60 to 65% homology at the amino acid level [167]. PON1 and PON3 reside on circulating HDL particles. PON2 is ubiquitously expressed and does not appear to be associated with HDL particles [168-170]. PON genes polymorphisms may affect the corresponding enzyme activity.

Two non-synonymous *PON1* polymorphisms with possible regulatory effects on enzyme activity [171], namely rs662 (c.575A>G or p.Gln192Arg) and rs854560 (c.163T>A or p.Leu55Met), have been extensively investigated as potential risk factors for atherosclerosis-related phenotypes, including coronary artery disease, peripheral arterial disease and IS [171-173]. Two previously published systematic reviews suggested that the G allele of rs662 is associated with a small increase (per-allele OR 1.12) in the risk of coronary artery disease, while no such association was found for rs854560 [172,173]. Inter-individual variability in PON1 levels is determined by the Q192R (Gln192Arg) and L55M (Leu55Met) coding region polymorphisms and by two described polymorphisms in the promoter of the PON1 gene, C(-107)T and G(-824)A. Five polymorphic sites were found in the promoter region of the PON1 gene: c.107C>T, c.126G>C, c.160G/A, c.824G>A, and c.907G>C. Specific polymorphisms are associated with the risk of acute IS.

According the literature data there are three meta–analysis [18,174,175] and 26 case–control studies [28,30,34,35,176-197] explored the association of PON1 polymorphisms and IS risk. A positive association of Gln192Arg PON1 polymorphism and IS was described in the meta–analyses and in five case-control studies [177,178,184,185,188], but this association was negative in all other reports. Only two studies in Turkish populations obtained evidence for a positive association of Leu55Met PON1 polymorphism and IS [188,193], in contrast to 12 where no evidence for this association was found [28,30,177-179,181,186,187,190-192,194].

Two recently published meta-analysis included the studies examined the association of two common polymorphisms in the coding region of PON1 gene (rs662 and rs854560) and the occurrence of IS. In meta-analysis [174] of 22 studies (7384 cases/11074 controls) PON1 polymorphism rs662 was associated with increased risk for IS (OR 1.10 per G allele copy, 95%CI 1.04–1.17, P=0.001), while no significant association of rs854560 was observed in meta-analysis of 16 eligible studies (OR 0.97 per T allele copy, 95% CI 0.90–1.04, P=0.37). The other meta-analysis [175] included 8 studies on rs854560 polymorphism and 9 studies on rs662 polymorphism. This analysis provides strong evidence that the rs662 polymorphism of PON1 gene is associated with IS (OR 1.21, 95%CI 1.02-1.43, P=0.03), and that the rs854560 gene polymorphism is not associated with IS (OR 1.12, 95%CI 0.96-1.31, P=0.13).

Man et al. [198] in 191 Han Chinese patients with acute IS, of whom 25% had concurrent stenosis found that genotype distributions of PON1 Q192R differed significantly between patients with stroke and controls, and that the presence of at least one R allele in PON1 Q192R was associated with concurrent stenosis.

Polymorphism c.107C>T is important because it contributes 23% of the variances in PON1 levels. Since the presence of T at position -107 of the PON1 gene disturbs a recognition sequence for stimulating protein-1 (Sp1), the TT genotype is associated with the lowest serum PON1 levels. Although the frequency of the T allele and TT genotype did not differ significantly between young adults with arterial IS and controls, the presence of the -107T allele was associated with an independent increase in the risk of arterial IS [197].

There are two common polymorphisms of the PON2 gene: A148G (Ala148Gly) and C311S (Ser311Cys)). Almost all research groups except one [192] agree that there is no significant association between IS and these polymorphisms [28,30,177,181,187,199] Four polymorphisms in the PON3 gene were examined in two studies on IS patients [181,187]. No evidence for an association was obtained. Whereas rs662 (c.575A>G or p.Gln192Arg) polymorphism of the PON1 gene could be regarded as a potential risk factor for IS, this does not seem to be the case for PON2 and PON3.

Although, Lazaros et al. [200] did not identified none of the PON polymorphisms (PON1(Q/R) 192, PON1(M/L) 55, and PON2(S/C) 311) as a risk factors for IS, they concluded that PON2 311C allele was significantly increased in patients with severe forms of IS and could be reviewed as a possible predisposing factor for severe cases of IS.

Large-scale multicenter-controlled prospective studies are warranted to further explore the effects of PON polymorphisms on stroke susceptibility and severity.

Low-density lipoprotein receptor (LDLR). LDLR is a cell surface receptor that plays an integral role in plasma lipoprotein metabolism, especially in cholesterol homeostasis. The LDLR gene is localized on chromosome 19, and comprises 45 kb with 18 exons. Mutations in this gene may lead to dysfunction of the receptor resulting in familial hypercholesterolemia and premature ischemic heart disease. The most frequently studied is A370T polymorphism (c.1171G>A in exon 8 that changes alanine to threonine at position 370 in the LDLR protein. The other described polymorphisms are *Nco*I, *Ava*II, c.1773C>T, and rs2738446 and rs2738450.

Only few studies explored the association of LDLR gene polymorphisms and IS. Guo et al. [201] investigated the relationship between *Nco*I and *Ava*II polymorphisms of the LDLR gene in Han Chinese patients with atherosclerotic cerebral infarction and concluded that the coexistence of A-A- and N+N+ genotypes significantly increases the risk of atherosclerotic cerebral infarction (RR 5.56, p<0.001). The data of Frikke-Schmidt et al. [202] support an association between c.370A>T polymorphism (370A allele) and increased risk of stroke. Two studies reported an association between rs2738450 and IS [135,203]. Recently published study [204], for the first time revealed the association of rs1122608 (located 58.7 kb upstream of the LDLR gene) and 530 IS patients in Chinese Han population.

Oxidized LDL that play a key role in the atherogenesis process exert most effects through the interaction with its major receptor lectin like oxidized low density lipoprotein receptor 1(LOX-1). LOX-1 is encoded by the lectin like oxidized low density lipoprotein receptor 1 (OLR1) gene, located in the p12.3–p13.2 region of human chromosome 12 and consists of 6 exons. Few SNPs located within introns 4, 5, and 3' untranslated region, are associated with higher risk of developing acute myocardial infarction. Polymorphism (rs11053646, G501C) located in exon 4, leads to a change from a lysine to an asparagine at position 167 (K167N). As the consequence, reduced binding and internalization of the oxLDL was noticed. Only one paper [205] relates G501C polymorphisms of the OLR1 gene and IS, with negative results. Except LOX-1 full receptor, LOXIN as an isoform lacking part of the functional domen was identified and it has a protective role by blocking LOX-1 activation. One recently published study examined the prevalence of OLR1 gene polymorphisms, IVS4-14 A/G and IVS4-73 C/T, which regulate the expression of LOXIN, in 43 patients with ischemic cerebrovascular diseases (ICVD). Patients with G homozygosity for IVS4-14 polymorphism and T homozygosity for IVS4-73 polymorphism have higher risk to develop ICVD [206]. Man et al. [198] in 191 Han Chinese patients with acute IS, of whom 25% had concurrent stenosis examined whether oxidized low-density lipoprotein receptor (OLR) 3' untranslated region (UTR) C > T (rs1050283) polymorphism and found that TT allele in OLR rs1050283 were associated with concurrent stenoses.

The association of LDLR and OLR1gene polymorphisms with IS should be further assessed in different populations and in wider series of patients.

Soluble epoxide hydrolase 2. Soluble (cytosolic) epoxide hydrolase (sEH) has two activities as epoxide hydrolase and phosphatase. It is an enyzme involved in conversion of epoxyeicosatrienoic acids (EETs) metabolites of arachidonic acid in less active corresponding diols. EETs functions as vasodilatators, have anti-inflammatory effects [207], and anti-trombotic effects [208,209]. EETs have been shown to regulate cerebral blood flow and, through their mitogenic properties, may contribute to angiogenesis in the brain. Hence, they may protect against IS [210-212]. It modifies blood pressure [213] or plasma lipid levels and composition of lipoprotein particles [214]. Soluble EH is encoded by EPHX2 gene located at chromosome 8 (8p21-p12). This gene contains 19 exons. It encodes 555 amino acids. Fourty four SNPs and one insertion/deletion polymoprhism [215] was identified in these gene. Substitutions Lys55Arg, Cys154Tyr and Glu470Gly resulted in an enzyme with increased epoxide hydrolase activity, until two other variants, the Arg287Gln substitution and the Ser402[Argins] insertion resulted in enzymes with reduced epoxide hydrolase activity.

Genetic studies links polymorphisms in the human EPHX2 gene with modified risk of IS in a number of human populations [216-218]. In the Fornage's study, a positive association was observed between the Glu470Gly variant and the incidence of IS in African American cohort [216].

Zhang et al. [218] examined potential associations between *EPHX2* G860A polymorphism and IS risk in Chinese population. The G860A polymorphism results in an amino acid substitution (R287Q, Arg287Gln) that alters enzyme stability and reduces enzyme activity [215,219]. They concluded that the presence of at least one A allele at position 860 of EPHX2 was independently associated with a decreased risk of IS. Gschwendtner et al. [217] in Caucasians found significant association between rs751141, rs7357432, rs2291635 and IS. The haplotype containing the associated alleles of the three SNPs showed an odds ratio of 1.59 (1.06-2.37, *P*=0.022) in the large-vessel subgroup and an odds ratio of 1.54 (0.96-2.41, *P*=0.062) in the subgroup of patients with undetermined etiology. Lee et al. [220] did not find positive association of three polymorphism in *EPHX2* gene (R103C, R287Q, and Arg[402-403ins]) and IS risk.

Fava's study [221] examined whether the EPHX2 missense K55R and R287Q, together with the –1452T>C (rs7003694) in the promoter region and the +1784A>G (rs1042032) in the 3'UTR polymorphisms, are associated with hypertension and with risk of cardiovascular events in middle-aged Swedes. They found no significant difference in the incidence of IS in carriers of different EPHX2 R287Q, EPHX2 –1452T>C genotypes, EPHX2 +1784A>G (P>0.05), until the higher incidence of IS was evident in male EPHX2 R-homozygotes versus male K-allele carriers.

2.2. Genome-wide association studies (GWAS) in ischemic stroke

The completion of the Human genome project, together with rapid improvements in laboratory techniques in this field, has enabled investigators to examine multiple genetic variants simultaneously in large study populations and it can be used for unlocking the genetic basis of complex human diseases [222,223]. The genetic variants that can be identified by GWAS are common SNP and have low effect size. By introducing GWAS a major limitation of the candidate gene study was overcame and candidate gene studies have now been largely superseded by the GWAS technique.

To date, GWAS of IS has been performed in 6 cohorts, resulting in 7 publications with somewhat inconsistent results. The initial step in a genome-wide genotyping study in patients with IS was performed in 2007 [224]. The analysis which compared 408,803 unique SNPs in 249 white patients with IS and 268 white neurologically normal controls in five US stroke centers do not suggest any single common genetic variant exerting a major risk for IS. The other recently published genome-wide association study [225] found a significant association between two SNPs rs11833579 and rs12425791on chromosome 12p13 with total, ischemic, and atherothrombotic stroke in white persons. The SNPs are located closed to the gene Ninjurin2 (nerve injury-induced protein 2-NINJ2) and WNK1- serine-threonine kinase that regulate ion channels involved in sodium and potassium transport. Finally, SNPs in paired-like homeodomain transcription factor 2 (*PITX2*) and zinc finger homeobox 3

(*ZFHX3*) were observed to be associated with cardioembolic stroke and atrial fibrillation in Icelandic population [226,227].

Three GWAS were performed in Japanese populations. Kubo et al. [228] found significant association of non-synonymous SNP (1425 G/A) in protein kinase C-eta (PRKCH) with lacunar infarction in the pathogenesis of IS. Hata et al. [229] found that SNP in the 5'-flanking region of angiotensin receptor like-1 (AGTRL1) gene (rs9943582, - 154G/A) to have a significant association with brain infarction. Also, rs9615362 of cell surface receptor CELSR1 (cadherin epidermal growth factor laminin A seven-pass G-type receptor 1) was associated with IS [230].

3. Conclusion

Genetics of IS represents a unique challenge. Among the most examined candidate genes in IS are those associated with lipid metabolism. Unfortunately, the results are complex and far from clear-cut. According the literature review in this chapter it can be concluded that genes (polymorphisms) that are the most likely to be associated with IS are: apoE (apo ε2/ε3/ε4) and PON1 gene (p.Gln192Arg). Insufficient or inconsistent data that neither supported nor excluded an association of some genes polymorphisms with IS apoAV (c.1131T>C), LPA (rs3798220), LPL (S447X), LDLR (c.370A>T), OLR1(IIVS4-14A/G, IVS4-73C/T) and EPHX2 (G860A). For other genes/polymorphisms that were reviewed in this paper, we are reasonably confident that an association with IS can be ruled out.

The reasons for contradictory results in the studies may be limited sample size, heterogeneity of study designs and endpoints, differences in inclusion and exclusion criteria, ethnically different patient populations, selection of control population, different stroke subtypes and age of stroke onset, type of statistical evaluation, covariates, correction for multiple testing etc. One of the limitations of multiple non-reproducible candidate gene studies was the restriction to a single or rather few genetic variants tested for association with disease in examined gene. Further, genetic variants of candidate genes with strong effects at the transcriptional level or others affecting the functionality of the protein may have escaped the test for association with disease risk. Thus, in retrospect, it is not surprising that the candidate gene approach resulted in only limited success in the elucidation of IS stroke genes.

Research in the field of IS should be directed towards facilitation of the characterization of IS pathogenesis at the molecular level and the development of genetic markers' panels for assessment of IS risk. Technological developments such as GWAS, NGS technology, transcription profiling and proteomics will provide huge amounts of genetic information and allow investigators to identify variants in patients with specific stroke subtype and to identify how they exert their effects at the molecular level. The replication in an independent study, in large and well-characterized groups of patients of different ethnic origin, is required to confirm previously obtained results. On the basis of genetic or genomic information the therapeutic outcome or side effects in stroke patients could be predicted, as the effectiveness and safety of applied therapy. Also, this approach may help in stroke prevention by identification of presymptomatic at-risk individuals, resulting in minimizing patients' morbidity and mortality and reducing health care costs associated with stroke.

Author details

Sanja Stankovic*, Milika Asanin and Nada Majkic-Singh
Clinical Center of Serbia, University of Belgrade, Serbia

4. References

[1] Shiber JR, Fontane E, Adewale A. A Stroke registry: hemorrhagic vs ischemic strokes. Am J Emerg Med 2010; 28(3):331-333.

[2] Baird AE. Genetics and genomics of stroke. J Am Coll Cardiol 2010;56(4):245-253.

[3] Adams HP, Bendixen BH, Kappelle LJ, Biller J, Love BB, Gordon DL, Marsh EE. Classification of subtype of acute ischemic stroke: definitions for use in a multicenter clinical trial. Stroke 1993;24(1):35-41.

[4] Bamford J, Sandercock PA, Dennis MS, Burn J, Warlow CP: Classification and natural history of clinically identifiable subtypes of brain infarction. Lancet 1991;337(8756):1521-1526.

[5] Bogousslavsky J, Van Melle G, Regli F: The Lausanne Stroke Registry: analysis of 1,000 consecutive patients with first stroke. Stroke 1988;19(9):1083-1092.

[6] Touboul PJ, Elbaz A, Koller C, Lucas C, Adrai V, Chedru F, Amarenco P, GENIC Investigators: Common carotid artery intima-media thickness and ischemic stroke subtypes: the GENIC case-control study. Circulation 2000;102(3):313-318.

[7] Whisnant JP. Modeling of risk factors for ischemic stroke. The Willis Lecture. Stroke 1997;28(9):1840-1844.

[8] Dichgans M. Genetics of ischaemic stroke. Lancet Neurol 2007;6(2):149-161.

[9] Schulz UG, Flossmann E, Rothwell PM. Heritability of ischemic stroke in relation to age, vascular risk factors, and subtypes of incident stroke in population-based studies. Stroke 2004;35(4):819-824.

[10] Jerrard-Dunne P, Cloud G, Hassan A, Markus HS. Evaluating the genetic component of ischemic stroke subtypes: a family history study. Stroke 2003;34(6):1364-1369.

[11] Brass LM, Isaacsohn JL, Merikangas AR. A study of twins and stroke. Stroke 1992;23(2):221-223.

[12] Touzé E, Rothwell PM. Sex differences in heritability of ischemic stroke: a systematic review and meta-analysis. Stroke 2008;39(1):16-23.

[13] Hassan A, Markus HS. Genetics and ischaemic stroke. Brain 2000;123 (Pt 9):1784-1812.

[14] Bersano A, Ballabio E, Bresolin N, Candelise L. Genetic polymorphisms for the study of multifactorial stroke.Hum Mutat 2008;29(6):776-795.

---

* Corresponding Author

[15] Stankovic S, Majkic-Singh N. Genetic aspects of ischemic stroke: coagulation, homocysteine, and lipoprotein metabolism as potential risk factors. Crit Rev Clin Lab Sci 2010;47(2):72-123.

[16] Casas JP, Hingorani AD, Bautista LE, Sharma P. Meta-analysis of genetic studies in ischemic stroke: thirty-two genes involving approximately 18,000 cases and 58,000 controls. Arch Neurol 2004; 61:1652-1662.

[17] Xin XY, Song YY, Ma JF, Fan CN, Ding JQ, Yang GY, Chen SD. Gene polymorphisms and risk of adult early-onset ischemic stroke: A meta-analysis. Thromb Res 2009;124(5):619-624.

[18] Xu X, Li J, Sheng W, Liu L. Meta-analysis of genetic studies from journals published in China of ischemic stroke in the Han Chinese population. Cerebrovasc Dis 2008;26(1):48-62.

[19] Ariyaratnam R, Casas JP, Whittaker J, Smeeth L, Hingorani AD, Sharma P. Genetics of ischaemic stroke among persons of non-European descent: a meta-analysis of eight genes involving approximately 32,500 Individuals. PLoS Med 2007;4:e131.

[20] Rao R, Tah V, Casas JP, Hingorani A, Whittaker J, Smeeth L, Sharma P. Ischaemic stroke subtypes and their genetic basis: a comprehensive meta-analysis of small and large vessel stroke. Eur Neurol 2009;61(2):76-86.

[21] Banerjee I, Veena Gupta V, Ganesh S. Association of gene polymorphism with genetic susceptibility to stroke in Asian populations: a meta-analysis. J Hum Genet 2007;52(3):205-219.

[22] McCarron MO, Delong D, Alberts MJ. ApoE genotype as a risk factor for ischemic cerebrovascular disease: a meta-analysis. Neurology 1999;53(6):1308-1311.

[23] Sudlow C, Martínez González NA, Kim J, Clark C. Does apolipoprotein E genotype influence the risk of ischemic stroke, intracerebral hemorrhage, or subarachnoid hemorrhage? Systematic review and meta-analyses of 31 studies among 5961 cases and 17,965 controls. Stroke 2006;37(2):364-370.

[24] Hamzi K, Tazzite A, Nadifi S. Large-scale meta-anlysis of genetic studies in ischemic stroke: Five genes involving 152797 individuals. Indian J Hum Genet 2011;17(3):212-217.

[25] Karttunen V, Alfthan G, Hiltunen L, Rasi V, Kervinen K, Kesaniemi YA, Hillbom M. Risk factors for cryptogenic ischaemic stroke. Eur J Neurol 2002;9(6):625-632.

[26] Szolnoki Z, Somogyvári F, Kondacs A, Szabó M, Fodor L. Evaluation of the interactions of common genetic mutations in stroke subtypes. J Neurol 2002;249(10):1391-1397.

[27] Szolnoki Z, Somogyvári F, Kondacs A, Szabó M, Fodor L, Bene J, Melegh B. Evaluation of the modifying effects of unfavourable genotypes on classical clinical risk factors for ischaemic stroke. J Neurol Neurosurg Psychiatry 2003;74(12):1615-1620.

[28] Zee RYL, Cook NR, Cheng S, Reynolds R, Erlich HA, Lindpaintner K, Ridker PM. Polymorphism in the P-selectin and interleukin-4 genes as determinants of stroke: a population-based, prospective genetic analysis. Hum Mol Genet 2004;13(4):389-396.

[29] Pezzini A, Grassi M, Del Zotto E, Archetti S, Spezi R, Vergani V, Assanelli D, Caimi L, Padovani A. Cumulative effect of predisposing genotypes and their interaction with modifiable factors on the risk of ischemic stroke in young adults. Stroke 2005;36(3):533-539.

[30] Lalouschek W, Endler G, Schillinger M, Hsieh K, Lang W, Cheng S, Bauer P, Wagner O, Mannhalter C. Candidate genetic risk factors of stroke: results of a multilocus genotyping assay. Clin Chem 2007;53(4):600-605.

[31] Berger K, Stögbauer F, Stoll M, Wellmann J, Huge A, Cheng S, Kessler C, John U, Assmann G, Ringelstein EB, Funke H. The glu298asp polymorphism in the nitric oxide synthase 3 gene is associated with the risk of ischemic stroke in two large independent case-control studies. Hum Genet 2007;121(2):169-178.

[32] Gao X, Yang H, ZhiPing T. Association studies of genetic polymorphism, environmental factors and their interaction in ischemic stroke. Neurosci Lett 2006;398(3):172-177.

[33] Kessler C, Spitzer C, Stauske D, Mende S, Stadlmüller J, Walther R, Rettig R. The apolipoprotein E and beta-fibrinogen G/A-455 gene polymorphisms are associated with ischemic stroke involving large-vessel disease. Arterioscler Thromb Vasc Biol 1997;17(11):2880-2884.

[34] Yamada Y, Metoki N, Yoshida H, Satoh K, Ichihara S, Kato K, Kameyama T, Yokoi K, Matsuo H, Segawa T, Watanabe S, Nozawa Y. Genetic risk for ischemic and hemorrhagic stroke. Arterioscler Thromb Vasc Biol 2006;26(8):1920-1925.

[35] Topić E, Šimundić AM, Štefanović M, Demarin V, Vuković V, Lovrenčić-Huzjan A, Žuntar I. Polymorphism of apoprotein E (APOE), methylenetetrahydrofolte reductase (MTHFR) and paraoxonase (PON1) genes in patients with cerebrovascular disease. Clin Chem Lab Med 2001;39(4):346-350.

[36] McIlroy SP, Dynan KB, Lawson JT, Patterson CC, Passmore AP. Moderately elevated plasma homocysteine, methylenetetrahydrofolate reductase genotype, and risk for stroke, vascular dementia, and Alzheimer disease in Northern Ireland. Stroke 2002;33(10):2351-2356.

[37] Mahieux F, Bailleul S, Fenelon R, Couderc R, Laruelle P, Gunel M. Prevalence of apolipoprotein E phenotypes in patients with acute ischemic stroke. Stroke 1990;21.I-115.

[38] Pedro-Botet J, Sentí M, Nogues X, Rubiés-Prat J, Roquer J, D'Olhaberriague L, Olivé J. Lipoprotein and apolipoprotein profile in men with ischemic stroke. Role of lipoprotein (a), triglyceride-rich lipoproteins, and apolipoprotein E polymorphism. Stroke 1992;23(11):1556-1562.

[39] Couderc R, Mahieux F, Bailleul S, Fenelon G, Mary R, Fermanian J. Prevalence of apolipoprotein E phenotypes in ischemic cerebrovascular disease. A case-control study. Stroke 1993;24(5):661-664.

[40] Coria F, Rubio I, Nuñez E, Sempere AP, SantaEngarcia N, Bayón C, Cuadrado N. Apolipoprotein E variants in ischemic stroke. Stroke 1995;26(12):2375-2376.

[41] De Andrade M, Thandi I, Brown S, Gotto A Jr, Patsch W, Boerwinkle E. Relationship of the apolipoprotein E polymorphism with carotid artery atherosclerosis. Am J Hum Genet 1995;56(6): 1379-1390.

[42] Kuusisto J, Mykkänen L, Kervinen K, Kesäniemi YA, Laakso M. Apolipoprotein E4 phenotype is not an important risk factor for coronary heart disease or stroke in elderly subjects. Arterioscler Thromb Vasc Biol 1995;15(9):1280-1286.

[43] Basun H, Corder EH, Guo Z, Lannfelt L, Corder LS, Manton KG, Winblad B, Viitanen M. Apolipoprotein E polymorphism and stroke in a population sample aged 75 years or more. Stroke 1996;27(8):1310-1315.

[44] Hachinski V, Graffagnino C, Beaudry M, Bernier G, Buck C, Donner A, Spence JD, Doig G, Wolfe BM. Lipids and stroke: a paradox resolved. Arch Neurol 1996;53(4):303-308.

[45] Ferrucci L, Guralnik JM, Pahor M, Harris T, Corti MC, Hyman BT, Wallace RB, Havlik RJ. Apolipoprotein E epsilon 2 allele and risk of stroke in the older population. Stroke 1997;28(12):2410-2416.

[46] Nakata Y, Katsuya T, Rakugi H, Takami S, Sato N, Kamide K, Ohishi M, Miki T, Higaki J, Ogihara T. Polymorphism of angiotensin converting enzyme, angiotensinogen, and apolipoprotein E genes in a Japanese population with cerebrovascular disease. Am J Hypertens 1997;10(12Pt1):1391-1395.

[47] Schmidt R, Schmidt H, Fazekas F, Schumacher M, Niederkorn K, Kapeller P, Weinrauch V, Kostner GM. Apolipoprotein E polymorphism and silent microangiopathy-related cerebral damage. Results of the Austrian Stroke Prevention Study. Stroke 1997;28(5):951-956.

[48] Yang G, Jinjin G, Jianfei N. The relationship between polymorphisms of apolipoprotein E gene and atherosclerotic cerebral infarction. Zhonghua Shen Jing Ge Za Zhi 1997;30:236–239.

[49] Wang DS, Jiang L, Dai YM. Primary study of ApoE gene polymorphism in patients with cerebral infarction. Zhong Feng Yu Shen Jing Ji Bing Za Zhi 1997;14:71–74.

[50] Zhu TB, Zhao SP, You XK. Effect of apolipoprotein E gene on plasma levels of lipids, lipoprotein, apolipoprotein and relation to cerebral infarction. Hu Nan Yi Xue 1997;14:265–266.

[51] Yan SK, Zhou X, Li XL. Relationship between gene polymorphism of apolipoprotein E and serum lipids, lipoproteins, and apolipoproteins in Chinese patients with atherothrombotic brain infarction. Zhong Guo Shen Jing Mian Yi Xue He Shen Jing Bing Xue Za Zhi 1997;4:16–21.

[52] Aalto-Setälä K, Palomäki H, Miettinen H, Vuorio A, Kuusi T, Raininko R, Salonen O, Kaste M, Kontula K. Genetic risk factors and ischaemic cerebrovascular disease: role of common variation of the genes encoding apolipoproteins and angiotensin-converting enzyme. Ann Med 1998;30(2):224–233.

[53] Ji Y, Urakami K, Adachi Y, Maeda M, Isoe K, Nakashima K. Apolipoprotein E polymorphism in patients with Alzheimer's disease, vascular dementia and ischemic cerebrovascular disease. Dement Geriatr Cogn Disord 1998;9(5):243-245.

[54] Margaglione M, Seripa D, Gravina C, Grandone E, Vecchione G, Cappucci G, Merla G, Papa S, Postiglione A, Di Minno G, Fazio VM. Prevalence of apolipoprotein E alleles in healthy subjects and survivors of ischemic stroke: an Italian case-control study. Stroke 1998;29(2):399-403.

[55] Cao W, Chen F, Teng L, Wang S, Fu S, Zhang G. The relationship between apolipoprotein E gene polymorphism and coronary heart disease and arteriosclerotic cerebral infarction. Zhonghua Yi Xue Yi Chuan Xue Za Zhi 1999;16:249–251.

[56] Peng DQ, Zhao SP, Wang JL. Lipoprotein (a), and apolipoprotein E4 as independent risk factors for ischemic stroke. J Cardiovasc Risk 1999;6(1):1-6.

[57] Liu WG, Li ZH. The relationship between polymorphisms of apolipoprotein E gene and atherosclerotic cerebral infarction in middle-aged and young adults. Lin Chuang Shen Jing Bing Xue Za Zhi 1999;12:134–136.

[58] Peng DQ, Zhao SP. Comparison of apolipoprotein E genotype distribution in two types of stroke. Zhong Guo Dong Mai Ying Hua Za Zhi 1999;7:34–36.

[59] Catto AJ, McCormack LJ, Mansfield MW, Carter AM, Bamford JM, Robinson P, Grant PJ. Apolipoprotein E polymorphism in cerebrovascular disease. Acta Neurol Scand 2000;101(6):399-404.

[60] Kokubo Y, Chowdhury AH, Date C, Yokoyama T, Sobue H, Tanaka H. Age-dependent association of apolipoprotein E genotypes with stroke subtypes in a Japanese rural population. Stroke 2000;31(6):1299-1306.

[61] McCarron MO, Muir KW, Nicoll JA, Stewart J, Currie Y, Brown K, Bone I. Prospective study of apolipoprotein E genotype and functional outcome following ischemic stroke. Arch Neurol 2000;57(10):1480-1484.

[62] Ding J, Zhu WB, Fan W. Association between apolipoprotein E polymorphisms and cerebral stroke. Zhong Guo Shen Jing Jing Shen Ji Bing Za Zhi 2000;26:371–372.

[63] Wang TG, He ZY, Li YQ. The relation between apolipoprotein E gene polymorphism and atherosclerotic cerebral infarction. Yi Chuan 2000;22:4–6.

[64] Chowdhury AH, Yokoyama T, Kokubo Y, Zaman MM, Haque A, Tanaka H. Apolipoprotein E genetic polymorphism and stroke subtypes in a Bangladeshi hospital-based study. J Epidemiol 2001;11(3):131-138.

[65] Frikke-Schmidt R, Nordestgaard BG, Thudium D, Moes Grønholdt ML, Tybjaerg-Hansen A. ApoE genotype predicts AD and other dementia but not ischemic cerebrovascular disease. Neurology 2001;56(2):194-200.

[66] MacLeod MJ, De Lange RP, Breen G, Meiklejohn D, Lemmon H, Clair DS. Lack of association between apolipoprotein E genoype and ischaemic stroke in a Scottish population. Eur J Clin Invest 2001;31(7):570-573.

[67] Serteser M, Visvikis S, Ozben T, Herbeth B, Balkan S, Siest G. Lipid profile and apolipoprotein E genotyping in stroke: a case-control study. Neuroscience-Net 2001;3, article 10015.

[68] Slooter AJC, Bots ML, Havekes LM, del Sol AI, Cruts M, Grobbee DE, Hofman A, Van Broeckhoven C, Witteman JC, van Dujn CM. Apolipoprotein E and carotid artery atherosclerosis: the Rotterdam study. Stroke 2001;32(9):1947-1952.

[69] Li YW, He X, Zhao LX. The relationship between polymorphisms of apolipoprotein E gene and cerebrovascular disorder. Xin Nao Xue Guan Bing Fang Zhi 2001;1:17–19.

[70] Li ZH, LiuWG, Zhao XY, Chen YQ. Risk factor for stroke and ApoE polymorphism in the young and middle-aged. Cu Zhong Yu Shen Jing Ji Bing 2001;8:326–329.

[71] Luthra K, Prasad K, Kumar P, Dwivedi M, Pandey RM, Das N. Apolipoprotein E gene polymorphism in cerebrovascular disease: a case-control study. Clin Genet 2002;62(1):39-44.

[72] Morrison AC, Ballantyne CM, Bray M, Chambless LE, Sharrett AR, Boerwinkle E. LPL polymorphism predicts stroke risk in men. Genet Epidemiol 2002;22(3):233-242.

[73] Shen LH, Ke KF, Li ZH. Research on apolipoprotein E gene polymorphism in patients with atherosclerotic cerebral infarction. Jiao Tong Yi Xue 2002;16:504-505.

[74] Xia Y, Li HL, Wang JL. Association between apolipoprotein E polymorphism and lipid metabolism in patients with cerebral infarction. Zhong Guo Bing Li Sheng Li Za Zhi 2002;18:826-829.

[75] Zhu L, Cui TP. The relation of apolipoprotein E gene polymorphism and cerebral infarction. Xue Shuan Yu Zhi Xue Xue 2002;8:14-15.

[76] Kolovou GD, Daskalova DCh, Hatzivassiliou M, Yiannakouris N, Pilatis ND, Elisaf M, Mikhailidis DP, Cariolou MA, Cokkinos DV. The epsilon 2 and 4 alleles of apolipoprotein E and ischemic vascular events in the Greek population-implications for the interpretation of similar studies. Angiology 2003;54(1):51-58.

[77] Slowik A, Iskra T, Turaj W, Hartwich J, Dembinska-Kiec A, Szczudlik A. LDL phenotype B and other lipid abnormalities in patients with large vessel disease and small vessel disease. J Neurol Sci 2003;214(1-2):11-16.

[78] Souza DR, Campos BF, de Arruda EF, Yamamoto LJ, Trinidane DM, Tognola WA. Influence of the polymorphism of apolipoprotein E in cerebral vascular disease. Arq Neuropsiquiatr 2003; 61(1):7-13.

[79] Um JY, Kim HM, Park HS, Joo JC, Kim KY, Kim YK, Hong SH. Candidate genes of cerebral infarction and traditional classification in Koreans with cerebral infarction. Int J Neurosci 2005;115(6):743-756.

[80] Wang XT, Huang HJ, Ju K. Apolipoprotein E gene polymorphism in people with cerebrovascular disease in the south of the Zhejiang province. Shen Jing Ji Bing Yu Jing Shen Wei Sheng 2003;3:17-19.

[81] Duzenli S, Pirim I, Gepdiremen A, Deniz O. Apolipoprotein E polymorphism and stroke in a population from eastern Turkey. J Neurogenet 2004;18(1):365-375.

[82] Jin ZQ, Fan YS, Ding J, Chen M, Fan W, Zhang GJ, Zhang BH, Yu SJ, Zhang YS, Ji WF, Zhang JG. Association of apolipoprotein E 4 polymorphism with cerebral infarction in Chinese Han population. Acta Pharmacol Sin 2004;25(3):352-356.

[83] Lin HF, Lai CL, Tai CT, Lin RT, Liu CK. Apolipoprotein E polymorpshim in ischemic cerebrovascular diseases and vascular dementia patients in Taiwan. Neuroepidemiology 2004;23(3):129-134.

[84] Pezzini A, Grassi M, Zotto ED, Bazzoli E, Archetti S, Assanelli D, Akkawi NM, Albertini A, Padovani A. Synergistic effect of apolipoprotein E polymorphisms and cigarette smoking on risk of ischemic stroke in young adults. Stroke 2004;35(2):438-442.

[85] Stanković S, Jovanović-Marković Z, Majkić-Singh N, Stanković A, Glišić S, Živković M, Kostic V, Alavantic D. Apolipoprotein E gene polymorphism as a risk factor for ischemic cerebrovascular disease. Jugoslov Med Biohem 2004;23(3):255-264.

[86] Cerrato P, Baima C, Grasso M, Lentini A, Bosco G, Cassader M, Gambino R, Cavallo Perin P, Pagano G, Fornengo P, Imperiale D, Bergamasco B, Bruno G. Apolipoprotein E polymorphism and stroke subtypes in an Italian cohort. Cerebrovasc Dis 2005;20(4):264-269.

[87] Zhou J, Xue YL, Guan YX, Yang YD, Fu SB, Zhang JC. Association study of apolipoprotein e gene polymorphism and cerebral infarction in type 2 diabetic patients. Yi Chuan 2005;27:35-38.

[88] Baum L, Ng HK, Wong KS, Tomlinson B, Rainer TH, Chen X, Cheung WS, Tang J, Tam WWS, Goggins W, Tong CSW, Kam D, Chan Y, Thomas GN, Chook P, Woo KS. Associations of apolipoprotein E exon 4 and lipoprotein lipase S447X polymorphisms with acute ischemic stroke and myocardial infarction. Clin Chem Lab Med 2006;44(3):274-281.

[89] Kang SY, Lee WI. Apolipoprotein e polymorphism in ischemic stroke patients with different pathogenetic origins. Korean J Lab Med 2006;26(3):210-216.

[90] Jiang ZQ, Liu H, Zhang GZ. Relationship between polymorphism of apolipoprotein E gene and atherosclerotic cerebral infarction, hypertensive intracerebral hemorrhage in the youth. J Gannan Med Univ 2006;26:331-334.

[91] Wang JH, Ning XJ, Lu HY. The study on apolipoprotein E gene polymorphism characteristics of cerebral infarction and intracerebral hemorrhage. Zhong Guo Man Xing Bing Yu Fang Yu Kong Zhi 2006;14:21-23.

[92] Giassakis G, Veletza S, Papanas N, Heliopoulos I, Piperidou H. Apolipoprotein E and first-ever ischaemic stroke in Greek hospitalized patients. J Int Med Res 2007;35(1):127-133.

[93] Lai CL, Liu CK, Lin RT, Tai CT. Association of apolipoprotein E polymorphism with ischemic stroke subtypes in Taiwan. Kaohsiung J Med Sci 2007;23(10):491-497.

[94] Parfenov MG, Nikolaeva TY, Sudomoina MA, Fedorova SA, Guekht AB, Gusev EI, Favorova OO. Polymorphism of apolipoprotein E (APOE) and lipoprotein lipase (LPL) genes and ischaemic stroke in individuals of Yakut ethnicity. J Neurol Sci 2007;255(1-2):42–49.

[95] Saidi S, Slamia LB, Ammou SB, Mahjoub T, Almawi WY. Association of apolipoprotein E gene polymorphism with ischemic stroke involving large-vessel disease and its relation to serum lipid levels. J Stroke Cerebrovasc Dis 2007;16(4):160-166.

[96] Abboud S, Viiri LE, Lütjohann D, Goebeler S, Luoto T, Friedrichs S, Desfontaines P, Gazagnes MD, Laloux P, Peeters A, Seeldrayers P, Lehtimaki T, Karhunen P, Pandolfo M, Laaksonen R. Associations of apolipoprotein E gene with ischemic stroke and intracranial atherosclerosis. Eur J Hum Genet 2008;16(8):955-960.

[97] Artieda M, Gañán A, Cenarro A, García-Otín AL, Jericó I, Civeira F, Pocoví M. Association and linkage disequilibrium analyses of APOE polymorphisms in atherosclerosis. Dis Markers 2008;24(2):65-72

[98] Tasdemir N, Tamam Y, Toprak R, Tamam B, Tasdemir MS. Association of apolipoprotein E genotype and cerebrovascular disease risk factors in a Turkish population. Int J Neurosci 2008;118(8):1109-1129.

[99] Wang B, Zhao H, Zhou L, Dai X, Wang D, Cao J, Niu W. Association of genetic variation in apolipoprotein E and low density lipoprotein receptor with ischemic stroke in Northern Han Chinese. J Neurol Sci 2009;276(1-2):118-122.

[100] Saidi S, Zammiti W, Slamia LB, Ammou SB, Almawi WY, Mahjoub T. Interaction of angiotensin-converting enzyme and apolipoprotein E gene polymorphisms in ischemic stroke involving large-vesssel disease. J Thromb Thrombolysis 2009;27(1):68-74.

[101] Tascilar N, Dursun A, Ankarali H, Mungan G, Sumbuloglu V, Ekem S, Bozdogan S, Baris S, Aciman E, Cabuk F. Relationship of apoE polymorphism with lipoprotein(a), apoA, apoB and lipid levels in atherosclerotic infarct. J Neurol Sci 2009;277(1-2):17-21.

[102] McCarron MO, Muir KW, Weir CJ, Dyker AG, Bone I, Nicoll JA, Lees KR. The apolipoprotein E epsilon4 allele and outcome in cerebrovascular disease. Stroke 1998;29(9):1882-1887.

[103] Wang L, Gu Y, Wu G, Wang W, Liu J, Liu J, Wu Z. A case control study on the distribution of apolipoprotein AI gene polymorphisms in the survivors of atherosclerosis cerebral infarction. Zhonghua Liu Xing Bing Xue Za Zhi 2000;21:22–25.

[104] Xia J, Yang Q, Yang Q, Xu H, Zhang L. The relationship of apolipoprotein H G1025C (Try316Ser) polymorphism with stroke and its effect on plasma lipid levels in Changsha Hans. Zhonghua Yi Xue Yi Chuan Xue Za Zhi 2003;20:114–118.

[105] Wang L, Gu Y, Wu G. The relation between polymorphisms of apolipoprotein B gene and atherosclerotic cerebral infarction. Zhonghua Yi Xue Za Zhi 1999;79:603–606.

[106] Stanković A, Stanković S, Jovanović-Marković Z, Zivković M, Djurić T, Glišić-Milosavljević S, Alavantić D. Apolipoprotein B gene polymorphisms in patients from Serbia with ischemic cerebrovascular disease. Arch Biol Sci 2007;59(4):303–309.

[107] Zhang L, Zeng Y, Ma M, Yang Q, Hu Z, Du X. Association study between C7673T polymorphism in apolipoprotein B gene and cerebral infarction with family history in a Chinese population. Neurol India 2009;57(5):584-588.

[108] Benn M, Nordestgaard BG, Jensen JS, Tybjaerg-Hansen A. Polymorphisms in apolipoprotein B and risk of ischemic stroke. J Clin Endocrinol Metab 2007;92(9):3611-3617.

[109] Havasi V, Szolnoki Z, Talián G, Bene J, Komlósi K, Maász A, Somogyvári F, Kondacs A, Szabó M, Fodor L, Bodor A, Melegh B. Apolipoprotein A5 gene promoter region T-1131C polymorphism associates with elevated circulating triglyceride levels and confers susceptibility of ischemic stroke. J Mol Neurosci 2006;29(2):177-183.

[110] Li J, Xu, Zhu XY. Association of APOA5 gene polymorphism with levels of lipids and atherosclerotic cerebral infarction in Chinese. Zhonghua Yi Xue Yi Chuam Xue Za Zhi 2007;24:576–578.

[111] Zhang K, Qiu F, Li L, Gu GY, Tao Y, Wang L, Luo XY, Xia YQ. The associated study on apolipoprotein A5 gene polymorphisms with carotid artherosclerosis in patients with cerebral infartion. Zhonghua Yi Xue Yi Chuan Xue Za Zhi 2008;25:284-288.

[112] Járomi L, Csöngei V, Polgár N, Szolnoki Z, Maász A, Horvatovich K, Faragó B, Sipeky C, Sáfrány E, Magyari L, Kisfali P, Mohás M, Janicsek I, Lakner L, Melegh B. Functional variants of glucokinase regulatory protein and apolipoprotein A5 genes in ischemic stroke. J Mol Neurosci 2010;41(1):121-128.

[113] Maász A, Kisfali P, Szolnoki Z, Hadarits F, Melegh B. Apolipoprotein A5 gene C56G variant confers risk for the development of large-vessel associated ischemic stroke. J Neurol 2008;255(5):649-654.

[114] Maasz A, Kisfali P, Jaromi L, Horvatovich K, Szolnoki Z, Csongei V, Safrany E, Sipeky C, Hadarits F, Melegh B. Apolipoprotein A5 gene IVS3+G476A allelic variant confers susceptibility for development of ischemic stroke. Circ J 2008;72(7):1065-1070.

[115] Li X, Su D, Zhang X, Zhang C. Association of apolipoprotein A5 gene promoter region -1131T>C with risk of stroke in Han Chinese. Eur J Intern Med 2011;22(1):99-102.

[116] Ogorelkova M, Kraft HG, Ehnholm C, Utermann G. Single nucleotide polymorphisms in exons of the apo(a) kringles IV types 6 to 10 domain affect Lp(a) plasma concentrations and have different patterns in Africans and Caucasians. Hum Mol Genet 2001;10(8):815-824.

[117] Trommsdorff M, Köchl S, Lingenhel A, Kronenberg F, Delport R, Vermaak H, Lemming L, Klausen IC, Faergeman O, Utermann G, Kraft HG. A pentanucleotide repeat polymorphism in the 5' control region of the apolipoprotein (a) gene is associated with lipoprotein (a) plasma concentrations in Caucasians. J Clin Invest 1995;96(1):150-157.

[118] Hu B, Zhou X, Shao H. Relationship between pentanucleotide repeat polymorphism of apolipoprotein (a) gene and atherosclerosis cerebral infarction in Han nationality. Zhonghua Shen Jing Ge Za Zhi 2000;33:172-175.

[119] Liu X, Sun L, Li Z, Gao Y, Hui R. Relation of pentanucleotide repeat polymorphism of apolipoprotein (a) gene to plasma lipoprotein (a) level among Chinese patients with myocardial infarction and cerebral infarction. Zhonghua Yi Xue Za Zhi 2002;82:1396-1400.

[120] Sun L, Li Z, Zhang H, Ma A, Liao Y, Wang D, Zhao B, Zhu Z, Zhao J, Zhang Z, Wang W, Hui R. Pentanucleotide TTTTA repeat polymorphism of apolipoprotein(a) gene and plasma lipoprotein(a) are associated with ischemic and hemorrhagic stroke in Chinese: a multicenter case-control study in China. Stroke 2003;34(7):1617-1622.

[121] Clarke R, Peden JF, Hopewell JC, Kyriakou T, Goel A, Heath SC, Parish S, Barlera S, Franzosi MG, Rust S, Bennett D, Silveira A, Malarstig A, Green FR, Lathrop M, Gigante B, Leander K, de Faire U, Seedorf U, Hamsten A, Collins R, Watkins H, Farrall M. Genetic variants associated with Lp(a) lipoprotein level and coronary disease. N Engl J Med 2009;361(26):2518-2528.

[122] Wang X, Cheng S, Brophy VH, Erlich HA, Mannhalter C, Berger K, Lalouschek W, Browner WS, Shi Y, Ringelstein EB, Kessler C, Luedemann J, Lindpaintner K, Liu L, Ridker PM, Zee RY, Cook NR. A meta-analysis of candidate gene polymorphisms and ischemic stroke in 6 study populations: association of lymphotoxin-alpha in nonhypertensive patients. Stroke 2009;40(3):683-695.

[123] Hopewell JC, Clarke R, Parish S, Armitage J, Lathrop M, Hager J, Collins R; Heart Protection Study Collaborative Group. Lipoprotein(a) genetic variants associated with coronary and peripheral vascular disease but not with stroke risk in the Heart Protection Study. Circ Cardiovasc Genet 2011;4(1):68-73.

[124] Chasman DI, Shiffman D, Zee RY, Louie JZ, Luke MM, Rowland CM, Catanese JJ, Buring JE, Devlin JJ, Ridker PM. Polymorphism in the apolipoprotein(a) gene, plasma lipoprotein(a), cardiovascular disease, and low-dose aspirin therapy. Atherosclerosis 2009;203(2):371-376.

[125] Ordovas JM, Cupples LA, Corella D, Otvos JD, Osgood D, Martinez A, Lahoz C, Coltell O, Wilson PW, Schaefer EJ. Association of cholesteryl ester transfer protein-TaqIB polymorphism with variations in lipoprotein subclasses and coronary heart disease risk: the Framingham study. Arterioscler Thromb Vasc Biol 2000;20(5):1323-1329.

[126] Brousseau ME, O'Connor JJ Jr, Ordovas JM, Collins D, Otvos JD, Massov T, McNamara JR, Rubins HB, Robins SJ, Schaefer EJ. Cholesteryl ester transfer protein TaqI B2B2 genotype is associated with higher HDL cholesterol levels and lower risk of coronary heart disease end points in men with HDL deficiency: Veterans Affairs HDL Cholesterol Intervention Trial. Arterioscler Thromb Vasc Biol 2002;22(7):1148-1154.

[127] Kuivenhoven JA, Jukema JW, Zwinderman AH, de Knijff P, McPherson R, Bruschke AV, Lie KI, Kastelein JJ. The role of a common variant of the cholesteryl ester transfer protein gene in the progression of coronary atherosclerosis. The Regression Growth Evaluation Statin Study Group. N Engl J Med 1998;338(2):86-93.

[128] Elosua R, Cupples LA, Fox CS, Polak JF, D'Agostino RA Sr, Wolf PA, O'Donnell CJ, Ordovas JM. Association between well-characterized lipoprotein-related genetic variants and carotid intimal medial thickness and stenosis: The Framingham Heart Study. Atherosclerosis 2006;189(1):222-228.

[129] Meguro S, Takei I, Murata M, Hirose H, Takei N, Mitsuyoshi Y, ishii K, Oguchi S, Shinohara J, Takeshita E, Watanabe K, Saruta T. Cholesteryl ester transfer protein polymorphism associated with macroangiopathy in Japanese patients with type 2 diabetes. Atherosclerosis 2001;156(1):151–156.

[130] Asselbergs FW, Moore JH, van den Berg MP, Rimm EB, de Boer RA, Dullaart RP, Navis G, van Gilst WH. A role for CETP TaqIB polymorphism in determining susceptibility to atrial fibrillation: a nested case control study. BMC Med Genet 2006;7:39.

[131] Zhuang Y, Wang J, Qiang H, Li Y, Liu X, Li L, Chen G. Cholesteryl ester transfer protein levels and gene deficiency in Chinese patients with cardio-cerebrovascular diseases. Chin Med J (Engl) 2002;115(3):371-374.

[132] Fidani L, Hatzitolios AI, Goulas A, Savopoulos C, Basayannis C, Kotsis A. Cholesterlyl ester transfer protein TaqI B and lipoprotein lipase Ser447Ter gene polymorphisms are not associated with ischaemic stroke in Greek patients. Neurosci Lett 2005;384(1-2):102-105.

[133] Quarta G, Stanzione R, Evangelista A, Zanda B, Sciarretta S, Di Angelantonio E, Marchitti S, Di Murro D, Volpe M, Rubattu S. A protective role of a cholesteryl ester transfer protein gene variant towards ischaemic stroke in Sardinians. J Int Med 2007;262(5):555-561.

[134] Enquobahrie DA, Smith NL, Bis JC, Carty CL, Rice KM, Lumley T, Hindorff LA, Lemaitre RN, Williams MA, Siscovick DS, Heckbert SR, Psaty BM. Cholesterol ester transfer protein, interleukin-8, peroxisome proliferator activator receptor alpha, and toll-like receptor 4 genetic variations and risk of incident nonfatal myocardial infarction and ischemic stroke. Am J Cardiol 2008;101(12):1683-1688.

[135] Hindorff LA, Lemaitre RN, Smith NL, Bis JC, Marciante KD, Rice KM, Lumley T, Enquobahrie DA, Li G, Heckbert SR, Psaty BM. Common genetic variation in six lipid-related and statin-related genes, statin use and risk of incident nonfatal myocardial infarction and stroke. Pharmacogenet Genomics 2008;18(8):677-682.

[136] Von Eckardstein A, Nofer JR, Assman G. High density lipoproteins and atherosclerosis. Role of cholesterol efflux and reverse cholesterol transport. Arterioscler Thromb Vasc Biol 2001;21(1):13-27.

[137] Braunham LR, Singaraja RR, Hayden MR. Variations on a gene: rare and common variants in ABCA1 and their impact on HDL cholesterol levels and atherosclerosis. Annu Rev Nutr 2006;26:105-129.

[138] Andrikovics H, Pongrácz E, Kalina E, Szilvási A, Aslanidis C, Schmitz G, Tordai I. Decreased frequencies of ABCA1 polymorphisms R219K and V771M in Hungarian patients with cerebrovascular and cardiovascular diseases. Cerebrovasc Dis 2006;21(4):254-259.

[139] Pasdar A, Yadegarfar G, Cumming A, Whalley L, St Clair D, MacLeod MJ. The effect of ABCA1 gene polymorphisms on ischaemic stroke risk and relationship with lipid profile. BMC Med Genetics 2007;8:30-36.

[140] Yamada Y, Metoki N, Yoshida H, Satoh K, Kato K, Hibino T, Yokoi K, Watanabe S, Ichihara S, Aoyagi Y, Yasunaga A, Park H, Tanaka M, Nozawa Y. Genetic factors for ischemic and hemorrhagic stroke in Japanese individuals. Stroke 2008;39(8):2211-2218.

[141] Hide WA, Chan L, Li WH. Structure and evolution of the lipase superfamily. J Lipid Res 1992;33(2):167-178.

[142] Mead JR, Irvine SA, Ramji DP. Lipoprotein lipase: structure, function, regulation, and role in disease. J Mol Med 2002;80(12):753-769.

[143] Goldberg IJ. Lipoprotein lipase and lipolysis: central roles in lipoprotein metabolism and atherogenesis. J Lipid Res 1996;37(4):693-707.

[144] Yla-Herttuala S, Lipton BA, Rosenfeld ME, Goldberg IJ, Steinberg D, Witztum JL. Macrophages and smooth muscle cells express lipoprotein lipase in human and rabbit atherosclerotic lesions. Proc Natl Acad Sci USA 1991;88(22):10143-10147.

[145] O'Brien KD, Gordon D, Deeb S, Ferguson M, Chait A. Lipoprotein lipase is synthesized by macrophage-derived foam cells in human coronary atherosclerotic plaques. J Clin Invest 1992;89(5):1544-1550.

[146] Lindqvist P, Ostlund-Lindqvist AM, Witztum JL, Steinberg D, Little JA. The role of lipoprotein lipase in the metabolism of triglyceride-rich lipoproteins by macrophages. J Biol Chem 1983;258(15):9086-9092.

[147] Mead JR, Ramji DP. The pivotal role of lipoprotein lipase in atherosclerosis. Cardiovasc Res 2002;55(2):261-269.

[148] Wittrup HH, Tybjaerg-Hansen A, Nordestgaard BG. Lipoprotein lipase mutations, plasma lipids and lipoproteins, and risk of ischemic heart disease. A meta-analysis. Circulation 1999;99(22):2901-2907.

[149] Kastelein JJ, Ordovas JM, Wittekoek ME, Pimstone SN, Wilson WF, Gagné SE, Larson MG, Schaefer EJ, Boer JM, Gerdes C, Hayden MR. Two common mutations (D9N, N291S) in lipoprotein lipase: a cumulative analysis of their influence on plasma lipids and lipoproteins in men and women. Clin Genet 1999;56(4):297-305.

[150] Chamberlain JC, Thorn JA, Oka K, Galton DJ, Stocks J. DNA polymorphisms at the lipoprotein lipase gene: associations in normal and hypertriglyceridaemic subjects. Atherosclerosis 1989;79(1):85-91.

[151] Gerdes C, Gerdes LU, Hansen PS, Faergeman O. Polymorphisms in the lipoprotein lipase gene and their associations with plasma lipid concentrations in 40-year-old Danish men. Circulation 1995;92(7):1765-1769.

[152] Shimo-Nakanishi Y, Urabe T, Hattori N, Watanabe Y, Nagao T, Yokochi M, Hamamoto M, Mizuno Y. Polymorphism of the lipoprotein lipase gene and risk of atherothrombotic cerebral infarction in the Japanese. Stroke 2001;32(7):1481-1486.

[153] Wang X, Cheng S, Brophy VH, Erlich HA, Mannhalter C, Berger K, Lalouschek W, Browner WS, Shi Y, Ringelstein EB, Kessler C, Luedemann J, Lindpaintner K, Liu L, Ridker PM, Zee RY, Cook NR. A meta-analysis of candidate gene polymorphisms and ischemic stroke in 6 study populations: association of lymphotoxin-alpha in nonhypertensive patients. Stroke 2009;40(3):683-95.

[154] Wang C, Sun T, Li H, Bai J, Li Y. Lipoprotein lipase Ser447Ter polymorphism associated with the risk of ischemic stroke: A meta-analysis. Thromb Res 2011;128(5):e107-e112.

[155] Cummings SR, Nevitt MC, Browner WS, Stone K, Fox KM, Ensrud KE, Cauley J, Black D, Vogt TM. Risk factors for hip fracture in white women. Study of Osteoporotic Fractures Research Group. N Engl J Med 1995;332(12):767-773.

[156] Zhao Y, Ma LY, Liu YX, Wang XY, Liu LS, Lindpaintner K. Relationship between alpha-ENaC gene Thr663Ala polymorphism and ischemic stroke. Zhongguo Yi Xue Ke Xue Yuan Xue Bao 2001;23:499-501.

[157] Huang P, Kostulas K, Huang WX, Crisby M, Kostulas V, Hillert J. Lipoprotein lipase gene polymorphisms in ischaemic stroke and carotid stenosis. Eur J Clin Invest 1997;27(9):740-742.

[158] Wittrup HH, Nordestgaard BG, Sillesen H, Schnohr P, Tybjaerg-Hansen A. A common mutation in lipoprotein lipase confers a 2-fold increase in risk of ischemic cerebrovascular disease in women but not in men. Circulation 2000;101(20):2393-2397.

[159] Myllykangas L, Polvikoski T, Sulkava R, Notkola IL, Rastas S, Verkkoniemi A, Tienari PJ, Niinistö L, Hardy J, Pérez-Tur J, Kontula K, Haltia M. Association of lipoprotein lipase Ser447Ter polymorphism with brain infarction: a population-based neuropathological study. Ann Med 2001;33(7):486-492.

[160] Zhao SP, Tong QG, Xiao ZJ, Cheng YC, Zhou HN, Nie S. The lipoprotein lipase Ser447Ter mutation and risk of stroke in the Chinese. Clin Chim Acta 2003;330(1-2):161-164.

[161] Guan GD, Xu E, Wang XJ, Xu YH, Qiu SD. Associations between Ser447Ter gene polymorphism of lipoprotein lipase and atherosclerotic cerebral infarction. Zhonghua Yi Xue Yi Chuan Xue Za Zhi 2006;23:519-522.

[162] Kostulas K, Brophy VH, Moraitis K, Manolescu A, Kostulas V, Gretarsdottir S, Cheng S, Hillert J. Genetic profile of ischemic cerebrovascular disease and carotid stenosis. Acta Neurol Scand 2008;118(3):146-152.

[163] Xu E, Li W, Zhan L, Guan G, Wang X, Chen S, Shi Y. Polymorphisms of the lipoprotein lipase gene are associated with atherosclerotic cerebral infarction in the Chinese. Neuroscience 2008;155(2):403-408.

[164] Tang X, Zhu YP, Li N, Chen DF, Zhang ZX, Dou HD, Hu YH. Genetic epidemiological study on discordant sib pairs of ischemic stroke in Beijing Fangshan District. Beijing Da Xue Xue Bao 2007;39:119-125.

[165] Aviram M, Billecke S, Sorenson R, Bisgaier C, Newton R, Rosenblat M, Erogul J, Hsu C, Dunlop C, La Du B. Paraoxonase active site required for protection against LDL

oxidation involves its free sulfhydryl group and is different from that required for Its arylesterase/paraoxonase activities: selective action of human paraoxonase allozymes Q and R. Arterioscler Thromb Vasc Biol 1998;18(10):1617-1624.

[166] Salonen JT, Ylä-Herttuala S, Yamamoto R, Butler S, Korpela H, Salonen R, Nyyssönen K, Palinski W, Witztum JL. Autoantibody against oxidised LDL and progression of carotid atherosclerosis. Lancet 1992;339(8798):883-887.

[167] Primo-Parmo SL, Sorenson RC, Teiber J, La Du BN. The human serum paraoxonase/arylesterase gene (PON1) is one member of a multigene family. Genomics 1996;33(3):498-507.

[168] Watson AD, Berliner JA, Hama SY, La Du BN, Faull KF, Fogelman AM, Navab M. Protective effect of high density lipoprotein associated paraoxonase. Inhibition of the biological activity of minimally oxidized low density lipoprotein. J Clin Invest 1995;96(6):2882-2891.

[169] Reddy ST, Wadleigh DJ, Grijalva V, Ng C, Hama S, Gangopadhyay A, Shih DM, Lusis AJ, Navab M, Fogelman AM. Human paraoxonase-3 is an HDL-associated enzyme with biological activity similar to paraoxonase-1 protein but is not regulated by oxidized lipids. Arterioscler Thromb Vasc Biol 2001;21(4):542-547.

[170] Ng CJ, Wadleigh DJ, Gangopadhyay A, Hama S, Grijalva VR, Navab M, Fogelman AM, Reddy ST. Paraoxonase-2 is a ubiquitously expressed protein with antioxidant properties and is capable of preventing cell-mediated oxidative modification of low density lipoprotein. J Biol Chem 2001;276(48):44444-44449.

[171] Mackness M, Mackness B. Paraoxonase 1 and atherosclerosis: is the gene or the protein more important? Free Radic Biol Med 2004;37(9):1317-1323.

[172] Wheeler JG, Keavney BD, Watkins H, Collins R, Danesh J. Four paraoxonase gene polymorphisms in 11212 cases of coronary heart disease and 12786 controls: meta-analysis of 43 studies. Lancet 2004;363(9410):689-695.

[173] Lawlor DA, Day IN, Gaunt TR, Hinks LJ, Briggs PJ, Kiessling M, Timpson N, Smith GD, Ebrahim S. The association of the PON1 Q192R polymorphism with coronary heart disease: findings from the British Women's Heart and Health cohort study and a meta-analysis. BMC Genet 2004;5:17.

[174] Dahabreh IJ, Kitsios GD, Kent DM, Trikalinos TA. Paraoxonase 1 polymorphisms and ischemic stroke risk: A systematic review and meta-analysis. Genet Med 2010;12(10):606-615.

[175] Banerjee I. Relationship between Paraoxonase 1 (PON1) gene polymorphisms and susceptibility of stroke: a meta-analysis. Eur J Epidemiol 2010;25(7):449-458.

[176] Cao H, Girard-Globa A, Serusclat A, Bernard S, Bondon P, Picard S, Berthezene F, Moulin P. Lack of association between carotid intima-media thickness and paraoxonase gene polymorphism in non-insulin dependent diabetes mellitus. Atherosclerosis 1998;138(2):361-366.

[177] Imai Y, Morita H, Kurihara H, Sugiyama T, Kato N, Ebihara A, Hamada C, Kurihara Y, Shindo T, Oh-hashi Y, Yazaki Y. Evidence for association between paraoxonase gene polymorphisms and atherosclerotic diseases. Atherosclerosis 2000;149(2):435-442.

[178] Voetsch B, Benke KS, Damasceno BP, Siqueira LH, Loscalzo J. Paraoxonase 192 Gln--
>Arg polymorphism: an independent risk factor for nonfatal arterial ischemic stroke
among young adults. Stroke 2002;33(6):1459-1464.

[179] Ueno T, Shimazaki E, Matsumoto T, Watanabe H, Tsunemi A, Takahashi Y, Mori M,
Hamano R, Fujioka T, Soma M, Matsumoto K, Kanmatsuse K. Paraoxonase1
polymorphism Leu-Met55 is associated with cerebral infarction in Japanese
populations. Med Sci Monit 2003;9(6):CR208-CR212.

[180] Chen JH, Zeng QX. Relationship between the paraoxonase gene 192 polymorphism
and atherosclerotic cerebral infarction. Zhong Guo Lin Chuang Kang Fu 2003;7:3036-
3037.

[181] Ranade K, Kirchgessner TG, Iakoubova OA, Devlin JJ, DelMonte T, Vishnupad P, Hui
L, Tsuchihashi Z, Sacks FM, Sabatine MS, Braunwald E, White TJ, Shaw PM, Dracopoli
NC. Evaluation of the paraoxonases as candidate genes for stroke: Gln192Arg
polymorphism in the paraoxonse 1 gene is associated with increased risk of stroke.
Stroke 2005;36(11):2346-2350.

[182] Huang Q, Liu YH, Yang Q. The association of PON1 Q192R gene polymorphism with
atherosclerotic cerebral infarction. Zhong Hua Shen Jing Ke Za Zhi 2005;38:454-455.

[183] Wu J, Zhao SP, Tan LM. The relationship between PON1-192 polymorphism and type
of cerebral infarction. Nao Yu Shen Jing Ji Bing Za Zhi 2005;13:253-255.

[184] Yu LT, Yu DC, Li L. The relationship between paraoxonase gene 192Gln/Arg
polymorphism and ischemic cerebrovascular disease. Zhong Hua Lao Nian Xin Nao
Xue Guan Bing Za Zhi 2005;7:254-256.

[185] Baum L, Ng HK, Woo KS, Tomlinson B, Rainer TH, Chen X, Cheung WS, Chan DK,
Thomas GN, Tong CS, Wong KS. Paraoxonase 1 gene Q192R polymorphism affects
stroke and myocardial infarction risk. Clin Biochem 2006;39(3):191-195.

[186] Huang Q, Liu YH, Yang QD, Xiao B, Ge L, Zhang N, Xia J, Zhang L, Liu ZJ. Human
serum paraoxonase gene polymorphisms, Q192R and L55M, are not associated with the
risk of cerebral infarction in Chinese Han population. Neurol Res 2006;28(5):549-554.

[187] Pasdar A, Ross-Adams H, Cumming A, Cheung J, Whalley L, St Clair D, MacLeod MJ.
Paraoxonase gene polymorphism and haplotype analysis in a stroke population. BMC
Medical Genetics 2006;7:28-33.

[188] Aydin M, Gencer M, Cetinkaya Y, Ozkok E, Ozbek Z, Kilic G, Orken C, Tireli H, Kara
I. PON1 55/192 polymorphism, oxidative stress, type, prognosis and severity of stroke.
IUBMB Life 2006;58(3):165-172.

[189] Chen WR, Xiao ZJ, Zhao SQ. The relationship between the gene polymorphism in
paraoxonase and lacunar infarction. Cu Zhong Yu Shen Jing Ji Bing 2006;13:75-78.

[190] Schiavon R, Turazzini M, De Fanti E, Battaglia P, Targa L, Del Colle R, Fasolin A,
Silvestri M, Biasioli S, Guidi G. PON1 activity and genotype in patients with arterial
ischemic stroke and in healthy individuals. Acta Neurol Scand 2007;116(1):26-30.

[191] Shin BS, Oh SY, Kim YS, Kim KW. The paraoxonase gene polymorphism in stroke
patients and lipid profile. Acta Neurol Scand 2008;117(4):237-243.

[192] Slowik A, Wloch D, Szermer P, Wolkow P, Malecki M, Pera J, Turaj W, Dziedzic T,
Klimkowicz-Mrowiec A, Kopec G, Figlewicz DA, Szczudlik A. Paraoxonase 2 gene

C311S polymorphism is associated with a risk of large vessel disese stroke in a Polish population. Cerebrovasc Dis 2007;23(5-6):395-400.

[193] Can Demirdöğen B, Türkanoğlu A, Bek S, Sanisoğlu Y, Demirkaya S, Vural O, Arinç E, Adali O. Paraoxonase/arylesterase ratio, PON1 192Q/R polymorphism and PON1 status are associated with increased risk of ischemic stroke. Clin Biochem 2008;41(1-2):1-9.

[194] Demirdöğen BC, Demirkaya S, Türkanoğlu A, Bek S, Arınç E, Adali O. Analysis of paraoxonase 1 (PON1) genetic polymorphisms and activities as risk factors for ischemic stroke in Turkish population. Cell Biochem Funct 2009;27(8):558-567.

[195] Xiao ZJ, Chen J, Sun Y, Zheng ZJ. Lack of association between the paraoxonase 1 Q/R192 single nucleotide polymorphism and stroke in a Chinese cohort. Acta Neurol Belg 2009;109(3):205-209.

[196] Schmidt R, Schmidt H, Fazekas F, Kapeller P, Roob G, Lechner A, Kostner GM, Hartung HP. MRI cerebral white matter lesions and paraoxonase PON1 polymorphisms: three-year follow-up of the Austrian Stroke Prevention Study. Arterioscler Thromb Vasc Biol 2000;20(7):1811-1816.

[197] Voetsch B, Benke KS, Panhuysen CI, Damasceno BP, Loscalzo J. The combined effect of paraoxonase promoter and coding region polymorphisms on the risk of arterial ischemic stroke among young adults. Arch Neurol 2004;61(3):351-356.

[198] Man BL, Baum L, Fu YP, Chan YY, Lam W, Hui CF, Leung WH, Wong KS. Genetic polymorphisms of Chinese patients with ischemic stroke and concurrent stenoses of extracranial and intracranial vessels. J Clin Neurosci 2010;17(10):1244-1247.

[199] Xu HW, Zhao Z, Yuan N, Xiao B, Yang XS, Tang BS. Relationship between single nucleotide polymorphisms of paraoxonase 2 and stroke. Zhonghua Yi Xue Yi Chuan Xue Za Zhi 2007;24:328-330.

[200] Lazaros L, Markoula S, Kyritsis A, Georgiou I. Paraoxonase gene polymorphisms and stroke severity. Eur J Neurol 2010;17(5):757-759.

[201] Guo Y, Guo J, Zheng D, Pan L, Li Q, Ruan G. Relationship between the Nco I, Ava II polymorphism of low density llpoprotein receptor gene and atherosclerotic cerebral infarction. Zhonghua Yi Xue Yi Chuan Xue Za Zhi 2002;19:209-212.

[202] Frikke-Schmidt R, Nordestgaard BG, Schnohr P, Tybjærg-Hansen A. Single nucleotide polymorphism in the low-density lipoprotein receptor is associated with a threefold risk of stroke. A case-control and prospective study. Eur Heart J 2004;25(11):943-951.

[203] Lee JD, Lin YH, Hsu HL, Huang YC, Wu CY, Ryu SJ, Lee M, Huang YC, Hsiao MC, Chang YJ, Chang CH, Lee TH. Genetic polymorphisms of low density lipoprotein receptor can modify stroke presentation. Neurol Res. 2010;32(5):535-540.

[204] Yang XC, Zhang Q, Li SJ, Wan XH, Zhong GZ, Hu WL, Li L, Yu SZ, Jin L, Wang XF. Association study between three polymorphisms and myocardial infarction and ischemic stroke in Chinese Han population. Thromb Res 2010;126(4):292-294.

[205] Hattori H, Sonoda A, Sato H, Ito D, Tanahashi N, Murata M, Saito I, Watanabe K, Suzuki N. G501C polymorphism of oxidized LDL receptor geen (OLR1) and ischemic stroke. Brain Res 2006;1121(1):246-249.

[206] Vietri MT, Molinari AM, Boggia M, Parisi M, Cioffi M.. IVS4-14 A/G and IVS4-73 C/T polymorphisms in OLR1 gene in patients with ischemic cerebrovascular diseases. Genet Test Mol Biomarkers 2010;14(1):9-11.

[207] Imig JD, Navar LG, Roman RJ, Reddy KK, Falck JR. Actions of epoxygenase metabolites on the preglomerular vasculature. J Am Soc Nephrol 1996;7(11):2364-2370.

[208] Heizer ML, McKinney JS, Ellis EF. 14,15-Epoxyeicosatrienoic acid inhibits platelet aggregation in mouse cerebral arterioles. Stroke 1991;22(11):1389-1393.

[209] Krötz F, Riexinger T, Buerkle MA, Nithipatikom K, Gloe T, Sohn H, Campbell WB, Pohl U. Membrane-potential-dependent inhibition of platelet adhesion to endothelial cells by epoxyeicosatrienoic acids. Arterioscler Thromb Vasc Biol 2004;24(3):595-600.

[210] Zhang W, Otsuka T, Sugo N, Ardeshiri A, Alhadid YK, Iliff JJ, DeBarber AE, Koop DR, Alkayed NJ. Soluble epoxide hydrolase gene deletion is protective against experimental cerebral ischemia. Stroke 2008;39(7): 2073-2078.

[211] Zhang L, Ding H, Yan J, Hui R, Wang W, Kissling GE, Zeldin DC, Wang DW. Genetic variation in cytochrome P450 2J2 and soluble epoxide hydrolase and risk of ischemic stroke in a Chinese population. Pharmacogenet Genomics 2008;18(1):45-51.

[212] Fornage M, Lee CR, Doris PA, Bray MS, Heiss G, Zeldin DC, Boerwinkle E. The soluble epoxide hydrolase gene harbors sequence variation associated with susceptibility to and protection from incident ischemic stroke. Hum Mol Genet 2005;14(19):2829-2837.

[213] Newman JW, Morisseau C, Hammock BD. Epoxide hydrolases: their roles and interactions with lipid metabolism. Prog Lipid Res 2005;44(1):1-51.

[214] Sato K, Emi M, Ezura Y, Fujita Y, Takada D, Ishigami T, Umemura S, Xin Y, Wu LL, Larrinaga-Shum S, Stephenson SH, Hunt SC, Hopkins PN. Soluble epoxide hydrolase variant (Glu287Arg) modifies plasma total cholesterol and triglyceride phenotype in familial hypercholesterolemia: intrafamilial association study in an eight-generation hyperlipidemic kindred. J Hum Genet 2004;49(1):29-34.

[215] Przybyla-Zawislak BD, Srivastava PK, Vázquez-Matiás H, et al. Polymorphism in human soluble epoxide hydrolase. J Mol Pharmacol 2003;64(2):482-490.

[216] Fornage M, Lee CR, Doris PA, Bray MS, Heiss G, Zeldin DC, Boerwinkle E. The soluble epoxide hydrolase gene harbors sequence variation associated with susceptibility to and protection from incident ischemic stroke. Hum Mol Genet 2005;14(19):2829-2837.

[217] Gschwendtner A, Ripke S, Freilinger T, Lichtner P, Müller-Myhsok B, Wichmann H, Meitinger T, Dichgans M. Genetic variation in soluble epoxide hydrolase (EPHX2) is associated with an increased risk of ischemic stroke in white Europeans. Stroke 2008;39(5):1593-1596.

[218] Zhang L, Ding H, Yan J, Hui R, Wang W, Kissling GE, Zeldin DC, Wang DW. Genetic variation in cytochrome P450 2J2 and soluble epoxide hydrolase and risk of ischemic stroke in a Chinese population. Pharmacogenet Genomics 2008;18(1):45-51.

[219] Sandberg M, Hassett C, Adman ET, Meijer J, Omiecinski CJ. Identification and functional characterization of human soluble epoxide hydrolase genetic polymorphisms. J Biol Chem 2000;275(37):28873-28881.

[220] Lee J, Dahl M, Grande P, Tybjaerg-Hansen A, Nordestgaard BG. Genetically reduced soluble epoxide hydrolase activity and risk of stroke and other cardiovascular disease. Stroke 2010,41(1):27-33.

[221] Fava C, Montagnana M, Danese E, Almgren P, Hedblad B, Engström G, Berglund G, Minuz P, Melander O. Homozygosity for the EPHX2 K55R polymorphism increases the long-term risk of ischemic stroke in men: a study in Swedes. Pharmacogenet Genomics 2010;20(2):94-103.

[222] Wang WYS, Barratt BJ, Clayton DG, Todd JA. Genome-wide association studies: theoretical and practical concerns. Nat Rev Genet 2005;6(2):109-118.

[223] Wellcome Trust Case Control Consortium. 2007. Genomewide association study of 14,000 cases of seven common diseases and 3,000 shared controls. Nature 2007;447(7145):661-678.

[224] Matarin M, Brown WM, Scholz S, Simon-Sanchez J, Fung HC, Hernandez D, Gibbs JR, De Vrieze FW, Crews C, Britton A, Langefeld CD, Brott TG, Brown RD Jr, Worrall BB, Frankel M, Silliman S, Case LD, Singleton A, Hardy JA, Rich SS, Meschia JF. A genome-wide genotyping study in patients with ischaemic stroke: Initial analysis and data release. Lancet Neurol 2007;6(5):414–420.

[225] Ikram MA, Seshadri S, Bis JC, Fornage M, DeStefano AL, Aulchenko YS, Debette S, Lumley T, Folsom AR, van den Herik EG, Bos MJ, Beiser A, Cushman M, Launer LJ, Shahar E, Struchalin M, Du Y, Glazer NL, Rosamond WD, Rivadeneira F, Kelly-Hayes M, Lopez OL, Coresh J, Hofman A, DeCarli C, Heckbert SR, Koudstaal PJ, Yang Q, Smith NL, Kase CS, Rice K, Haritunians T, Roks G, de Kort PL, Taylor KD, de Lau LM, Oostra BA, Uiterlinden AG, Rotter JI, Boerwinkle E, Psaty BM, Mosley TH, van Duijn CM, Breteler MM, Longstreth WT Jr, Wolf PA. Genomewide association studies of stroke. N Engl J Med 2009;360(17):1718-1728.

[226] Gretarsdottir S, Thorleifsson G, Manolescu A, Styrkarsdottir U, Helgadottir A, Gschwendtner A, Kostulas K, Kuhlenbaumer G, Bevan S, Jonsdottir T, Bjarnason H, Saemundsdottir J, Palsson S, Arnar DO, Holm H, Thorgeirsson G, Valdimarsson EM, Sveinbjornsdottir S, Gieger C, Berger K, Wichmann HE, Hillert J, Markus H, Gulcher JR, Ringelstein EB, Kong A, Dichgans M, Gudbjartsson DF, Thorsteinsdottir U, Stefansson K. Risk variants for atrial fibrillation on chromosome 4q25 associate with ischemic stroke. Ann Neurol 2008;64(4):402-409.

[227] Gudbjartsson DF, Holm H, Gretarsdottir S, Thorleifsson G, Walters GB, Thorgeirsson G, Gulcher J, Mathiesen EB, Njolstad I, Nyrnes A, Wilsgaard T, Hald EM, Hveem K, Stoltenberg C, Kucera G, Stubblefield T, Carter S, Roden D, Ng MC, Baum L, So WY, Wong KS, Chan JC, Gieger C, Wichmann HE, Gschwendtner A, Dichgans M, Kuhlenbaumer G, Berger K, Ringelstein EB, Bevan S, Markus HS, Kostulas K, Hillert J, Sveinbjornsdottir S, Valdimarsson EM, Lochen ML, Ma RC, Darbar D, Kong A, Arnar DO, Thorsteinsdottir U, Stefansson K. A sequence variant in zfhx3 on 16q22 associates with atrial fibrillation and ischemic stroke. Nat Genet 2009;41(8):876-878.

[228] Kubo M, Hata J, Ninomiya T, Matsuda K, Yonemoto K, Nakano T, Matsushita T, Yamazaki K, Ohnishi Y, Saito S, Kitazono T, Ibayashi S, Sueishi K, Iida M, Nakamura Y,

Kiyohara Y. A nonsynonymous snp in prkch (protein kinase c eta) increases the risk of cerebral infarction. Nat Genet 2007;39(2):212-217.

[229] Hata J, Matsuda K, Ninomiya T, Yonemoto K, Matsushita T, Ohnishi Y, Saito S, Kitazono T, Ibayashi S, Iida M, Kiyohara Y, Nakamura Y, Kubo M. Functional snp in an sp1-binding site of agtrl1 gene is associated with susceptibility to brain infarction. Hum Mol Genet 2007;16(6):630-639.

[230] Yamada Y, Fuku N, Tanaka M, Aoyagi Y, Sawabe M, Metoki N, Yoshida H, Satoh K, Kato K, Watanabe S, Nozawa Y, Hasegawa A, Kojima T. Identification of celsr1 as a susceptibility gene for ischemic stroke in japanese individuals by a genome-wide association study. Atherosclerosis 2009;207(1):144-149.

Plasma Lipoproteins in Brain Inflammatory and Neurodegenerative Diseases

Armando Sena, Carlos Capela, Camila Nóbrega,
Véronique Férret-Sena, Elisa Campos and Rui Pedrosa

Additional information is available at the end of the chapter

1. Introduction

Functions of the central nervous system (CNS) are mainly performed by neurons and glial cells (astrocytes, oligodendrocytes and microglia). Microglia or microcytes have macrophage-like immune related functions; oligodendrocytes are the myelinating cells in the CNS; and astrocytes have diverse roles in synaptogenesis, neurotransmission, myelination and reactive mechanisms to injury. CNS tissue is separated from blood circulation by specialized cell barriers, the most extensive being the endothelium of the so-called blood-brain barrier (BBB).

Brain cholesterol and lipid homeostasis is largely independent of plasma lipoproteins because the BBB restricts the transport of these molecules. In consequence, lipoprotein fractions and compositions in the CNS are different from those in the blood, and consist mainly of high-density lipoproteins (HDL)-like particles. Glial cells (in particular astrocytes) are the main source of cholesterol and HDL-like particles in the CNS [1]. This specialized scenario is reflected on the analysis of cerebrospinal fluid (CSF). The CSF contains apolipoproteins similar to those of plasma, including apoE, apoA-I and A-II, apoC-I, C-II and C-III, apoJ and apoD, but not apoB; apoE and apo A-I are the most abundant. Importantly, while HDL-cholesterol and apoA-I in blood influence its levels in the CSF, this is not the case for apoE, apoJ and apoD, which are synthetized by glial cells [1-2]. In vitro studies have suggested apoA-I expression by brain endothelium and that plasma HDL (containing apoA-I) is transcytosed across the BBB [3]. In consequence, the implications of plasma lipoprotein metabolism in brain physiology and pathological states have been controversial. Nevertheless, many CNS disorders are associated with disturbances of the plasma lipoprotein profile and there is increasingly evidence for pathogenic and clinical relevance of these alterations.

In this chapter we do not pretend to make an exhaustive review on the vast literature related to this theme. Rather, we intend to incorporate some relevant studies in a comprehensive framework addressed to open new avenues of research. With this purpose, we will mainly focus on two frequent and disabling conditions, multiple sclerosis (MS) and Alzheimer disease (AD), and discuss the involvement of plasma lipoproteins in brain inflammatory and neurodegenerative mechanisms. With this approach we expect that useful insights may emerge regarding the contribution of plasma lipoproteins in CNS physiology and pathological states.

2. Multiple sclerosis

MS is a demyelinating inflammatory and neurodegenerative disease of the CNS with heterogeneous pathology (see below) and clinical outcomes. More than 80% of MS patients present initially with acute attacks (relapses) of neurological dysfunction (follow by variable degree of recovery and periods of "remission"), characterizing the relapsing-remitting phenotype (RR-MS). Most of these patients develop a disabling progressive course independently of eventual relapses (secondary progressive MS). A small percentage of patients (10-15%) presents initially with a progressive disease course (primary progressive MS). Patients with clinical isolated syndrome (CIS) have an isolated episode suggestive of MS. The investigation of CIS patients is of special theoretical and practical interest because of their increased risk to develop the disease.

A possible involvement of plasma lipoprotein in MS pathogenesis was suggested in 1953 by the work of Swank [4]. This author presented evidence for a favorable disease course in patients taking a diet poor in animal fat [5]. Sinclair, in 1956, called attention for the importance of a deficiency in polyunsaturated fatty acids and remarked similar epidemiological aspects of MS and cerebrovascular disease [6]. A landmark work, providing a clear potential involvement of plasma lipoproteins in disease was published by Shore *et al*, in 1987 [7]. These authors studied the animal model of MS, experimental autoimmune encephalomyelitis (EAE) and concluded that *"major changes in apoE-containing lipoproteins are undoubtedly significant in the altered immune function in EAE"*. Supporting this prediction, it was observed higher plasma apoE concentrations in MS patients during relapses in comparison to remission states and lower levels in patients under remission in comparison to normal controls [8-10]. Studying EAE induction in apoE-deficient female mice, Karussis in 2003, found that apoE deficiency might be connected with a defective neuronal repair mechanism and enhanced immune reactivity and worse course of the disease [11]. These results could indicate that plasma apoE may have an immunosuppressive role in MS [9]. In agreement with this concept, our group observed that lower levels of plasma apoE might promote immune reactivity in these patients [12].

In their work, Shore *et al* observed higher concentrations of total LDL and HDL cholesterol after onset of clinical symptoms. Giubilei *et al*, in 2002, studied plasma lipoproteins and magnetic resonance imaging (MRI) in patients with a first clinical episode suggestive of MS (CIS), supporting the findings in EAE [13]. These authors observed high total and HDL-cholesterol in these patients and a significant correlation between disease activity (as assessed by MRI) and both total and LDL-cholesterol levels. Jamroz-Wisniewska *et al* found high total cholesterol levels in patients (RR and progressive forms) and also higher LDL-cholesterol in RR patients in remission and in progressive forms than in healthy subjects [14]. Serum paraoxonase 1 (a HDL associated enzyme) activity in relapses was significantly lower in RR patients in comparison to other MS groups. An epidemiological survey based on almost 9000 patients with MS found that the presence of hypercholesterolemia, among other vascular co-morbidities, increased the risk of a more rapid disabling progression of the disease [15]. Recently, Weinstock-Guttman *et al* studied the serum lipid profiles in association with clinical disability and MRI measures in 492 MS patients [16]. They found that worsening disability was associated with higher total and LDL cholesterol, and triglycerides. Higher HDL levels were associated with lower probability for the presence of acute inflammatory lesions (assessed by MRI). Other authors have found higher HDL-cholesterol (and total blood homocysteine) levels in MS patients during a phase of clinical inactivity in comparison to normal controls [17].

The possible influence of apoE allele polymorphism in MS susceptibility and disease severity has been addressed in many studies. Overall, literature does not suggest a role of apoE alleles as risk factor of developing MS [18-19]. An association of apoE polymorphism with disease severity in MS patients has been more controversial. Using MRI methodology, some studies have shown an association between the apoE4 isoform and more severe brain tissue destruction in these patients [20-22]. However, an influence of this isoform on the clinical course of the disease is not established and the interaction with potential confounders should be considered. For example, it was suggested that an influence of apoE polymorphism on the clinical course, and even the risk of MS, could particularly exist in women [23]. Our group has provided evidence for an influence of cigarette smoking in apoE4-carriers, in modulating the clinical severity of RR-MS patients [24]. Some studies have suggested an association of apoE4 allele and apoA1 promoter polymorphism with cognitive impairment in these patients, which may occur very early in the clinical course of the disease [25-26].

As mentioned above, MS is a heterogeneous clinical entity. RR-MS has a higher prevalence in women (which is increasing) and the course of the disease is in general more disabling in men. It is not unreasonable to hypothesize that gender-related and other genetic influences could implicate different impacts of lipoprotein metabolism in MS. Few studies have analyzed the influence of MS therapies in plasma lipoproteins of these patients. However, these studies could provide useful insights on the pathogenic role of this metabolism. Our group first suggested that interferon beta therapy changes this metabolism in RR-MS patients. In particular, we found that at 12 month of therapy, lower apoA1 and higher apoE levels were associated with the presence of relapses and/or progression of the disease [27]. Others authors have found that MS therapy is associated with a decrease of plasma total cholesterol [28-29]. Overall, the reviewed data strongly support a role of plasma lipoproteins metabolism in the pathophysiology of the disease, as discussed below.

2.1. Pathophysiological mechanisms

A major link between plasma lipoproteins and MS concerns the immune system. It is well known that immune reactivity interacts with adaptive alterations of lipoprotein metabolism [30-31]. Recent reports have showed that distinct metabolic programs are essential for survival and functional specialization of different lymphocyte cell populations. For example, lipid oxidation is essential for Treg generation while Th1 differentiation and cytokine production by differentiated Th1, Th2 and Th17 cells are suppressed by lipids and require glucose metabolism [32]. Although the immunopathogenesis of MS lesions (demyelinating plaques) is heterogeneous and may differ in different patients, an imbalance favoring a Th1 effector cell activation is generally accepted [33]. Therefore, it would not be unexpected if an abnormal lipid modulation of immune functions could contribute for MS pathogenesis. However, a primary role of lymphocytes (T cells and B cells) in mediating CNS injury in this disease (at least in all patients) is controversial [32]. Myeloid cells play a pivotal role in the regulation of infiltrating lymphocyte cell activities and are involved in myelin breakdown and axonal injury [33-34]. Macrophages of M1 phenotype are characterized by high production of pro-inflammatory mediators and are crucial in Th1 cell response, while M2 phenotypes are associated with tissue remodeling/repair and expression of anti-inflammatory molecules [35]. In MS lesions, myelin phagocytosis by myeloid cells induces a foamy appearance. Foamy macrophages are originated from resident myeloid cells (microglia) and infiltrating monocytes and are suggested to be of M2-type macrophages and to contribute to the resolution of brain inflammation [36-37].

Macrophage polarization is modulated by different factors. For example, the M2 antinflammatory phenotype is induced by HDLs and apoE [35, 38], and fatty acid and phospholipid synthesis is essential for phagocytic differentiation of human monocytes [39]. ApoE is one ligand for the LDL-receptor-related-protein-1 (LRP1). Quite interesting, LRP1 mediates the downregulation of microglial inflammatory activity by apoE [40] and is essential for phagocytosis of degraded myelin in mice with EAE [41]. Moreover, LRP1 is also expressed in neurons and astrocytes and regulates BBB permeability [42]. This scenario is consistent with a reduction of inflammatory infiltrates and clinical disability by apoE-derived peptides in EAE [43] and immunosuppressive and neuroprotective effects of plasma apoE in EAE [11] and MS patients [8-10, 12].

Among the transcriptional factors regulating macrophage polarization, peroxisome proliferator-activated receptor (PPAR) γ is known to promote M2 macrophages [35]. This is of potential interest in the context of preliminary evidence implicating PPARs in MS pathogenesis and as therapeutic targets for the disease [44-45].

In brain, apoE is associated with HDL-like particles, also containing the second major apolipoprotein, apoA-I. These apolipoproteins are primarily located on separated lipoproteins particles [1]. Although apoE in the brain is predominantly synthetized by glial cells, plasma HDL/apoA-I may cross the BBB and influence its levels in the brain [2]. HDL effects include an inhibition of cytocine-induced expression of adhesion molecules in endothelial cells, which could further depress brain parenchyma immune reactivity [46]. As mentioned, higher levels of plasma HDL were found in CIS and RR-MS and were associated with a lower probability in development of acute inflammatory lesions in these patients [13, 16-17]. Recently, preliminary evidence from our group suggests that higher plasma HDL levels are associated with an increased intrathecal IgG synthesis in these patients [47]. Because low plasma HDL-cholesterol is associated with a predominance of pro-inflammatory phenotype of monocyte-derived macrophage [48], these findings suggest an immunosuppressive role of HDL in the development of MS lesions. This interpretation is further supported by the beneficial therapeutical effects of fingolimod in the disease [49].

Fingolimod (FTY720) is a structural analog of sphingosine, which down modulates sphingosine 1-phosphate (S1P) receptors. S1P is a major component of HDL, including in the CNS and induces an anti-inflammatory phenotype in macrophages. SP1 receptors are widespread in CNS cells and a defect of sphingolipid and phospholipid metabolism is observed early in normal appearing white and grey matter in MS patients. Moreover, S1P is reduced in affected white matter and is increased in CSF of these patients [49]. Importantly, FTY720 treatment has been shown to have neuroprotective effects independent of immunomodulatory mechanisms [50]. These data suggest a protective role of endogenous HDL components not only in the genesis of acute inflammatory lesions but also in the neurodegenerative process of MS.

An involvement of oxidative stress in MS, including of lipid peroxidation has recently received much support [51]. Newcombe et al, in 1994, demonstrated for the first time the presence of oxidized LDL (ox-LDL) and their peroxidative end-products in early and actively demyelinating plaques in post-mortem MS brain [36]. They suggested that plasma LDL enters (through a damaged BBB) the parenchyma and is oxidatively modified in the lesions. More recent data supports an important involvement of oxidative damage including oxidized phospholipids in myelin and axon injury in MS [52]. Several studies have also demonstrated that measures of oxidative stress and lipid peroxidation are consistently increased in the blood of these patients [51]. Our group reported increased levels of serum oxLDL in RR-MS patients in remission in comparison to normal controls and higher levels during relapses [53]. These findings are consistent with a contribution of plasma ox-LDL in promotion BBB permeability and acute inflammatory CNS lesions in the disease. However, increased plasma lipid peroxidation or oxidative stress is probably not associated with disability progression in these patients [54]. The pathophysiology of acute lesions (MS plaques) and disability progression are indeed thought to be mediated by different mechanisms. In fact, it was suggested that low oxygen radical formation in peripheral leukocytes may be associated with a increased severity of the disease [55]. These findings indicate that the role of oxidative stress in MS is complex. An oral formulation of dimethylfumarate (BG-12) activates the Nrf2 antioxidant pathway and was recently observed to be of clinical benefit in RR-MS patients, possibly in disease progression also [56]. These recent promising results should stimulate future research to clarify the involvement of lipid peroxidation in the disease. It should be noted that this involvement further supports a role of plasma HDL in disease pathogenesis, as discussed above. Plasma HDL-associated α-tocopherol is transcytosed across the BBB and may have antioxidant as well as anti-inflammatory effects [3].

Ludewig and Laman (2004) remarked the similarities that may exist between the atherosclerotic plaque development and MS lesions and suggested: *"Systematic comparison of these two diseases involving foam cells in chronic lesions may prove fruitful"* [57]. As we have reviewed, recent research clearly supports this prediction. Moreover, patients with MS have several vascular abnormalities and a higher risk for ischemic stroke [58]. In 2003, our group first reported a pilot trial suggesting a benefit of statin monotherapy in the pathogenesis process (assessed by MRI) and clinical activity of RR-MS patients [59]. These beneficial effects were confirmed by Vollmer *et al* trial in 2004 [60] and in a long-term follow-up of our patients [61]. Very recently, beneficial effects of statin monotherapy were reported in patients with a first clinical episode (CIS) suggestive of MS [62-63]. A synthesis of some shared pathophysiological factors involved in MS and atherosclerosis is presented on Table 1. As we will discuss below, the presence of similar mechanisms involving plasma lipoprotein metabolism in the pathogenesis of atherosclerosis/ischemic and demyelinating lesions may be extensive to other chronic inflammatory and neurodegenerative pathologies.

Pathophysiology	Comment	References
Lesions		
Foam Cells Plaques	Macrophage lipid uptake in early lesion formation	[36-37]
Lipoprotein Related		
Total and LDL cholesterol	Promotion of lesion formation and/or progression	[13-16]
HDL-Cholesterol	Protective of lesion formation and/or progression	[13, 16-17, 47]
ApoA-I	Immunosuppressive and protective	[26-27]
ApoE	Immunosuppressive and protective	[8-10, 12]
Sphingosine-1-phosphate	Modulation of immune reactivity and lesion formation	[49-50]
Oxidative stress and oxLDL	Lipid peroxidation in lesion formation	[51-53]
Secretory phospholipase A_2	Increased expression	[115]
Immunopathogenesis		
M_2-Macrophages	Anti-inflammatory, phagocytic cells in lesions	[37]
T-Cells	Promoting lesion formation	[33]
Inflammatory cytokines	Promoting lesion formation	[33]
Interleukin-10	Protective of lesion formation	[116]

Pathophysiology	Comment	References
Adipocytokines	Leptin and Adiponectin involvement in immune dysfunction	[117-118]
MMP-9	Upregulation associated with lesion formation	[42]
Others		
Statins	Pathological and clinical benefits	[59, 63]
PPARs	PPAR-gamma agonists protective	[45]
Estrogens	Protective of lesion formation and/or progression	[119]
Homocysteine	Increased levels associated with lesion formation	[17, 58]
Platelets	Increased adhesiveness and aggregation	[58, 120]
Smoking	Promotes lesion formation	[121]
Ischemic events	Increased risk associated with atherogenesis and MS pathogenesis	[58]

Table 1. Some pathogenic similarities between multiple sclerosis and atherosclerosis.

3. Alzheimer disease

Possession of the apoE4 allele is the major genetic risk factor for sporadic late-onset AD [64-65]. This observation led to a large body of research on cholesterol and lipid metabolism in patients and animal models of AD during the last two decades. However, the investigation of this metabolism in patients with clinical AD is not sufficient to clarify its role in the pathogenesis of the dementia. It is generally accepted that the pathogenic processes in AD begin many decades before the appearance of evident symptoms. More recently, a major focus of interest has been on longitudinal studies addressing the association between lipoprotein profiles and clinical evolution of cognitive normal subjects or patients with mild cognitive impairment (MCI). A large percentage of patients with the diagnosis of MCI by the 6th decade are known to develop AD later in life. Therefore prospective studies are crucial for development of efficient preventive or therapeutic measures.

Amyloid-β (Aβ) deposition in plaques (AP) (also known senile or neuritic plaques (NP)) and neurofibrillary tangles (NFT), characterized by hyperphosphorylated tau protein aggregates, are pathologic hallmarks of AD [66-67]. The association between plasma lipoprotein profiles and risk of development of clinical manifestations of dementia has been controversial. An association between high cholesterol levels in midlife and an increased risk for dementia in old age has been suggested by several publications [68], but it was not confirmed by a recent large population study [69]. Instead, this study found that low cholesterol levels in late life were predictive of subsequent dementia. Supporting this conclusion, another study in elderly individuals found that low HDL-cholesterol and low total and Non-HDL cholesterol were associated with higher AD risk [70]. These authors suggested a protective effect of late life total cholesterol level on the risk for mild cognitive impairment and AD. Low HDL-cholesterol levels were also associated with decline of memory in middle-aged adults [71]. Within this framework, decreased plasma apoA-I levels have also been found in AD as well as in vascular dementia, and higher apoA-I levels associated with decreased risk of dementia [72-73]. Few studies have investigated the association of lipid profiles with AD-related pathology. A recent work has found that high total cholesterol, LDL-cholesterol and non-HDL-cholesterol levels were associated with risk of development of AP, but not NFT [74]. However, as we will discuss below, the genesis of pathological hallmarks of AD is not invariably associated with clinical manifestations of cognitive impairment and dementia.

The apoE4 allele is an established risk factor for the development of sporadic AD; it is associated with an early age at onset of dementia in an allele dose-dependent manner; and with increased Aβ burden. Moreover, in MCI it predicts conversion to AD. In contrast, apoE2 allele is associated with delayed age of onset of AD [66]. Recent data have provided evidence for an important role of apoE protein levels, independently of the genotype. In one study, middle-aged offspring with familial history of AD were found to have lower plasma apoE levels when compared with offspring without familial history of AD, independent of APOE genotype [75]. In other study, plasma apoE levels were found to be lower in patients with AD and decreased with Aβ load [76].

Overall the reviewed data strongly support a role of plasma lipoprotein metabolism in the pathogenesis of AD, as discussed in more detail below.

3.1. Pathophysiological mechanisms

As already mentioned, NP and NFT are the hallmarks of AD pathology. However, these aggregates are present in a variable extend in about 30% of cognitively normal elderly subjects. In AD, synaptic structural and functional alterations also occur early and are more pronounced than in normal ageing individuals. ApoE-containing lipoproteins, mainly derived from astrocytes, may influence these pathogenic processes in several ways. Cholesterol associated with these lipoproteins is necessary for neurons and to stimulate axonal growth and synaptogenesis. Lipidate-apoE contributes for clearing out Aβ from the brain, a process mediated by apoE receptors (especially LRP1) present in glial cells, neurons and in endothelium of the BBB. Pathways for Aβ clearance also include proteolytic degradation and oligomerization in the aggregates of amyloid plaques, mechanisms also modulated by apoE. For all these processes the isoform apoE4 (which is in general associated with less secreted production of the protein) is less efficient and promotes synaptic dysfunction, toxicity of soluble Aβ and NP deposits. Moreover, it is suggested that apoE4 fragments induce mitochondrial dysfunction and neurotoxicity and that cholesterol levels may regulate Aβ production [65, 66]. Supporting this important role of apoE for AD and the harmful effects of Aβ on cognitive functions, cognitive performance in normal older adults was associated with Aβ load (PET), mainly in ε4 carriers [77].

The above findings do not exclude the contribution of other apolipoproteins for Aβ pathology. For example, apoJ and apoD (see below) also modulate Aβ deposition, a deficiency of apoA-I promotes cognitive impairment and polymorphisms of all these apolipoproteins were associated with risk for AD [1]. Interestingly, increased plasma levels of apoJ (clusterin) are not present before the development of AD but are indeed associated with the severity and progression of the disease, supporting a neuroprotective role [78].

The link between a lipoprotein dysregulation and tau pathology (NFT deposition), in contrast, is not well understood. Beyond the involvement of cholesterol and apolipoproteins, AD is associated with disturbances of sphingolipids and phospholipid metabolism that may contribute for its pathogenesis [67]. Moreover, cognitive impairment and dementia, including AD, are frequently associated with markers of systemic and brain inflammatory activity [79], vascular atherogenic [80] and white-matter (myelin) pathology [81]. An underlying dysregulation of lipoprotein metabolism could be linked to all these pathogenic pathways.

The scenario briefly described above is clearly consistent with the observations that low plasma apoE may be associated with increased risk of AD and correlates with Aβ load, as assessed by PET [75-76]. As remarked, the last studies emphasized the importance of total apoE levels, independently of the genotype. Supporting this concept, it was recently reported in AD mouse models a stimulation of Aβ clearance and cognitive function by inducing apoE expression [82]. After apoE-mediated transport through the BBB, plasma Aβ transport is accomplished by triglyceride-rich lipoproteins (TRL) rich in apoE, for uptake in liver [83]. These findings are also consistent with the risk conferred by low plasma apoE levels. Low plasma apoE levels could also promote systemic immune reactivity and atherogenic pathology in these patients.

Although no relation exists between plasma and brain apoE levels, a strong correlation was found between HDL-cholesterol and apoA-I in serum and in CSF lipoproteins (which are HDL-like particles) [2]. This scenario could contribute to the risk of cognitive impairment and AD conferred by low plasma HDL-cholesterol and apoA-I levels [70-73]. On one hand, these deficiencies could be linked to an increased systemic inflammatory and oxidative status and promotion of atherogenesis. On the other hand, low HDL and apoA-I levels would provide less neurotrophic and immunosuppressive abilities to the brain [84]. If high total and LDL or non-HDL cholesterol in plasma cannot influence its levels in the brain, how could they be associated in some studies with an increased AD risk and Aβ load (NPs)? Experimental studies suggest that plasma cholesterol levels do not normally regulate production of brain Aβ [85]. One possibility resides in the fact that high non-HDL cholesterol in these patients may be associated with low HDL, apoE and apoA-I levels, a pro-inflammatory systemic status and increased atherogenic/ischemic pathology. Supporting this hypothesis, in animal models, cognitive impairment following high fat diet consumption was associated with brain inflammation [86]. Among other markers of inflammation [79], serum levels of adipocytokines have been associated with cognitive impairment and progression of AD, as well as atherogenic/ischemic disease [87-88]. Metabolic syndrome [89] and insulin resistance and type 2 diabetes [90] are associated not only with higher risk of vascular disease but also with risk of dementia, including AD. All these conditions may promote the development of dementia also by affecting myelin integrity and white-matter connective functions.

It should be noted that clinical overt cognitive impairment and dementia do not depend solely on the severity of neurodegenerative and vascular pathologies. Human brain is provided with potential compensatory or plastic mechanism, which may mitigate the clinical impact of ageing-associated pathologies [91-92]. This means that in old age, risk factors for dementia may not have the same significance they have in previous decades. Those factors may include high total and non-HDL cholesterol plasma levels, which may have a major impact in promoting atherogenesis/ischemic/inflammatory processes and AD-related pathology in middle-life, but not in neuroplastic mechanisms increasingly required with advancing age. Lower total and LDL cholesterol have indeed been associated with a poor prognosis in the ischemic stroke [93] and in elderly individuals, as observed above, this profile may increase the risk for overt dementia. Increased body mass index (BMI) in middle life appears to be a risk factor for latter development of cognitive decline and AD, but in late life the burden of cerebral amyloid and tau is associated with lower BMI in cognitively normal and MCI subjects [94-95]. These facts could contribute to the inconsistent results regarding the benefits of statins on prevention and treatment of AD, despite in vitro and animal studies demonstrating an effect in decreasing Aβ formation [96].

In Figure 1 are presented some of the suggested implications of lipoproteins in the pathogenesis of AD and MS.

4. Other brain disorders

Results concerning an association between high serum cholesterol levels and the risk for **Parkinson disease (PD)** have been conflicting. However, this possible association may not exist in older subjects (≥55 years). As for AD, low serum total and LDL-cholesterol levels may increase the risk of PD with advancing age [97-99]. In this context, it is intriguing that hyperlipidemia probably has also a protective role on the neurodegenerative process of **amyotrophic lateral sclerosis (ALS)** [100-101]. PD and ALS also involve inflammatory processes and it has been noted some convergence in the mechanisms underlying neurodegeneration in these disorders, in AD and MS. [102]. An abnormal brain cholesterol homeostasis may also contribute to the pathophysiology of **Huntington's disease** [103]. In what concerns PD, and in contrast to AD, an involvement of apoE genotypes is not clarified. Several studies reported no influence of apoE4 allele in the development of PD or in dementia associated with the disease, which is in contrast to its established role in AD pathogenesis [104]. However, apoE and LRP1 were found to be increased in brain from PD patients, suggesting an involvement in the deposition of α-sinuclein aggregates (Lewy bodies) typical of this disease [105].

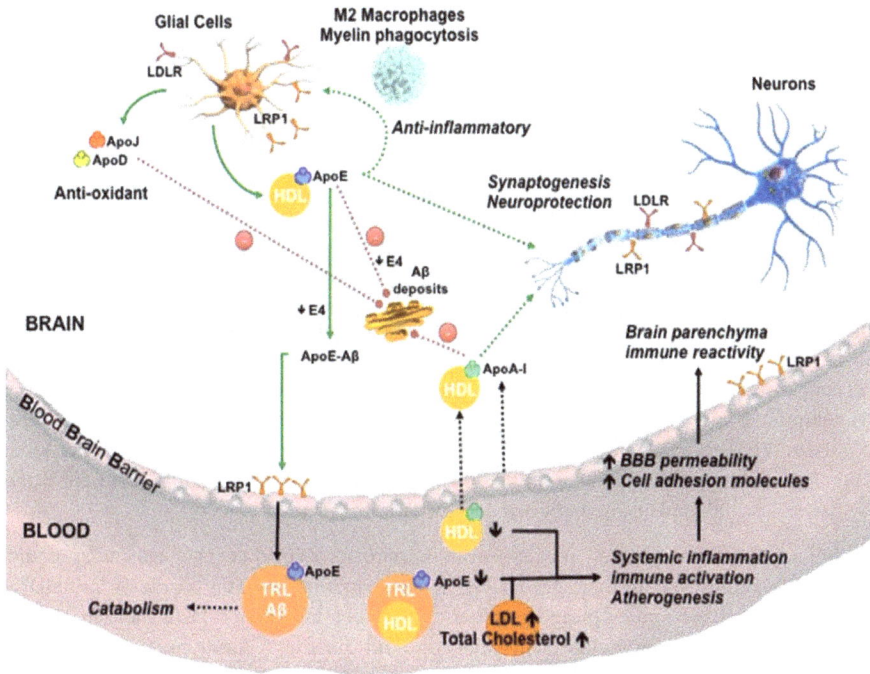

Figure 1. Some putative implications of lipoproteins for the pathogenesis of Multiple Sclerosis and Alzheimer disease (see text for interpretation).

As mentioned above, apolipoprotein D in the CNS is normally synthetized by glial cells (astrocytes and oligodendrocytes). Although present in the CSF in lower concentrations than apoE, A-I and J, some studies have suggested a possible neuroprotective role of apoD in neuropathological states [1,106]. ApoD is a member of the lipocalin family of proteins that are involved in the transport of small hydrophobic ligands. Among several proposed ligands (cholesterol, progesterone, pregnenolone, bilirubin), apoD can bind with high affinity arachidonic acid (AA). Inflammatory responses and oxidative stress associated with brain insults are known to mobilize AA from membranes. Therefore, apoD could have a neuroprotective role by controlling oxidative damage [106-107]. In fact, higher levels of apoD have been found in brain or CSF of AD and other neuropathologies [108]. Curiously, an increase of apoD has also been reported in plasma and certain brain regions of patients with **schizophrenia** and **bipolar disorder**. In these conditions, a disturbance of phospholipid metabolism has been proposed and apoD could represent a response addressed to stabilize membrane AA or bind free AA [106]. The fact that atypical antipsychotics such as clozapine up-regulate apoD expression supports neurotrophic effects of this protein [106]. It should be noted that other apolipoproteins have been implicated in these neuropsychiatric disorders, including apoE, and apoA-I [106, 109-110]. Interestingly, as observed for AD and MS, lower serum apoA-I levels were found in schizophrenia [110].

Overall, these data emphasizes the relevance of plasma lipoprotein metabolism in brain physiology and the convergence of similar dysfunctions of this metabolism associated with several neuropathologies.

5. Conclusion

This review has addressed MS and AD as a strategy to explore the potential relevance of plasma lipoproteins in CNS inflammatory and neurodegenerative disorders. Despite quite different in their demographics, clinical and pathological characteristics, some similarities in their inflammatory and neurodegenerative components have been noted previously [102].

In MS as in AD, the genesis of brain pathology is thought to begin many years before the clinical overt disease. Despite the occurrence of widespread lesions, brain plastic compensatory mechanisms may maintain those disorders clinically silent, delay their symptoms or modify their clinical evolution. Molecular mechanisms underlying grey and white matter plasticity are of outstanding neurobiological and medical importance and are currently poorly understood [111]. This review suggests that an involvement of lipoprotein metabolism in brain plasticity mechanisms is highly plausible and deserves much future research.

Clinical signs of MS very rarely first appear in individuals after 60 years of age and sporadic AD rarely manifest before that age. However, it is remarkable that a profile of low HDL-cholesterol, apoE and apoA-I plasma levels and elevated total and non-HDL cholesterol may promote the risk or progression of disability in both disorders. As discussed, this profile could be associated with both the genesis of lesions in the CNS and the systemic immune-related or metabolic alterations implicated in their pathophysiology (Table 1, Figure 1). It is to note that disturbances in brain cholesterol transport (that may occur in MS, AD and other neuropathologies) can lead to alterations in cholesterol uptake from plasma to brain and decrease plasma HDL levels (112). In MS as in AD, this lipoprotein profile may promote foam cell plaque formations. In young individuals genetically susceptible to MS, this profile may promote the genesis of demyelinating plaques; instead with advanced age, atheroma plaques formation prevails, contributing to AD, in genetically susceptible subjects. Supporting this speculation, MS pathogenesis may share many lipoprotein-related and inflammatory mechanisms underlying atherogenesis (Table1). In addition, with aging, this lipoprotein profile could have a convergent impact for the maintenance of the typical CNS lesions occurring in MS and AD. In fact, advanced ageing may be associated with lower recruitment of anti-inflammatory and phagocytic macrophages and other blood-derived factors to the CNS [113]. This situation, on one hand, favors lower capacity of β-amyloid clearance, oligodendrocyte toxicity and myelin lesions, early present in incipient AD. On the other hand, it restricts remyelination capacities in MS, which are more accentuated with advancing ageing in these patients. The presence of age-related changes in blood circulation has recently been noted of possible relevance for MS and AD [114]. These relevant age-related changes should comprise circulating lipoprotein metabolism.

Despite the similarities of lipoproteins involvement in these two disorders, including the neuroprotective, immunosuppressive and vascular/ischemic protective functions of HDL-cholesterol and associated apolipoproteins (Fig. 1), distinctive implications on their pathogenesis are expected. In MS, a participation of lymphocyte infiltration is certainly important while this is not the case for AD. For example, sphingosine-1-phosphate component of HDL could be special relevant for the immune dysfunction and the abnormal sphingosine metabolism associated with the genesis of demyelinating plaques and neurodegenerative processes in MS. In AD, triglyceride-rich plasma lipoproteins and apoE4 isoform are especially relevant in the clearance of Aβ and genesis of amyloid plaques. It should be emphasized that MS and AD are pathological and clinical heterogeneous diseases. For example, the immunopathogenesis of MS differ among patients even with similar clinical profiles and prominent atherosclerosis lesions are absent in some patients with AD. Therefore, the contribution of plasma lipoprotein metabolism for the pathogenesis of these disorders may be variable and this could explain discrepancies among some studies. Future work aimed to clarify the roles of plasma lipoproteins in these diseases should address clinical homogeneous patient populations, include concomitant pathological and immunological markers and consider potential environmental confounders. Ideally, laboratory data should be correlated with neuroimaging measures. Finally, MS and AD are clear examples of complex conditions for which multiple genetic risk factors for developing and progression are to be expected. Selected genetic typing of the study population is therefore convenient, because lipoprotein alterations may not have the same significance and the same therapeutical implications in different genetic backgrounds.

In sum, the available reviewed data suggest that plasma lipoproteins metabolism is a fruitful "window" to an improved understanding of MS and AD and other neurological diseases. Of outstanding interest, plasma lipoproteins may represent useful targets for discovering preventive and therapeutical strategies for these common disabling human conditions.

A very recent paper from Dr Lawrence Steinman group at Stanford University highlights the importance of lipids in the pathogenesis of MS and the therapeutic potential of lipid-based strategies for the disease (Science Transl Med 2012; 8 (137); E-pub 2012 6 Jun).

Author details

Armando Sena, Carlos Capela, Camila Nóbrega and Elisa Campos
Centro de Estudos de Doenças Crónicas (CEDOC), Faculdade de Ciências Médicas, UNL, Campo Mártires da Pátria, Lisboa, Portugal

Armando Sena, Carlos Capela, Camila Nóbrega and Rui Pedrosa
Serviço de Neurologia, Hospital dos Capuchos, Centro Hospitalar de Lisboa-Central, Lisboa, Portugal

Armando Sena, Camila Nóbrega and Véronique Férret-Sena
Interdisciplinary Centre of Research Egas Moniz (CRiEM), Cooperativa Egas Moniz, Monte da Caparica, Portugal

Acknowledgement

The authors thank to nurses Cristina Araújo and Ana Mendes (Neurology Service) for helping in research and dedicated assistance to our patients and to Merck-Serono, Biogen Idec, Bayer HealthCare, Lundbeck, Octapharma, Teva Pharma and Sanofi-Aventis for supporting our research.

6. References

[1] Hayashi H (2011) Lipid Metabolism and Glial Lipoproteins in the Central Nervous System. Biol. pharm. bull. 34: 453-461.

[2] Fagan AM, Younkin LH, Morris JC et al (2000) Differences in the Aβ40/ Aβ42 Ratio Associated with Cerebrospinal Fluid Lipoproteins as a Function of Apolipoprotein E Genotype. Ann. neurol. 48: 201-210.

[3] Balazs Z, Panzenboeck u, Hammer A et al (2004) Uptake and Transport of High-Density Lipoproteins (HDL) and HDL-Associated α-Tocopherol by an *in vitro* Blood-Brain Barrier Model. J. neurochem. 89: 939-950.

[4] Swank RL (1953) Treatment of Multiple Sclerosis with Low-Fat Diet. Arch. neurol. psych. 69: 91-103.

[5] Swank RL (1970) Multiple Sclerosis. Twenty Years on Low-Fat Diet. Arch. neurol. 23: 460-474

[6] Sinclair HM (1956) Deficiency of Essential Fatty Acids and Atherosclerosis, etcetera. Lancet, 270: 381-383.

[7] Shore VG, Smith ME, Perret V et al (1987) Alterations in Plasma Lipoproteins and Apolipoproteins in Experimental Allergic Encephalomyelitis. J. lipid res. 28: 119-129.

[8] Rifai N, Christenson RH, Gelman BB et al (1987) Changes in Cerebrospinal Fluid IgG and Apoliporotein E Indices in Patients with Multiple Sclerosis during Demyelination and Remyelination. Clin. chem. 33: 1155-1157.

[9] Gelman BB, Rifai N, Christenson RH, et al (1988) Cerebrospinal Fluid and Plasma Apolipoproteins in Patients with Multiple Sclerosis. Ann. clin. lab. science. 18: 46-52.

[10] Carlsson J, Armstrong VW, Reiber H et al (1991) Clinical Relevance of the Quantification of Apolipoprotien E in Cerebrospinal Fluid. Clin. chim. acta, 196: 167-176.

[11] Karussis D, Michaelson DM, Grigoriadis N et al (2003) Lack of Apolipoprotein-E Exarcebates Experimentally Allergic Encephalomyelitis. Mult. scler. 9: 476-480.

[12] Sena A, Bendtzen K, Cascais MJ et al (2010) Influence of Apolipoprotein E Plasma Levels and Tobacco Smoking on the Induction of Neutralising Antibodies to Interferon-Beta. J. neurol. 257: 1703-1707.

[13] Giubilei F, Antonioni G, Di Legge S et al (2002) Blood Cholesterol and MRI Activity in First Clinical Episode Suggestive of Multiple Sclerosis. Acta neurol. scand. 106: 109-112.

[14] Jamroz-Wisniewska A, Beltowski J, Stemasiak Z et al (2009) Paraoxonase 1 Acitivity in Different Types of Multiple Sclerosis. Mult.scler. 15: 399-402.

[15] Marrie RA, Rudick R, Horwitz R et al (2010) Vascular Comorbility is Associated with More Rapid Disability Progression in Multiple Sclerosis. Neurology 74: 1041-1047.

[16] Weinstock-Guttman B, Zivadinov R, Mahfooz N et al (2011) Serum Lipid Profiles are Associated with Disability and MRI Outcomes in Multiple Sclerosis. J. neuroinflammation 8:127-133

[17] Salemi G, Gueli MC, Vitale F, et al (2010) Blood Lipids, Homocysteine, Stress Factors, and Vitamins in Clinically Stable Multiple Sclerosis Patients. Lipids in health and disease 9:19-21.

[18] Burwick RM, Ramsay PP, Haines JL et al (2006) ApoE Epsilon Variation in Multiple Sclerosis Susceptibility and Disease Severity. Neurology 66: 1373-1383.

[19] Pinholt M, Frederiksen JL, Christiansen M (2006) The Association Between Apolipoprotein E and Multiple Sclerosis. Eur. j. neur. 13: 573-580.

[20] Enzinger C, Ropele S, Strasser-Fuchs S et al (2003) Lower Levels of N-Acetylaspartate in Multiple Sclerosis Patients with the Apolipoprotein E ε4 Allele. Arch. neurol. 60:65-70.

[21] Enzinger C, Ropele S, Smith S et al (2004) Accelerated Evolution of Brain Atrophy and "Black Holes" in MS Patients with ApoE- ε4. Ann. neurol. 55:563-569.

[22] De Stefano N, Bartolozzi ML, Nacmias B et al (2004) Influence of Apolipoprotein E ε4 Genotype on Brain Tissue Integrity in Relapsing-Remitting Multiple Sclerosis. Arch. neurol. 61: 536-540.

[23] Kantarci OH, Hebrink DD, Achenbach SJ et al (2004) Association of ApoE Polymorphisms with Disease Severity in MS is Limited to Women. Neurology 62: 811-814.

[24] Sena A, Couderc R, Ferret-Sena V et al (2009) Apolipoprotein E Polymorphisms Interacts with Cigarette Smoking in Progression of Multiple Sclerosis. Eur. j. neur. 16: 832-837.

[25] Shi J, Zhao CB, Vollmer TL et al (2008) ApoE ε4 Allele is Associated with Cognitive Impairment in Patients with Multiple Sclerosis. Neurology, 70: 185 190.

[26] Koutsis G, Panas M, Giogkaraki E et al (2009) An ApoAI Promoter Polymorphism is Associated with Cognitive Performance in Patientswith Multiple Sclerosis. Mult. scler. 15: 174-179.

[27] Sena A, Pedrosa R, Ferret-Sena V et al (2000) Interferon β1a Therapy Changes Lipoprotein Metabolism in Patients with Multiple Sclerosis. Clin. chem. lab. med. 38: 209-213.

[28] Morra BV, Coppola G, Orefice G et al (2004) Interferon β Treatment Decreases Cholesterol Plasma Levels in Multiple Sclerosis Patients. Neurology 62: 829-830.

[29] Coppola G, Lanzillo R, Florio C et al (2006) Long-Term Clinical Experience with Weekley Interferon β-1a in Relapsing Multiple Sclerosis. Eur. j. neurol. 13: 1014-1021.

[30] Hansson GK (2007) Light Hits the Liver. Science 316: 206-207.

[31] Glass CK, Saijo, K (2008) Oxysterols Hold T Cells in Check. Nature 455:40-41.

[32] Michalek RD, Gerriets VA, Jacobs SR et al (2011) Cutting Edge: Distinct Glycolytic and Lipid Oxidative Metabolic Programs Are Essential for Effector and Regulatory CD4+ T Cells Subsets. J. immunol. 186: 3299-3303.

[33] Henderson APD, Barnett MH, Parratt JDE et al (2009) Multiple Sclerosis – Distribution of Inflammatory Cells in Newly Forming Lesions. Ann. neurol. 66: 739-753.

[34] Weber MS, Prod'homme T, Ypussef S. et al (2007) Type II Monocytes Modulate T Cell-Mediated Central Nervous System Autoimmune Disease. Nat. med. 13: 935-943.

[35] Chinetti-Gbaguidi G, Staels B (2011) Macrophage Polarization in Metabolic Disorders: Functions and Regulation. Curr. opin. lipidol. 22: 365-372.

[36] Newcombe J, Li H, Cuzner ML (1994) Low Density Lipoprotein Uptake by Macrophages in Multiple Sclerosis Plaques: Implications for Pathogenesis. Neuropathol. appl. neurobiol. 20: 152-162.

[37] Boven LA, Van Mars M, Van Zwam M et al (2006) Myelin-Laden Macrophages Are Anti-Inflammatory with Foam Cells in Multiple Sclerosis. Brain 129: 517-526.

[38] Baitsch D, Bock HH, Engel T et al (2011) Apolipoprotein E Induces Antiinflammatory Phenotype in Macrophages. Arterioscler. thromb. vasc. biol. 31: 1160-1168.

[39] Ecker J, Liebisch G, Englmaier M et al (2010) Induction of Fatty Acids Synthesis is a Key Requirement for Phagocytic Differentiation of Human Monocytes. Proc. natl. acad. sci. USA 107: 7817-7822.

[40] Pocivavsek A, Michailenko I, Strickland DK et al (2009), Microglial Low-Density Lipoprotein Receptor-Related Protein 1 Modulates c-Jun N-Terminal Kinase Activation. J. neuroimmunol. 214: 25-32.

[41] Gaultier A, Wu X, Le Moan N et al (2008) Low-Density Lipoprotein Receptor-Related Protein 1 Is An Essencial Receptor for Myelin Phagacytosis. J. cell sci. 122: 1155-1162.

[42] Yepes M, Sandkvist M, Moore EG et al (2003) Tissue-Type Plasminogen Activator Opening of the Blood-Brain Barrier Via the LDL Receptor-RelatedProtein. J. clin. invest. 112:1533-1540.

[43] Li F-Q, Sempowski GD, McKenna SE et al (2006) Apolipoprotein E-derived Peptides Ameliorate Clinical Disability and Inflammatory Infiltrates into the Spinal Cord in a Murine Model of Multiple Sclerosis. J pharmacol. exp. ther. 318: 956-965.

[44] Sena A, Tavares A, Ferret-Sena V et al (2008) Peroxisome Proliferator-Activated Receptors (PPARs) in Relapsing-Remitting Multiple Sclerosis Patients. Mult. scler. 14: S244.

[45] Shukla DK, Kaiser CC, Stebbins GT et al (2010) Effects of Poliglitazone on Diffusion Tensor Imaging Indices in Multiple Sclerosis Patients. Neuroscience Letters 472: 153-156.

[46] Cockerill GW, Rye KA, Gamble JR et al (1995) High-Density Lipoproteins Inhibit Cytocine-Induced Expression of Endothelial Cell Adhesion Molecules. Arterioscler. thromb. vasc. biol. 15: 1987-1994.

[47] Nobrega C, Capela C, Gorjon A et al (2011) Plasma Lipoproteins and Intrathecal Immunoglobulin Synthesis in Multiple Sclerosis. J. neurol. 258 (Suppl 1): S202.

[48] Sarov-Blat, L, Kiss RS, Haidar B et al (2007) Predominance of a Proinflammatory Phenotype in Monocyte-Derived Macrophages from Subjects with Low Plasma HDL-Cholesterol. Arterioscler. thromb. vasc. biol. 27: 1115-1122.

[49] Mehling M, Johnson TA, Antel J (2011) Clinical Immunology of the Sphongosine 1-Phosphate Receptor Modulator Fingolimod (FTY720) in Multiple Sclerosis. Neurology, 76 (Suppl 3): S20-S27.

[50] Norimatsu Y, Ohmori T, Kimura A et al (2012) FTY720 Improves Functional Recovery after Spinal Cord Injury by Primarily Nonimmunomodulatory Mechanisms. Am. J. pathol. 180: 1625-1635.

[51] Ferretti G, Bacchetti T (2011) Peroxidation of Lipoproteins in Multiple Sclerosis. J. neurol. sci. 311: 92-97

[52] Haider L, Fisher MT, Frischer JM et al (2011) Oxidative Damage in Multiple Sclerosis Lesions. Brain 134: 1914-1924.

[53] Sena A, Pedrosa R, Roque R et al (2006) Oxidised Low Density Lipoprotein in Serum of Relapsing-Remitting Multiple Sclerosis Patients. Mult. scler. 12 (Suppl 1): S168-S169.

[54] Koch M, Mostert J, Arutjunyan AV et al (2007) Plasma Lipid Peroxidation and Progression of Disability in Multiple Sclerosis. Eur. j. neurol. 14: 529-533.

[55] Mossberg N, Movitz C, Hellstrand K et al (2009) Oxygen Radical Production in Leukocytes and Disease Severity in Multiple Sclerosis. J. immunol. 213: 131-134.

[56] Linker RA, Lee D-H, Ryan S et al (2011) Fumaric Acid Esters Exert Neuroprotective Effects in Neuroinflammation Via Activation of the Nrf2 Aantioxidante Pathway. Brain 134: 678-692.

[57] Ludewig B, Laman JD (2004) The In and Out of Monocytes in Atherosclerotic Plaques: Balancing Inflammation through Migration. Proc. natl. acad. sci. USA 101: 11529-11530.

[58] D'haesseleer M, Cambron M, Vanopdenbosch L et al (2011) Vascular Aspects of Multiple Sclerosis. Lancet neurol. 10: 657-666.

[59] Sena A, Pedrosa R, Morais MG (2003) Therapeutical Potential of Lovastatin in Multiple Sclerosis. J. neurol. 250: 754-755.

[60] Vollmer T, Key L, Durkalski V et al (2004) Oral Simvastatin Treatment in Relapsing-Remitting Multiple Sclerosis. Lancet, 363: 1607-1608.

[61] Sena A, Pedrosa R, Morais MG (2007) Beneficial Effect of Statins in Multiple Sclerosis: Is It Dose-Dependent? Atherosclerosis, 191: 462.

[62] Tskiri A, Lakkenbach K, Fuglø D et al (2011) Simvastatin Improves Final Visual Outcome in Acute Optic Neuritis: a Randomized Study. Mult. scler. j. 18: 72-81.

[63] Waubant E, Pelletier D, Mass M et al (2012) Randomized Controlles Trial of Atorvastatin in Clinically Isolated Syndrome. Neurology 78:1171-1178.

[64] Strittmatter WJ, Saunders AM, Schmechel D et al (1993) Apolipoprotein E: High-Avidity Binding to β-Amyloid and Increased Frequency of Type 4 Allele in Late-Onset Alzheimer Disease. Proc. natl. acad. sci. USA 90: 1977-1981.

[65] Corder EH, Saunders AM, Strittmatter WJ (1993) Gene Dose of Apolipoprotein E Type 4 Allele and the Risk of Alzheimer's Disease in Late-Onset Families. Science 261:921-931.

[66] Bu G, (2009) Apolipoprotein E and its Receptors in Alzheimer's Disease: Pathways, Pathogenesis and Therapy. Nat. rev. neuroscience 10: 333-344.

[67] Di Paolo G, Kim T-W (2011) Linking Lipids to Alzheimer's Disease. Cholesterol and Beyond. Nat. rev. neuroscience 12: 284-296.

[68] Haan MN (2010) Midlife Cholesterol Level and Dementia 32 Years Later. Is There a Risk? Neurology 75: 1862-1863.

[69] Mielke MM, Zandi PP, Shao H (2010) The 32-Year Relationship Between Cholesterol and Dementia from Midlife to Late Life. Neurology 75: 1888-1895.

[70] Reitz C, Tang M-X, Schupf N et al (2010) Association of Higher Levels of High-Density Lipoprotein Cholesterol in Elderly Individuals and Lower Risk of Late-Onset Alzheimer Disease. Arch. neurol. 67: 1491-1497.

[71] Singh-Manoux A, Gimeno D, Kivimaki M et al (2008) Low HDL Cholesterol Is a Risk Factor for Deficit and Decline in Memory in Midlife. The Whitehall II Study. Arterioscler. thromb. vasc. biol. 28: 1556-1562.

[72] Merched A, Xia Y, Visvikis S et al (2000) The Relation Between Apolipoprotein AI and Dementia. The Honolulu-Asia Aging Study. Am. j. epidemiol. 165: 985-992.

[73] Kutiyama M, Takahashi K, Yamano T et al (1994) Low Levels of Serum Apolipoprotein AI and AII in Senile Dementia. Jpn j. psychiatry neurol. 48: 589-593.

[74] Matsuzaki T, Sasaki K, Hata J et al (2011) Association of Alzheimer Disease Pathology with Abnormal Lipid Metabolism. The Hisayama Study. Neurology 77: 1068-1075.

[75] Van Vliet p, Westendorp RGJ, Eikelenboom P et al (2009) Parental History of Alzheimer Disease Associated with Lower Plasma Apolipoprotein E Levels. Neurology 73: 681-687.

[76] Gupta VB, Laws SM, Villemagne VL et al (2011) Plasma Apolipoprotein E and Alzheimer Disease Risk. The AIBL Study of Aging. Neurology 76: 1091-1098.

[77] Kantarci K, Lowe V, Przybelski SA et al (2012) ApoE Modifies the Association Between Aβ Load and Cognition in Cognitively Normal Older Adults. Neurology 78: 232-240.

[78] Schrijvers EM, Koudstaal PJ, Hofman A et al (2011) Plasma Clusterin and the Risk of Alzheimer Disease. Jama 305: 1322-1326.

[79] Holmes C, Cunningham C, Zotova E et al (2009) Systemic Inflammation and Disease Progression in Alzheimer Disease. Neurology 73: 768-774.

[80] Iadecola C (2003) Atherosclerosis and Neurodegeneration. Unexpected Conspirators in Alzheimer's Dementia. Arterioscler. thromb. vasc. biol. 23: 1951-1953.

[81] Desai MK, Mastrangelo MA, Ryan DA et al (2010) Early Oligodendrocyte/Myelin Pathology in Alzheimer's Disease Mice Constitutes a Novel Therapeutic Target. Am. J. pathol. 177: 1422-1435.

[82] Cramer PE, Cirrito JR, Wesson DW et al (2012) ApoE-Directed Therapeutics Rapidly Clear β-Amyloid and Reverse Deficits in AD Mouse Models. Science 335: 1503-1506.

[83] Takechi, R, Galloway S, Pallebage-Gamarallage MMS et al (2008) Chylomicron Amyloid-Beta in the Aetiology of Alzheimer's Disease. Atherosclerosis (Suppl 9) 19-25.

[84] Kontush A, Chapman MJ (2008) HDL: Close to Our Memories? Arterioscler. thromb. vasc. biol. 28: 1418-1420.

[85] Elder GA, Cho JY, English DF et al (2007) Elevated Plasma Cholesterol. Does Not Affect Brain Aβ in Mice Lacking the Low-Density Lipoprotein Receptor. J. neurochem. 102: 1220-1231.

[86] Pistell PJ, Morrison CD, Gupta S et al (2010) Cognitive Impairment Following High Fat Diet Consumption Is Associated with Brain Inflammation. J. immunol. 219: 25-32.

[87] Lieb W, Beiser AS, Vasan RS et al (2009) Association of Plasma Leptin Levels with Incident Alzheimer Disease and MRI Measures of Brain Aging. Jama 302: 2565-2572.

[88] Une K, Takei A, Tomita N et al (2010) Adiponectin in Plasma and Cerebrospinal Fluid in MCI and Alzheimer's Disease. Eur. j. neurol. 18: 1006-1009.

[89] Yaffe K, Weston AL, Blackwell T et al (2009) The Metabolic Syndrome and Development of Cognitive Impairment Among Older Women. Arch. neurol. 66: 324-328.

[90] Craft S (2009) The Role of Metabolic Disorders in Alzheimer Disease and Vascular Dementia. Arch. neurol. 66: 300-305.

[91] Brayne C, Ince PG, Keage H et al (2010) Education, the Brain and Dementia: Neuroprotection or Compensation? Brain 133: 2210-2216.

[92] Belleville S, Clément F, Mellah S et al (2011) Training-Related Brain Plasticity in Subjects at Risk of Development Alzheimer's Disease. Brain 134: 1623-1634.

[93] Cuadrado-Godia E, Jiménez-Conde J, Ois A et al (2009) Sex Differences in the Prognostic Value of the Lipid Profile After the First Ischemic Stroke. J. neurol. 256: 989-995.

[94] Hughes TF, Borenstein AR, Schofield E et al (2009) Association Between Late-Life Body Mass Index and Dementia. The Kame Project. Neurology 72: 1741-1746.

[95] Vidoni ED, Townley RA, Honea RA et al (2011) Alzheimer Disease Biomarkers Are Associated with Body Mass Index. Neurology 77: 1913-1920.

[96] Shepardson NE, Shankar GM, Selkoe DJ (2011) Cholesterol Level and Statin Use in Alzheimer Disease. I. Review of Epidemiological and Preclinical Studies. Arch. neurol. 68: 1239-1244.

[97] De Lau LML, Koudstaal PJ, Hofman A et al (2006) Serum Cholesterol Levels and the Risk of Parkinson's Disease. Am. j. epidemiol. 164: 998-1002.

[98] Huang X, Chen H, Miller WC et al (2007) Lower Low-Density Lipoprotein Cholesterol Levels Are Associated with Parkinson's Disease. Mov. disord. 22: 377-381.

[99] Hu G, Antikainen R, Jousilahti P et al (2008) Total Cholesterol and the Risk of Parkinson Disease. Neurology 70: 1972-1979.

[100] Dupuis L, Corcia P, Fergani A et al (2008) Dyslipidemia Is a Protective Factor in Amyotrophic Lateral Sclerosis. Neurology 70: 1004-1009.

[101] Chiò A, Calvo A, Ilardi A et al (2009) Lower Serum Lipids Are Related to Respiratory Impairment in Patients with ALS. Neurology 73: 1681-1685.

[102] Glass CK; Saijo K, Winner B et al (2010) Mechanisms Underlying Inflammation in Neurodegeneration. Cell140: 918-934.

[103] Valenza M, Cattaneo E (2011) Emerging Roles for Cholesterol in Huntington's Disease. Trends in neurosci. 34: 474-486.

[104] Whitehead AS, Bertrandy S, Finnan F et al (1996) Frequency of the Apolipoprotein E ε4 Allele in a Case-Control Study of Early Onset Parkinson's Disease. J. neurol. neurosurg. psychiatry (1996) 61: 347-351.

[105] Wilhelmus MMM, Bol JGJM, Rozemuller AJM et al (2011) Apolipoprotein E and LRP1 Increase Early in Parkinson's Disease Pathogenesis. Am. J. pathol. 179: 2152-2156.

[106] Thomas EA, Sutcliffe JG (2002) The Neurobiology of Apoliporoteins in Psichiatric Disorders. Mol. Neurobiol. 26: 369-388.

[107] Muffat J, walker DW, Benzer S (2008) Human ApoD, an Up-Regulated in Neurodegenerative Diseases, Extends Lifespan and Increases Stress Resistance in *Drosophila*. Proc. natl. acad. sci. USA 105: 7088-7093.

[108] Terrise L, Poirier J, Bertrand P et al (1998) Increased Levels of Apolipoprotein D in Cerebrospinal Fluid and Hippocampus of Alzheimer's Patients. J. neurochem. 71: 1643-1650.

[109] Digney A, Keriakous D, Scarr E et al (2005) Differential Changes in Apolipoprotein E in Schizophrenia and Bipolar I Disorder. Biol. Psychiatry 57: 711-715.

[110] Huang JT-J, Wang L, Prabakaran S et al (2008) Independent Protein-Profiling Studies Show a Decrease in Apolipoprotein AI Levels in Schizophrenia CSF, Brain and Peripheral Tissues. Mol. psychiatry 13: 1118-1128.

[111] Zatorre RJ, Fields RD, Johansen-Berg H et al (2012) Plasticity in Gray and White: Neuroimaging Changes in Brain Structure During Learning. Nat. rev. neurosci. 15: 528-536.

[112] Karasinska JM, Rinninger F, Lütjohann et al (2009) Specific Loss of Brain ABCA1 Increases Brain Cholesterol Uptake and Influences Neuronal Structure and Function, J. neurosci., 29: 3579-3589.

[113] Ruckh JM, Zhao J-W, Shadrach JL et al (2012) Rejuvenation of Regeneration in the Aging Central Nervous System. Cell stem cell 10: 96-103.

[114] Redmond SA, Chan JR (2012) Revitalizing Remyelination–the Answer Is Circulating. Science 336: 161-162.

[115] Cunnigham TJ, Yao L, Oetinger M et al (2006) Secreted Phospholipase A2 Activity in Experimental Autoimmune Encephalomyelitis and Multiple Sclerosis. J. neuroinflammation 3: 26-33.

[116] Hesse D, Krakauer M, Lund H et al (2011) Disease Protection and Interleukin-10 Induction by Endogenous Interferon-β in Multiple Sclerosis? Eur. j. neurol. 18: 266-272.

[117] Matarese G, Procaccini C, De Rosa V, (2008) The Intricate Interface Between Imuune and Metabolic Regulation: a Role for Leptin in the Pathogenesis of Multiple Sclerosis? J. leukoc. biol. 84: 893-899.

[118] Hietaharju A, Kuusisto H, Nieminen R et al (2010) Elevated Cerebrospinal Fluid Adiponectin and Adipsin Levels in Patients with Multiple Sclerosis: a Finnish Co-Twin Study. Eur. j. neurol. 17: 332-334.

[119] Sena A, Couderc R, Vasconcelos JC et al (2012) Oral Contraceptive Use and Clinical Outcomes in Patients with Multiple Sclerosis. J. neurol. sci. 317(1-2): 47-51.

[120] Neu IS, Prosiegel M, Pfaffenrath V (1982) Platelet Aggregation and Multiple Sclerosis. Acta neurol. scandinav. 66:497-504.

[121] Hawkes CH, (2007) Smoking is a Risk Factor for Multiple Sclerosis: a metanalysis. Mult. scler. 13: 610-615.

Lipoproteins and Cancer

Chapter 11

Lipoproteins and Cancer

Caryl J. Antalis and Kimberly K. Buhman

Additional information is available at the end of the chapter

1. Introduction

Circulating lipoproteins perform vital functions, including the transport of fatty acids and cholesterol from intestine and liver throughout the body. However, in well-fed Western societies, elevated concentrations of lipoproteins in blood have long been recognized to convey increased risk for cardiovascular disease. High fat diets, obesity, and heredity can all contribute to hyperlipidemia. More recently, there has been concern for the possible effects of hyperlipidemia on risk for or progression of cancers, which have a far greater demand for lipids than normal tissues. For example, obesity is now an established risk factor for certain types of cancer and is also found to affect the prognosis for cancer patients (Calle and Kaaks 2004; Cleary and Grossmann 2009). While the association of obesity with cancer is complex, higher circulating lipids may be a contributing element. Similarly type 2 diabetes, a condition of multiple co-morbidities including hyperlipidemia, is associated with the incidence of and mortality from cancer (Faulds and Dahlman-Wright 2012).

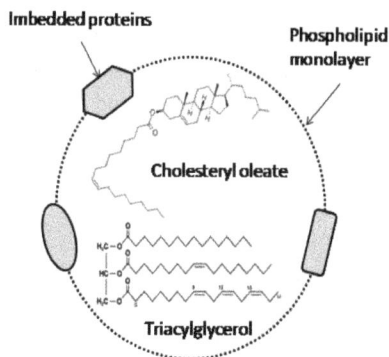

Figure 1. Cytoplasmic lipid droplets consist of an oily core of TAG and CE surrounded by a phospholipid monolayer, specific coat proteins, and other proteins.

The association of hyperlipidemia with cancer began with early observations of an accumulation of cholesterol in tumors (reviewed, (Mulas, Abete et al. 2011)). Higher levels of cholesterol and cholesteryl esters (CE) in malignant compared to less malignant tumors and normal tissues were first measured chemically (Yasuda and Bloor 1932). The accumulation of lipids in tumors was subsequently noted in tumor sections through histological examination and staining for lipid droplets (also called lipid bodies) (Freitas, Pontiggia et al. 1990). Lipid droplets are cellular organelles that store neutral lipids triacylglycerol (TAG) and CE (**Fig. 1**). Adipocytes store lipids in a single, large lipid droplet. Most other cell types have fewer, smaller lipid droplets except under pathological conditions when increased numbers and amounts of lipid may be present (Bozza and Viola 2010). Lipid droplets were detected *in vivo* in tumors with proton magnetic resonance (Delikatny, Chawla et al. 2011), and more recently, *in vivo* and *in vitro* with coherent anti-Stokes Raman scattering microscopy (Le, Huff et al. 2009). Unlike adipocyte lipid droplets, tumor cell lipid droplets contain significant quantities of CE (Tosi and Tugnoli 2005); therefore as these tumors grow and accumulate cholesterol, they may be expected to affect whole body cholesterol homeostasis and circulating cholesterol levels.

The observation of changes in plasma cholesterol in cancer patients constitutes the second line of evidence in the association of lipoproteins with cancer. It appeared in multiple studies over many years that lower plasma cholesterol was associated with a higher risk of cancer (Rose and Shipley 1980). This was a concern because lowering plasma cholesterol is a goal in cardiovascular disease prevention. The relationship between plasma cholesterol and cancer was examined in many population-based studies. Although total plasma cholesterol (total-C) measurements were used in many studies, determinations of individual lipoprotein cholesterol fractions were increasingly included. Plasma cholesterol resides primarily in low density lipoproteins (LDL) and high density lipoproteins (HDL), the lipoproteins that transport cholesterol to cells and collect excess cholesterol from cells, respectively. High HDL-C is a protective factor against atherosclerosis, while high LDL-C is positively associated with risk of atherosclerosis.

Two trends ultimately emerged from the data. First, total-C concentrations were lower two to six years prior to a cancer diagnosis, suggesting reverse causation: i.e., the early stages of the tumor led to lower circulating cholesterol (Sharp and Pocock 1997). Second, the plasma cholesterol fraction associated with tumor-caused decreases was primarily HDL-C, although the trend was detectable in total-C values also (Ahn, Lim et al. 2009). These conclusions were supported by data showing an increase in HDL-C when the patient was in remission (Dessi, Batetta et al. 1995).

The observations above suggest that in some types of cancer, tumor cells accumulate cholesterol as CE in lipid droplets and efflux less cholesterol to HDL, resulting in lower circulating HDL-C, detectable even before the tumor can be diagnosed. There is also some indication that low HDL-C levels may contribute to the development of cancer (Mondul, Weinstein et al. 2011). HDL has antioxidant and anti-inflammatory properties in addition to its role in reverse cholesterol transport (Kwiterovich 2000), and low HDL-C is a defining characteristic of the metabolic syndrome which has already been linked to cancer risk (Faulds and Dahlman-Wright 2012). Although lower HDL-C can have multiple etiologies, it can be one indicator of the presence of a tumor. If some tumors accumulate cholesterol, then it might be reasonable to ask if LDL-C fuels the development of this type of tumor.

In this chapter, we will review the evidence that LDL-C, which is usually highly correlated to total-C, is positively associated with the risk of some types of cancer. We will also review the growing body of data on what mechanisms may be involved in tumor cholesterol accumulation and what markers may be useful to identify tumors that are stimulated by cholesterol. We will address the questions: does higher circulating cholesterol increase the risk of or prognosis for certain cancers, and should lowering LDL-C be a goal in the prevention or management of some types of cancer?

2. Clinical and epidemiological evidence for an association of LDL with cancer

The presence of cancer can affect whole body cholesterol homeostasis, leading to the observation of low plasma HDL-C in cancer patients as described above. Plasma LDL-C levels in cancer may be confounded by the increased catabolism of LDL by a known or undiagnosed tumor, leading to an apparent association of low LDL-C with some types of cancer (Vitols, Gahrton et al. 1985). These apparent interactions of synchronous lipoprotein levels with cancer make it difficult to distinguish a tumor-promoting effect of lipoproteins from a tumor-induced effect on lipoproteins. Prospective studies that include a baseline measurement of blood cholesterol levels and a sufficient follow-up period could reveal if there was a positive association of hypercholesterolemia with the incidence of cancer, or in cancer patients, with prognosis or survival. Such studies have been conducted and the results have been somewhat inconsistent, which may be partially explained by the fact that tumors vary greatly by tissue of origin and even by sub-types of tumor arising from the same tissue.

Additional insight has been gained from studies of statins and statin users. Statins (inhibitors of 3-hydroxy-3-methylglutaryl-CoA reductase (HMGCR)), the rate limiting step in cholesterol biosynthesis) are considered to have pleiotropic effects against cancer due to the multiple biosynthetic products downstream of HMGCR (Gazzerro, Proto et al. 2012). However, pharmacokinetic data suggests that the peripheral tissues do not have access to high enough concentrations of therapeutic statins to effect other pathways and that the major effect of statins is through the reduction of cholesterol biosynthesis in the liver (Solomon and Freeman 2008). Statins lower plasma total-C, which reflects a large reduction in LDL-C (up to 50% or more), a lesser reduction of VLDL-C and minor effects on HDL-C. The reduction of circulating LDL-C, a major consequence of statin use, is likely the primary anti-cancer action of statins.

The largest prospective study to date on cholesterol and cancer was done in Korean adults enrolled in the Korean National Health Insurance Corporation (NHIC); participants (n = > one million) underwent biennial medical evaluations where a baseline fasting total-C measurement was obtained and follow-up data was collected for up to 14 years (Kitahara, Berrington de Gonzalez et al. 2011). The study identified cancer types that had a positive trend with quintiles of total-C in men (prostate, P = 0.002, and colon, P = 0.05) and women (breast, P = 0.003, and colon, P = 0.004), as well as those that had a negative trend in men (esophageal, stomach, liver, and lung) and women (liver). The results were adjusted for multiple factors including BMI, and excluded cancers diagnosed in the first 5 years of follow-up. This study identified the hormone-related cancers and colon cancer as having the greatest association with total-C. These cancers are also the most heavily studied with respect to the effects of total-C, statins, or dietary fat.

Prostate cancer. Early stage prostate cancer (PrC) is stimulated by circulating testosterone through over-expression of the androgen receptor (AR). AR signaling regulates the expression of the PrC marker prostate specific antigen (PSA); androgen-deprivation (castration) therapies block AR signaling, providing an effective treatment and reducing PSA levels. However, over time advanced PrC emerges which is resistant to castration therapies (androgen-independent), although the AR may still play a role in tumor progression (Taplin and Balk 2004). Testosterone is synthesized from cholesterol in the testes, but also in advanced prostate tumor cells, providing a rationale for an effect of cholesterol availability on prostate tumorigenesis (Mostaghel, Solomon et al. 2012).

Several large prospective studies in the USA showed an association between higher baseline plasma total-C and the development of high-grade (Gleason sum $\geq$ 7), but not total or low-grade PrC. In the Health Professionals Follow-Up Study, 18,018 men provided a baseline blood sample and were followed for up to 7 years (Platz, Clinton et al. 2008). Men with low total-C had a reduced incidence of high-grade PrC (odds ratio (OR) = 0.61, 95% CI, 0.39-0.98), and the association persisted after excluding men who were diagnosed within 2 years of blood draw. In the Prostate Cancer Prevention Trial (7 years), 5586 men in the placebo arm with a lower baseline total-C measurement had a reduced incidence of Gleason 8-10 PrC (OR = 0.41, 95% CI, 0.22-0.77) (Platz, Till et al. 2009). In the CLUE II study, 6816 men in Washington County, Maryland were followed for a mean of 12 years (Mondul, Clipp et al. 2010). Those with a baseline total-C in the desirable or borderline range had a reduced incidence of high grade PrC (hazard ratio (HR) = 0.68, 95% CI, 0.40-1.18), which was more pronounced in men with a higher BMI (HR = 0.36, 95% CI, 0.16-0.79). Excluding users of cholesterol-lowering drugs or cases diagnosed within two years of follow-up did not change the results.

The differential effects of total-C on high-grade PrC were supported in several studies conducted outside the USA. In the Alpha Tocopherol, Beta Carotene Cancer Prevention Study cohort, baseline fasting total-C and HDL-C were obtained for >29,000 Finnish male smokers who were enrolled between 1985 and 1988. After long-term follow-up (still ongoing) in 2006, and excluding the first 10 years from baseline, it was found that men with higher total-C had increased risk of overall (HR = 1.22, 95% CI, 1.03-1.44) and advanced (HR = 1.85, 95% CI, 1.13-3.03) PrC (Mondul, Weinstein et al. 2011). The Midspan studies (begun in the 1960s and 1970s in Scotland, UK) had a median follow-up period of 24 years after a baseline plasma total-C measurement (Shafique, McLoone et al. 2012). In 12,926 men diagnosed with PrC >5 years after entry into the study (n = 650), the HR for the risk of high-grade disease (Gleason score ≥ 8) in those with cholesterol levels in the second highest quintile or the highest two quintiles combined compared to the lowest quintile was 1.75 (95% CI, 1.03-2.97) and 1.88 (95% CI, 1.08-3.27), respectively. The use of statins was not available. The Nijmegen Biomedical Study in the Netherlands reported that among 2118 men followed for a median period of 6.7 years who had never used cholesterol-lowering drugs (and excluding those diagnosed in the first year), those with higher baseline total-C had increased risk for PrC (HR = 1.39, 95% CI, 1.03–1.88) and aggressive PrC (HR = 1.65, 95% CI, 1.10–2.47 (Kok, van Roermund et al. 2011). An even stronger association was seen for LDL-C levels and PrC (HR = 1.42, 95% CI, 1.00–2.02) and aggressive PrC (HR = 1.83, 95% CI, 1.15–2.90).

Some studies did not support a role for cholesterol in PrC. No association of baseline plasma total-C or HDL-C with incident, advanced, or fatal PrC was found in the HUNT 2 study where a cohort of 29,364 Norwegian men were followed for a mean 9.3 years (Martin, Vatten et al. 2009). A stated limitation of the study was the small number of advanced or fatal cases. Similarly, no association of total-C with incidence of PrC was found in the Apolipoprotein MOrtality RISk (AMORIS) study, which followed 200,660 Swedish men for a mean of 8 years (Van Hemelrijck, Garmo et al. 2011). In this study no information was available on tumor severity, precluding a finding of a differential effect based on tumor grade.

Other types of studies have contributed evidence for the effects of blood cholesterol on PrC. In a cross-sectional cohort study of 531 American men, the incidence of benign prostate hyperplasia was 4-fold greater in those with diabetes who were in the highest compared to the lowest quartile of LDL-C; this effect was not seen in those without diabetes (Parsons, Bergstrom et al. 2008). A positive diagnosis of PrC in African-American (AA) men (n = 521), but not non-AA men (n = 451), undergoing biopsy was >3-fold higher for those in the highest quartile of LDL-C compared to the lowest (Moses, Abd et al. 2009). In a case-control study in 1294 Italian men <75 years of age with incident PrC compared to 1451 men hospitalized with acute, non-neoplastic conditions, the odd ratio (OR) for prostate cancer was 1.54 (95% CI, 1.26-1.89) for those with hypercholesterolemia (Pelucchi, Serraino et al. 2011). A post hoc analysis of the REDUCE study (which evaluated the anti-testosterone dutasteride in men with high prostate specific antigen (PSA) values but no PrC) examined the association of coronary artery disease (CAD) with PrC risk (Thomas, Gerber et al. 2012). In 6729 men who underwent at least one biopsy, those with CAD had an increased risk of PrC diagnosis (OR = 1.35, 95% CI, 1.08–1.67), suggesting common risk factors.

The benefit of statins in PrC prevention or treatment is still under evaluation, but observational studies have demonstrated reduced risk of PrC in statin users (reviewed, (Solomon and Freeman 2008; Marcella, David et al. 2011)). Statin use was recently shown to reduce the risk if death from PrC in a case-control study; cases were residents of New Jersey, USA ages 55 to 79 years who died from PrC between 1997 and 2000 (n = 380) and controls from the population were matched by 5-year age group and race. The unadjusted OR for death from PrC was 0.49 (95% CI, 0.34-0.70) for any exposure to statins and decreased to 0.37 (P < .0001) after multivariate adjustment (Marcella, David et al. 2011). Users of high-potency statins had about 2.5 times more protection compared with users of low-potency statins; the authors suggest that this points to cholesterol-lowering as the mechanism of protection. A positive association between LDL-C and PSA was demonstrated in a longitudinal study of 1214 American veterans undergoing statin treatment between 1990 and 2006 (Hamilton, Goldberg et al. 2008). After a relatively short period of statin use (< 1 year), there was a near-linear relationship between changes in LDL-C and changes in PSA values. After adjustment for multiple factors, for every 10% change in LDL-C, PSA changed by 1.64% (95% CI, 0.64% to 2.65%, P = .001). This relationship held over increases or decreases in the values, although the mean and median changes in LDL and PSA were -26% and -4.1%, respectively (Hamilton, Goldberg et al. 2008). A subsequent study showed that statin use dose-dependently lowered the risk of a PSA recurrence in men who underwent a radical prostectomy (n = 1319) (30% lower risk of PSA recurrence (HR = 0.70, 95% CI, 0.50-0.97) (Hamilton, Banez et al. 2010). Median follow-up time was 24 months for statin users (n = 236, 18%), 36 months for non-users.

Breast cancer. Epidemiological studies showing a higher incidence of breast cancer (BrC) in Westernized countries led to a focus on the role of dietary fat in BrC risk (Kelsey 1993). Although dietary fat may affect circulating cholesterol levels, the specific contribution of plasma lipoproteins to BrC has received less attention. In addition, the relationship between circulating cholesterol and BrC risk may be complicated by the fact that, as for testosterone, cholesterol is a biosynthetic estrogen precursor and structurally similar to estrogen. Estrogen lowers plasma LDL by increasing the expression of the LDLR (Kovanen, Brown et al. 1979; Hulley, Grady et al. 1998), but stimulates breast tumor growth through over-expression of estrogen receptor alpha (herein referred to as ER). Obesity and menopausal status can affect circulating lipids, estrogen levels, and BrC risk.

The Nurses' Health Study of >70,000 female, married, American nurses used self-reported serum cholesterol levels to analyze the association of blood cholesterol with risk of invasive BrC during up to 12 years of follow-up (Eliassen, Colditz et al. 2005). In that study, BrC incidence was not affected by cholesterol levels or use of statins or other lipid-lowering drugs. In a 10-year follow-up of postmenopausal Korean women (n = 170,374), a positive trend for quartiles of baseline fasting serum total-C and BrC incidence was found (HR = 1.31, 95% CI, 1.06-1.61); however, after adjustment for BMI the trend was no longer significant (Ha, Sung et al. 2009). In contrast, 157 of 5865 peri/postmenopausal Swedish women in the Malmö Preventive Project developed BrC over a mean of 6.6 years; relative risk was increased by quartiles of baseline fasting total-C (P for trend, 0.05) (Manjer, Kaaks et al. 2001). This effect was not seen among the 112/3873 premenopausal women who developed BrC over a mean of 9.6 yrs. BMI was not a factor in the risk of BrC in either group.

Because BrC has multiple types with distinct and recognizable patterns of gene expression, different treatments and prognoses, it may be more useful to examine BrC types separately

(Hu, Fan et al. 2006). Expression of the ER is an important discriminating factor among BrC types, with ER- BrC having fewer treatment options and a worse prognosis. A number of studies have shown differences in cholesterol metabolism between ER+ and ER- BrC. LDLR and ER content were determined (by ligand binding) in tumors from 72 Swedish patients who had undergone mastectomy (Rudling, Stahle et al. 1986). Interesting, LDLR content was negatively, while ER content was positively correlated with survival in months. LDLR content strongly and independently predicted a worse prognosis in these patients (Rudling, Stahle et al. 1986). This finding is consistent with more recent data on tumor gene expression, where LDLR mRNA expression was generally higher in ER- as compared to ER+ human breast tumors in multiple studies (P < 0.05, oncomine.org).

Circulating cholesterol may affect severity, recurrence, or outcome of BrC. In a prospective study of Canadian women diagnosed with early stage BrC (n = 520) and followed for a median period of 8.7 years, a trend toward higher risk of recurrence was seen in women with a higher fasting baseline total-C or LDL-C (Bahl, Ennis et al. 2005). Unfortunately, women with preexisting hyperlipidemia were excluded from the study, leaving a population with a smaller range of cholesterol levels in the evaluation. In 24,329 Norwegian women, a higher baseline non-fasting total-C level was not associated with BrC incidence (Vatten and Foss 1990), but those in the highest quartile did have an increased the risk of death from BrC (HR = 2.0, 95% CI, 1.1 – 3.7) (Vatten, Foss et al. 1991). In the Women's Intervention Nutrition Study (WINS), women with BrC counseled for a low-fat diet (20% of calories) and followed for a median period of 5 years had a 24% lower risk of recurrence (n = 96/975, HR = 0.76, 95% CI, 0.60 to 0.98) as compared to the control group (n = 181/1462); interestingly, the effect was even stronger in those whose tumor was ER- (n = 28/205, HR = 0.58, 95% CI, 0.37 to 0.91) as compared to those whose tumor was ER+ (n = 59/273) (Chlebowski, Blackburn et al. 2006). Although neither total-C nor LDL-C were reported, serum fatty acid analysis showed a reduction in saturated fats in the diet group, and saturated fats are known to increase circulating cholesterol levels (Blackburn and Wang 2007).

A number of clinical trials are underway to evaluate statins for the prevention or treatment of breast cancer. Large scale prospective studies on the association of statin use with risk of breast cancer have had mixed results (Cauley, McTiernan et al. 2006; Jacobs, Newton et al. 2011), but beneficial effects of statins on disease recurrence have been documented. In a prospective cohort study of all female residents in Denmark diagnosed with stage I-III invasive BrC between 1996 and 2003 (n = 18,769), users of simvastatin (a lipophilic statin) had a 10% lower risk of recurrence (95% CI, -11% to -8%) as compared with nonusers of statins (Ahern, Pedersen et al. 2011). No reduced risk was observed in users of hydrophilic statins. In 703 American women treated for stage II/III breast cancer between 1999 and 2005 and followed until 2008, users of statins (n = 156) had a reduced risk of recurrence in multivariate analysis (HR = 0.40, 95% CI, 0.24–0.67) (Chae, Valsecchi et al. 2011). No effect was seen on overall survival. Interestingly, a retrospective analysis of BrC patients in the Kaiser Permanente Cancer Registry in California (n = 2141) found that those who had used statins for one year or more had fewer aggressive ER-/PR- tumors and were more likely to have low grade and less invasive tumors (Kumar, Benz et al. 2008).

In a small study of women with newly diagnosed BrC (chemotherapy and radiotherapy naïve, n = 17) who were postmenopausal and normal weight, it was found that oxidized LDL (oxLDL) (P < 0.001), total-C (P = 0.001) and LDL-C (P = 0.001) were higher compared to a matched control group (n = 30) (Delimaris, Faviou et al. 2007). While LDL-C may contribute to cancer risk or prognosis, as in cardiovascular disease oxLDL may also play a role. OxLDL is present as a small percentage of total LDL in normal individuals, but the percentage of oxLDL may increase in pathological states (Holvoet, Lee et al. 2008; Mello, da Silva et al. 2011). An oxLDL receptor (OLR1) and was recently identified experimentally as part of gene signature responsible for transformation, tumor growth, and proliferation in multiple cancer cell lines (Hirsch, Iliopoulos et al. 2010). There is evidence that oxLDL is higher in hypercholesterolemic subjects, and that lowering total LDL with statins will result in lower oxLDL (Stojakovic, Claudel et al. 2010; Tavridou, Efthimiadis et al. 2010).

Ovarian cancer. Ovarian cancer (OvC) has a much lower incidence than BrC, but is more deadly as most tumors are highly advanced at diagnosis. OvC is not stimulated by estrogen, but there is some evidence that circulating cholesterol affects outcomes. In a prospective study of 132 American women with stage III or IV OvC, serum banked at the time of diagnostic surgery was analyzed for total-C, HDL-C, and TAG (LDL was calculated; statin users were excluded) (Li, Elmore et al. 2010). Disease-specific survival was longer in patients with normal LDL as compared to those with elevated LDL-C (59 and 51 months, respectively, P = 0.04). In another study at the same site, statin use was found to be an independent positive prognostic factor in 126 women with stage III/IV OvC, 17 of whom were taking statins at the time of initial surgery (Elmore, Ioffe et al. 2008). Mean progression-free survival, as well as overall survival, was longer for statin users (24 months compared to 16 months, P = 0.007) as compared to statin non-users (62 months compared to 46 months, P = 0.04). Serum was not available to determine actual levels of lipoproteins. In a small study, women with OvC (n = 15) compared to a matched control group (n = 30) had higher oxLDL (P = 0.006) and there was a trend toward higher LDL-C (P = 0.076) (Delimaris, Faviou et al. 2007). The women had not yet received any chemotherapy or radiotherapy at the time of blood collection.

Colorectal cancer. Colon cancer risk was associated with baseline total-C in the Korean NHIC data (Kitahara, Berrington de Gonzalez et al. 2011). Other studies have had mixed results. In the European Prospective Investigation into Cancer and Nutrition, 1238 incident cases of colorectal cancer (CRC) and matched controls were analyzed for an association of CRC risk with serum lipoproteins (van Duijnhoven, Bueno-De-Mesquita et al. 2011). No significant trend for quintiles of total-C or LDL-C with CRC incidence was detected; a negative trend for HDL-C with colon cancer was seen, even when excluding the first two years of follow-up. No correction for the use of statins, aspirin or other medications was possible in this study. In the Japan Collaborative Cohort Study for Evaluation of Cancer Risk, the association of oxLDL and autoantibodies to oxLDL (oLAB) with the incidence of CRC was examined (Suzuki, Ito et al. 2004). A positive trend was found for oxLDL and CRC, even after multiple adjustments (P = 0.038, n = 119 cases, 316 controls); the trend for oLAB was not significant. The adjusted OR for the highest compared to the lowest quartile of oxLDL was 3.10, 95% CI, 1.04-9.23. Although total-C was not different between cases and controls, oxLDL was strongly associated with total-C (P < 0.001, n = 304).

Plasma cholesterol may affect the progression of colon cancer to a more aggressive disease. The fasting lipid profiles of Italian men and women with metastatic CRC (n = 22) had higher synchronous total cholesterol, LDL-cholesterol and LDL/HDL ratios compared to those without metastases (n = 62) (P = 0.03, 0.01, and 0.002, respectively) (Notarnicola, Altomare et al. 2005). These results were independent of BMI. The authors hypothesized that LDL is beneficial for the proliferation and invasion steps of tumor progression. The effect of statin use on CRC incidence is unsettled due to mixed results from several retrospective analyses (Poynter, Gruber et al. 2005; Flick, Habel et al. 2009; Singh, Mahmud et al. 2009). There is hopeful data that statins may lower the recurrence rate of CRC , and a large-scale clinical trial is currently examining the potential of statin therapy to reduce the relapse rate in colon cancer in patients who have had surgery for early stage colon cancer (Hede 2011).

Other cancers. There is little consistent evidence to date from large prospective studies for the positive association of total-C or LDL-C with the incidence of other cancers. However, retrospective case control and observational studies showing a reduced risk of cancer in statin users are suggestive that lowering LDL-C may be an effective preventative strategy for a wider range of cancer types. For example, renal clear cell carcinoma (the most prevalent renal cell carcinoma) is known to accumulate large amounts of CE (Gebhard, Clayman et al. 1987), and a large case control study in American veterans (n = 1446 cases) found a 48% reduction in risk for this cancer in statin users (Khurana, Caldito et al. 2008). In the same population, a 55% reduction in the incidence of lung cancer in statin users compared to nonusers was found (n = 7280 cases) (Khurana, Bejjanki et al. 2007).

The evidence cited in this section suggests that higher circulating cholesterol can have the strongest effects on more advanced tumors. The question of whether more advanced or aggressive tumors accumulate more cholesterol as compared to early stage tumors *in vivo* has not been specifically addressed, although there is some evidence to suggest that this is the case (Tosi and Tugnoli 2005). Experimental data in the next section provide more support for the association of exogenous cholesterol with more aggressive cancer, as well as insight into how and why cancer cells accumulate cholesterol against normal homeostatic mechanisms.

3. Experimental and mechanistic evidence for role of LDL in cancer

Cholesterol homeostasis. If cholesterol homeostasis is altered in cancer cells to meet a greater demand for cholesterol, an understanding of the mechanisms involved will open up new targets against cancer. In normal cells, free cholesterol in cells is closely regulated to maintain adequate membrane cholesterol but prevent free cholesterol toxicity. Excess cholesterol is stored in the form of neutral cholesteryl esters (CE) that are available to the cell through the CE cycle (Brown, Ho et al. 1980), or is effluxed to circulating HDL for transport back to the liver (Fielding and Fielding 2001). In cholesterol-accumulating tumors, there is more CE storage and less efflux of cholesterol to HDL. Is this cholesterol newly synthesized

Study (Country)	Years of follow-up	n (n for cases)	Sex	Type of cancer	Association with risk of cancer for:			Reference
					Total-C	LDL-C	HDL-C	
National Health Insurance Corp. enrollees (South Korea)	Up to 14	1,189,719 (M:53,944 F: 24,475)	M,F	All	Positive for PrC (M), BrC (F), CRC (M,F); negative for stomach, liver (M,F), lung (M)	Not measured	Not measured	{Kitahara, 2011}
Health Professionals Follow-Up (USA)	Up to 7	18,018 (698)	M	PrC	Positive for high-grade PrC	Not measured	Not measured	{Platz, 2008}
Prostate Cancer Prevention Trial (USA)	Up to 7	5,586 (1,251)	M	PrC	Positive for high-grade PrC			{Platz, 2009}
CLUE II (USA)	Mean of 11.9	6,816 (438)	M	PrC	Positive for high-grade PrC	Not measured	Not measured	{Mondul, 2010}
Alpha-Tocopherol, Beta-Carotene Cancer Prevention (smokers, Finland)	>10	29,093 (2,041)	M	PrC	Positive for aggressive and advanced PrC	Not measured	Negative trend	{Mondul, 2011}
Midspan (Scotland, UK)	Up to 37	12,926 (650)	M	PrC	Positive for high-grade PrC	Not measured	Not measured	{Shafique, 2012}
Nijmegen Biomedical (Netherlands)	Mean of 6.6	2,118 (43)	M	PrC	Positive for total and aggressive PrC	Positive for total and aggressive PrC	Positive for non-aggressive PrC	{Kok, 2011}
HUNT 2 (Norway)	Mean of 9.3	29,364 (687)	M	PrC	None	Not measured	None	{Martin, 2009}
Apolipoprotein MOrtality RISk (Sweden)	Mean of 7.0 - 8.3	200,660 (5,112)	M	PrC	None			{Van Hemelrijck, 2011}
Nurses' Health (self-reported serum cholesterol) (USA)	6 - 12	79,994 (3177)	F	BrC	None	Not measured	Not measured	{Eliassen, 2005}
Postmenopausal public servants (South Korea)	Up to 10	170,374 (714)	F	BrC	Positive trend	Not measured	Not measured	{Ha, 2009}
Malmö Preventive Project (Sweden)	Up to 20	9,738 (269)	F	BrC	Positive for postmenopausal; none for premenopausal	Not measured	Not measured	{Manjer, 2001}
National Health Screening Service (Norway)	11 - 14	24,329 (242)	F	BrC	Negative (pre-menopausal); none (post-menopausal)	Not measured	Not measured	{Vatten, 1990}
EPIC and Nutrition (nested case-control)	Mean of 3.8	521,448 (1238)	M,F	CRC	None	Not measured	Positive for colon cancer	{van Duijn-hoven, 2011}

Table 1. Large, prospective studies with a baseline total cholesterol measurement and long-term follow-up for cancer incidence. M, male; F, female; PrC, prostate cancer; BrC, breast cancer; CRC, colorectal cancer.

or obtained from LDL, and what determines this? Normal cells obtain cholesterol primarily through endocytosis of circulating LDL through the LDLR, but have the capacity for endogenous synthesis via the mevalonate pathway; both mechanisms are tightly controlled for cholesterol homeostasis (Goldstein, DeBose-Boyd et al. 2006). The expressions of both LDLR and HMGCR are regulated by the transcription factors sterol response element binding proteins (SREBP1/2), whose processing and maturation proceed in response to decreased intracellular cholesterol (Brown and Goldstein 1997). The observed accumulation of CE in some tumors, the positive association of total-C with the risk of some types of cancer, and the demand for cholesterol for membrane building in growing cells, all suggest that the expression of these proteins and other components of the cholesterol homeostatic response system are altered in cancer.

Cholesterol biosynthesis in cancer. In order to obtain sufficient cholesterol, proliferating cells may accelerate the rate of cholesterol biosynthesis. Oncogenes that transform cells and dysregulate growth activate anabolic and biosynthetic pathways leading to *de novo* cholesterol and fatty acid synthesis. This is accomplished by a greatly increased flux of glucose into cells and through the glycolytic pathway to produce energy, and transport of TCA cycle citrate from the mitochondria to the cytosol for lipid biosynthesis (Vander Heiden, Cantley et al. 2009). The cytosolic enzyme ATP citrate lyase converts citrate to acetyl-CoA, the basic building block for both fatty acids and cholesterol. Growth factor activation of tyrosine kinase receptors and downstream PI3K/AKT and MAP-kinase signaling pathways increase expression and activation of the SREBPs (Kotzka, Muller-Wieland et al. 2000; Porstmann, Griffiths et al. 2005; Krycer, Sharpe et al. 2010), which control many lipid biosynthetic enzymes. Interesting, it was recently demonstrated that a mutated form of the cell cycle regulator p53, common in many tumors, bound to the promoter regions of the SREBPs and increased the expression of mevalonate pathway genes in BrC cells (Freed-Pastor, Mizuno et al. 2012).

A high enough rate of *de novo* biosynthesis may not always be possible; for example in solid tumors, expansion and insufficient vascularization may limit the delivery of glucose and oxygen. If oxygen is limited, activation of the hypoxia inducible factor 1 (HIF1) pathway can increase survival but divert pyruvate to lactate, reducing production of citrate (Gordan, Thompson et al. 2007). If glucose is limited, reducing ATP production, the AMP activated protein kinase (AMPK) pathway can inactivate key biosynthetic enzymes by phosphorylation (Shackelford and Shaw 2009). If biosynthesis becomes constrained, cells would have an advantage by being able to obtain lipids exogenously from circulating lipoproteins.

Cholesterol uptake in cancer. Uptake of cholesterol from LDL is primarily through the LDLR, although several scavenger receptors may also contribute. Over-expression of LDLR without feedback regulation by cholesterol has been observed in many types of cancer cells (Chen, Li et al. 1988; Hirakawa, Maruyama et al. 1991; Chen and Hughes-Fulford 2001; Antalis, Uchida et al. 2011). Although the role of SREBPs in feedback regulation of LDLR expression is well understood (Goldstein, DeBose-Boyd et al. 2006), there is evidence that cell signaling pathways also contribute to LDLR up-regulation in cancer. In BrC cells, LDLR mRNA expression was 3-5-fold higher in ER- as compared to ER+ cell lines; PKC activation was strongly associated with increased LDLR expression in ER+ BrC cells, and to a lesser extent, even in ER- cells (Stranzl, Schmidt et al. 1997). Activation of the p42/44 (MAPK) cascade was sufficient to induce LDLR transcription in human hepatoma HepG2 cells expressing oncogenic Raf-1 kinase (Kapoor, Atkins et al. 2002). In glioblastoma cells, chronic activation of the EGF receptor tyrosine kinase, or other mechanisms which ultimately activated the PI3K/AKT pathway, led to increased expression of SREBP1 and the LDLR and to LDL-responsive proliferation (Guo, Reinitz et al. 2011).

Increased dietary cholesterol has been shown to promote tumorigenesis in animal models. A Western-type high cholesterol diet compared to a chow diet increased tumor incidence and metastasis in a mouse model of PrC (Llaverias, Danilo et al. 2010). The same group, using similar diets, showed an increase in tumor formation and more aggressive tumors in a mouse model of BrC (Llaverias, Danilo et al. 2011). In both studies, plasma total-C was reduced following tumor development, suggesting utilization of circulating cholesterol by the tumor and similarity to what is observed in people with cancer.

Role of cholesterol esterification. Whether tumor cells obtain the needed cholesterol endogenously or exogenously, it would be imperative to have a way to manage the increased flux of cholesterol so as to meet the dual goals of ensuring a ready supply and avoiding toxicity. Cholesterol toxicity is prevented by effluxing the excess free cholesterol to an extracellular acceptor or converting free cholesterol to non-toxic esters of fatty acids. The observed low HDL-C in cancer patients, combined with the observed increased cholesterol content in tumors suggest that efflux mechanisms are reduced and esterification is increased. Synthesis and storage of CE in lipid droplets not only reduces toxicity but provides an accessible depot of cholesterol for future cell needs.

The enzyme responsible for cholesterol esterification is acyl-CoA:cholesterol acyltransferase 1 (ACAT1/SOAT1), a constitutive resident of the endoplasmic reticulum. ACAT1 esterifies cholesterol obtained from LDL and also from endogenous synthesis (Chang, Li et al. 2009). ACAT1 is frequently found to be over-expressed in cancer vs. normal tissues in human tumor gene expression analyses, including cancers of brain, breast, cervix, esophagus, head and neck, kidney, and testis ($P < 0.05$, oncomine.org). Over-expression of ACAT1 has been specifically associated with cholesterol accumulation in renal clear cell carcinoma, a tumor type characterized by 35-fold more CE as compared to normal kidney (Gebhard, Clayman et al. 1987).

ACAT activity has been associated with proliferation in cancer cells. The CE content of lymphocytes from patients with acute or chronic lymphocytic leukemia (n = 30) was 6-fold higher as compared to lymphocytes from healthy age-matched controls (n = 15), and plasma HDL was >40% reduced in the leukemia patients compared to the controls (Mulas, Abete et al. 2011). Phytohemaglutinin (PHA)-stimulated proliferation of the isolated leukemic cells was positively correlated to esterification of oleate to cholesterol, and inhibition of ACAT greatly reduced PHA-induced proliferation (Mulas, Abete et al. 2011). Cholesterol esterification and ACAT1 expression were also studied in leukemia cell lines. Cells with a greater ability to esterify cholesterol and with lower cholesterol efflux (CEM) had a higher rate of proliferation as compared to cells with a greater ability to synthesize cholesterol *de novo* (MOLT4) (Dessi, Batetta et al. 1997). Further work demonstrated that the faster-growing CEM cells expressed more ACAT1 and less HMGCR mRNA as compared to the slower-growing MOLT4 cells (Batetta, Pani et al. 1999).

In BrC, we showed that more aggressive basal-like ER- BrC cells had more lipid droplets and a much higher ratio of CE to TAG in stored neutral lipids as compared to less aggressive ER+ BrC cells; this was associated with higher expression of ACAT1 (Antalis, Arnold et al. 2010). The cell line differences were mirrored in gene expression analyses of human breast tumors, where higher expression of ACAT1/SOAT1 is characteristic of basal-like ER- tumors (Antalis, Arnold et al. 2010). We further showed that ER- cells took up more LDL as compared to ER+ cells, and that LDL dose-responsively increased proliferation only of ER- cells and in an ACAT-sensitive manner. In a follow-up study, we examined the effect of lipoprotein deprivation on chemotactic migration of the highly motile basal-like ER- cell line MDA-MB-231. We showed that lipid droplets were depleted and migration was reduced 85% when cells were grown in medium without lipoproteins, and that adding back LDL or fatty acids restored migration in an ACAT-sensitive manner (Antalis, Uchida et al. 2011). In addition, LDLR expression in these cells was not affected by exogenous LDL but was reduced 75% in the presence of an ACAT inhibitor, suggesting that high ACAT1 expression permitted continued high expression of the LDLR.

What mediates the over-expression of ACAT1 in cancer is not completely understood. Although ACAT1 is a critical component of intracellular cholesterol homeostasis, its expression is not known to be regulated by the SREBPs (Goldstein, DeBose-Boyd et al. 2006). In monocytes and macrophages, ACAT1 expression was up-regulated by interferon γ and all-*trans*-retinoic acid via STAT1 (Yang, Duan et al. 2001) and by dexamethasone via a glucocorticoid response element in its promoter (Yang, Yang et al. 2004). ACAT1 has also been shown to have an NFκB binding element in its proximal promoter and to be up-regulated in response to TNFα signaling through NFκB (Lei, Xiong et al. 2009). Cholesterol acts as an allosteric activator of ACAT1 activity (Liu, Chang et al. 2005).

The LXR pathway. The transcription factor LXR is a major regulator of fatty acid and cholesterol metabolism in cells. When cellular free cholesterol levels are high, some cholesterol is oxidized to form oxysterols, which act as endogenous ligands for LXR; thus LXRs are considered "cholesterol sensors"(Tontonoz 2011). LXR has an absolute requirement for RXRα as a dimerization partner. RXRα expression is highly regulated by both transcription and protein degradation (Boudjelal, Wang et al. 2000; Lefebvre, Benomar et al. 2010). RXRα availability is also affected by competition with its other binding partners, including PPAR, RAR, VDR, TR and FXR. LXR/RXRα is a permissive heterodimer, being stimulated by agonists of either partner (Tontonoz 2011).

LXR signaling is known to have dual roles: up-regulation of genes of fatty acid biosynthesis (including fatty acid synthase and stearoyl-CoA desaturase 1/2) and repression of NFκB controlled inflammatory genes (including IL-6, COX-2, and nitric oxide synthase) (Joseph, Castrillo et al. 2003). In addition, LXR/RXRα controls the transcription of key genes in cholesterol homeostasis: MYLIP/IDOL, the E3-ligase that ubiquitinates the LDLR leading to its degradation, ABCA1 and ABCG1, transporters involved in cholesterol efflux to APOA1 and HDL, and others (Tontonoz 2011). The demonstrated control of ACAT1 by NFκB suggests that its transcription could be antagonized by LXR activity. LXR signaling may have the ability to mediate the balance between lipid biosynthesis/efflux mechanisms and uptake/storage mechanisms. **Fig. 2** and **Fig. 3** illustrate how key factors in cellular cholesterol homeostasis may be affected by the activity of LXR and its target genes.

The uptake of exogenous LDL through LDLR leads to increased cellular free cholesterol, reduced maturation of SREBPs and reduced transcription of LDLR. When LXR/RXRα is active (**Fig. 2**), LDLR protein is degraded by MYLIP and cholesterol efflux mechanisms are increased (Beltowski 2008). ACAT1 transcription may be reduced by the inhibitory effect of LXR/RXRα on NFκB transactivation activity, blocking cholesterol accumulation. Similarly ApoA1, the apolipoprotein acceptor for cholesterol efflux, which under some conditions is repressed by NFκB, could be increased (Mogilenko, Dizhe et al. 2009). As a result, normal cellular cholesterol homeostasis is enforced.

When LXR/RXRα is less active (**Fig. 3**), and under the influence of cytokines, a different pattern of gene expression predominates. Cholesterol efflux is reduced and thus free cholesterol is maintained at a high enough level in bilayer membranes that maturation of SREBPs is not triggered. More free cholesterol is esterified and stored in lipid droplets, due to a possible induction of ACAT1. LDLR protein degradation is reduced, allowing the cell to maintain high LDLR expression and unrestrained uptake of LDL. In this way, cellular cholesterol homeostasis is perturbed in the direction of LDL uptake and cholesterol accumulation.

The pathways described in **Figs. 2 and 3** are hypothesized to explain the observed cholesterol accumulation in some tumors and cancer cell lines. LDLR is placed at the center of the process of LDL uptake and accumulation, with LXR pathway inactivation being the key factor allowing cholesterol accumulation. No doubt the situation is more complicated than shown, as it does not account for scavenger receptor participation. However, the central role of LXR makes it a potential target in cancer.

LXR agonists have been tested in experimental models of cancer. In glioblastoma cells over-expressing the EGFR, EGF stimulated PI3K/Akt-driven up-regulation of SREBP1 and LDLR (Guo, Reinitz et al. 2011). An LXR agonist induced MYLIP/IDOL-mediated degradation of LDLR, ABCA1-mediated cholesterol efflux, and cell death both *in vitro* and in an animal model. In OvC cells, oxLDL stimulated proliferation and secretion of the cytokine cardiotrophin 1 (Scoles, Xu et al. 2010). An LXR agonist blocked both the cytokine secretion and the proliferation induced by oxLDL; the authors attribute the response to increased cholesterol efflux and decreased inflammatory effects of the LXR agonist. In an athymic model of PrC, progression of androgen-dependent tumors to androgen-independent tumors after castration was accompanied by decreases in expression of LXR target genes in the tumor, and treatment with an LXR agonist delayed the progression for about 4 weeks (Chuu, Hiipakka et al. 2006).

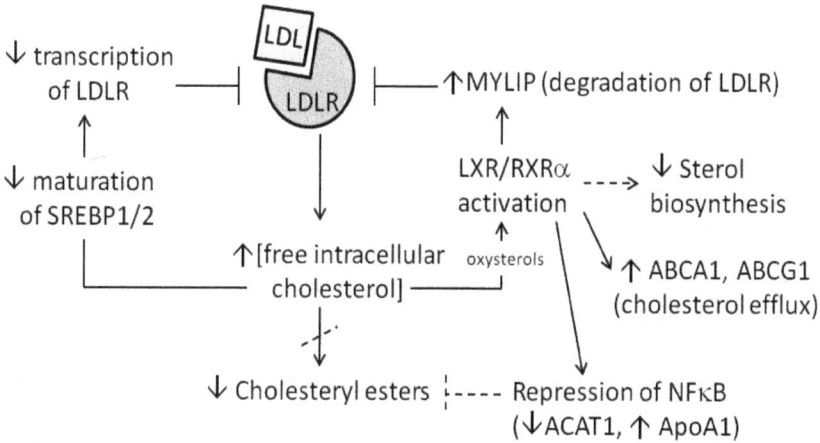

Figure 2. LXR transcriptional targets control intracellular cholesterol concentrations. Dotted line indicates pathways not proven.

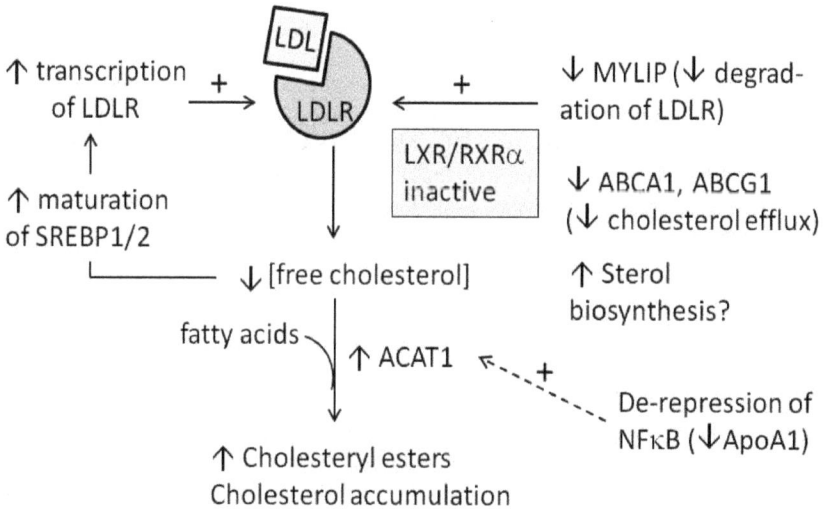

Figure 3. Reduced LXR signaling allows increased LDL uptake and intracellular cholesterol accumulation. Dotted line indicates pathways not proven.

Cholesterol and tumorigenesis. The question remains as to the role that CEs may play in the survival, proliferation and metastasis of cancer cells. We and others have proposed that accumulation of CE spares energy needed for *de novo* sterol synthesis, allowing greater proliferation and migration and perhaps a quicker return to growth after a period of stasis (Batetta, Pani et al. 1999; Antalis, Arnold et al. 2010; Antalis, Uchida et al. 2011). The process of cholesterol esterification was linked to proliferation in multiple studies in different cancer cell lines (Batetta, Pani et al. 1999; Peiretti, Dessi et al. 2007; Paillasse, de Medina et al. 2009; Antalis, Arnold et al. 2010; Mulas, Abete et al. 2011), implying a complex network of signaling pathways and gene expression that ties cholesterol accretion to tumorigenesis. However, the exact role of CE in tumorigenesis remains to be determined.

PrC is a unique case considering the slow growth characteristics of this malignancy. The lipid raft concept has been proposed to account for the tumorigenic effects of cholesterol (Freeman, Cinar et al. 2007), and a higher level of cholesterol in PrC cells has been linked to membrane lipid raft-induced oncogenic cell signaling (Hager, Solomon et al. 2006). A connection between LXR signaling and lipid raft-associated signaling was demonstrated in androgen-responsive LnCAP cells, where an LXR agonist down-regulated Akt signaling in a cholesterol- and lipid raft-dependent manner, resulting in apoptosis of cells and xenograft tumors (Pommier, Alves et al. 2010). In addition, a relationship between androgens and cholesterol metabolism was demonstrated in PrC cells. It was first noted that androgen stimulation caused a dramatic increase in lipid droplets in LNCap cells. The induced neutral lipids included both TAG (33-fold) and CE (7-fold increase), most of which originated from new lipid synthesis (Swinnen, Van Veldhoven et al. 1996). This was later found to be due to an up-regulation of the SREBPs and lipid biosynthetic genes (Nelson, Clegg et al. 2002). The androgen-independent PC-3 cells had a higher content of CE and but not higher ACAT1 activity or expression as compared to LNCap cells (Locke, Wasan et al. 2008). In both an androgen-independent cell line and a mouse xenograft model of PrC progression, changes in cholesterol metabolism and homeostasis were associated with initiation of tumoral androgen production and expression of the AR and PSA (Locke, Wasan et al. 2008; Leon, Locke et al. 2010). These data, along with the clinical data cited in **Section 2,** suggest that in PrC cholesterol accumulation may be important for androgen synthesis, which is closely involved with PrC progression even under castration therapy.

Another function of LDL and other lipoproteins is the provision of essential fatty acids. Mammalian cells are not able to make polyunsaturated fatty acids; the essential n-6 and n-3 fatty acids are derived from the diet and carried to cells by lipoproteins. Human glioma, one of the deadliest types of cancer, was found to contain up to 100-fold more CE compared to control tissue, and the fatty acid composition of the tumor CEs indicated an LDL origin (Nygren, von Holst et al. 1997). The n-6 fatty acid arachidonic acid is necessary for synthesis of second messengers such as the prostaglandin PGE_2, a tumor promoter (Wang and Dubois 2006). In androgen-independent PrC PC-3 cells, PGE_2 production increased >3-fold in response to LDL (Chen and Hughes-Fulford 2001). Thus the fatty acids esterified to cholesterol and other lipids may be important for the effect of LDL on cancer cells.

Finally, although lower plasma HDL-C in cancer patients may be due to reduced efflux of cholesterol to HDL from the tumor, there is evidence that some cancer cells can take up CE from circulating HDL, providing another explanation for low HDL. Recent investigations with the CEM-CCRF lymphoblastic cell line into the source of intracellular CE showed that

HDL-CE were taken up and stored without hydrolysis and re-esterification, while LDL-CE were hydrolyzed and re-esterified (Uda, Accossu et al. 2012). Although the mechanism was not clear, the data implied that HDL as well as LDL could be a source of CE for leukemic cells. A previous study in BrC cells showed that either HDL or LDL dose-dependently stimulated proliferation of ER- cell lines, but only HDL had the effect on ER+ cells lines (Rotheneder and Kostner 1989). In an animal model of PrC, a diet high in fat and cholesterol resulted in increased tumor incidence and increased tumor expression of scavenger receptor B1, the receptor responsible for selective uptake of HDL-C (the major form of circulating cholesterol in mice) by cells (Llaverias, Danilo et al. 2010). The question of whether HDL can supply cholesterol to tumor cells *in vivo* in humans remains open.

4. Conclusions and future directions

The heterogeneous nature of cancer and the changes that accompany tumor progression make it very difficult to draw overall conclusions about the effects of circulating cholesterol on cancer incidence or progression. However, large scale prospective studies have shown that higher plasma total-C and LDL-C can increase the risk for some cancers, with the hormone-related cancers in men and women being especially affected. Data also point to a more potent effect of exogenous cholesterol on more aggressive cancers. These conclusions are supported by data on the effect of statins, which have been shown to reduce both the risk and the progression of some cancers. As more clinical trial data emerges, we will have a clearer picture of the usefulness of cholesterol reduction and statins in cancer and what types of cancer respond to these therapies.

Individualized approaches are the future for cancer therapy. Gene and protein expressions may serve as biomarkers to identify tumors that are stimulated by LDL. The genes/proteins expected to be more expressed as a result of LXR/RXRα pathway activation, i.e. MYLIP and ABCA1, and those expected to be more expressed as a result of LXR/RXRα pathway inactivation, i.e. ACAT1/SOAT1 and LDLR, may be used to distinguish tumors that are cholesterol-accumulating. The cholesterol and CE content of tumor biopsies determined by chemical or enzymatic methods could also be used as biomarkers. Imaging methods such as magnetic resonance (Delikatny, Chawla et al. 2011) and coherent anti-Stokes Raman scattering (Le, Huff et al. 2009) have the potential to allow *in vivo* visualization of lipids in tumors. These kinds of data will help to substantiate and clarify the association of CE accumulation with types of cancer.

If it can be shown that a tumor has the markers of higher cholesterol uptake and accumulation, treatments to lower circulating lipids and affect intracellular cholesterol homeostasis are available. Existing drugs developed for prevention or treatment of cardiovascular disease or metabolic syndrome, such as statins and metformin (an AMPK activator), are being "repurposed" for the treatment of cancer. ACAT inhibitors that did not have the expected result of reducing atherosclerotic plaques in clinical trials may find a new use in cholesterol-accumulating cancers. A new ACAT1-specific inhibitor was effective in killing glioma cells in *in vitro* studies (Bemlih, Poirier et al. 2010). LXR pathway modulators that can increase cholesterol efflux and HDL-C levels without stimulating lipid biosynthesis in the liver, needed to treat cardiovascular disease and metabolic syndrome, may also be useful in cancer (Ratni, Blum-Kaelin et al. 2009). Dietary regimens targeting fat and cholesterol reduction in those with hyperlipidemia, with known benefits in preventing and treating heart disease, may be recommended to decrease the risk or recurrence of some types of cancer.

Author details

Caryl J. Antalis and Kimberly K. Buhman
Indiana University School of Medicine & Purdue University, USA

5. References

Ahern, T. P., L. Pedersen, et al. (2011). "Statin prescriptions and breast cancer recurrence risk: a Danish nationwide prospective cohort study." *J Natl Cancer Inst* 103(19): 1461-1468.

Ahn, J., U. Lim, et al. (2009). "Prediagnostic total and high-density lipoprotein cholesterol and risk of cancer." *Cancer Epidemiol Biomarkers Prev* 18(11): 2814-2821.

Antalis, C. J., T. Arnold, et al. (2010). "High ACAT1 expression in estrogen receptor negative basal-like breast cancer cells is associated with LDL-induced proliferation." *Breast Cancer Res Treat* 122(3): 661-670.

Antalis, C. J., A. Uchida, et al. (2011). "Migration of MDA-MB-231 breast cancer cells depends on the availability of exogenous lipids and cholesterol esterification." *Clin Exp Metastasis* 28(8): 733-741.

Bahl, M., M. Ennis, et al. (2005). "Serum lipids and outcome of early-stage breast cancer: results of a prospective cohort study." *Breast Cancer Res Treat* 94(2): 135-144.

Batetta, B., A. Pani, et al. (1999). "Correlation between cholesterol esterification, MDR1 gene expression and rate of cell proliferation in CEM and MOLT4 cell lines." *Cell Prolif* 32(1): 49-61.

Beltowski, J. (2008). "Liver X receptors (LXR) as therapeutic targets in dyslipidemia." *Cardiovasc Ther* 26(4): 297-316.

Bemlih, S., M. D. Poirier, et al. (2010). "Acyl-coenzyme A: cholesterol acyltransferase inhibitor Avasimibe affect survival and proliferation of glioma tumor cell lines." *Cancer Biol Ther* 9(12): 1025-1032.

Bjorge, T., A. Lukanova, et al. (2011). "Metabolic risk factors and ovarian cancer in the Metabolic Syndrome and Cancer project." *Int J Epidemiol* 40(6): 1667-1677.

Blackburn, G. L. and K. A. Wang (2007). "Dietary fat reduction and breast cancer outcome: results from the Women's Intervention Nutrition Study (WINS)." *Am J Clin Nutr* 86(3): s878-881.

Boudjelal, M., Z. Wang, et al. (2000). "Ubiquitin/proteasome pathway regulates levels of retinoic acid receptor gamma and retinoid X receptor alpha in human keratinocytes." *Cancer Res* 60(8): 2247-2252.

Bozza, P. T. and J. P. Viola (2010). "Lipid droplets in inflammation and cancer." *Prostaglandins Leukot Essent Fatty Acids* 82(4-6): 243-250.

Brown, M. S. and J. L. Goldstein (1997). "The SREBP pathway: regulation of cholesterol metabolism by proteolysis of a membrane-bound transcription factor." *Cell* 89(3): 331-340.

Brown, M. S., Y. K. Ho, et al. (1980). "The cholesteryl ester cycle in macrophage foam cells. Continual hydrolysis and re-esterification of cytoplasmic cholesteryl esters." *J Biol Chem* 255(19): 9344-9352.

Calle, E. E. and R. Kaaks (2004). "Overweight, obesity and cancer: epidemiological evidence and proposed mechanisms." *Nat Rev Cancer* 4(8): 579-591.

Cauley, J. A., A. McTiernan, et al. (2006). "Statin use and breast cancer: prospective results from the Women's Health Initiative." *J Natl Cancer Inst* 98(10): 700-707.

Chae, Y. K., M. E. Valsecchi, et al. (2011). "Reduced risk of breast cancer recurrence in patients using ACE inhibitors, ARBs, and/or statins." *Cancer Invest* 29(9): 585-593.

Chang, T. Y., B. L. Li, et al. (2009). "Acyl-coenzyme A:cholesterol acyltransferases." *Am J Physiol Endocrinol Metab* 297(1): E1-9.

Chen, J. K., L. Li, et al. (1988). "Altered low density lipoprotein receptor regulation is associated with cholesteryl ester accumulation in Simian virus 40 transformed rodent fibroblast cell lines." *In Vitro Cell Dev Biol* 24(4): 353-358.

Chen, Y. and M. Hughes-Fulford (2001). "Human prostate cancer cells lack feedback regulation of low-density lipoprotein receptor and its regulator, SREBP2." *Int J Cancer* 91(1): 41-45.

Chlebowski, R. T., G. L. Blackburn, et al. (2006). "Dietary fat reduction and breast cancer outcome: interim efficacy results from the Women's Intervention Nutrition Study." *J Natl Cancer Inst* 98(24): 1767-1776.

Chuu, C. P., R. A. Hiipakka, et al. (2006). "Inhibition of tumor growth and progression of LNCaP prostate cancer cells in athymic mice by androgen and liver X receptor agonist." *Cancer Res* 66(13): 6482-6486.

Cleary, M. P. and M. E. Grossmann (2009). "Minireview: Obesity and breast cancer: the estrogen connection." *Endocrinology* 150(6): 2537-2542.

Delikatny, E. J., S. Chawla, et al. (2011). "MR-visible lipids and the tumor microenvironment." *NMR Biomed* 24(6): 592-611.

Delimaris, I., E. Faviou, et al. (2007). "Oxidized LDL, serum oxidizability and serum lipid levels in patients with breast or ovarian cancer." *Clin Biochem* 40(15): 1129-1134.

Dessi, S., B. Batetta, et al. (1997). "Role of cholesterol synthesis and esterification in the growth of CEM and MOLT4 lymphoblastic cells." *Biochem J* 321 (Pt 3): 603-608.

Dessi, S., B. Batetta, et al. (1995). "Clinical remission is associated with restoration of normal high-density lipoprotein cholesterol levels in children with malignancies." *Clin Sci (Lond)* 89(5): 505-510.

Eliassen, A. H., G. A. Colditz, et al. (2005). "Serum lipids, lipid-lowering drugs, and the risk of breast cancer." *Arch Intern Med* 165(19): 2264-2271.

Elmore, R. G., Y. Ioffe, et al. (2008). "Impact of statin therapy on survival in epithelial ovarian cancer." *Gynecol Oncol* 111(1): 102-105.

Faulds, M. H. and K. Dahlman-Wright (2012). "Metabolic diseases and cancer risk." *Curr Opin Oncol* 24(1): 58-61.

Fielding, C. J. and P. E. Fielding (2001). "Cellular cholesterol efflux." *Biochim Biophys Acta* 1533(3): 175-189.

Flick, E. D., L. A. Habel, et al. (2009). "Statin use and risk of colorectal cancer in a cohort of middle-aged men in the US: a prospective cohort study." *Drugs* 69(11): 1445-1457.

Freed-Pastor, W. A., H. Mizuno, et al. (2012). "Mutant p53 disrupts mammary tissue architecture via the mevalonate pathway." *Cell* 148(1-2): 244-258.

Freeman, M. R., B. Cinar, et al. (2007). "Transit of hormonal and EGF receptor-dependent signals through cholesterol-rich membranes." *Steroids* 72(2): 210-217.

Freitas, I., P. Pontiggia, et al. (1990). "Histochemical probes for the detection of hypoxic tumour cells." *Anticancer Res* 10(3): 613-622.

Gazzerro, P., M. C. Proto, et al. (2012). "Pharmacological actions of statins: a critical appraisal in the management of cancer." *Pharmacol Rev* 64(1): 102-146.

Gebhard, R. L., R. V. Clayman, et al. (1987). "Abnormal cholesterol metabolism in renal clear cell carcinoma." *J Lipid Res* 28(10): 1177-1184.

Goldstein, J. L., R. A. DeBose-Boyd, et al. (2006). "Protein sensors for membrane sterols." *Cell* 124(1): 35-46.

Gordan, J. D., C. B. Thompson, et al. (2007). "HIF and c-Myc: sibling rivals for control of cancer cell metabolism and proliferation." *Cancer Cell* 12(2): 108-113.

Guo, D., F. Reinitz, et al. (2011). "An LXR agonist promotes GBM cell death through inhibition of an EGFR/AKT/SREBP-1/LDLR-dependent pathway." *Cancer Discov* 1(5): 442-456.

Ha, M., J. Sung, et al. (2009). "Serum total cholesterol and the risk of breast cancer in postmenopausal Korean women." *Cancer Causes Control* 20(7): 1055-1060.

Hager, M. H., K. R. Solomon, et al. (2006). "The role of cholesterol in prostate cancer." *Curr Opin Clin Nutr Metab Care* 9(4): 379-385.

Hamilton, R. J., L. L. Banez, et al. (2010). "Statin medication use and the risk of biochemical recurrence after radical prostatectomy: results from the Shared Equal Access Regional Cancer Hospital (SEARCH) Database." *Cancer* 116(14): 3389-3398.

Hamilton, R. J., K. C. Goldberg, et al. (2008). "The influence of statin medications on prostate-specific antigen levels." *J Natl Cancer Inst* 100(21): 1511-1518.

Hede, K. (2011). "Hints that statins reduce colon cancer risk finally being put to the test." *J Natl Cancer Inst* 103(5): 364-366.

Hirakawa, T., K. Maruyama, et al. (1991). "Massive accumulation of neutral lipids in cells conditionally transformed by an activated H-ras oncogene." *Oncogene* 6(2): 289-295.

Hirsch, H. A., D. Iliopoulos, et al. (2010). "A transcriptional signature and common gene networks link cancer with lipid metabolism and diverse human diseases." *Cancer Cell* 17(4): 348-361.

Holvoet, P., D. H. Lee, et al. (2008). "Association between circulating oxidized low-density lipoprotein and incidence of the metabolic syndrome." *JAMA* 299(19): 2287-2293.

Hu, Z., C. Fan, et al. (2006). "The molecular portraits of breast tumors are conserved across microarray platforms." *BMC Genomics* 7: 96.

Hulley, S., D. Grady, et al. (1998). "Randomized trial of estrogen plus progestin for secondary prevention of coronary heart disease in postmenopausal women. Heart and Estrogen/progestin Replacement Study (HERS) Research Group." *JAMA* 280(7): 605-613.

Jacobs, E. J., C. C. Newton, et al. (2011). "Long-term use of cholesterol-lowering drugs and cancer incidence in a large United States cohort." *Cancer Res* 71(5): 1763-1771.

Joseph, S. B., A. Castrillo, et al. (2003). "Reciprocal regulation of inflammation and lipid metabolism by liver X receptors." *Nat Med* 9(2): 213-219.

Kapoor, G. S., B. A. Atkins, et al. (2002). "Activation of Raf-1/MEK-1/2/p42/44(MAPK) cascade alone is sufficient to uncouple LDL receptor expression from cell growth." *Mol Cell Biochem* 236(1-2): 13-22.

Kelsey, J. L. (1993). "Breast cancer epidemiology: summary and future directions." *Epidemiol Rev* 15(1): 256-263.

Khurana, V., H. R. Bejjanki, et al. (2007). "Statins reduce the risk of lung cancer in humans: a large case-control study of US veterans." *Chest* 131(5): 1282-1288.

Khurana, V., G. Caldito, et al. (2008). "Statins might reduce risk of renal cell carcinoma in humans: case-control study of 500,000 veterans." *Urology* 71(1): 118-122.

Kitahara, C. M., A. Berrington de Gonzalez, et al. (2011). "Total cholesterol and cancer risk in a large prospective study in Korea." *J Clin Oncol* 29(12): 1592-1598.

Kok, D. E., J. G. van Roermund, et al. (2011). "Blood lipid levels and prostate cancer risk; a cohort study." *Prostate Cancer Prostatic Dis*.

Kotzka, J., D. Muller-Wieland, et al. (2000). "Sterol regulatory element binding proteins (SREBP)-1a and SREBP-2 are linked to the MAP-kinase cascade." *J Lipid Res* 41(1): 99-108.

Kovanen, P. T., M. S. Brown, et al. (1979). "Increased binding of low density lipoprotein to liver membranes from rats treated with 17 alpha-ethinyl estradiol." *J Biol Chem* 254(22): 11367-11373.

Krycer, J. R., L. J. Sharpe, et al. (2010). "The Akt-SREBP nexus: cell signaling meets lipid metabolism." *Trends Endocrinol Metab* 21(5): 268-276.

Kumar, A. S., C. C. Benz, et al. (2008). "Estrogen receptor-negative breast cancer is less likely to arise among lipophilic statin users." *Cancer Epidemiol Biomarkers Prev* 17(5): 1028-1033.

Kwiterovich, P. O., Jr. (2000). "The metabolic pathways of high-density lipoprotein, low-density lipoprotein, and triglycerides: a current review." *Am J Cardiol* 86(12A): 5L-10L.

Le, T. T., T. B. Huff, et al. (2009). "Coherent anti-Stokes Raman scattering imaging of lipids in cancer metastasis." *BMC Cancer* 9: 42.

Lefebvre, B., Y. Benomar, et al. (2010). "Proteasomal degradation of retinoid X receptor alpha reprograms transcriptional activity of PPARgamma in obese mice and humans." *J Clin Invest* 120(5): 1454-1468.

Lei, L., Y. Xiong, et al. (2009). "TNF-alpha stimulates the ACAT1 expression in differentiating monocytes to promote the CE-laden cell formation." *J Lipid Res* 50(6): 1057-1067.

Leon, C. G., J. A. Locke, et al. (2010). "Alterations in cholesterol regulation contribute to the production of intratumoral androgens during progression to castration-resistant prostate cancer in a mouse xenograft model." *Prostate* 70(4): 390-400.

Li, A. J., R. G. Elmore, et al. (2010). "Serum low-density lipoprotein levels correlate with survival in advanced stage epithelial ovarian cancers." *Gynecol Oncol* 116(1): 78-81.

Liu, J., C. C. Chang, et al. (2005). "Investigating the allosterism of acyl-CoA:cholesterol acyltransferase (ACAT) by using various sterols: in vitro and intact cell studies." *Biochem J* 391(Pt 2): 389-397.

Llaverias, G., C. Danilo, et al. (2011). "Role of cholesterol in the development and progression of breast cancer." *Am J Pathol* 178(1): 402-412.

Llaverias, G., C. Danilo, et al. (2010). "A Western-type diet accelerates tumor progression in an autochthonous mouse model of prostate cancer." *Am J Pathol* 177(6): 3180-3191.

Locke, J. A., K. M. Wasan, et al. (2008). "Androgen-mediated cholesterol metabolism in LNCaP and PC-3 cell lines is regulated through two different isoforms of acyl-coenzyme A:Cholesterol Acyltransferase (ACAT)." *Prostate* 68(1): 20-33.

Manjer, J., R. Kaaks, et al. (2001). "Risk of breast cancer in relation to anthropometry, blood pressure, blood lipids and glucose metabolism: a prospective study within the Malmo Preventive Project." *Eur J Cancer Prev* 10(1): 33-42.

Marcella, S. W., A. David, et al. (2011). "Statin use and fatal prostate cancer: A matched case-control study." *Cancer*.

Martin, R. M., L. Vatten, et al. (2009). "Components of the metabolic syndrome and risk of prostate cancer: the HUNT 2 cohort, Norway." *Cancer Causes Control* 20(7): 1181-1192.

Mello, A. P., I. T. da Silva, et al. (2011). "Electronegative low-density lipoprotein: origin and impact on health and disease." *Atherosclerosis* 215(2): 257-265.

Mogilenko, D. A., E. B. Dizhe, et al. (2009). "Role of the nuclear receptors HNF4 alpha, PPAR alpha, and LXRs in the TNF alpha-mediated inhibition of human apolipoprotein A-I gene expression in HepG2 cells." *Biochemistry* 48(50): 11950-11960.

Mondul, A. M., S. L. Clipp, et al. (2010). "Association between plasma total cholesterol concentration and incident prostate cancer in the CLUE II cohort." *Cancer Causes Control* 21(1): 61-68.

Mondul, A. M., S. J. Weinstein, et al. (2011). "Serum total and HDL cholesterol and risk of prostate cancer." *Cancer Causes Control* 22(11): 1545-1552.

Moses, K. A., T. T. Abd, et al. (2009). "Increased low density lipoprotein and increased likelihood of positive prostate biopsy in black americans." *J Urol* 182(5): 2219-2225.

Mostaghel, E. A., K. R. Solomon, et al. (2012). "Impact of circulating cholesterol levels on growth and intratumoral androgen concentration of prostate tumors." *PLoS One* 7(1): e30062.

Mulas, M. F., C. Abete, et al. (2011). "Cholesterol esters as growth regulators of lymphocytic leukaemia cells." *Cell Prolif* 44(4): 360-371.

Nelson, P. S., N. Clegg, et al. (2002). "The program of androgen-responsive genes in neoplastic prostate epithelium." *Proc Natl Acad Sci U S A* 99(18): 11890-11895.

Notarnicola, M., D. F. Altomare, et al. (2005). "Serum lipid profile in colorectal cancer patients with and without synchronous distant metastases." *Oncology* 68(4-6): 371-374.

Nygren, C., H. von Holst, et al. (1997). "Increased levels of cholesterol esters in glioma tissue and surrounding areas of human brain." *Br J Neurosurg* 11(3): 216-220.

Paillasse, M. R., P. de Medina, et al. (2009). "Signaling through cholesterol esterification: a new pathway for the cholecystokinin 2 receptor involved in cell growth and invasion." *J Lipid Res* 50(11): 2203-2211.

Parsons, J. K., J. Bergstrom, et al. (2008). "Lipids, lipoproteins and the risk of benign prostatic hyperplasia in community-dwelling men." *BJU Int* 101(3): 313-318.

Peiretti, E., S. Dessi, et al. (2007). "Modulation of cholesterol homeostasis by antiproliferative drugs in human pterygium fibroblasts." *Invest Ophthalmol Vis Sci* 48(8): 3450-3458.

Pelucchi, C., D. Serraino, et al. (2011). "The metabolic syndrome and risk of prostate cancer in Italy." *Ann Epidemiol* 21(11): 835-841.

Platz, E. A., S. K. Clinton, et al. (2008). "Association between plasma cholesterol and prostate cancer in the PSA era." *Int J Cancer* 123(7): 1693-1698.

Platz, E. A., C. Till, et al. (2009). "Men with low serum cholesterol have a lower risk of high-grade prostate cancer in the placebo arm of the prostate cancer prevention trial." *Cancer Epidemiol Biomarkers Prev* 18(11): 2807-2813.

Pommier, A. J., G. Alves, et al. (2010). "Liver X Receptor activation downregulates AKT survival signaling in lipid rafts and induces apoptosis of prostate cancer cells." *Oncogene* 29(18): 2712-2723.

Porstmann, T., B. Griffiths, et al. (2005). "PKB/Akt induces transcription of enzymes involved in cholesterol and fatty acid biosynthesis via activation of SREBP." *Oncogene* 24(43): 6465-6481.

Poynter, J. N., S. B. Gruber, et al. (2005). "Statins and the risk of colorectal cancer." *N Engl J Med* 352(21): 2184-2192.

Ratni, H., D. Blum-Kaelin, et al. (2009). "Discovery of tetrahydro-cyclopenta[b]indole as selective LXRs modulator." *Bioorg Med Chem Lett* 19(6): 1654-1657.

Rose, G. and M. J. Shipley (1980). "Plasma lipids and mortality: a source of error." *Lancet* 1(8167): 523-526.

Rotheneder, M. and G. M. Kostner (1989). "Effects of low- and high-density lipoproteins on the proliferation of human breast cancer cells in vitro: differences between hormone-dependent and hormone-independent cell lines." *Int J Cancer* 43(5): 875-879.

Rudling, M. J., L. Stahle, et al. (1986). "Content of low density lipoprotein receptors in breast cancer tissue related to survival of patients." *Br Med J (Clin Res Ed)* 292(6520): 580-582.

Scoles, D. R., X. Xu, et al. (2010). "Liver X receptor agonist inhibits proliferation of ovarian carcinoma cells stimulated by oxidized low density lipoprotein." *Gynecol Oncol* 116(1): 109-116.

Shackelford, D. B. and R. J. Shaw (2009). "The LKB1-AMPK pathway: metabolism and growth control in tumour suppression." *Nat Rev Cancer* 9(8): 563-575.

Shafique, K., P. McLoone, et al. (2012). "Cholesterol and the risk of grade-specific prostate cancer incidence: evidence from two large prospective cohort studies with up to 37 years' follow up." *BMC Cancer* 12: 25.

Sharp, S. J. and S. J. Pocock (1997). "Time trends in serum cholesterol before cancer death." *Epidemiology* 8(2): 132-136.

Singh, H., S. M. Mahmud, et al. (2009). "Long-term use of statins and risk of colorectal cancer: a population-based study." *Am J Gastroenterol* 104(12): 3015-3023.

Solomon, K. R. and M. R. Freeman (2008). "Do the cholesterol-lowering properties of statins affect cancer risk?" *Trends Endocrinol Metab* 19(4): 113-121.

Stojakovic, T., T. Claudel, et al. (2010). "Low-dose atorvastatin improves dyslipidemia and vascular function in patients with primary biliary cirrhosis after one year of treatment." *Atherosclerosis* 209(1): 178-183.

Stranzl, A., H. Schmidt, et al. (1997). "Low-density lipoprotein receptor mRNA in human breast cancer cells: influence by PKC modulators." *Breast Cancer Res Treat* 42(3): 195-205.

Suzuki, K., Y. Ito, et al. (2004). "Serum oxidized low-density lipoprotein levels and risk of colorectal cancer: a case-control study nested in the Japan Collaborative Cohort Study." *Cancer Epidemiol Biomarkers Prev* 13(11 Pt 1): 1781-1787.

Swinnen, J. V., P. P. Van Veldhoven, et al. (1996). "Androgens markedly stimulate the accumulation of neutral lipids in the human prostatic adenocarcinoma cell line LNCaP." *Endocrinology* 137(10): 4468-4474.

Taplin, M. E. and S. P. Balk (2004). "Androgen receptor: a key molecule in the progression of prostate cancer to hormone independence." *J Cell Biochem* 91(3): 483-490.

Tavridou, A., A. Efthimiadis, et al. (2010). "Simvastatin-induced changes in circulating oxidized low-density lipoprotein in different types of dyslipidemia." *Heart Vessels* 25(4): 288-293.

Thomas, J. A., 2nd, L. Gerber, et al. (2012). "Prostate Cancer Risk in Men with Baseline History of Coronary Artery Disease: Results from the REDUCE Study." *Cancer Epidemiol Biomarkers Prev*.

Tontonoz, P. (2011). "Transcriptional and Posttranscriptional Control of Cholesterol Homeostasis by Liver X Receptors." *Cold Spring Harb Symp Quant Biol*.

Tosi, M. R. and V. Tugnoli (2005). "Cholesteryl esters in malignancy." *Clin Chim Acta* 359(1-2): 27-45.

Uda, S., S. Accossu, et al. (2012). "A lipoprotein source of cholesteryl esters is essential for proliferation of CEM-CCRF lymphoblastic cell line." *Tumour Biol* 33(2): 443-453.

van Duijnhoven, F. J., H. B. Bueno-De-Mesquita, et al. (2011). "Blood lipid and lipoprotein concentrations and colorectal cancer risk in the European Prospective Investigation into Cancer and Nutrition." *Gut* 60(8): 1094-1102.

Van Hemelrijck, M., H. Garmo, et al. (2011). "Prostate cancer risk in the Swedish AMORIS study: the interplay among triglycerides, total cholesterol, and glucose." *Cancer* 117(10): 2086-2095.

Vander Heiden, M. G., L. C. Cantley, et al. (2009). "Understanding the Warburg effect: the metabolic requirements of cell proliferation." *Science* 324(5930): 1029-1033.

Vatten, L. J. and O. P. Foss (1990). "Total serum cholesterol and triglycerides and risk of breast cancer: a prospective study of 24,329 Norwegian women." *Cancer Res* 50(8): 2341-2346.

Vatten, L. J., O. P. Foss, et al. (1991). "Overall survival of breast cancer patients in relation to preclinically determined total serum cholesterol, body mass index, height and cigarette smoking: a population-based study." *Eur J Cancer* 27(5): 641-646.

Vitols, S., G. Gahrton, et al. (1985). "Hypocholesterolaemia in malignancy due to elevated low-density-lipoprotein-receptor activity in tumour cells: evidence from studies in patients with leukaemia." *Lancet* 2(8465): 1150-1154.

Wang, D. and R. N. Dubois (2006). "Prostaglandins and cancer." *Gut* 55(1): 115-122.

Yang, J. B., Z. J. Duan, et al. (2001). "Synergistic transcriptional activation of human Acyl-coenzyme A: cholesterol acyltransferase-1 gene by interferon-gamma and all-trans-retinoic acid THP-1 cells." *J Biol Chem* 276(24): 20989-20998.

Yang, L., J. B. Yang, et al. (2004). "Enhancement of human ACAT1 gene expression to promote the macrophage-derived foam cell formation by dexamethasone." *Cell Res* 14(4): 315-323.

Yasuda, M. and W. R. Bloor (1932). "Lipid Content of Tumors." *J Clin Invest* 11(4): 677-682.

Chapter 12

Structural Origin of ELOA Toxicity – Implication for HAMLET-Type Protein Complexes with Oleic Acid

Vladana Vukojević and Ludmilla A. Morozova-Roche

Additional information is available at the end of the chapter

1. Introduction

Self-assembled proteinaceous complexes with oleic acid (OA) acquire distinct properties that are not characteristic of the native protein. Most notably, the newly obtained features include the ability to specifically kill tumor cells while sparing the healthy, normally functioning ones, as it is the case with human or bovine α-lactalbumin made lethal to tumor cells (HAMLET or BAMLET) [1,2] or to indiscriminately induce cell death in all tested cell lines, as it is the case with equine lysozyme (EL) complex with oleic acid (ELOA) [3,4]. While extensive information has been accumulated on the structural, functional and therapeutic properties of protein complexes with OA, many questions remain still unanswered, such as what is the structural origin of their toxicity, what are the specific targets at the cell surface and/or the cellular interior, what are the mechanisms of cellular uptake?

In this chapter, we summarize our current understanding of the structure and function of HAMLET-type protein complexes with oleic acid, using ELOA as an example.

2. Origin of HAMLET - Human α-lactalbumin made lethal for tumor cells

Complexes of human α-lactalbumin with OA were discovered by Catharina Svanborg and co-workers about two decades ago [5]. Initially, Håkansson et al. [5,6] and Svensson et al. [7] discovered that a multimeric human α-lactalbumin derivative isolated from the casein fraction of milk was a potent Ca^{2+}-elevating and apoptosis-inducing agent with a broad, yet selective cytotoxic activity. It was found that the apoptosis-inducing fraction of α-lactalbumin contained oligomers of α-lactalbumin that have undergone a conformational change towards a molten globule-like state [7]. Oligomerization appeared to have conserved α-lactalbumin in a state with molten globule-like properties under physiological conditions. Multimeric α-lactalbumin was shown to bind to the cell surface, enter the cytoplasm and accumulate in cell nuclei [7]. Multimeric α-lactalbumin was also shown to increase the rate of respiration in isolated mitochondria by exerting an uncoupling effect, which was abolished completely by bovine serum albumin. Multimeric α-lactalbumin accumulated in the nuclei of sensitive cells rather than in the cytoplasm, vesicular fraction, or ER-Golgi complex [6]. Nuclear uptake was shown to occur rapidly in cells that are susceptible to an apoptosis-inducing effect, but not in nuclei of resistant cells. Nuclear uptake was shown to proceed through the nuclear pore complex and was critical for the induction of DNA fragmentation. Ca^{2+} was required for induction of DNA fragmentation by multimeric α-lactalbumin but nuclear uptake of multimeric α-lactalbumin is independent of Ca^{2+}.

Similar cytotoxic activity was observed by in vitro produced HAMLET complexes, in which human α-lactalbumin was converted into the apoptosis-inducing tumoricidal folding variant by binding OA [8-10]. The formation of HAMLET was carried out in chromatography ion exchange columns preconditioned with fatty acids. It was also identified that HAMLET formation is governed by stereo-specific lipid-protein interactions and that only unsaturated C16-C20 fatty acids in cis conformation, but not other fatty acids could induce HAMLET [11]. Among such complexes, only HAMLET complex with OA and cis vaccenic acid complexes were shown to kill tumor cells efficiently, while the C16 or C20 cis fatty acid complexes with α-lactalbumin showed low or intermediate activity [11].

HAMLET's remarkable tumor-selective cytotoxicity correlated with the conformational change of the protein that has taken place upon complex formation, i.e. conversion to molten globule-like state. However, α-lactalbumin in a molten globule state without OA does not possess such activity per se, indicating that the presence of both components is required. As partially unfolded α-lactalbumin can revert easily to its native state upon Ca^{2+} binding in natural cell culture media or within cells, the D87A Ca^{2+}- binding site mutant of α-lactalbumin was produced [12], which was lacking Ca^{2+}-binding property and remained partially unfolded at physiological conditions. Such mutant formed a tumoricidal HAMLET-like complex with OA, but the partially unfolded protein alone did not kill tumor cells. Another non-native α-lactalbumin variant with all amino acids building disulfide bridges substituted by Ala residues also did not exhibit cytotoxic activity in the absence of OA, while its HAMLET-like form displayed strong tumoricidal activity against lymphoma and carcinoma cell lines [13]. Together, these experiments consistently confirmed that both molten globule like protein conformation and specific fatty acids are required for the tumoricidal activity of the investigated complexes.

It has been suggested that naturally occurring HAMLET may have a protective function. In the stomach of nursing children low pH can induce the release of Ca^{2+} from the high-affinity Ca^{2+}-binding site of α-lactalbumin and activate lipases hydrolyzing free fatty acids from milk triglycerides, thereby providing naturally occurring conditions that favor the formation of α-lactalbumin lethal to tumors [14]. This could be important for lowering the incidence of cancer in breast-fed children by purging tumor cells from the gut of the neonate.

3. Equine lysozyme (EL) as a structural homologue of α-lactalbumin

The protein component of ELOA is equine lysozyme (EL), a protein that is abundant in mare milk and kumys (a fermented beverage produced from mare milk that is widely used in Middle Asia). EL belongs to an important calcium-binding sub-family within the extended family of lysozymes, *i.e.* in contrast to common c-type lysozyme EL possesses high affinity calcium binding site, resembling with this regards α-lactalbumins. Lysozymes and α-lactalbumins are characterized by not more than 35-40% in sequence homology, but share remarkably similar tertiary folds. EL serves as an evolutionary bridge between lysozymes and α-lactalbumins, combining the structural and folding properties of both. These are rather small molecules of about 14.6 kDa, consisting of two sub-domains – α-helical and β-sheet rich domains separated by a deep cleft. Lysozyme active site is located in this cleft (absent in α-lactalbumins). Calcium is coordinated by a loop positioned at the bottom of the cleft and important for the structural integrity of the protein, yet the physiological function of calcium binding to EL and other calcium-binding lysozymes is still unclear. The calcium-binding usually increases the protein stability against denaturing treatments, however in the case of EL, the significantly lower stability and cooperatively was observed compared to non-calcium-binding lysozymes even in its holo-form, while in the apo-form its thermodynamic stability is closer of α-lactalbumins than to c-type lysozymes [15,16]. EL forms a wide range of partially folded states under equilibrium conditions similar to these of α-lactalbumins [16,3,17,18]. However, EL molten globule is much more structured compared to the "classical" molten globules of α-lactalbumins, possessing an extended native-like hydrophobic core stabilised by interactions between three major α-helices (A, B and D-helices) in the α-domain [17,18]. Like c-type lysozymes, during refolding kinetics EL forms an ensemble of well-defined transient kinetic intermediates, possessing very persistent structures [19]. Importantly, the rapidly formed kinetic intermediate of EL (2.5 ms refolding time) is characterised by the same extended core structure as its equilibrium molten globule analogues populated under acidic conditions, indicating that the hydrophobic collapse into molten globule-like state is an essential step in protein folding. Given its distinct structural properties, EL may be used as an invaluable research object in revealing the general mechanism and role of intermediate states in protein folding.

4. Controlled ELOA production using ion-exchange chromatography

Similar to HAMLET, ELOA was produced at the solid-liquid interface in an ion-exchange chromatography column preconditioned with OA (Figure 1).

ELOA was eluted as a strong peak by using a 0-1.5 M NaCl gradient. In the absence of OA, free EL was eluted as a narrow peak at a low NaCl concentration [4]. EL was subjected to column chromatography without decalcification as it has been performed with human α-lactalbumin during original HAMLET production, indicating a difference in the generic properties of EL and α-lactalbumin. ELOA complex remains stable in its lyophilized form suitable for long storage as well as it can be kept in solution for up to a week. It is also important to note, that co-incubation of a 50 fold excess of OA mixed with EL in solution at

270

Lipoproteins: Current Concepts

room temperature did not lead to ELOA formation as was evident from the lack of characteristic ELOA conformational transitions monitored by near-UV CD [4]. Thus, the application of a solid-liquid interface facilitating protein self-assembly and protein-OA interactions proved to be an efficient approach in production of both ELOA and HAMLET complexes. By comparison, the complex of hen egg white lysozyme with oleic acid was also produced under the same conditions, but it was very low populated, unstable and OA can be easily depleted from its structure. Hen egg white lysozyme is much more stable than EL and it is evident that the hydrophobic interface in the column chromatography is not sufficient to cause its partial unfolding and interactions with OA molecules.

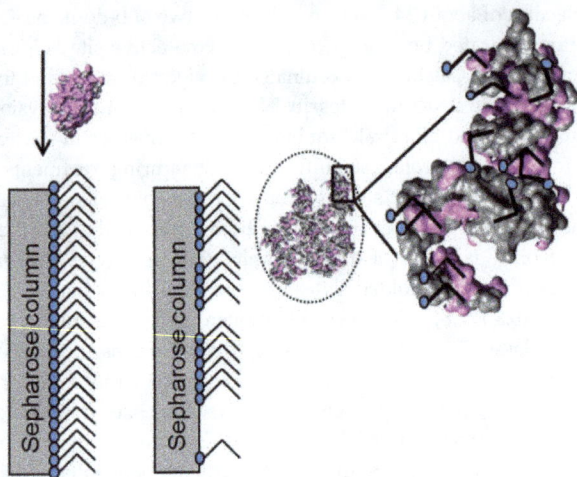

Figure 1. Schematic presentation of ELOA formation at the solid–liquid interface in a Sepharose chromatography column. The positively charged Sepharose matrix is preloaded with oleic acid (the hydrophilic carboxyl group is denoted by a blue circle and the aliphatic chain by a gray line with a "kink" at the position of the double bond). When folded, EL molecules (shown in space-filling representation, with exposed hydrophilic residues in purple and buried hydrophobic residues in grey) are added to the column some hydrophobic residues become exposed and interact with oleic acid molecules forming ELOA.

Indeed, hydrophobic and charged surfaces often facilitate the self-assembly processes by recruiting proteins and modifying their interactions [20]. Within the ion-exchange matrix bound OA molecules constitute an extended surface, facilitating both charged and hydrophobic interactions with EL molecules, while in solution OA, like many other small aliphatic molecules, would be present as a micelle. In addition, the solid-liquid interface may induce EL partial unfolding and expose its hydrophobic surfaces buried in the native state; this can also be critical for ELOA complex formation. It is worth noting, that hydrophobic interactions within the column chromatography may effectively model the interactions, which can take place at the hydrophobic and charged surfaces in biological systems. For example, the interactions with cell lipid membranes may be able to induce protein-ligand complexation otherwise not occurring in solution.

5. EL conformation in ELOA

Similar to human α-lactalbumin in the HAMLET complex, EL in ELOA acquires a partially folded state as evident from spectroscopic and NMR measurements [4]. The ELOA near-UV CD spectrum at room temperature shows the presence of less structure than in the native holo-state and even in the EL molten globule at 57 °C, i.e. the characteristic CD peaks are largely overlapped and the magnitude of the ellipticity is diminished at all wavelengths. ELOA spectra in the far-UV CD region recorded at both 25 °and 57 °C exhibit the same shape as the EL molten globule spectrum at 57 °C, which together with the near UV CD data demonstrate disordering of the tertiary interactions, but preservation of the secondary structure.

Consistently with molten globule conformation, the 1D ^{1}H NMR spectrum of ELOA at pH 9.0 exhibit very broad aromatic and aliphatic resonances, indicative of conformational mobility in a millisecond time scale, and a complete absence of resolved methyl peaks in the up-field region of 2.5 - 0.5 ppm [4,21]. This is in contrast to the NMR spectrum of native EL characterized by well-dispersed resonances in both the aromatic and aliphatic regions. Examination of the 1D ^{1}H NMR spectrum of ELOA showed up-field shifts of the resonance of bound OA compared with the resonances of free OA. This unequivocally demonstrates that OA molecules are integrated in ELOA. Specifically, OAs interact directly with EL aromatic residues as manifested in the presence of cross-peaks between the protons of aromatic residues and OA observed in the ^{1}H NOESY spectrum of ELOA. Due to the poor chemical shift dispersion of the ELOA spectrum, it is impossible to assign the positive NOE cross-peaks to specific aromatic residues, nonetheless this is an absolutely clear indication that EL aromatic residues are directly involved in OA binding.

Similar to typical molten globule states, ELOA binds the hydrophobic dye ANS, which is commonly used to examine the partially folded protein conformations. Interaction with ELOA resulted in ca. 10-fold increase in dye fluorescence compared with free ANS in solution. A shorter wavelength shift of the spectrum maximum of ANS (from 515 to 495 nm) indicates that ANS in its bound form is involved into a more hydrophobic environment. These results demonstrate that the ELOA complex is characterized by some exposed hydrophobic surfaces, which attract hydrophobic ANS molecules.

The surface dynamics and exposure of aromatic residues of ELOA were also probed by photochemically induced dynamic nuclear polarization (photo-CIDNP) spectroscopy [21]. CIDNP method evaluates the surface structure of proteins and complexes by means of a laser induced photochemical reaction, which takes place only if the aromatic side-chains of histidine, tyrosine and tryptophan residues are accessible to a photosensitizer [22]. The ELOA CIDNP spectra were compared with those of holo EL and EL molten globule.

The CIDNP spectra of the native EL at several pH 4.5, 6.9 and 9.0 are well-resolved and were assigned by comparison with NMR chemical shifts [17,18,22]. ELOA and EL molten display less resolved CIDNP spectra, consistent with their millisecond conformational fluctuations, although it is still straight forward to distinguish tyrosine and tryptophan/histidine residues based on their emissive (negative) and absorptive (positive) polarizations, respectively. In the EL molten globule state the characteristic emissive peak corresponding to the Tyr ε protons is the dominant feature, while the other peaks observed for the native state are not present, indicating that these residues are not surface-accessible. The ELOA spectrum contains the same emissive Tyr peak (≈ 6.75 ppm) as seen for the molten globule state. In addition, at ca. 7.7 ppm narrow absorptive photo-CIDNP signals assigned to either His 114 or Trp 63 or potentially both are present in ELOA spectrum, but not in the spectrum of EL molten globule. Both of these residues occur close to the EL inter-domain cleft, indicating limited conformational mobility in this region compared to the rest of the protein and hence that this region is affected by the presence of OA and may be an OA-binding site. Occupation of this cleft by OA may induce further large-scale changes in the relative positions of EL α- and β-domains, possibly lowering the affinity of the calcium-binding site. Hence, although there are clear similarities with EL molten globule, ELOA is characterized by some more structured regions arising from OA binding. These structural changes may be also related to ELOA functional activity as exposed hydrophobic residues in these regions may promote the ELOA interactions with the hydrophobic environment in lipid bilayers and cell membranes.

The thermal unfolding transition of ELOA, monitored by far-UV CD ellipticity at 222 nm, was manifested in an overall decrease of the CD signal and occurred over a very board range of temperatures from 30 °C and up to 80 °C. In EL alone, dissolved at both pH 9.0 and pH 7.0, two unfolding transitions were observed over the same temperature range, however, these transitions were not distinguished in ELOA. This indicates that the conformational changes in ELOA and EL may have different structural origin. It is interesting to note, that HAMLET is less stable towards thermal denaturation than holo human α-lactalbumin, while exhibiting the same stability towards urea denaturation [23]. This demonstrates that OA may produce some destabilizing effect on proteinaceous compounds in both ELOA and HAMLET, but to different degree and with different manifestation in their thermal unfolding transitions.

6. ELOA stoichiometry and comparison with EL amyloid oligomers

The question which is still debated concerns how many protein and OA molecules can be involved in the HAMLET-type complex formation and which conditions can affect this process. Firstly, in the case of ELOA the analysis of 1D ¹H NMR spectrum enabled us to determine the amount of bound OA per protein molecule by comparing the peak areas of the bound OA, reflecting the contribution of 2 olefinic protons, with the peaks corresponding to EL aromatic proton resonances [4]. This value can vary from 4 to 48 OA molecules per EL molecule, depending on the specific chromatographic conditions during ELOA formation. In general, increasing saturation of the column with OA resulted in the formation of ELOA with higher OA content. Secondly, the number of EL molecules in ELOA was determined by pulsed-field gradient NMR diffusion measurements and estimated to be 4–9 in most cases [4]. Thus, both number of OA and protein molecules can vary significantly within the ELOA complexes and the largest ELOA lies at the upper scale among the HAMLET-type complexes.

At the same time, the size of ELOA complexes tends to decrease upon dilution to micromolar concentration range. Their dimensions were characterized by AFM using the volume measurements of the round-shaped particles naturally attached to the mica surface, under the assumption that they acquire a shape of spherical cup due to their interactions with mica [24,22]. At the concentrations used in this study (< 1.5 μM) ELOA was predominantly present in the form of low molecular weight complexes - monomers to pentamers, while some larger aggregates were observed in a lower quantity. This finding was corroborated by fluorescence correlation spectroscopy (FCS). By comparing the average residence time of fluorescently labeled ELOA (*ca.* 120 μs) to the residence time of the reference fluorescent dye (*ca.* 35 μs), FCS indicated that ELOA is predominantly present as a low molecular weight complex (20-30 kDa) in diluted solutions [22].

It is interesting to draw comparison between ELOA and EL amyloid oligomers since they display some common properties. Amyloid oligomers attracted particular attention among protein self-assembled complexes due to their critical involvement in several amyloid and conformational diseases [25-27]. Oligomerisation precedes the amyloid fibril formation and oligomers may serve as nuclei for fibrillar growth. It has also been suggested that oligomers, rather than the apparently inert amyloid fibrils are major cytotoxic agents in amyloid diseases. Both α-lactalbumins and lysozymes form amyloids *in vitro* [28,29] and the lysozymes amyloid formation is associated with systemic amyloidosis in the body [30]. Under EL self-assembly both amyloid oligomers and ELOA become well populated, providing a unique opportunity to compare them directly.

Both ELOA and EL amyloid oligomers exhibited very similar stochiometry with 4 to 20-30 EL molecules involved [4,22,24]. Both ELOA and the amyloid species of corresponding size display the cytotoxic apoptotic activity, clearly absent in EL itself. ELOA and EL amyloid oligomers were characterized by spherical morphology examined by AFM and both tended to self-assemble into donut-like circular structures with very similar diameters of ca. 30 nm as measured by AFM [4,31]. In addition, ELOA and EL amyloid oligomers possess characteristic amyloid tinctorial properties such as binding of thioflavin-T dye, which is known as an amyloid specific marker. Thus, ELOA has some common structural and cytotoxic features with both HAMLET and amyloid oligomers and their further studies may shed light on both these phenomena and potential link between them. It is important that ELOA complexes are stable enough to be amenable for structural characterization at atomic resolution, whereas the amyloid oligomers are often transient in nature and tend to associate into larger aggregates or split into monomers. Amyloid oligomers are also not well populated and attempts have been made to stabilize them by using fatty acids and surfactants [32-35], which extend further the comparison between HAMLET-type complexes and amyloid species.

7. Live cell study of ELOA interaction with the plasma membrane

Molecular mechanisms of protein complexes interaction with living cells and their primary targets at the cell surface remain largely unknown and disputed [20,36]. Methods with single molecule sensitivity, Fluorescence Correlation Spectroscopy (FCS) and Confocal Laser Scanning Microscopy (CLSM) imaging by avalanche photodiodes (APD), so called APD imaging [37], which enable quantitative and nondestructive studies of molecular interactions and mobility in living cells, revealed that ELOA primarily acts on the plasma membrane of PC12 cells, inflicting damage and eventually causing plasma membrane rupture (Figure 2 A and B) followed by a rapid influx and distribution of ELOA inside the already dead cell (Figure 2 C) [21].

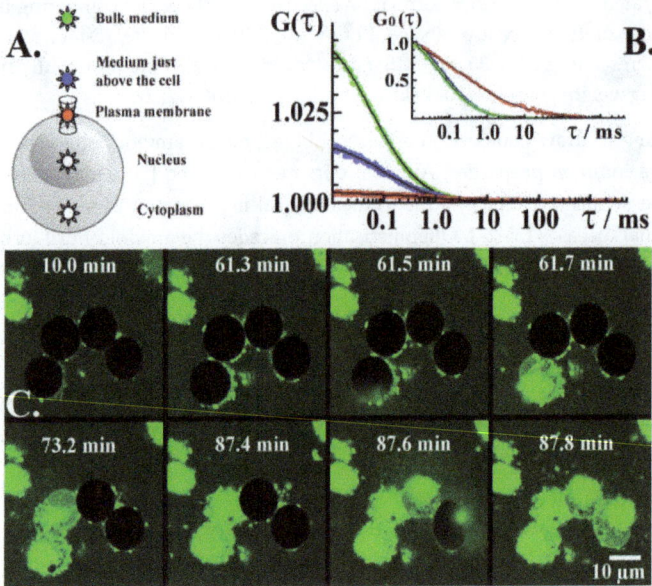

Figure 2. ELOA interaction with live PC12 cells studied by Fluorescence Correlation Spectroscopy (FCS) and APD imaging. (A) Schematic presentation of different locations – the cell culturing medium, PC12 cell plasma membrane and nucleus, at which FCS measurements were performed. **(B)** FCS measurements show that the concentration of ELOA in the bulk medium (c_{medium}^{bulk} = 240 nM) is lower than in the immediate vicinity of the cell ($c_{medium}^{cell\ surr.}$ = 670 nM) and the plasma membrane ($c_{medium}^{plasma\ membrane.}$ = 2.5 µM). **Insert:** Autocorrelation curves normalized to the same amplitude show that lateral mobility of ELOA in the plasma membrane (red) is significantly slower than in the medium (blue and green), as evident from the shift of the autocorrelation curve recorded at the plasma membrane (red) towards longer characteristic times. ELOA was neither detected in the cell nucleus nor in the cytoplasm. FCS measurements were taken 40-45 min after exposing PC12 cells to fluorescently labeled ELOA. **(C)** APD imaging shows that fluorescently labeled ELOA complexes are not gradually taken up by PC12 cells. Instead, the concentration of ELOA complexes in the immediate cellular surroundings progressively increases, reaching a local concentration that is several times higher than the concentration in the bulk medium. At a critical time-point (61.5 min), the plasma membrane ruptures. Only then the ELOA complexes "stream in" and swiftly distribute in the cellular interior, preferring particularly the cell nucleus. The scale bar is 10 µm.

8. Putative mechanism of ELOA-induced cellular toxicity

Local rearrangements of lipid organization in the plasma membrane of PC12 cells (Figure 3) observed using a general lipophilic marker that differently partitions between the ordered and disordered phase of the lipid bilayer (1,1'-dioctadecyl-3,3,3',3'-tetramethylind ocarbocyanine perchlorate dye, DiIC18(5)) [21,38], are consistent with the hypothesis that ELOA may form transient pores in the plasma membrane.

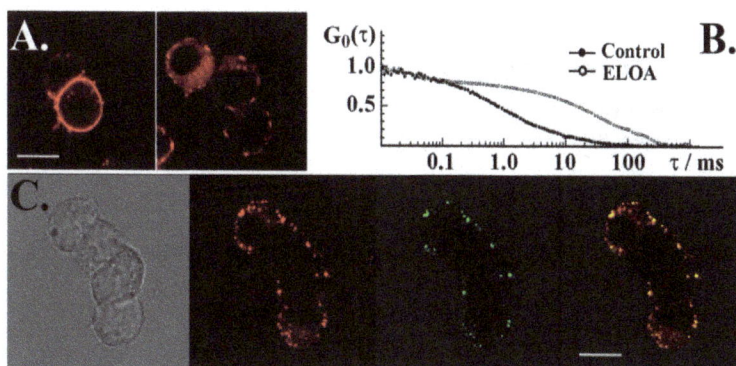

Figure 3. Lipid marker distribution in the plasma membrane of live PC12 treated with ELOA. (A)
Representative image showing uniform distribution of the fluorescent lipid marker DiIC18(5) in PC12 cells not exposed to ELOA (left). In cells exposed to ELOA the distribution of DiIC18(5) becomes patchy, and regions of local accumulation could be observed (right). **(B)** DiIC18(5) partitioning between different regions in the plasma membrane is also affected, as evident from the shifting of the autocorrelation curve to longer characteristic times in cells treated with ELOA. (C) Transmitted-light and APD images of PC12 cells taken 40 min after exposure to ELOA show that the plasma membrane marker DiIC18(5) (red) colocalizes with the fluorescently labeled ELOA complexes (green). The scale bar is 10 μm.

9. Future development and prospective applications

Recently, complexes of bovine β-lactoglobulin and pike parvalbumin with OA were produced and classified as HAMLET-type complexes [39]. These proteins are neither structurally related to α-lactalbumins nor to lysozymes. Nevertheless, their complexes with OA displayed cytotoxic activity that bears a resemblance to the cytotoxic activity of HAMLET [39]. This suggests that protein self-assembly may be mediated by oleic acid and more oleic acid-protein complexes can be discovered in future. Their putative ability to eliminate specifically rapidly divided cells, such as cancer cells, has a significant therapeutic potential. The mechanisms of their toxic activity are still debated. Our research provides first insight at a single cell level that ELOA interactions with the cellular membrane play critical role in cytotoxicity, leading to membrane permeability and even rupture. There are obvious differences in the composition and structure of protein-oleic acid complexes arising due to differences in the structure and dynamics of the protein component and differences in the conditions of complex formation. The common feature of these complexes can be related to the fact that they all serve as cargo vessels delivering oleic acid to the cells and facilitating its penetration into cell membrane and cell interior. HAMLET is the first example of proteinaceous complexes with oleic acid effectively used in combating various cancer conditions and other complexes can be also potentially used for this purpose if their properties will be well-understood and controlled.

Author details

Vladana Vukojević
Department of Clinical Neuroscience, Karolinska Institute, Stockholm, Sweden

Ludmilla A. Morozova-Roche
Department of Medical Biochemistry and Biophysics, Umeå University, Umeå, Sweden

Acknowledgement

Support from the Swedish Medical Research Council, Insamlingsstiftelsen, Umeå and the Kempe foundation is gratefully acknowledged.

10. References

[1] Mossberg AK, Hun Mok K, Morozova-Roche LA, Svanborg C (2010) Structure and function of human α-lactalbumin made lethal to tumor cells (HAMLET)-type complexes. FEBS J. 277:4614-4625.

[2] Lisková K, Kelly AL, O'Brien N, Brodkorb A (2010) Effect of denaturation of alpha-lactalbumin on the formation of BAMLET (bovine alpha-lactalbumin made lethal to tumor cells). J Agric Food Chem. 58:4421-4427.

[3] Morozova-Roche LA (2007) Equine lysozyme: the molecular basis of folding, self-assembly and innate amyloid toxicity. FEBS Lett. 581: 2587-2592.

[4] Wilhelm K, Darinskas A, Noppe W, Duchardt E, Mok KH, Vukojević V, Schleucher J, Morozova-Roche LA (2009) Protein oligomerization induced by oleic acid at the solid-liquid interface--equine lysozyme cytotoxic complexes. FEBS J. 276:3975-3989.

[5] Håkansson A, Zhivotovsky B, Orrenius S, Sabharwal H, Svanborg C (1995) Apoptosis induced by a human milk protein. Proc Natl Acad Sci U S A. 92:8064-8068.

[6] Håkansson A, Andréasson J, Zhivotovsky B, Karpman D, Orrenius S, Svanborg C (1999) Multimeric alpha-lactalbumin from human milk induces apoptosis through a direct effect on cell nuclei. Exp Cell Res. 246:451-460.

[7] Svensson M, Sabharwal H, Håkansson A, Mossberg AK, Lipniunas P, Leffler H, Svanborg C, Linse S (1999) Molecular characterization of alpha-lactalbumin folding variants that induce apoptosis in tumor cells. J Biol Chem. 274:6388-6396.

[8] Svensson M, Håkansson A, Mossberg AK, Linse S, Svanborg C (2000) Conversion of alpha-lactalbumin to a protein inducing apoptosis. Proc Natl Acad Sci U S A. 97:4221-4226.

[9] Svanborg C, Agerstam H, Aronson A, Bjerkvig R, Düringer C, Fischer W, Gustafsson L, Hallgren O, Leijonhuvud I, Linse S, Mossberg AK, Nilsson H, Pettersson J, Svensson M

(2003) HAMLET kills tumor cells by an apoptosis-like mechanism--cellular, molecular, and therapeutic aspects. Adv Cancer Res. 88:1-29.

[10] Gustafsson L, Hallgren O, Mossberg AK, Pettersson J, Fischer W, Aronsson A, Svanborg C (2005) HAMLET kills tumor cells by apoptosis: structure, cellular mechanisms, and therapy. J Nutr. 135:1299-1303.

[11] Svensson M, Mossberg AK, Pettersson J, Linse S, Svanborg C (2003) Lipids as cofactors in protein folding: stereo-specific lipid-protein interactions are required to form HAMLET (human alpha-lactalbumin made lethal to tumor cells). Protein Sci. 12:2805-2814.

[12] Svensson M, Fast J, Mossberg AK, Düringer C, Gustafsson L, Hallgren O, Brooks CL, Berliner L, Linse S, Svanborg C (2003) Alpha-lactalbumin unfolding is not sufficient to cause apoptosis, but is required for the conversion to HAMLET (human alpha-lactalbumin made lethal to tumor cells). Protein Sci. 12:2794-2804.

[13] Pettersson-Kastberg J, Aits S, Gustafsson L, Mossberg A, Storm P, Trulsson M, Persson F, Mok KH, Svanborg C (2009) Can misfolded proteins be beneficial? The HAMLET case. Ann Med. 41:162-176.

[14] Svensson M, Håkansson A, Mossberg AK, Linse S, Svanborg C (2000) Conversion of alpha-lactalbumin to a protein inducing apoptosis. Proc Natl Acad Sci U S A. 97:4221-4226.

[15] Morozova L, Haezebrouck P, Van Cauwelaert F (1991) Stability of equine lysozyme. I. Thermal unfolding behaviour. Biophys Chem. 41:185-191.

[16] Van Dael H, Haezebrouck P, Morozova L, Arico-Muendel C, Dobson CM (1993) Partially folded states of equine lysozyme. Structural characterization and significance for protein folding. Biochemistry. 32:11886-11894.

[17] Morozova LA, Haynie DT, Arico-Muendel C, Van Dael H, Dobson CM (1995) Structural basis of the stability of a lysozyme molten globule. Nature Struct. Biol. 10:171- 175.

[18] Morozova-Roche LA, Arico-Muendel C, Haynie DT, Emelyanenko VI, Van Dael H, Dobson CM (1997) Structural characterisation and comparison of the native and A-states of equine lysozyme. J. Mol. Biol. 268: 903-921.

[19] Morozova-Roche LA, Jones JA, Noppe W, Dobson CM (1999) Independent nucleation and heterogeneous assembly of structure during folding of equine lysozyme. J. Mol. Biol. 289: 1055-1073.

[20] Stefani M (2007) Generic cell dysfunction in neurodegenerative disorders: role of surfaces in early protein misfolding, aggregation, and aggregate cytotoxicity. Neuroscientist. 13:519-531.

[21] Vukojević V, Bowen AM, Wilhelm K, Ming Y, Ce Z, Schleucher J, Hore PJ, Terenius L, Morozova-Roche LA (2010) Lipoprotein complex of equine lysozyme with oleic acid (ELOA) interactions with the plasma membrane of live cells. Langmuir. 26:14782-14787.

[22] Mok KH, Hore PJ (2004) Photo-CIDNP NMR methods for studying protein folding. Methods. 34:75-87.

[23] Fast J, Mossberg AK, Svanborg C, Linse S (2005) Stability of HAMLET-a kinetically trapped alpha-lactalbumin oleic acid complex. Protein Sci. 14:329-340.

[24] Malisauskas M, Ostman J, Darinskas A, Zamotin V, Liutkevicius E, Lundgren E, Morozova-Roche LA (2005) Does the cytotoxic effect of transient amyloid oligomers from common equine lysozyme in vitro imply innate amyloid toxicity? J Biol Chem. 280:6269-6275.

[25] Campioni S, Mannini B, Zampagni M, Pensalfini A, Parrini C, Evangelisti E, Relini A, Stefani M, Dobson CM, Cecchi C, Chiti F (2010) A causative link between the structure of aberrant protein oligomers and their toxicity. Nat Chem Biol. 6:140-147.

[26] Chiti F, Dobson CM (2006) Protein misfolding, functional amyloid, and human disease. Annu. Rev. Biochem. 75:333-366.

[27] Morozova-Roche LA, Malisauskas M (2007) A false paradise - mixed blessings in the protein universe: the amyloid as a new challenge in drug development. Curr. Med. Chem. 14:1221–1230.

[28] Morozova-Roche LA, Zurdo J, Spencer A, Noppe W, Receveur V, Archer DB, Joniau M, Dobson CM (2000) Amyloid fibril formation and seeding by wild-type human lysozyme and its disease-related mutational variants. J Struct Biol. 130:339-351.

[29] Goers J, Permyakov SE, Permyakov EA, Uversky VN, Fink AL (2002) Conformational prerequisites for alpha-lactalbumin fibrillation. Biochemistry. 41:12546–12551.

[30] Harrison RF, Hawkins PN, Roche WR, MacMahon RF, Hubscher SG, Buckels JA (1996) 'Fragile' liver and massive hepatic haemorrhage due to hereditary amyloidosis. Gut. 38:151-152.

[31] Malisauskas M, Zamotin V, Jass J, Noppe W, Dobson CM, Morozova-Roche LA (2003) Amyloid protofilaments from the calcium-binding protein equine lysozyme: formation of ring and linear structures depends on pH and metal ion concentration. J Mol Biol. 330:879-890.

[32] Nagarajan S, Ramalingam K, Neelakanta Reddy P, Cereghetti DM, Padma Malar EJ, Rajadas J (2008) Lipid-induced conformational transition of the amyloid core fragment Abeta(28-35) and its A30G and A30I mutants. FEBS J. 275:2415-2427.

[33] Otzen DE, Sehgal P, Westh P (2009) Alpha-Lactalbumin is unfolded by all classes of surfactants but by different mechanisms. J Colloid Interface Sci. 329:273-283.

[34] Sharon R, Bar-Joseph I, Frosch MP, Walsh DM, Hamilton JA, Selkoe DJ (2003) The formation of highly soluble oligomers of alpha-synuclein is regulated by fatty acids and enhanced in Parkinson's disease. Neuron. 37:583–595.

[35] Otzen DE, Nesgaard LW, Andersen KK, Hansen JH, Christiansen G, Doe H, Sehgal P (2008) Aggregation of S6 in a quasi-native state by sub-micellar SDS. Biochim. Biophys. Acta 1784:400–414.

[36] Cecchi C, Baglioni S, Fiorillo C, Pensalfini A, Liguri G, Nosi D, Rigacci S, Bucciantini M, Stefani M (2005) J. Cell Sci. 118:3459-3470.

[37] Vukojević V, Heidkamp M, Ming Y, Johansson B, Terenius L, Rigler R (2008) Quantitative single-molecule imaging by confocal laser scanning microscopy. Proc Natl Acad Sci U S A. 105:18176-1081.

[38] Loura LM, Fedorov A, Prieto M (2000) Partition of membrane probes in a gel/fluid two-component lipid system: a fluorescence resonance energy transfer study. Biochim Biophys Acta. 1467:101-112.

[39] Permyakov SE, Knyazeva EL, Khasanova LM, Fadeev RS, Zhadan AP, Roche-Hakansson H, Håkansson AP, Akatov VS, Permyakov EA (2012) Oleic acid is a key cytotoxic component of HAMLET-like complexes. Biol Chem. 393:85-92.

Chapter 13

Role of Lipoproteins in Carcinogenesis and in Chemoprevention

Adebowale Bernard Saba and Temitayo Ajibade

Additional information is available at the end of the chapter

1. Introduction

Lipoproteins are complex aggregates of lipids and proteins that render endogenous lipids compatible with the aqueous environment of body fluids (Brown, 2007). The major physiological role of lipoproteins is to transport water-insoluble lipids from their point of origin to their respective destinations. Lipoproteins are synthesised mainly in the liver and intestines. They are in a state of constant flux, in circulation, changing in composition and physical structure as the peripheral tissues take up the various components before the remnants are returned to the liver. The most abundant lipid constituents of lipoproteins are free cholesterol, cholesterol esters, triacylglycerols and phospholipids, though fat-soluble vitamins and anti-oxidants are also found in lipoproteins (Kwiterovich, 2000).

Classification of lipoproteins

Lipoproteins are classified as chylomicrons (CM), very-low-density lipoproteins (VLDL), low-density lipoproteins (LDL) and high-density lipoproteins (HDL), based on the relative densities of the aggregates on ultracentrifugation. These classes are further refined by improved separation procedures, and intermediate-density lipoproteins (IDL) and subdivisions of the HDL (e.g. HDL_1, HDL_2, HDL_3 etc) are often defined. Density of lipoproteins is determined largely by the relative concentrations of triacylglycerols and proteins and by the diameters of the broadly spherical particles, which vary from about 6000Å in CM to 100Å or less in the smallest HDL. An alternative nomenclature is based on the relative mobilities on electrophoresis on agarose gels. Thus, α, pre-β and β lipoproteins correspond to HDL, VLDL and LDL, respectively (Lacko et al., 2007).

Chylomicrons

Chylomicrons, the largest and least dense of the lipoproteins are formed in the intestinal cell walls from dietary fat and cholesterol. Their main task is to carry triglycerides from the intestine to the tissues where they are needed as a source of energy. In the circulation, triglycerides are removed from chylomicrons via the action of lipoprotein lipase (LPL), an enzyme present in the capillaries of many tissues. If present in large amounts, such as after a fatty meal, chylomicrons cause the plasma to appear milky.

Very low density lipoproteins

Very low density lipoproteins (VLDLs) are synthesised in the liver. Much like chylomicrons, they function primarily to distribute triglycerides to target sites such as adipose tissue and skeletal muscle where they are used for storage and energy. The manner in which triglycerides are removed from the circulation is the same as that of chylomicrons. Gradually with removal of triglycerides and protein, VLDLs are converted to LDL. High plasma levels of VLDL are associated with familial hypertriglyceridaemia, diabetes mellitus and underactive thyroid.

Low density lipoproteins

Low density lipoproteins are cholesterol-rich particles. About 70% of plasma cholesterol occurs in this form. LDLs are chiefly involved in the transport of the cholesterol manufactured in the liver to the tissues, where it is used. Uptake of cholesterol into cells occurs when lipoprotein binds to LDL receptors on the cell surface. LDL is then taken into the cell and broken down into free cholesterol and amino acids. Disorders involving a defect in or lack of LDL receptors are usually characterised by high plasma cholesterol levels. The cholesterol cannot be cleared efficiently from the blood and therefore accumulates.

High density lipoproteins

The high-density lipoproteins (HDLs) are small, dense, and spherical lipid-protein complexes which are normally considered to consist of those plasma lipoprotein particles which fall into the density range of $1.063–1.210\,g/mL$. HDL particles are composed of an outer layer containing free cholesterol, phospholipid, and various apolipoproteins (Apo), which covers a hydrophobic core consisting primarily of triglycerides and cholesterol esters (Barter et al., 2003). The major proteins are Apo A-I (Mr 28,000) and Apo A-II (Mr 17,000). Apo A-I, the primary protein constituent of these particles, accounts for about 60% of the protein content of HDL. Apo A-I is synthesized in the intestines and liver and is thought to be largely responsible for the antiatherogenic effects of HDL. Some HDL particles carry only Apo A-I, whereas others contain both Apo A-I and Apo A-II (Shah et al., 2001). Other apolipoprotein species found in HDL particles include Apo A-IV, Apo C (C-I, C-II, and C-III), and Apo E.

High density lipoprotein subtypes

Several subtypes of HDL particles have been identified on the basis of density, electrophoretic mobility, particle size, and apolipoprotein composition (Albers et al., 1984). Differences in particle size are mainly the result of the number of apolipoprotein particles and the volume of the cholesterol ester in the core of the particle. HDL can also be classified into larger, less dense HDL2 or smaller, denser HDL3 which falls within the density ranges 1.063–1.125 and $1.125–1.210\,g/mL$, respectively. Although the major proportion of HDL is normally present in HDL3, individual variability in HDL levels in human populations usually reflects different amounts of HDL2 (Skinner, 1994). HDL2 is richer in particles containing Apo A-I without Apo A-II, whereas HDL3 is richer in particles containing both Apo A-I and Apo A-II (Gotto, 2001).

Synthesis, Metabolism and Regulation of plasma lipoprotein concentration

Normal metabolism and homeostasis of carbohydrates, amino acids and lipids *in vivo* depend on integrated liver function. Most plasma apolipoproteins and endogenous lipids and lipoproteins, including apolipoprotein(a) (apo(a)) and lipoprotein(a) (Lp(a)), are synthesized in the liver. The apolipoprotein(a) is a high molecular weight glycoprotein, of 250-838 kD (Bowden, 1994) and a total of 34 different apo(a) isoforms have been identified in populations (Marcovina *et al.*, 1993). The core components of Lp(a) are neutral lipid and an apoB-100 molecule, which are covalently connected by a disulfide-bond bridge and surrounded by hydrophilic apo(a) (Byrne, 1994). The heterogeneity of apo(a) determines the changes in plasma Lp(a) concentrations, and there is a negative correlation between the molecular weight of apo(a) and the plasma Lp(a) concentration (Wade, 1993)

Lp(a) has a simple Mendelian dominant inheritance, which is controlled by the alleles Lpa and Lp0 (Utermann *et al.*, 1988). Plasma Lp(a) concentration is controlled by three alleles., i.e., LpA, Lpa and Lp0 (Hasstedt *et al.* 1986). Pedigree analysis indicated that the size polymorphism of Lp(a) is controlled by a series of alleles of a single point (Utermann, 1988). The apo(a) gene which is located in q26-27 of chromosome 6 in humans has a linkage to the plasminogen (PGN) gene, and is inherited in a codominant Mendelian model (Amemiya, 1996). Apolipoprotein(a) mRNA (14 kb) encodes for a mature protein of 4529 amino acid residues in the presence of a signal peptide with 19 amino acid residues (McLean, 1987) while a high-degree homology exists between the molecular structures of the apo(a) gene and the PGN gene. The high-degree homology between the apo(a) gene and the PGN gene determines the biological actions of Lp(a) (Romics *et al.*, 1996).

2. Role of lipoprotein in cancer

Cancerous cells generally have high requirements for cholesterol as they are rapidly dividing cells. The Low density lipoproteins (LDL) which are cholesterol-rich particles have been especially found to play significant role in the pathogenesis of a large number of cancers. For instance, increased LDL requirement and receptor activity have been reported in cancer of the prostate gland (Chen and Hughes-Fulford, 2001); colon (Niendorf *et al.*, 1995); adrenal gland (Nakagawa *et al.*, 1995); hormone unresponsive breast tumors (Stranzl *et al.*, 1997), cancers of gynecological origin, tumors of lung tissues (Vitols *et al.*, 1992), leukemia (Tatidis *et al.*, 2002), and malignant brain tumors (Rudling *et al.*, 1990). In contrast, high density lipoproteins (HDL) have been reported significantly lowered in patients with primary or metastatic liver cancer (Moorman *et al.*, 1998). Hoyer and Engohm (1992) observed an inverse association between serum HDL-cholesterol and risk of breast cancer in a cohort of 5,207 Danish women, who participated in the Glostrup population studies of breast cancer.

Lymphoma patients often exhibit abnormal lipid metabolism. Numerous clinical studies of lymphoma patients have reported lipid abnormalities that are similar to the dyslipidemia observed in inflammatory and infectious diseases that are believed to develop secondary to circulating cytokines and the accompanying acute-phase response (Blackman *et al.*, 1993). Spiegel *et al.* (1989) investigated plasma lipids and lipoproteins at presentations in 25 patients with acute leukemia and non-Hodgkin's lymphoma and reported that all patients demonstrated an abnormality in at least one plasma lipid fraction and most exhibited a predictable pattern of lipid alterations that consisted of extremely low levels of HDL-cholesterol, elevated triglyceride, and elevated very low density lipoprotein (VLDL). The degree of lipid abnormality was directly related to the underlying tumor burden and particularly to the presence of bone marrow involvement. Therefore, low levels of circulating HDL-cholesterol in lymphoma patients may occur before the clinical onset of cancer and may serve as a marker for inflammation-induced lymphomagenesis, rather than a consequence of lymphoma-induced acute-phase responses.

2.1. Role of lipoprotein(a) in pathogenesis of cancer

Lipoprotein[a] is an intriguing molecule consisting of a low-density lipoprotein core and a covalently bound apolipoprotein[a]. Apolipoprotein[a] possess an inactive protease domain which is a single copy of the plasminogen kringle 5 and multiple repeats of domains homologous to the plasminogen kringle 4. The plasminogen kringle 5 (K5) domain, which is distinct from angiostatin, possesses potent anti-angiogenic properties on its own, which can be exploited in cancer therapy. The angiostatic effect and novel proinflammatory role of the K5 protein is via its ability to recruit tumor-associated neutrophils and NKT lymphocytes, leading to a potent antitumor response (Perri *et al.*, 2007).

Recently, anti-angiogenic agents have been found to promote leucocyte-vessel wall interaction as part of their anti-tumorigenic effects. Studies on animal models have indicated that the proteolytic break-down products of apolipoprotein[a] may posses anti-angiogenic and anti-tumorigenic effects both *in vitro* and *in vivo*. This is a convenient premise to develop novel therapeutic modalities which may efficiently suppress tumor growth and metastasis (Giuseppe *et al.*, 2007). Significant decrease in Lipoprotein[a] levels have been reported in liver cancer patients by Samonakis *et al.*, 2004. Although the liver plays an important role in lipid metabolism, several non-hepatic factors such as hormones, cytokines, genetics and nutrition are also involved in different ways. For example, several inflammatory and tumoral diseases are characterized by the production and delivery of cytokines influencing serum Lipoprotein[a] levels.

The mechanisms by which cancers induce cachexia involve inflammatory cytokine production; which is responsible for a wide number of metabolic disorders, essentially involving lipid metabolism (Langstein and Norton, 1991) and serum Lipoprotein[a] level changes during inflammatory disease. Liver damage has been linked to reduce Lipoprotein[a] serum levels (Malaguarnera *et al.*, 1996). Geiss *et al.*(1996) observed marked increase in Lipoprotein[a] concentration from 7 mg/dl in acute stage to 32 mg/dl in convalescence in hepatitis patients.The lipoprotein[a] half-life is short *in vivo*, 3.3~3.9 d, (Krempler *et al.*, 1983), and influenced early by liver function alterations (Malaguarnera *et al.*, 1994), The high involvement of Lipoprotein[a] in lipid and protein metabolism have been suggested to be a sensitive and early marker of liver malfunction, therefore Lipoprotein[a] may supply useful additional information for a more complete assessment and monitoring of the liver function in patients with hepatocellular carcinoma and liver cirrhosis (Uccello *et al.*, 2011).

2.1.1. Role of high density lipoprotein (HDL) in the pathogenesis of cancer

The origin and fate of HDL are less well understood than other lipoproteins. HDL may be formed both in the intestine and in the liver. It is also formed during lipolysis of Triglycerides-rich lipoproteins. Apolipoprotein AI (ApoAI) and apolipoprotein AII (apoAII) are the major apolipoproteins of HDL and the production rate of apoA-I is an important determinant of the variability of plasma HDL concentrations. Production rate of apoA-I is however, influenced by many factors and apoA-I transcriptional regulation has an impact on plasma HDL concentrations.

A large number of cellular lipid transporters and receptors including a spectrum of HDL and intermediates of HDL participate in the transport of excess cholesterol from peripheral cells to the liver. At one end of the spectrum are lipid-free or lipid-poor apoA-I particles, referred to as pre-β HDL. These particles are secreted by the liver and small intestine or generated from surface material from partially lipolyzed chylomicrons or from HDL2 in the periphery by the action of cholesteryl ester transfer protein (CETP), hepatic lipase, or phospholipid transfer protein (PTP) (Dullaart *et al.*, 2001). Lipolyzed chylomicrons are HDL precursors that accept unesterified cholesterol and phospholipids transferred by ATP-binding cassette transporter A1 (ABCA1) on peripheral cells, giving rise to discoidal lipoproteins containing apoA-I (Yokoyama, 2005). The acquired unesterified cholesterol of the apoA-I is esterified by the plasma enzyme lecithin:cholesterol acyltransferase (LCAT) to form cholesteryl ester which is packaged into the hydrophobic core of the discoidal particle, converting it to spherical HDL3. HDL3 can continue to accept unesterified cholesterol and phospholipids from the class B, type I scavenger receptor (SR-BI); through continued action of LCAT (Fig. 1). This causes the hydrophobic core of the discoidal lipoprotein particle to expand and the size increases, thereby forming HDL2.

High density lipoprotein (HDL) plays a key role in the reverse cholesterol transport pathway (RCTP) (Genest *et al.*, 1990) and various fractions of HDL have been shown to offer a new approach to study liver diseases (Cooper *et al.*, 1996). Plasma HDL-C, HDL-PL and HDL-C/HDL-PL have been reported to be lower in hepatocellular carcinoma patients than those in normal patients (Li *et al.*, 1993). In a study on 40 patients with hepatocellular carcinoma, LDL-C level was found to be significantly lower in the hepatocellular carcinoma patients than in the controls, but HDL-C did not show a statistically significant difference to the controls (Motta *et al.*, 2001). HDL-C itself has also been reported significantly decreased in patients with primary or metastatic liver cancer (Kanel *et al.* 1983). Therefore, variations in the level of plasma lipids and lipoproteins may assist in describing the nature of cirrhosis and hepatocellular carcinoma (Ahaneku *et al.*, 1992).

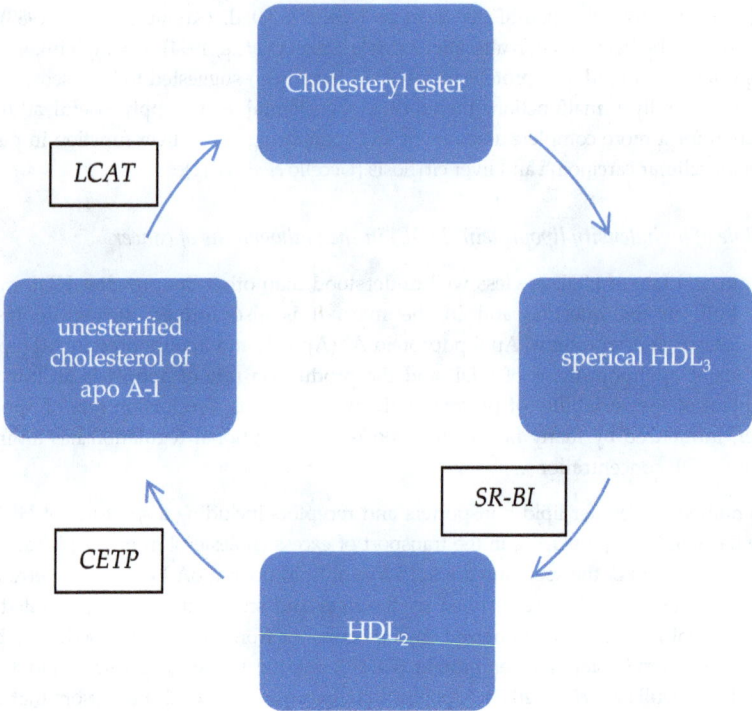

Figure 1. Formation of HDL2 from unesterified cholesterol

Furthermore, an inverse and significant association exists between levels of HDL cholesterol and the risk of incident cancer, according to the results of a recent study (Jafri *et al.*, 2010). Exceptionally, a few cancer risk factors are associated with increased levels of HDL. For instance, many breast cancer risk factors are associated with high HDL-C and the relationship between breast cancer and HDL-C is independent of other risk factors (Moorman *et al.*, 1998).

HDL could play a role in carcinogenesis through its influence on cell cycle entry, via a mitogen-activated protein kinase-dependent pathway or regulation of apoptosis. Specifically, an inverse association exists between serum HDL and risk of breast cancer, and several studies have reported lower levels of HDL in breast cancer patients. Further studies have also shown that tumor progression from localized to metastatic disease is associated with declining HDL levels. At least two population-based screening surveys involving Norwegian women have established low HDL, as part of the metabolic syndrome associated with increased post-menopausal breast cancer risk. The risk of post-menopausal breast cancer among overweight and obese women in the highest serum HDL-cholesterol quartile was one-third the risk of women in the lowest serum HDL-cholesterol quartile (Furberg *et al.*, 2004).

2.2. Serum lipid profile and incidence of cancer

Cancer patients often present altered serum lipid profile including changes of HDL level. Case-control studies of newly diagnosed lung cancer have shown that HDL levels are reduced in lung cancer cases. Patients with advanced nonresectable lung cancer also often have decreased serum HDL. Although the biological mechanisms that might link low plasma levels of HDL with cancer are not well understood, the HDL regulation of cell cycle entry through a mitogen activated protein kinase-dependent pathway and apoptosis, modulation of cytokine production, and antioxidative function (Perletti et al., 1996) has been suggested biologically plausible.

The association between higher HDL and lower overall cancer incidence observed in the ATBC cohort is biologically plausible, as HDL has anti-inflammatory properties (Perletti et al., 1996). However, it is also plausible that this association reflects the effect of factors such as inflammation, which are associated with both HDL and risk of cancer.

2.3. Effect of inflammation on HDL levels in the body

Inflammation reduces HDL and likely increases risk of lung cancer (Mendez et al., 1991). Studies have shown that chronic inflammation is known to reduce both serum HDL-cholesterol levels and its anti-inflammatory properties. The lipoprotein abnormalities seen in patients with inflammatory diseases are thought to develop secondary to circulating cytokines and the accompanying acute-phase response (Blackman et al., 1993). Low HDL-cholesterol may therefore be a marker for the severity of systemic inflammation and inflammation-induced non-Hodgkin's lymphoma risk. Conversely, high HDL-cholesterol itself may be protective against non-Hodgkin's lymphoma. High-density lipoprotein-cholesterol seems to modulate inflammatory responses independent of non-HDL cholesterol levels by suppressing chemotactic activity of monocytes and lymphocytes and inhibiting cytokine-induced expression of endothelial cell adhesion molecules.

2.4. Association between antioxidant enzymes and HDL

There are a number of enzymes associated with HDL that have antioxidant properties, including paraoxonase, platelet-activating factor acetylhydrolase, and glutathione peroxidase. Paraoxonase-1 (PON1), the enzyme primarily responsible for HDL's antioxidant function, is closely bound to the HDL particle. PON1's enzymatic activity is highly regulated by environmental factors such as diet and physical activity, by certain drugs, and by genetic factors, especially certain genetic polymorphisms in the paraoxonase-1 gene, PON1. PON1 is synthesized in the mammalian liver and circulates in blood bound to HDL apolipoprotein (apo) A-1 and apo J. There are 2 other proteins in the same family as PON1 that probably also have antioxidant actions. These are PON2 and PON3. PON2 is ubiquitously expressed within cells, whereas PON3 exhibits a basal constitutive antioxidant activity and is essentially bound to HDL (Reddy et al., 2001; Ng et al., 2001). These enzymes have the ability to prevent the formation of proinflammatory oxidized phospholipids and to block the activity of those already formed; however, these oxidized lipids negatively regulate the activities of the HDL-associated enzymes. During an acute-phase response in rabbits, mice, and humans, there seems to be an increase in the formation of these oxidized lipids that results in the inhibition of the HDL-associated enzymes and an association of acute-phase proteins with HDL that renders HDL pro-inflammatory rather than being anti-inflammatory.

2.5. Role of lipoproteins in cancer chemotherapy

The efficacy of cancer chemotherapy is often limited by severe cytotoxic effects induced by anticancer drugs, on healthy and cancerous cells (Sehouli *et al.*, 2002). In addition, most of the available dosage forms perform with less than optimal efficiency because of poor solubility and limited accessibility to target tissues (Rosen and Abribat, 2005). Furthermore, it is difficult to eradicate cancer cells *in vivo* because they share the same biochemical machinery with normal cells.

In spite of the overwhelmingly large number of anticancer drugs that have been developed, none is completely selective for cancer cells. Consequently, all anticancer drugs presently in use induce significant dose-limiting toxic side effects. For this reason, there has been increased emphasis on selective delivery of drugs to tumours in ways that bypass normal body tissues.

2.5.1. Lipoproteins as special anticancer drug delivery agent

The cytotoxic effects of cancer chemotherapeutic drugs on healthy organs can be significantly diminished by employing special drug delivery systems targeted specifically to cancer cells (Minko *et al.*, 2004). Targeting is especially important in circumstances where a localized tumor is removed surgically, and chemotherapy is prescribed as a follow-up preventive against potential metastases (Dharap *et al.*, 2005). Among the vehicles that can be used for special anticancer drug delivery are lipoproteins.

Lipoproteins have been considered appropriate drug-delivery vehicles for anticancer drugs (Braschi *et al.*, 1999), owing to their structural features, biocompatibility and targeting capability via receptor mediated mechanisms (Nikanjam *et al.*, 2007). The basic structure of lipoproteins, which comprises of an outer protein–phospholipid shell with a lipophilic surface and an interior hydrophobic compartment, positions them as ideal transporters of hydrophobic drugs, including anticancer agents. Due to their biocompatible, lipoproteins have considerable advantages over the conventional carrier systems currently used in cancer chemotherapy in that they provide the opportunity for targeted delivery of the anticancer drug they carry through endocytosis by receptor mediated uptake or by selective uptake of core components (Pathania, *et al.*, 2003).

The main advantages of lipoprotein-based formulations are their biocompatible components, their relative stability in the blood circulation and their track record of having already been safely injected into human subjects (Bisoendial *et al.*, 2002). In addition, lipoprotein-based formulations have vast targeting potential via receptor-mediated mechanisms that are overexpressed in cancer cells compared with normal cells. It has been shown that anticancer drugs can be targeted to cancer cells that generally express a high level of lipoprotein receptors, by encapsulation into reconstituted high density lipoprotein nanoparticles which are similar to the endogenous lipoprotein particles responsible for shuttling of hydrophobic molecules to different parts of the body.

Lipoprotein complexes are ideal for loading and targeted delivery of cancer therapeutic and diagnostic agents because they can mimic the shape and structure of endogenous lipoproteins, and as such, remain in circulation for an extended period of time, while largely evading the reticuloendothelial cells in the body's defenses. The small size (less than 30 nm) of the low-density and high-density classes of lipoproteins allows them to maneuver deeply into tumors. Furthermore, lipoproteins can be targeted to their endogenous receptors, especially, when the receptors are implicated in cancer. Although the lipophilic character of certain pharmaceuticals may be a disadvantage during intravenous therapy, this can be advantageous in anticancer drug deliveries as the highly lipid compounds are ideally suited for incorporation into lipoproteins. Most of the 'orphaned' anticancer drugs that have failed primarily due to their poor water solubility can also be made to progress faster through development process by incorporating them into lipoproteins.

2.5.2. Low density lipoprotein in cancer chemotherapy

Low density lipoprotein (LDL) has been found to represent a suitable carrier for cytotoxic drugs that may target them to cancer (Kader and Pater, 2002). This is because the low-density lipoprotein receptor (LDLR) has been found to be over-expressed in numerous cancers. The upregulated levels of low-density receptor in these cancers are believed to provide the cancer cells with the necessary lipid substrates needed for active membrane synthesis. In fact, sequestration of plasma LDL cholesterol in cancers has been suggested to explain the low levels of circulating total and LDL cholesterol observed in patients with malignancies. These findings have led many researchers to investigate the possibility of exploiting LDL as a delivery vehicle for cancer diagnostics and therapeutics (Corbin and Zheng, 2007).

The high requirement of LDL by cancer cells and thus the overexpression of LDL receptor can be utilized for developing a novel targeted drug delivery system. This can be achieved by targeting of the LDL particle and allowing the anticancer drugs to be transferred to the natural LDL inside of the body. LDL will function as a secondary carrier of anticancer molecules and deliver these molecules selectively to cancerous cells via elevated LDL receptors. This approach requires the anticancer molecules to have affinity for the LDL particle endogenously and to have certain special physicochemical properties.

There are at least three different ways in which diagnostic or anticancer agents can be incorporated into LDL. The first of these is protein labeling in which the anticancer drug is covalently attached to the amino acid residues of apolipoprotein (apo) B-100 protein of LDL; the second is surface labeling which involves intercalation of the diagnostic or anticancer agent into the phospholipid monolayer of LDL; the third is reconstitution core loading via substitution of anticancer agents into the lipid core of LDL. A large number of researchers have shown that various anticancer agents could be actively incorporated into LDL through intercalation or reconstitution methods. Moreover, these novel LDL–drug complexes were shown to be more efficacious against cancer cells than their conventional counterparts.

Coupling of doxorubicin to human LDL to form a LDL-doxorubicin complex injected to mice resulted in greater accumulation of LDL-doxorubicin, in the liver, than free doxorubicin. In contrast, LDL- doxorubicin was less accumulated in heart than free doxorubicin. This suggests LDL could be used as carriers to conjugate anti-cancer drugs.

2.5.3. Problems limiting the use of LDL as delivery vehicle

Although a lot of studies have produced promising results, progress towards utilizing LDL as a delivery vehicle in the clinical setting has been impeded by the need to isolate LDL from fresh donor plasma. Relying on donor plasma to acquire LDL is problematic because LDL samples vary from batch to batch, methods for isolating LDL are lengthy and large quantities of LDL are difficult to attain. Furthermore, isolated LDL can only be stored for finite periods before aggregation and degradation processes compromise the integrity of the LDL sample. As a result of these limitations, attempts have been made to prepare synthetic LDL-like particles consisting of phospholipid/cholesterol ester microemulsions and apoB-100 (the LDLR-binding component of LDL). Difficulties also plagued this endeavor owing to the size and complexity of the apoB-100 protein which is one of the largest monomeric proteins known consisting of over 4500 amino acids with a molecular weight of 550 kDa. Furthermore, apoB-100 is highly insoluble in aqueous solutions, making it difficult to work with. In addition, these problems are compounded still by the difficulties of having to isolate apoB-100 from donor plasma and other approaches to working with apoB-100 are therefore needed.

2.5.4. High-density Lipoprotein in cancer chemotherapy

The targeted delivery of anticancer agents via lipoprotein carriers is based on the concept that cancer cells have a higher expression of lipoprotein receptors (Lacko et al., 2002; Cao et al., 2004) due to their increased need for cholesterol to promote rapid proliferation. Clinical studies have shown that HDL cholesterol levels, like LDL levels, are lower in cancer patients, including those with haematological malignancies (Fiorenza et al., 2000). Unfortunately, the targeting of chemotherapeutic agents via HDL is daunting because overwhelming efforts are required for the isolation HDL from human plasma and a lot of biosafety concerns are attached with the injection of human-blood-derived products (Adams et al., 2003). Consequently, the focus of future studies is likely to be on synthetic/reconstituted lipoproteins with favourable drug-carrying capacity, and the exploitation of their potential for targeting tumour cells and tissues.

HDL transports cholesterol to liver cells, where they are recognized and taken up via specific receptors. Cholesteryl esters within HDL are selectively uptaken by hepatocytes via the scavenger receptor class B type I (SR-BI). An interesting feature of SR-BI is that the receptor selectively translocates HDL-cholesteryl esters from the lipoprotein particle to the cytosol of the liver parenchymal cells without a parallel uptake of the apolipoproteins and this property may allow for the delivery of its loaded drugs avoiding lysosomal degradation (Lou et al., 2005). The high affinity of cancer cells for HDL has made them useful as carriers for delivering anticancer drugs into hepatoma cells to treat HCC. Anti-cancer drug-HDL complexes work as efficient drug delivery vehicles due to the ability of cancer cells to acquire HDL core components (Wasan et al., 1996). Complexing of anti-cancer drugs with HDL does not influence characteristics of the anticancer drugs (Kader et al., 2002) and administration of anti-cancer drug-HDL complex may reduce toxic side-effects during the chemotherapy(Lacko, et al.,2002).

In a cell culture system, cellular uptake of recombinant HDL-aclacinomycin by the SMMC-7721 hepatoma cells was significantly higher than that of free aclacinomycin at the concentration range of 0.5–10 μg/mL. Cytotoxicity of recombinant HDL- aclacinomycin to the hepatoma cells was significantly higher than that of free aclacinomycin at concentration range of less than 5 μg/mL just as IC50 of recombinant HDL-aclacinomycin was lower than IC50 of free aclacinomycin (Lou et al., 2005). These results strongly suggest that HDL could be used as carriers to conjugate water-insoluble anti-cancer drugs in order to achieve higher therapeutic concentrations of the drugs in the microenvironment of the cancer cells.

3. Conclusion

Lipoproteins are complex endogenous aggregates of lipids and proteins that function primarily for the transport of water insoluble lipids from their point of origin to their respective destinations. Lipoproteins are classified as chylomicrons, very low density lipoproteins, low density lipoproteins and high density lipoproteins, based on the relative densities of the aggregates on ultracentrifugation.

Lipoproteins, which are cholesterol-rich particles, have been especially found to play significant roles in the pathogenesis of a large number of cancers because rapidly dividing cancerous cells generally have high requirements for cholesterol. This is exemplified by the increased LDL-requirement associated with a number of cancers including cancers of the prostate, colon, adrenal gland etc.

The efficacy of cancer chemotherapy is often limited by severe deleterious effects induced by anticancer drugs, on healthy and cancerous cells because of lack of specificity for the cancerous cells. In addition most available anticancer drugs do not perform optimally because of limited accessibility to target tissues. The deleterious effects of anticancer drugs on healthy organs can be markedly diminished by employing special drug delivery systems that specifically target cancer cells, using lipoproteins as carriers.

The main advantages of lipoproteins as anti-cancer drug carriers are: (1) lipoproteins are spherical particles consisting of a core of apolar lipids surrounded by a phospholipid monolayer, in which cholesterol and apoproteins are embedded. Therefore, highly lipophilic drugs can be incorporated into the apolar core without affecting lipoprotein receptor recognition; (2) lipoproteins are completely bio-degradable, do not trigger immunological responses, escape from recognition and elimination by the reticuloendothelial system, and have a relatively long half-life in the circulation; (3) lipoproteins can be recognized and taken up via specific receptors, and can mediate cellular uptake of the carried drugs ; and (4) many cancer cells show a high ability of lipoprotein uptake and therefore high therapeutic levels of the conjugated drugs can be rapidly attained at the target site(s).

Author details

Adebowale Bernard Saba and Temitayo Ajibade
Department of Veterinary Physiology,
Biochemistry and Pharmacology, University of Ibadan, Ibadan, Nigeria

4. References

Adams, K. M.; Lamber, N.C.; Heimfeld S.; Tylee, T. S. & Pang J. M. (2003). Male DNA in female donor apheresis and CD34-enriched products. *Blood*   102: 3845-3847. *doi:10.1182/blood-2003-05-1570.*

Ahaneku, J.E.; Taylor, G.O.; Olubuyide, I.O. & Agbedana, E.O. (1992). Abnormal lipid and lipoprotein patterns in liver cirrhosis with and without hepatocellular carcinoma. *J Pak Med Assoc*, 42(11):260-263. PMID:1336073

Albers, J. J.; Tollefson J. H.;Chen, C. H. & Steinmetz A. (1984) "Isolation and characterization of human plasma lipid transfer proteins," *Arteriosclerosis*; 4 (1): 49–58. PMID:6691846

Amemiya, H.; Arinami, T.; Kikuchi S.; Yamakawa-Kobayashi K.; Li L.; Fujiwara H. (1996). Apolipoprotein(a) and pentanucleotide repeat polymorphisms are associated with the degree of atherosclerosis in coronary heart disease. *Atherosclerosis*. 123:181-191. PMID:8782849

Barter, P.J.; Kastelein, & Kastelein J. (2003). "High density lipoproteins (HDLs) and atherosclerosis; the unanswered questions," *Atherosclerosis*.   168 (2) 195–211. PMID:12801602

Bisoendial, R.J.; Hovingh, G.K & de Groot, E. (2002). Measurement of subclinical atherosclerosis: beyond risk factor assessment. *Curr Opin Lipidol* 13:595-603. PMID:12441883

Blackman, J.D.; Cabana, V.G. & Mazzone, T. (1993). The acute-phase response and associated lipoprotein abnormalities accompanying lymphoma. *J Intern Med*. 233(2):201-4. PMID: 8433082

Bowden, J.F.; Pritchard, P.H.; Hill, J.S. & Frohlich, J.J. (1994). Lp(a) concentration and apo(a) isoform size. Relation to the presence of coronary artery disease in familial hypercholesterolemia. *Arterioscler Thromb* 14:1561-1568. PMID:7918305

Braschi, S.; Neville, T.A.; Vohl, M.C.& Sparks, D.L. (1999). Apolipoprotein A-I charge and conformation regulate the clearance of reconstituted high density lipoprotein *in vivo. J Lipid Res.*4 0(3):522–532. PMID:10064741

Brown, W.V. (2007). High-density lipoprotein and transport of cholesterol and triglyceride in blood. *J. Clin. Lipidology*, 1, 7-19. PMID: 21291664

Byrne, C.D. & Lawn, R.M. (1994). Studies on the structure and function of the apolipoprotein(a) gene. *Clin Genet*; 46:34-41. PMID:7988075

Cao, W.M.; Murao, K. & Imachi H. (2004). A mutant high-density lipoprotein receptor inhibits proliferation of human breast cancer cells. *Cancer Res*; 64: 1515-1521. PMID:14973113

Chen, Y. & Hughes-Fulford M. (2001). Human prostate cancer cells lack feedback regulation of low-density lipoprotein receptor and its regulator, SREBP2. *Int J Cancer*. 2001;91:41–45. PMID:11149418

Cooper, M.E., Akdeniz A. & Hardy K.J. (1996): Effects of liver transplantation and resection on lipid parameters: a longitudinal study. *Aust N Z J Surg*; 66(11):743-746. PMID:8918381

Corbin, I.R. & Zheng G. (2007). Mimicking Nature's Nanocarrier: Synthetic Low-density Lipoprotein-like Nanoparticles for Cancer-drug Delivery. *Nanomedicine* ;2(3):375-380. PMID:17716181

Dharap, S.S.; Wang, Y.; Chandna, P.; Khandare, J.J.; Qiu, B; Gunaseelan, S.; Sinko, P.J.; Stein, S; Farmanfarmaian, A. & Minko, T. (2005). Tumor-specific targeting of an anticancer drug delivery system by LHRH peptide. *PNAS* 102 36 12962-12967. PMID:16123131

Dullaart, R.P.F. & van Tol, A. (2001). Role of phospholipid transfer protein and pre β-high density lipoproteins in maintaining cholesterol efflux from Fu5AH cells to plasma from insulin-resistant subjects. *Scand. J. Clin. Lab. Invest.*;61:69–74. PMID:11300613

Fiorenza, A.M.; Branchi, A. & Sommariva, D. (2000). Serum lipoprotein profile in patients with cancer. A comparison with non-cancer subjects. *Int J Clin Lab Res*; 30:141-145. PMID:11196072

Furberg, A.; Marit, B.V.; Wilsgaard, T.; Bernstein, L.& Thune, .I (2004). Serum High-Density Lipoprotein Cholesterol, Metabolic Profile, and Breast Cancer Risk. *JNCI J Natl Cancer Inst* 96 (15): 1152-1160. PMID: 15292387

Geiss, H.C.; Ritter, M.M.; Richter, W.O.; Schwandt, P.& Zachoval, R. (1996). Low lipoprotein (a) levels during acute viral hepatitis. *Hepatology*.;24(6):1334–1337. PMID:8938156

Genest, J.J.; McNamara, J.R.; Ordovas, J.M.; Martin-Munley, S.; Jenner, J.L.; Millar, J.;Salem, D.N. & Schaefer, E.J. (1990): Effect of elective hospitalization on plasma lipoprotein cholesterol and apolipoproteins A-I, B and Lp(a). *Am J Cardiol*, 65(9):677-679. PMID:2106773

Giuseppe, L.; Massimo F.; Gian L.S. & Gian CG (2007). Lipoprotein[a] and cancer: Anti-neoplastic effect besides its cardiovascular potency. *Cancer Treatment Reviews* 33, 427-436. doi:10.1016/j.ctrv.2007.02.006

Gotto, A.M. (2001). "Low high-density lipoprotein cholesterol as a risk factor in coronary heart disease: a working group report," *Circulation*; 103 (17) 2213–2218. PMID:11331265

Hasstedt, S.J.;   Ash, K.O. & Williams R.R. (1986). A re-examination of major locus hypotheses for high density lipoprotein cholesterol level using 2,170 persons screened in 55 Utah pedigrees. *Am J Med Genet*;24:57-67. PMID:3706413

Hoyer, A. P.; & Engholm G. (1992). "Serum lipids and breast cancer risk: a cohort study of 5,207 Danish women," Cancer Causes and Control; 3(5) 403–408. PMID:1525320

Jafri, H.; Alsheikh-Ali, A.A. & Karas, R.H. (2010). Baseline and on-treatment high-density lipoprotein cholesterol and the risk of cancer in randomized controlled trials of lipid-altering therapy. *J Am Coll Cardiol*; 55:2846-54. PMID:21173414

Kader, A. & Pater, A. (2002). Loading anticancer drugs into HDL as well as LDL has little affect on properties of complexes and enhances cytotoxicity to human carcinoma cells. *J Control Release*; 80(1–3):29-44. PMID:11943385

Kanel, G.C.; Radvan, G. & Peters RL (1983). High-density lipoprotein cholesterol and liver disease. *Hepatology*; 3: 343-348. PMID:6840679

Krempler. F.; Kostner, G.M.; Roscher, A.; Haslauer, F.;Bolzano, K. &Sandhofer, F. (1983). Studies on the role of specific cell surface receptors in the removal of lipoprotein (a) in man. *J Clin Invest*.;71:1431–1441. PMID: 6304146

Kwiterovich, P.O. (2000). "The metabolic pathways of high-density lipoprotein, low-density lipoprotein, and triglycerides: a current review". *The American journal of cardiology* 86 (12A): 5L–10L. PMID: 11374859

Lacko, A.G.; Nair, M.; Paranjape, S.; Johnson, S. & McConathy, W.J. (2002). High density lipoprotein complexes as delivery vehicles for anticancer drugs. *Anticancer Res.* 22(4):2045-9. PMID:12174882

Lacko, A.G.; Nair, M.; Prokai, L.& McConathy, W.J. (2007).Prospects and challenges of the development of lipoprotein-based formulations for anti-cancer drugs. *Expert Opin Drug Deliv*.;4(6):665-75. PMID:17970668

Langstein, H.N. & Norton, J.A. (1991). Mechanisms of cancer cachexia. *Hematol Oncol Clin North Am*. 5:103–123. PMID: 2026566

Li, W.X. (1993). [Serum cholesterol and cancer mortality: eleven-year prospective cohort study on more han nine thousand persons]. *Zhonghua Liu Xing Bing Xue Za Zhi*. 14(1):6-9. PMID:8504456

Lou, B.; Liao, X.L.; Wu, M.P.; Cheng, P.F.; Yin, C.Y. & Fei, Z. (2005). High-density lipoprotein as a potential carrier for delivery of a lipophilic antitumoral drug into hepatoma cells. *World J Gastroenterol*; 11(7):954-959. PMID:15742395

Malaguarnera, M.; Trovato, G.; Restuccia, S.; Giugno, I., Franze, C.M., Receputo, G., Siciliano, R.; Motta, M. & Trovato, B.A. (1994). Treatment of nonresectable hepatocellular carcinoma: review of the literature and meta-analysis. *Adv Ther*; 11:303–319. PMID:10150270

Malaguarnera, M.; Giugno, I.; Trovato, B.A.; Panebianco, M.P.; Restuccia, N. & Ruello, P. (1996). Lipoprotein(a) in cirrhosis. A new index of liver functions? *Curr Med Res Opin*; 13:479–485. PMID:8840366

Marcovina, S.M.; Zhang, Z.H.; Gaur, V.P. & Albers, J.J. (1993). Identification of 34 apolipoprotein(a) isoforms: differential expression of apolipoprotein(a) alleles between American blacks and whites. *Biochem Biophys Res Commun*;191:1192-1196. PMID:8466495

McLean, J.W.; Tomlinson, J.E.; Kuang, W.J.; Eaton, D.L.; Chen, E.Y. & Fless, G.M. (1987). cDNA sequence of human apolipoprotein(a) is homologous to plasminogen. *Nature*;330:132-137. PMID:3670400

Mendez, A.J.; Oram, J.F.& Bierman, E.L. (1991). Protein kinase C as a mediator of high density lipoprotein receptor-dependent efflux of intracellular cholesterol. *J Biol Chem*; 266. PMID:1645339

Minko, T.; Dharap, S.S.; Pakunlu, R.I.& Wang, Y. (2004). Molecular targeting of drug delivery systems to cancer. *Curr. Drug Targets*; 5 389-406. PMID: 15134222

Moorman, P.G.; Hulka, B.S.; Hiatt, R.A.; Krieger, N.; Newman, B.; Vogelman, J.H. & Orentreich, N. (1998). Association between high-density lipoprotein cholesterol and breast cancer varies by menopausal status. *Cancer Epidemiol Biomarkers Prev*;7(6):483-8. PMID: 9641492

Motta, M.; Giugno, I.; Ruello, P.; Pistone, G.; Di Fazio, I. & Malaguarnera, M. (2001). Lipoprotein(a) behaviour in patients with hepatocellular carcinoma. *Minerva Med.*;92:301–305. PMID:11675573

Niendorf, A.; Nagele, H.; Gerding, D.; Meyer-Pannwitt, U. & Gebhardt, A (1995) Increased LDL receptor mRNA expression in colon cancer is correlated with a rise in plasma cholesterol levels after curative surgery. *Int J Cancer*, 61: 461–464. PMID:7759150

Ng, C.J.; Wadleigh, D.J.; Gangopadhyay, A.; Hama, S.; Grijalva, V.R. & Navab, M. (2001). Paraoxonase-2 is a ubiquitously expressed protein with antioxidant properties and is capable of preventing cell-mediated oxidative modification of low density lipoprotein. *J Biol Chem* ; 276:44444-9. PMID:11579088.

Nikanjam, M.; Gibbs, A.R.; Hunt, C.A.; Budinger, T.F. & Forte, T.M. (2007). Synthetic nano-LDL with paclitaxel oleate as a targeted drug delivery vehicle for glioblastoma multiforme. *J Control Release, 124:163–171.* PMID:17964677

Pathania, D.; Millard, M. & Neamati N (2009). Opportunities in discovery and delivery of anticancer drugs targeting mitochondria and cancer cell metabolism. *Adv Drug Deliv.*,30;61(14):1250-75. PMID:19716393

Perletti, G.; Tessitore, L.; Sesca, E.; Pani, P.; Dianzani, M.U. & Piccinini, F. (1996). Epsilon PKC acts like a marker of progressive malignancy in rat liver, but fails to enhance tumorigenesis in rat hepatoma cells in culture. *Biochem Biophys Res Commun*; 221: 688-691. PMID: 8630022

Perri, S.R.; Martineau, D.; François, M.; Bisson, L.; Durocher, Y. & Galipeau, J. (2007). Plasminogen Kringle 5 blocks tumor progression by antiangiogenic and proinflammatory pathways. *Mol Cancer Ther*;6 (2):441–9. doi:10.1158/1535-7163.MCT-06-0434

Reddy, S.T.; Wadleigh, D.J.; Grijalva, V., Ng, C., Hama, S. & Gangopadhyay A (2001). Human paraoxonase-3 is an HDL-associated enzyme with biological activity similar to paraoxonase-1 protein but is not regulated by oxidized lipids. *Arterioscler Thromb Vasc Biol*;21:542-7. PMID:11304470

Romics, L.; Nemesánszky, E.; Szalay, F.; Császár, A.; Tresch, J. & Karádi, I. (1996). Lipoprotein(a) concentration and phenotypes in primary biliary cirrhosis. *Clin Chim Acta*;255:165-171. PMID:8937759

Rosen, H. & Abribat T (2005). The rise and rise of drug delivery. *Nat Rev Drug Discov.* ;4:381–5. PMID: 15864267

Rudling, M.J.; Angelin, B.; Peterson, C.O. & Collins, V.P. (1990). Low density lipoprotein receptor activity in human intracranial tumors and its relation to the cholesterol requirement. *Cancer Res.* 50, 483-487. Available from http://cancerres.aacrjournals.org/content/50/3/483#related-urls

Samonakis, D.N.; Koutroubaki, I.E.; Sfiridaki, A; Malliaraki, N.; Antoniou, P.& Romanos J (2004). Hypercoagulable states in patients with hepatocellular carcinoma. *Dig Dis Sci;* 49:854-858. PMID: 15259509

Sehouli, J.; Stengel, D.; Elling, D.; Ortmann, O.; Blohmer, J. Riess, H. &Lichtenegger W (2002) *Gynecol. Oncol.,* 85 , 321-326. PMID: 11972395

Shah, P.K.; Kaul, S.; Nilsson, J. & Cercek, B. (2001). "Exploiting the vascular protective effects of high-density lipoprotein and its apolipoproteins: an idea whose time for testing is coming, part I," *Circulation,* 104 (19). 2376–2383. PMID:11696481

Skinner, E. R. (1994). "High-density lipoprotein subclasses," *Current Opinion in Lipidology,* 5 (3). 241–247. doi:10.1155/2011/496925

Spiegel, D.; Bloom, J.R.; Kramer, H.C. & Gottheil, E. (1989). Effect of treatment on the survival of patients with metastasic breast cancer. *Lancet,* 2, 888–891. PMID: 9885092

Uccello, M.; Malaguarnera, G.; Pelligra, E.M.; Biondi, A.; Basile, F. & Motta M (2011). Lipoprotein(a) as a potential marker of residual liver function in hepatocellular carcinoma. *Indian J Med Paediatr Oncol.;* 32(2): 71–75. PMID:22174493

Utermann, G; Duba, C.& Menzel, H.J. (1988). Genetics of the quantitative Lp(a) lipoprotein trait. II. Inheritance of Lp(a) glycoprotein phenotypes. *Hum Genet;*78:47-50. PMCID: PMC287411

Vitols, S.; Peterson, C.; Larsson, 0.; Holm, P. & Aberg B (1992). Elevated uptake of low density lipoproteins by human lung cancer tissue in vivo. *Cancer Res.,* 52, 6244-6247. 0007-0920/95

Wade, D.P.; Clarke, J.G.; Lindahl, G.E.; Liu, A.C.; Zysow, B.R.& Meer, K. (1993) l. 5' control regions of the apolipoprotein(a) gene and members of the related plasminogen gene family. *Proc Natl Acad Sci* U S A;90:1369-1373. PMID:7679504

Wasan, K.M. & Morton RE (1996): Differences in lipoprotein concentration and composition modify the plasma distribution of free and liposomal annamycin. *Pharm Res.,* 13(3):462-468. DOI: 10.1021/js960495j

Yokoyama, S. (2005). Assembly of high density lipoprotein by the ABCA1/apolipoprotein pathway. *Curr. Opin. Lipidol.;*16:269–279. Downloaded from www.jlr.org

Lipoproteins in Inflammatory and Infectious Diseases

Chapter 14

Dyslipoproteinemia in Chronic HCV Infection

Yoshio Aizawa, Hiroshi Abe, Kai Yoshizawa, Haruya Ishiguro, Yuta Aida, Noritomo Shimada and Akihito Tsubota

Additional information is available at the end of the chapter

1. Introduction

Hepatitis C virus (HCV) is a unique virus whose life cycle is closely associated with lipoprotein metabolism [1, 2, 3]. Assembly of HCV particles, formation of HCV-virions, is closely connected to the formation of lipid droplets in hepatic cells that may serve as an assembly platform [1, 4]. In addition, the production of HCV particles is tightly linked to the very low-density lipoprotein (VLDL) production pathway [5, 6]. HCV particles circulating in the blood during chronic HCV infection form lipo-viral particles (LVP) that are rich in triglycerides (TG), apoB-100 and apoE, with physiochemical similarity to VLDL particles and are highly infectious [7, 8]. In contrast, denser HCV particles are less infectious. These data strongly suggest that both viral particles and VLDL are integral components of LVPs with high infective capability. Although LVPs are thought to be assembled in liver cells by association with host lipoproteins prior to secretion, association between HCV and VLDL in the circulation after secretion from the liver cannot be ruled out.

Studies to date have indicated that the process of HCV assembly and secretion largely utilizes the VLDL pathway. Therefore, suppression of apoB-100 or apoE also inhibits secretion of HCV. Inhibition of microsomal triglyceride transfer protein (MTP), a critical protein for the initial step of VLDL assembly by co-translational lipidation of apoB-100 [9, 10], inhibits HCV secretion.

As HCV depends on VLDL pathways for its assembly and secretion, the lipid-rich environment of the liver cell combined with reduced VLDL secretion may be required for efficient assembly and secretion of HCV virions by ensuring the feasibility of co-assembly with VLDL. Hypobetalipoproteinemia, reduced activity of MTP with negative correlation to hepatic steatosis and viral load, is observed in HCV-G3 chronic infection [11]. Secretion of apoB-100 was reduced by HCV nonstructural proteins using the HCV subgenomic replicon expression system and interaction between the HCV NS5A and apoB-100 was observed [12].

In addition, there has been accumulating evidence that the HCV core protein is induced upon the redistribution of lipid droplets, affecting the assembly of both HCV and VLDL [13, 14]. The function of core protein in lipid metabolism has been widely examined including in models of steatosis involving HCV core protein transgenic mice [15]. Findings include reduction in activity of MTP [9] and Tyr164Phe substitution in relation to marked steatosis in HCV-G3 infection [16]. Although the participation of Arg70Gln/His substitution in steatosis of hepatocytes has been proposed [17], the precise connection between hepatic steatosis and HCV-G1b and/or HCV-G2 infection remains unclear.

As mentioned earlier, HCV particles in peripheral blood may associate not only with VLDL, but also with other lipoproteins, especially LDL, since circulating LVPs span a wide range of buoyant gravity and physicochemical characteristics [8, 18]. The association of lipoproteins with HCV particles may be beneficial for HCV through protection against anti-HCV neutralizing antibodies, as the antigenicity of HCV surface proteins is hidden beneath the associated lipoprotein particle [18]. Lipoprotein particles isolated from sera of HCV patients displayed differentially modulated lipid synthesis in human monocyte-derived macrophages in comparison to lipoproteins obtained from normal subjects, suggesting that HCV infection influences the biochemical composition of lipoproteins, thus revealing an alternative influence on lipid metabolism [19]. HCV entry into liver cells may occur through many receptors, including CD81 (direct binding to HCV E2 protein) and claudin-1, both of which act during the later steps of HCV entry [20]. The predominant role of LDL-receptors or remnant receptors is to catch VLDL-derived lipoprotein particles. Meanwhile, SR-BI (a receptor for HDL and oxidized LDL) directly binds HCV E2 protein [21]. Very recently, Nieman-Pick C1-like 1 cholesterol absorption receptor has been reported as a new factor for HCV entry to hepatic cells [22]. Interestingly, lipoprotein lipase, which hydrolyzes VLDL, is reported to increase the binding of LVP to hepatic cells while simultaneously decreasing infection levels of hepatic cells [23].

These findings suggest that examination of serum lipid profiles in chronic HCV infection may be important for understanding the biological features of HCV infection. Compared to normal subjects, low levels of TC, high-density lipoprotein cholesterol (HDL-C) and LDL-C was reported in chronic HCV-G3a infection [24]. However, lipoprotein profiles in infections of genotypes other than HCV-G3 have not been fully described and the data are somewhat conflicted. Moriya et al. indicated that TC levels and apoB, CII and CIII were reduced in HCV-G1b compared with chronic HCV-G2a or hepatitis B virus (HBV) infection [25], while others have not reported such a distinction. In addition, distortion of serum lipid levels has been widely observed in connection with virological outcome of IFN-based antiviral therapy, especially in HCV-G1 infection. Lower LDL-C, HDL-C, TC and/or TG was reported to be a possible predictor for unfavorable response to IFN-based therapy [24, 26, 27]. However, after the discovery of a genetic polymorphism near the human IL28B gene as the most potent predictor of the outcome of IFN-based therapy, the distortion of serum lipid levels is no longer thought to be an independent factor, but rather a confounding variable for predicting therapeutic efficacy [28].

In this chapter, we described the lipoprotein profiles in chronic HCV-G1b infection (the most common genotype in Japan) compared with that in chronic HCV-G2 infection (the second most common genotype in Japan). In addition, the influence of the genetic polymorphism near the human IL28B gene and aa substitutions in the core and NS5A regions of HCV on lipoprotein profiles in chronic HCV-G1b infection was determined.

To examine the serum lipid profiles of many patients, ultracentrifugation was unsuitable. We instead measured serum lipoprotein using HPLC in addition to a conventional laboratory method involving measurement of apolipoproteins. To examine serum LDL-C levels, the Friedwald equation can be used as an indirect calculation method yielding "total cholesterol minus HDL-C minus 0.2 × TG. However, the precision of this equation has not been determined in pathological conditions such as chronic HCV infection.

2. Methods

2.1. Patients and materials

Fasting sera of patients who were diagnosed as having chronic HCV infection were collected and stored at lower than -30 degrees centigrade until examination of apolipoproteins and/or lipoproteins. At the time of serum collection, TC, TG and HDL-C were measured using a routine laboratory kit. Serum LDL-C was measured by a direct assay using an LDL-cholesterol kit (Sekisui Medical, Japan). A good correlation between the direct assay and indirect measurement using the Friedwald equation (r=0.96) in 96 healthy adults whose TG level was lower than 400 mg/dl was observed.

Detection of HCV infection was made using a commercial real-time PCR Kit. Detection of HCV genotype was performed by a PCR method based on the 5′ non-coding sequence. Some patients diagnosed as having HCV-G1b infection, aa substitutions at core 70/91, interferon sensitivity determined region (ISDR) and/or IFN RBV resistance determining region (IRRDR) were examined. All patients examined were Japanese without ongoing treatment of IFN-based antiviral therapy.

All patients were confirmed not to be co-infected with HIV, HTLV, tuberculosis or other chronic bacterial infections. In addition, patients who were diagnosed with hepatic cirrhosis were excluded from the study. Study protocols were approved by the review board of each institution and written informed consent was obtained prior to study enrollment.

2.2. Detection of serum apolipoproteins

Serum levels of apolipoproteins (apoA1, apoA11, apoB, apoCII, apoCIII and apoE) were analyzed by immunonephelometry using an apolipoprotein detection kit (Sekisui Medical, Japan). Serum apoB-48 was assayed by chemiluminescent enzyme immunoassay [29] using an apoB-48 CLEIA kit (Fujirebio, Japan). Serum apoB-100 level was determined as apoB minus apoB-48.

2.3. Detection of lipoprotein fractions by HPLC based method

Fasting serum lipoprotein profiles were analyzed using an HPLC system with on-line enzymatic dual detection of cholesterol and TG as described previously (LipoSEARCH, Skylight Biotech, Japan) [30]. Briefly, 10 μl of whole serum sample was injected into two connected columns (300 × 7.8 mm) of TSKgel LipopropakXL (Tosoh, Japan) and eluted by TSKeluent Lp-1 (Tosoh). The eluent from the columns was divided by a micro splitter and continuously monitored at 550 nm after an online enzymatic reaction with a commercial kit, Determiner L TC (Kyowa Medex, Japan) and Determiner L TG (Kyowa Medex). Then, the cholesterol and TG concentrations were calculated by the computer program, which was designed to process complex chromatograms with a modified Gaussian curve fitting for resolving overlapping peaks by mathematical treatment.

Lipoprotein particles were fractionated into four major lipoproteins according to particle diameter as follows: >80 nm classified as chylomicrons; 30 to 80 nm as VLDL; 16 to 30 nm as LDL; and 8 to 16 nm as HDL. Next, the concentration of cholesterol and TG was measured in each major lipoprotein fraction and the ratio of cholesterol:TG concentration (C:T ratio) was calculated. This system has been successfully applied elsewhere in clinical research with excellent reproducibility [31]. Although freezing and thawing may affect the lipoprotein fraction, the influence of freezing and thawing on the measurement of cholesterol and TG concentration in healthy samples is fairly low, as described on homepage of Skylight Biotech (http://www.lipo-search.com).

Figure 1 illustrates chromatographic pattern of cholesterol and triglycerides derived LipoSEARCH was illustrated.

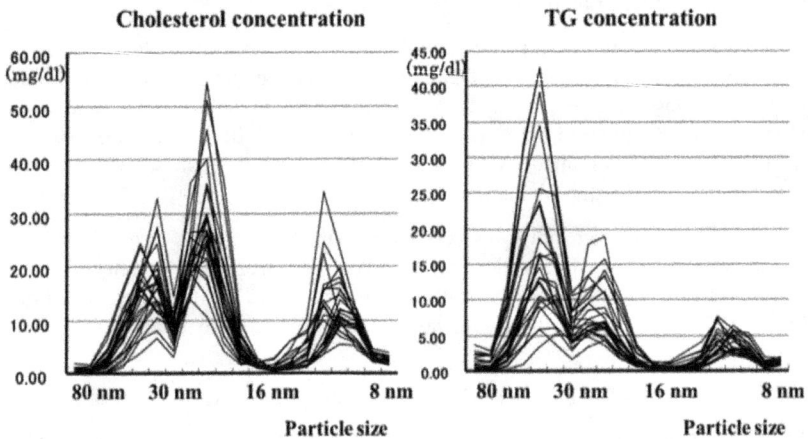

Figure 1. Chromatographic pattern of fasting serum cholesterol and TG derived HPLC-based method (LipoSEARCH).

2.4. Detection of amino acid substitutions at aa 70/91 in core region of HCV-G1b

HCV RNA was extracted from serum samples and reverse transcribed with random primers and MMLV reverse transcriptase (Takara Shuzo, Japan). Based on the method of Akuta et al. [32], nucleotide sequences of the core region were analyzed by direct sequencing after nested PCR. Subsequently, the aa substitutions at position 70 (arginine, Arg70; or glutamine/histidine, Gln70/His70) and at position 91 (leucine, Leu91; or methionine Met91) were determined.

2.5. Detection of amino acid substitutions related to ISDR in NS5A of HCV-G1b

Nucleotide sequences of NS5A-ISDR were analyzed by direct sequencing after double-round PCR according to Enomoto et al. [33]. Then, the aa sequence from 2209-2248, termed the ISDR region, in NS5A was determined and the numbers of aa substitutions defined.

2.6. Detection of amino acid substitutions related to IRRDR in NS5A of HCV-G1b

Nucleotide sequences of NS5A-IRRDR were analyzed by direct sequencing after double-round PCR according to El-Shamy et al. [34] and the numbers of aa substitutions within this region (2334-2376) determined. Moreover, the particular position of aa substitutions were evaluated in connection to dyslipoproteinemia, found in chronic HCV-G1b patients.

3. Results

Difference in lipoprotein profiles between patients with chronic HCV-G2 and HCV-G1b infection, between HCV-G1b patients with favorable and unfavorable response to PEG-IFN plus RBV combination therapy, and between minor (non-responder) genotypes and major (responder) genotypes of IL28B (rs8099917)

Serum lipoprotein profiles are clearly distorted in patients with chronic HCV-G3 infection and this is characterized by a decrease in apoB-100-related cholesterol [24]. However, serum lipoprotein disturbances in other genotypes are an issue that has been under discussion. In addition, disturbances in lipoprotein profiles have been reported in patients who were not responsive to IFN-based antiviral therapy in comparison to responsive patients with chronic HCV-G1 infection [26, 27]. However, whether or not dyslipoproteinemia is an independent factor affecting the efficacy of IFN-based therapy remains controversial. The latest interpretation seems to be that dyslipoproteinemia in HCV-G1 patients may be a confounding factor of the host genotype of IL28B that is the strongest predictor for virological outcome following PEG-IFN plus RBV therapy [28].

Initially, we compared serum levels of apolipoproteins in chronically HCV-G2- and G1b-infected patients paying special attention to virological outcome of PEG-IFN plus RBV therapy in the HCV-G1b patients. Of the pre-treatment fasting sera taken from 42 HCV-G1b patients, 23 achieved a sustained viral response (SVR; negative for HCV RNA at 6 months after standard 48 weeks of therapy); 8 had a transient viral response (TVR; HCV RNA-negative during therapy, but reappearing after therapy); and 11 had a non-viral response (NVR; HCV RNA-positive during therapy). 24 HCV-G2 patients were also examined.

There were no differences in the concentration of apoAI, apoAII, apoCIII, apoE and apoB-48 (138.2±27.4 mg/dl vs. 142.4±32.4 mg/dl, 30.4±6.5 mg/dl vs. 29.9±4.8 mg/dl, 5.7±2.0 mg/dl vs. 6.2±2.4 mg/dl, 4.4±1.1 mg/dl vs. 4.3±1.0 mg/dl and 3.3±4.2 mg/dl vs. 3.1±1.4 mg/dl, respectively) between HCV-G1b and G2 patients. In addition, no significant changes were observed among patients with HCV-G1b showing different outcome of PEG-IFN plus RBV therapy. However, there were substantial differences in apoB and apoCII levels according to the response to PEG-

IFN plus RBV therapy in HCV-G1b patients. In HCV-G1b infection, apoB and apoCII levels were significantly higher in SVR patients than in NVR patients. The levels of apoB and apoCII in HCV-G2 patients were similar to those in HCV-G1b patients who achieved SVR (Figure 2).

These data suggest that the apolipoprotein profile in HCV-G1b patients is basically indistinguishable from that in HCV-G2 patients. Meanwhile, the profile differed in relation to different outcomes of PEG-IFN plus RBV therapy among chronic HCV-G1b infection. From these observations, we presumed that a decrease in LDL and/or VLDL may be a feature of dyslipoproteinemia in HCV-G1b patients who failed to respond to PEG-IFN plus RBV therapy. ApoB-100 (about 95% of apoB is apoB-100 in fasting sera) is an indicator of total VLDL, intermediate-density lipoprotein (IDL) and LDL particles in the blood because each particle of VLDL, IDL or LDL is composed of one molecule of apoB-100. Moreover, the majority of apoCII, which activates the enzyme lipoprotein lipase in capillaries, is associated with VLDL. Therefore, a decrease in apoCII could be related to a decrease in VLDL. Taking into consideration these observations, we concluded that a characteristic of the lipoprotein profile during the pre-treatment period of chronic HCV-G1b patients who subsequently fail antiviral IFN-based therapy is a decline in VLDL levels. Although lower serum apoE has recently been reported to be related to a favorable response to IFN-based therapy [35], we did not find any differences in serum apoE levels.

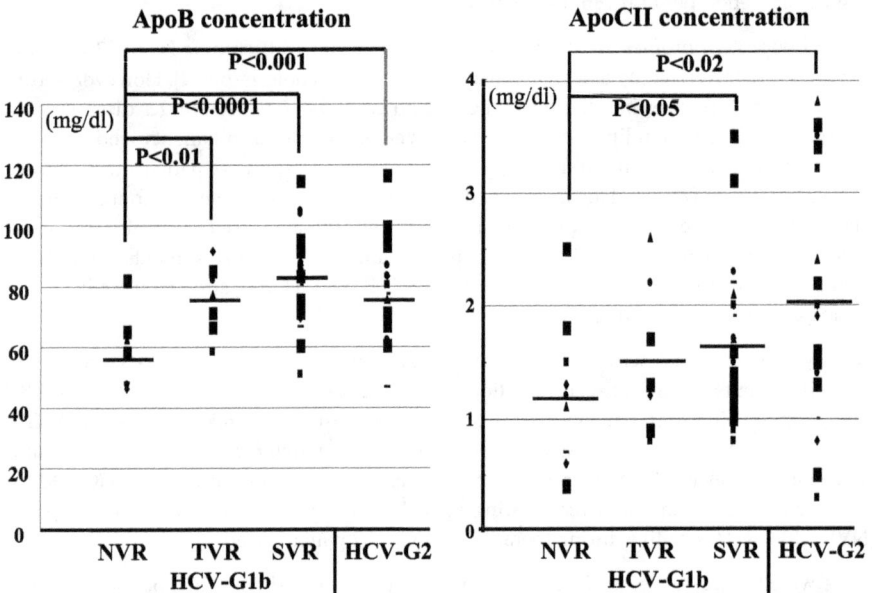

Figure 2. Serum apoB and apoCII levels in patients with chronic HCV-G1b or HCV-G2 infection. All HCV-G1b patients were treated with standard PEG-IFN plus RBV combination therapy and clarified according to the virological response. (NVR; non-viral response, TVR; transient viral response, SVR; sustained viral response)

Next, we examined the profiles of serum lipoproteins in 32 patients who were chronically infected with HCV-G2 and 111 patients with HCV-G1b, along with genotyping of the region near the IL28B gene (SNP ID rs8099917), the strongest predictor of outcome to PEG-IFN plus RBV therapy. Recently, we reported that serum apoB-100 levels are prescribed by genetic polymorphism of rs8099917, and a minor genotype of rs8099917 (TG or GG; also known as non-responder genotype of PEG-IFN plus RBV combination therapy) may be related to a decrease in serum apoB-100 in chronic HCV-G1b infection [36]. In chronic HCV-G1b infections in Japan, the minor genotype was found in 25-35% of cases, whereas it was only 10-20% in chronic HCV-G2 infections. We examined SNP of rs8099917 and the lipoprotein profiles obtained by LipoSEARCH were compared among 29 HCV-G2 patients who had the major genotype of IL28B (rs8099917), 75 HCV-G1b patients who had the major IL28B genotype and 36 HCV-G1b patients who had the minor IL28B genotype.

There were no differences related to gender, age, fibrosis score of liver biopsy, BMI, serum ALT level, viral load or platelet count among these three groups (data not shown). We did not cite the chylomicron fraction because this fraction was too small to analyze precisely.

Lipoprotein fraction	TC	VLDL-C	LDL-C	HDL-C
HCV-G2, IL28B major	167.92±31.20	38.22±17.65	81.18±22.25	44.79±14.49
HCV-G1b, IL28B major	169.65±30.52	45.52±18.64	72.80±17.27*	46.54±12.22
HCV-G1b, IL28B minor	161.69±30.87	36.22±15.18***	68.61±18.40*	52.20±16.64**

(mg/dl)

Lipoprotein fraction	TG	VLDL-TG	LDL-TG	HDL-TG
HCV-G2, IL28B major	104.46±59.41	52.48±36.43	27.22±10.17	16.22±5.91
HCV-G1b, IL28B major	93.30±36.74	43.52±20.99	24.65±7.37	16.82±6.20
HCV-G1b, IL28B minor	92.92±42.21	41.53±20.32	22.10±6.69*	19.14±8.25

(mg/dl)

*$P>0.05$ vs HCV-G2, **$P>0.05$ vs HCV-G1b with IL28B major, ***$P>0.01$ vs HCV-G1b with IL28B major

Table 1. Differences in lipoprotein fractions among patients with chronic HCV-G1b and G2 infection with different IL28 genotype

In patients with HCV-G1b having the IL28B major (responder) genotype, the concentration of cholesterol in the LDL fraction (LDL-C) was significantly lower than that in patients with HCV-G2. In addition, the tendency of a reciprocal increase of VLDL-C ($P=0.055$) in HCV-G1b patients having the responder genotype was observed. As a result, the level of LDL-C plus VLDL-C was similar between these two groups. In contrast, a reciprocal increase of VLDL-C was not observed in G1b patients who had the non-responder genotype. Decrease of VLDL-C and increase of HDL-C is a feature of the non-responder genotype compared with the responder genotype in chronic HCV-G1b patients.

We also examined serum lipoprotein profiles of five patients who had been infected with HCV-G1b, but achieved SVR by PEG-IFN plus RBV therapy (HCV-free status continuing for longer than 6 months). In these cured patients, low levels of VLDL-C and high levels of LDL-C and HDL-C were observed (Table 2). The lipoprotein profiles of these cured patients were quite normal with a normal composition of cholesterol and TG in each lipoprotein fraction when assessed by C:T ratio. When compared with cured patients, a relative decrease of TG in the VLDL fraction and a relative increase of TG in the LDL and HDL fractions were noted in chronic HCV infection.

In chronic HCV infection, we unexpectedly found that the serum levels of LDL-C measured by an HPLC system (LipoSEARCH) were considerably lower than those measured directly using a conventional method (HCV-G1b: 93.8±26.76 mg/dl; HCV-G2: 101.21±34.19 mg/dl), or measured indirectly by the Friedwald equation. In place of the decreased LDL fraction, the VLDL fraction was increased. This pattern is somewhat reminiscent of hyperlipidemic (high TG) samples [31]. However, in chronic HCV infection, TG levels are not much different from those in cured patients. Therefore, this finding cannot be explained by an increase of TG.

Lipoprotein fraction	TC	VLDL-C	LDL-C	HDL-C
HCV-G1b, achieved SVR	177.49±35.63	26.34±7.24	89.41±31.61	61.28±15.14

(mg/dl)

	TG	VLDL-TG	LDL-TG	HDL-TG
HCV-G1b, achieved SVR	79.15±25.33	44.11±18.77	22.08±4.96	10.47±2.64

(mg/dl)

Table 2. Serum lipid profiles of patients who achieved SVR by PEG-IFN plus RBV therapy for chronic HCV-G1b infection (sera were obtained at least 6 months after HCV was completely eradicated)

An increase of the VLDL fraction in chronic HCV infection could be explained by reduced enzymatic activity of lipoprotein lipase, which may facilitate HCV cell entry [23] while delaying the conversion of VLDL to LDL. Alternatively, discrepancy between chemically determined LDL (conventional measurement method) and levels determined by particle size (HPLC-based method) may be explained by the existence of LDL-associated LVPs in the blood during chronic HCV infection. These particles may have the physicochemical surface nature of LDLs, but particle sizes larger than 55 nm because the diameter of the HCV particle is about 55 nm, and hence must be eluted in the VLDL fraction. Although we must take into consideration that lipoprotein particles could theoretically become fused together during freezing and thawing, thus seriously distorting the lipoprotein fraction pattern determined by the HPLC-based method, freezing and thawing has reportedly been found not to seriously affect lipoprotein profiles (Skylight Biotech, http://www.lipo-search.com).

A recent study by Nishimura et al. [37] suggested that diminished VLDL-TG/non-VLDL-TG is a key feature of chronic HCV infection. They detected VLDL-TG based on the chemical nature of VLDL. Their findings do not conflict with our data. Our results indicate a relative decrease of TG in the VLDL fraction, but a relative increase of TG in the LDL and HDL fractions. Thus, their findings of decreased VLDL-TG/non-VLDL-TG appear to be consistent with our results.

As this kind of lipid abnormality is not easily determined by conventional methodology, the HPLC-based method is extraordinarily useful for the study of lipoprotein profiles in chronic

HCV infection. However, close attention to sample handling is needed because lipoprotein particles are fairly unstable.

The tentative conclusions are as follows: 1. Decrease of apoB (apoB-100) and apoB-100-related cholesterol might be a main feature of dyslipoproteinemia in chronic HCV-G1b infection with unfavorable response to PEG-IFN plus RBV combination therapy; 2. Existence of abnormally large particles (LPVs) eluted in the VLDL fraction by HPLC, in spite of the chemical characteristics of LDL, and/or reduced activity of lipoprotein lipase with delayed VLDL dissociation, may be a feature of dyslipoproteinemia in chronic HCV infection; 3. Relative increase of TG in the LDL and HDL fractions with a relative decrease of TG in the VLDL fraction may be a feature of chronic HCV infection. These conclusions are partially in concordance with those reported by Mawatari et al. [38].

4. Do viral factors participate in the dyslipoproteinemia seen during chronic HCV-G1b infection?

HCV-G1 infection is widely distributed worldwide and the most common genotype in the world, while it is one of the most resistant genotypes to IFN-based therapy. In Japan, almost all G1 subtypes are 1b, contributing to more than 70% of chronic HCV infection cases. Viral factors participating in the response to IFN-based therapies have been extensively studied, especially with regard to HCV-G1b in Japan. Among them, core protein substitution at aa 70/91 [32] and aa substitutions in the ISDR [33], IRRDR [34] and in NS5A are widely accepted as candidates. Among them, substitution at core protein 91, Leu91Met, did not affect serum levels of apoB-100 as reported earlier by us [36]. We further examined the significance of substitution at aa 70, Arg70Gln/His, aa substitutions in the ISDR and IRRDR and aa substitutions at particular positions within the IRRDR.

In that former study, we determined that substitution of Arg70 to Gln/His70 was a distinctive factor participating in the regulation of serum apoB-100 levels in chronic HCV-G1b patients, independent from the IL28B genotype. To clarify the lipoprotein profiles according to substitution at aa 70, we examined the lipoprotein profiles of fasting sera from 113 chronic HCV-G1b patients (68 were Arg70 and 45 were Gln/His70) by LipoSEARCH, as described earlier (Table 3).

Lipoprotein fraction	TC	VLDL-C	LDL-C	HDL-C
HCV-G1b Arg70	169.80±30.75	43.16±18.39	74.23±18.68	48.09±14.03
HCV-G1b Gln/His70	158.54±28.44	40.02±17.40	65.99±15.51*	47.57±14.09

(mg/dl)

Lipoprotein fraction	TG	VLDL-TG	LDL-TG	HDL-TG
HCV-G1b Arg70	89.49±33.68	42.15±20.80	23.53±6.71	16.24±5.31
HCV-G1b Gln/His70	95.95±41.99	42.13±19.50	18.14±7.83	19.42±8.31*

(mg/dl)

(*: $p > 0.05$)

Table 3. Differences in lipoprotein profiles related to the substitution at aa 70 in core region (Arg70 to Gln/His70) in patients with chronic HCV-G1b infection

There were no differences related to gender, age, fibrosis score of liver biopsy, BMI, serum ALT level, viral load or platelet count between Arg70 and Gln/His70. However, a significant difference was found in the distribution of the IL28B genotype (SNP of rs8099917). In Arg70, 49 patients were of the major (responder) genotype while 15 were of the minor (non-responder) genotype. In Gln/His70, 22 were major and 20 were minor. A total of 7 patients remained undetermined. The difference in the IL28B genotype distribution between these two groups was significant ($P=0.0005$ by chi-square test with Yate's correction). Therefore, the influence of the IL28B genotype was not excluded in this study. However, the pattern of dyslipoproteinemia seen in Gln/His70 cases is dissimilar to that in the IL28B minor genotype, which was described in detail. In core 70 mutants (Gln/His70), a significant decrease in LDL-C and increase in HDL-TG levels was demonstrated without a decrease in VLDL-C. TC levels tended to be lower than that in the core Arg70 cases ($P=0.052$). These findings may indicate that the core mutation at aa 70 is an important viral feature in relation to the dyslipoproteinemia seen in HCV-G1b, functioning mainly through decreasing LDL-C. As a result of the substitution at aa 70, the nature of the amino acid is substantially changed. Therefore, the configuration and the biological activity of the core protein may be significantly disturbed, which may lead to the disruption of lipid metabolism. However, the precise mechanism of the consequence of aa 70 substitution on lipid metabolism is a matter to be solved in the future.

To exclude the influence of the IL28B genotype, we further compared the lipoprotein profiles in 49 patients with core 70 wild-type, and 22 patients with core 70 mutant phenotypes in whom the IL28B genotype was major (Table 4).

Lipoprotein fraction	TC	VLDL-C	LDL-C	HDL-C
HCV-G1b Arg70	173.37±30.83	45.80±18.29	76.41±18.28	46.38±12.37
HCV-G1b Gln/His70	155.44±24.13*	41.01±17.02	65.71±13.34*	45.72±12.45

(mg/dl)

Lipoprotein fraction	TG	VLDL-TG	LDL-TG	HDL-TG
HCV-G1b Arg70	92.33±35.95	42.57±22.27	25.50±6.77	15.87±5.28
HCV-G1b Gln/His70	93.44±36.33	41.36±17.84	26.12±8.72	18.28±7.38

(mg/dl)

(*: $p>0.05$)

Table 4. Differences in lipoprotein profiles related to substitution at aa 70 in the core region (Arg70 to Glun/His70) in patients with chronic HCV-G1b having IL28B major (responder) genotype

Although a relatively small-sized study, the features of the dyslipoproteinemia seen in patients with the core aa 70 mutation was clearly elucidated as a decrease of TC due to a decrease of LDL-C. Even after exclusion of the influence of the IL28B genotype, the core 70 aa substitution was found to have a role in dyslipoproteinemia that may be critical.

We also examined aa substitutions in the NS5A region in relation to disturbance of serum lipid/lipoprotein levels, since NS protein may inhibit the secretion of apoB-100 in vitro [12]. Moreover, a polypeptide comprised of aa residues 2135 to 2419 within the NA5A protein co-precipitated with apoB, suggesting a possible interaction between NS5A protein and apoB-

100. Thus, we examined aa substitutions in particular regions of NS5A to elucidate the possibility of a viral factor being the determinant of lipid metabolism. We compared the aa sequence 2209-2248 (ISDR) with the sequence of HCV-J and the number of aa substitutions was classified as wild-type (0 or 1) or non-wild-type (≥2). According to the numbers of aa substitutions in the ISDR [33], 102 of 117 subjects were judged to be wild-type and 15 non-wild-type. No significant differences in serum apoB and lipid concentrations were found between wild-type and non-wild-type ISDR (Table 5).

Lipid profile	T.C.	TG	LDL-C*	apoB
ISDR wild	172.8±32.0	104.3±54.1	92.5±26.6	80.9±19.8
ISDR non-wild	167.1±28.0	82.2±26.2	92..9±22.1	78.2±13.9

(mg/dl)

*measured directly using commercial kit. Note the substantial differences of LDL-C level measured by HPLC system shown in Table 1, Table 2 and Table 3.

Table 5. Lipid profiles of HCV-G1b patients with ISDR wild and ISDR non-wild

Although not described in Table 5, there was no statistical difference between patients with wild-type and non-wild-type ISDR in terms of serum levels of apoAI, apoAII, apoCII, apoCIII, apoE and apoB-48. We also compared lipid profiles between patients with a substitution number of 0 (N=80) and ≥1 (N=37), and found no significant difference (data not shown).

We very recently examined the aa substitution number and the place of substitution in the IRRDR (aa 2334-2376) in 105 patients who were chronically infected with HCV-G1b. By comparison with the HCV-J sequence, the number of aa substitutions was determined. A high degree (≥6 substitutions) of sequence variation in the IRRDR, which is thought to be a useful marker for predicting SVR [34], was found in 34 patients, whereas a less diverse (≤5 substitutions) IRRDR sequence (predictive of non-SVR) was found in 71 patients.

The clinical background is illustrated in Table 6.

Gender (M/F)	42/63	TG	89.8±40.2 (mg/dl)
Age	62.9±12.0 (years)	LDL-C	95.5±29.2 (mg/dl)
ALT	53.5±38.5 (U/L)	apoB-100	73.0±21.1 (mg/dl)
Albumin	4.1±0.4 (g/dl)	rs8099917 (Major/Minor)	65/40
Plt	16.0±6.3x10⁴	aa 70 (Wild/Mutant)	64/41
HCV-RNA	6.4±0.6 (logIU/ml)	aa 91 (Wild/Mutant)	63/42
TC	171.3±32.7 (mg/dl)	ISDR (Wild / Mutant)	92/12

Table 6. Clinical characteristics of 105 HCV-G1b patients whose IRRDR sequences were examined

The number of aa substitution detected in this study was illustrated in Figure 3.

As shown in Figure 3, the substitution number in the IRRDR was widely distributed. There was no difference in lipid profile between the two groups (substitution number ≥6 vs. substitution number ≤5) along with other clinical backgrounds except for the distribution of the IL28B genotype (Table 7).

Number of patient

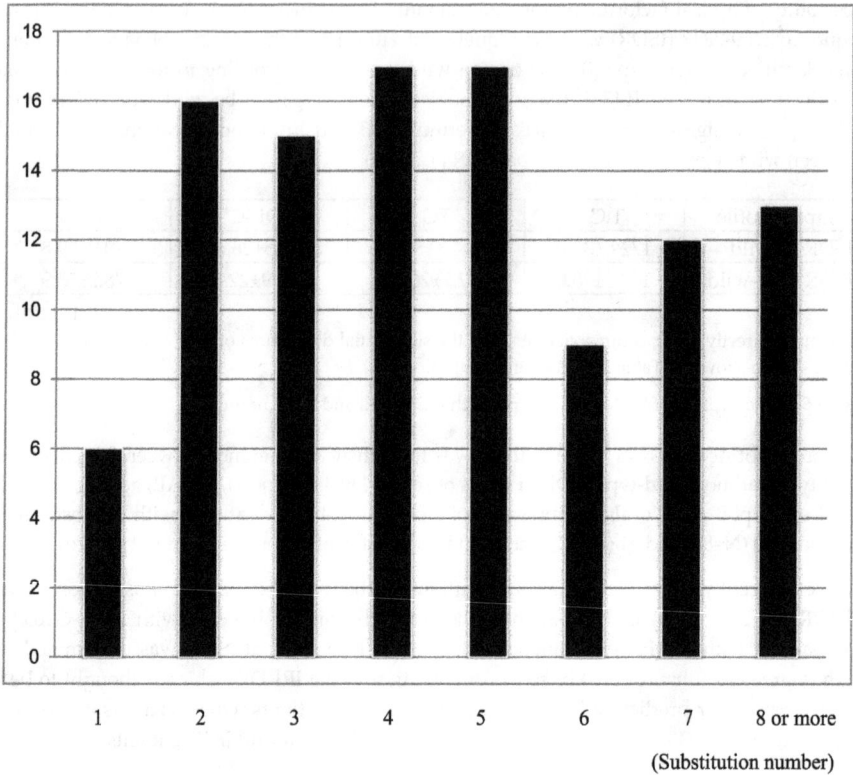

(Substitution number)

Figure 3. Distribution of aa substitution number in the IRRDR among 105 chronic HCV-G1b patients

	Substitutions (≤5)	Substitutions (≥6)	P value
TC	171 (115-253)	160 (96-268)	0.10
TG	81(42-267)	82 (37-207)	0.83
LDL-C	93 (36-193)	86 (50-172)	0.40
apoB-100	75 (39-131)	72 (44-135)	0. 23
IL28-B (TT/nonTT)	39/35	20/5	0.024

Data were expressed as median (range)

Table 7. Differences in lipid profile according to the substitution number in the IRRDR

Next, we examined the relationship between each aa substitution in the IRRDR and dyslipoproteinemia. As illustrated in Figure 4, substitution at aa 2356 may impact serum lipid profiles. There was no difference in lipid profile related to substitutions other than aa 2356.

A substitution at aa 2356 from Gly to Glu, Lys or Ala may be critical for distortion of the serum lipid profile. This substitution was previously shown to be a key substitution determining virological outcome of PEG-IFN plus RBV therapy in HCV-G1b patients [39]. The therapeutic outcome of 63 patients treated with PEG-IFN plus RBV is indicated in the lower right of Figure 4. Although a clear difference was not observed in the outcome of the therapy, NVR tended to be frequent in patients with non-Gly at aa 2356. The substitution at aa 2356 of Gly to Glu or Gly to Lys caused a drastic change in the nature of the amino acid that may influence the nature of the protein and in turn affect the biochemical interaction of apoE or apoB-100 with NA5A protein resulting in the decrease of serum apoB-100 and LDL-C. However, the change in serum apoB-100 level was minor. As aa substitution in the IRRDR is somewhat related to aa substitution in other regions such as the core aa 70, further examination is needed to establish the importance of aa 2356 substitution on lipoprotein metabolism in chronic HCV-G1b infection.

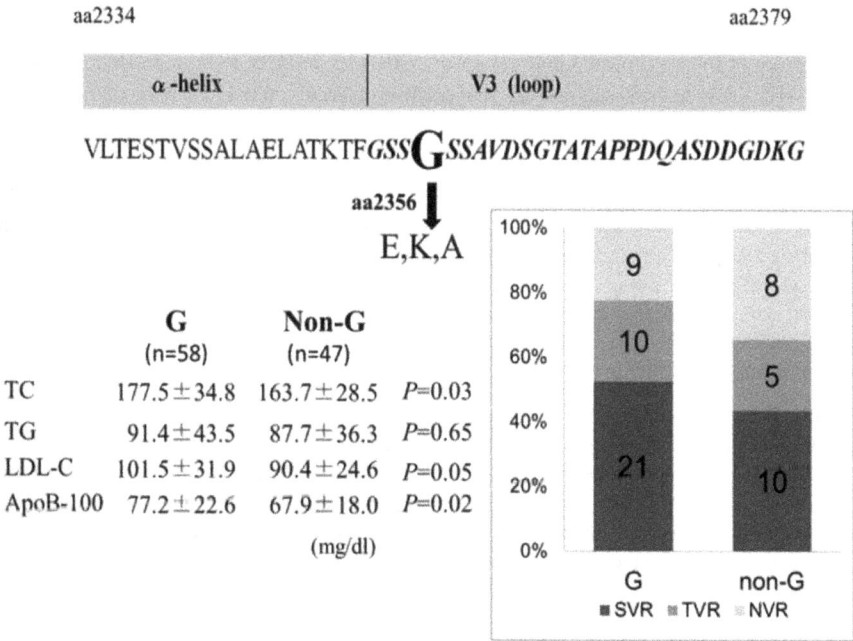

Figure 4. Consequence of substitution at aa 2356 from G to E, K or A in the IRRDR. Serum levels of TC, LDL-C and apoB-100 were significantly higher in patients who had G2356 than in patients who had not. At lower right, the virological response to PEG-IFN plus RBV therapy in 63 patients with HCV-G1b was summarized.

5. Are lipid/lipoprotein profiles independent predictors of therapeutic outcome in chronic HCV- G1b infection?

Chronic HCV infection has been treated with PEG-IFN and RBV combination therapy. HCV-G2 and HCV-G3 responded fairly well to this combination therapy. However, the response was weaker in HCV-G1 patients with only half of the patients achieving a sustained virological response. Before the discovery of the IL28B genotype, a decrease in TC, LDL-cholesterol or apoB-100-related cholesterol had been considered as an independent predictor of the outcome of PEG-IFN plus RBV therapy in HCV-G1. However, after the discovery that the genotype near the IL28B gene was a potent predictor, serum lipid levels have since been considered to be a confounding factor of the IL28B genotype. Actually, as we described earlier in this chapter and also in an earlier manuscript [36], the IL28B genotype may profoundly affect serum lipid levels in HCV-G1b infection.

Our latest knowledge, based on a prospective study examining predictive factors for the outcome of PEG-IFN plus RBV therapy, indicates that an increase in serum apoB-100 levels is an independent factor predicting rapid virological response (decline of serum HCV RNA at 4 weeks) to PEG-IFN plus RBV therapy. However, it does not appear to be an independent predictor of the final outcome of virological response in chronic HCV-G1b infection (data currently under consideration for contribution). These findings may partly indicate that disturbance of the lipid profile may reflect the efficiency of HCV replication due to efficient entry of HCV into hepatic cells and efficient production/secretion of HCV along with VLDL.

Although the evidence described in this chapter is not based on in vitro studies using HCV-secreting cells, our observations based on clinical samples may contribute to furthering the understanding of HCV-lipid metabolism interaction. From this viewpoint, it is noteworthy that lipoprotein lipase may act not only in the conversion of VLDL to LDL, but also in inhibiting HCV entry into liver cells [23]. Interestingly, in in vitro studies using naturally HCV-secreting cells, the enzymatic activity of lipoprotein lipase may be reduced in HCV-infected cells and may act to promote cell entry of HCV. This may contribute to an increase in the number of large-size (suitable for VLDL particle size) LVP having a chemical nature of LDL as suggested in our HPLC-based study.

Finally, hepatic steatosis may be associated with dyslipoproteinemia in chronic HCV infection and has been extensively studied with special attention to its relation to the IL28B genotype [40] and substitution at aa 70 [17] in HCV-G1b. However, in our experience, the relationship between hepatic steatosis and these host and/or viral factors was equivocal, perhaps because there are other environmental and host factors strongly affecting hepatic steatosis other than the factors discussed in this manuscript.

6. Summary

The features of dyslipoproteinemia in chronic HCV infection have been described. Serum lipid/lipoprotein profiles were in part HCV-genotype specific. The HCV core protein of HCV-G1b is closely associated with lipoprotein metabolism, and substitution of amino acid (aa 70) in the core protein may precipitate dyslipoproteinemia, in addition to substitution of aa 2356 in NS5A. In addition, dyslipoproteinemia in chronic HCV-G1b infection is largely affected by the genotype near the human IL28B gene. Therefore, dyslipoproteinemia in chronic HCV infection may involve complicated interplay between viral and host factors that could affect human lipid metabolism. Further study of lipids/lipoproteins in chronic HCV infection will be valuable to clarify the interaction of HCV and host lipid metabolism in detail. Dyslipoproteinemia in chronic HCV-G1b infection may play a role in the rapid decline of HCV during PEG-IFN plus RBV therapy. However, the clinical utility of dyslipoproteinemia as a predictor of final response to PEG-IFN plus RBV treatment remains controversial.

Author details

Yoshio Aizawa*, Hiroshi Abe, Kai Yoshizawa, Haruya Ishiguro and Yuta Aida
Jikei University Katsushika Medical Center, Division of Gastroenterology and Hepatology, Internal Medicine, Aoto, Katsushika-ku, Tokyo, Japan

Noritomo Shimada
Shinmatsudo Chuo General Hospital, Department of Gastroenterology and Hepatology, Shin-Matsudo, Matsudo City, Chiba, Japan

Akihito Tsubota
Jikei University Kashiwa Hospital, Institute of Clinical Medicine and Research (ICMR), Kashiwa City, Chiba, Japan

Acknowledgement

We thank professor Enomoto and doctor Maekawa, University of Yamanashi for kind examination of aa substitution within IRRDR.

7. References

[1] Target-Adams P, Boulant s, Douglas MW, McLauchlan J. (2010) Lipid metabolism and HCV infection. Viruses 2: 1195-1217.

[2] Syed GH, Amako Y, Siddqui A. (2010) Hepatitis C virus hijacks host lipid metabolism. Trends Endocrinol Metab 21:33-40.

[3] Popescu CI, Dubuisson J. (2009) Role of lipid metabolism in hepatitis C virus assembly and entry. Biol Cell 102:63-74

[4] Alvisi G, Madan V, Bartenschlager R. (2011) Hepatitis C virus and host cell lipids: an intimate connection. RNA Biol 8:258-269.

[5] Gastaminza P, Cheng G, Wieland S, Zhong J, Liao W et al. (2008) Cellular determinants of hepatitis C virus assembly, maturation, degradation and secretion. J Virol 82:2120-2129.

[6] Huang H, Sun F, Owen DM, Li W, Chen Y et al. (2007) Hepatitis C virus production by human hepatocytes dependent on assembly and secretion of very low-density lipoproteins. Proc Natl Acad Sci USA 104:5848-5853.

[7] Sabahi A, Marsh KA, Dahari H, Corcoran P, Lamora JM et al. (2010) The rate of hepatitis C virus infection initiation in vitro is directly related to particle density. Virology 407:110-119.

---

* Corresponding Author

[8] Nielsen SU, Bassendine MF, Burt AD, Martin C, Pumeechockchai W et al. (2006) Association between hepatitis C virus and very-low-density lipoprotein (VLDL)/LDL analyzed in iodixanol density gradients. J Virol 80:2418-2428.

[9] Popescu CI, Dubuisson J. (2010) Role of lipid metabolism in hepatitis C virus assembly and entry. Biol cell 102:63-74.

[10] Chang K.S, Jiang J, Cai Z, Luo G (2007) Human apolipoprotein e is required for infectivity and production of hepatitis C in cell culture. J Virol 81:13783-13793

[11] Zampino R, Ingrosso D, Durante-Mangoni E, Capasso R, Tripodi MF et al. (2008) Microsomal triglyceride transfer protein (MTP) -493G/T gene polymorphism contributes to fat liver accumulation in HCV genotype 3 infected patients. J Viral Hepat 15:740-746

[12] Domitrovich AM, Felmlee DJ, Siddiqui A. (2005) Hepatitis C nonstructural proteins inhibit apolipoprotein B100 secretion. J Biol Chem 280:39802-39808.

[13] Counihan NA, Rawlinson SM, Lindenbach BD. (2011) Trafficking of hepatitis C virus core protein during virus particle assembly. PLoS Pathog 7: e1002302.

[14] Roingeard P, Depla M. (2011) The birth and life of lipid droplets: learning from the hepatitis C virus. Biol Cell 103:223-231.

[15] Koike K, Tsutsumi T, Yotsuyanagi H, Moriya K. (2010) Lipid metabolism and liver disease in hepatitis C viral infection. Oncology 78 Suppl 1:24-30.

[16] Hourioux C, Patient R, Morin A, Blanchard E, Moreau A et al. (2007) The genotype 3-specific hepatitis C virus core protein residue phenylalanine 164 increases steatosis in an in vitro cellular model. Gut 56:1302-1308.

[17] Sumida Y, Kanemasa K, Hara T, Inada Y, Sakai K et al. (2011) Impact of amino acid substitutions in hepatitis C virus genotype 1b core region on liver steatosis and glucose tolerance in non-cirrhotic patients without overt diabetes. J Gastroenterol Hepatol 26:836-842.

[18] André P, Komurian-Pradel F, Deforges S, Perret M, Berland JL et al. (2002) Characterization of low- and very-low-density hepatitis C virus RNA-containing particles. J Virol 76:6919-6928.

[19] Napolitano M, Giuliani A, Alonzi T, Mancone C, D'Offizi G et al. (2007) Very low density lipoprotein and low density lipoprotein isolated from patients with hepatitis C infection induce altered cellular lipid metabolism. J Med Virol 79:254-258.

[20] Meredith LW, Wilson GK, Fletcher NF, McKeating JA. (2012) Hepatitis C virus entry: beyond receptors. Rev Med Virol doi: 10.1002/rmv.723.

[21] Dao Thi VL, Dreux M, Cosset FL. (2011) Scavenger receptor class B type I and the hypervariable region-1 of hepatitis C virus in cell entry and neutralisation. Expert Rev Mol Med 13:e13.

[22] Sainz Jr B, Barretto N, Martin DN, Hiraga N, Imamura M et al. (2012) Identification of the Nieman-Pick C1-like 1 cholesterol absorption receptor as a new hepatitis C virus entry factor. Nature Med doi:10,1038/nm.2581.

[23] Maillard P, Walic M, Meuleman P, Roohvand F, Huby T et al. (2011) Lipoprotein lipase inhibits hepatitis C virus (HCV) infection by blocking virus cell entry. PLoS One 6:e26637.

[24] Sheridan DA, Price DA, Schmid ML, Toms GL, Donaldson P et al. (2009) Apolipoprotein B-associated cholesterol is a determinant of treatment outcome in patients with chronic hepatitis C virus infection receiving anti-viral agents interferon-alpha and ribavirin. Aliment Pharmacol Ther 29:1282-1290.

[25] Moriya K, Shintani Y, Fujie H, Miyoshi H, Tsutsumi T et al. (2003) Serum lipid profile of patients with genotype 1b hepatitis C viral infection in Japan. Hepatol Res 25:371-376.

[26] Akuta N, Suzuki F, Kawamura Y, Yatsuji H, Sezaki H ET AL. (2007) Predictors of viral kinetics to peginterferon plus ribavirin combination therapy in Japanese patients infected with hepatitis C virus genotype 1b. J Med Virol 79:1686-1695.

[27] Ramcharran D, Wahed AS, Conjeevaram HS, Evans RW, Wang T et al. (2010) Association between serum lipids and hepatitis C antiviral treatment efficacy. Hepatology 52:854-863.

[28] Li JH, Lao XQ, Tillmann HL, Rowell J, Patel K et al. (2010) Interferon-lambda genotype and low-density lipoprotein cholesterol levels in patients with chronic hepatitis C infection. Hepatology 51:1904-1911.

[29] Kinoshita M, Kojima M, Matsushima T, Teramoto T (2005) Determination of apolipoprotein B-48 in serum by a sandwich ELISA. Clin Chim Acta 351:115-120.

[30] Usui S, Hara S, Hosaki S, Okazaki M. (2002) A new on-line dual enzymatic method for simultaneous quantification of cholesterol and triglycerides in lipoproteins by HPLC. J Lipid Res 43: 805-814.

[31] Okazaki M, Usui S, Ishigami M, Sasaki N, Nakamura T et al. (2005) Identification of Unique Lipoprotein Subclasses for Visceral Obesity by Component Analysis of Cholesterol Profile in High-Performance Liquid Chromatography. S

[32] Akuta N, Suzuki F, Kawamura Y, Yatsuji H, Sezaki H et al. (2007) Predictive factors of early and sustained responses to peginterferon plus ribavirin combination therapy in Japanese patients infected with hepatitis C virus genotype 1b: amino acid substitutions in the core region and low-density lipoprotein cholesterol levels. J Hepatol 46:403-410.

[33] Enomoto N, Sakuma I, Asahina Y, Kurosaki M, Murakami T et al. Mutations in the nonstructural protein 5A gene and response to interferon in patients with chronic hepatitis C virus 1b infection. (1996) N Engl J Med 1996; 334: 77-81.

[34] El-Shamy A, Nagano-Fujii M, Sasase N, Imoto S, Kim SR et al. (2008) Sequence variation in hepatitis C virus nonstructural protein 5A predicts clinical outcome on pegylated interferon/ribavirin combination therapy. Hepatology 48: 38-47.

[35] Sheridan DA, Bridge SH, Felmlee DJ, Crossey MM, Thomas HC et al. (2012) Apolipoprotein-E and hepatitis C lipoviral particles in genotype 1 infection: Evidence for an association with interferon sensitivity. J Hepatol Mar 10. [Epub ahead of print]

[36] Aizawa Y, Yoshizawa K, Aida Y, Ishiguro H, Abe H et al. (2012) Genotype rs8099917 near the IL28B gene and amino acid substitution at position 70 in the core region of the hepatitis C virus are determinants of serum apolipoprotein B-100 concentration in chronic hepatitis C. Mol Cell Biochem 360:9-14.

[37] Mawatari H, Yoneda M, Fujita K, Nozaki Y, Shinohara Y et al. (2010) Association between lipoprotein subfraction profile and the response to hepatitis C treatment in Japanese patients with genotype 1b. J Virol Hepatitis 17:274-279.

[38] ElHefnawi MM, Zada S, El-Azab IA. (2010) Prediction of prognostic biomarkers for interferon-based therapy to hepatitis C virus patients: a meta-analysis of the NS5A protein in subtypes 1a, 1b, and 3a. Virol J Jun 15;7:130.

[39] Ohnishi M, Tsuge M, Kohno T, Zhang Y, Abe H et al. (2012) IL28B polymorphism is associated with fatty change in the liver of chronic hepatitis C patients. J Gastroenterol. 2012 Feb 18. [Epub ahead of print].

Chapter 15

Adiponectin: A Perspective Adipose Tissue Marker with Antiinflammatory and Antiaterogenic Potential

Dalibor Novotný, Helena Vaverková and David Karásek

Additional information is available at the end of the chapter

1. Introduction

Adipose tissue as a substantial part of the human body contains about 10 % of body mass. It serves both as a reservoir of the energy storage and the active endocrine tissue producing many proactive substances including adipokines. These molecules have many important metabolic effects [1]. Adiponectin is an adipose tissue-derived adipokine which circulates at relatively high concentrations in blood. It has protective role in the initiation and progression of atherosclerosis through its antiinflammatory and antiatherogenic effects. Adiponectin serum levels are decreased in obesity, type 2 diabetes, and patients with coronary artery disease, etc [2]. The level of circulating adiponectin correlates positively with HDL cholesterol, and negatively with inflammatory markers, markers of insulin resistance, triglyceride-rich lipoprotein particles, and other adipokines. Adiponectin disposes of protective actions on development of various obesity-linked diseases. The antiinflammatory properties may be the major component of its beneficial effects on cardiovascular and metabolic disorders including atherosclerosis and insulin resistance. In addition, adiponectin displays a direct biological activity through the induction of a classical pathway of complement activation.

2. Adiponectin and atherosclerosis

Human adiponectin is a protein containing 244 amino acids. It is produced by apM1 cDNA transcripts. Adiponectin consists of two structurally distinct domains and the C-terminal part is likely to be involved in protection against atherosclerosis (Figure 1).

As a member of the soluble collagen superfamily, adiponectin has a structural homology with collagen type VIII, X, complement C1q and tumor necrosis factor alpha family [2]. In human plasma, adiponectin is present in a variety of heterogeneous isoforms, from large multimeric structures of high molecular weight to trimeric isoform. Monomeric one is present only in adipose tissue. The biological activity of various multimeric isoforms are not fully known yet, but it appears different isoforms have varying effects in different diseases. Although some studies have proposed that the ratio of high molecular weight (HMW) form to the other forms may serve as a better indicator of metabolic disorders, the majority of studies that have linked adiponectin to metabolic diseases have used assays for total adiponectin.

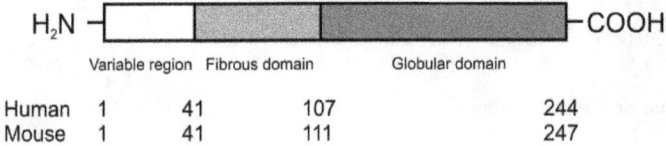

Figure 1. Schematic presentation of adiponectin structure (adapted from [2]).

Adiponectin gene is located on chromosome 3q27 and contains 3 exons and 2 introns. In 2003, DNA sequences encoding two receptors for adiponectin Adipo R1 and Adipo R2 were identified [3]. They are localized on chromosome 1 (1p36.13-q41) and 12 (12p13.31), respectively, with expression in most organs (AdipoR1 in skeletal muscle, AdipoR2 in the liver, in particular). Adiponectin gene is polymorphic, located in the region that contains susceptible loci for type 2 diabetes mellitus and metabolic syndrome. A number of single nucleotide polymorphisms (SNPs) and missense mutations were observed, especially in exons 2, 3 and the gene promoter.

Figure 2 schematically depicts some of the antiatherogenic properties of adiponectin towards different types of cells that have been established in experimental models. Adiponectin negatively regulates the expression of TNF alpha and C-reactive protein (CRP) in adipose tissue. On the contrary, its expression is negatively regulated by TNF alpha and interleukin 6 (IL 6). Adiponectin reduces expression of vascular and intracellular adhesion molecules (VCAM 1, ICAM 1), E-selectin, interleukin 8, and monocyte adhesion to human aortal endothelial cells after their stimulation with TNF alpha [4]. The proliferation and migration of smooth muscle cells induced by platelet growth factor (PDGF) is abolished or diminished by adiponectin action as inhibition of activation of nuclear factor kappa B in endothelial cells. This effect is partially mediated by its ability to support the action of cyclic adenosine monophosphate - proteinkinase A system (cAMP-PKA).

In endothelial cells, adiponectin inhibits the production of reactive oxygen species (ROS) induced by high levels of glucose via above mentioned the cAMP-PKA system. Adiponectin inhibits macrophage transformation to foam cells and reduces the intracellular content of cholesterol esters via suppression of expression of scavenger receptors, class A (SR-A). In these cells, adiponectin reduces lipopolysaccharides stimulated TNF alpha production. Recent clinical trials show a positive correlation of plasma levels of adiponectin and IL 10 [5]. In accordance with these findings, adiponectin has an antiatherogenic properties in mice models. Adenovirus-mediated supplementation of adiponectin inhibits the formation of atherosclerotic lesions and reduces the levels of mRNA of SR-A, TNF alpha and VCAM 1 in the vascular wall [6]. It is interesting that in these models, adiponectin has no effect on glucose and lipid parameters. The authors conclude that adiponectin affects atherogenesis through a series of antiinflammatory effects on macrophages and vascular endothelium.

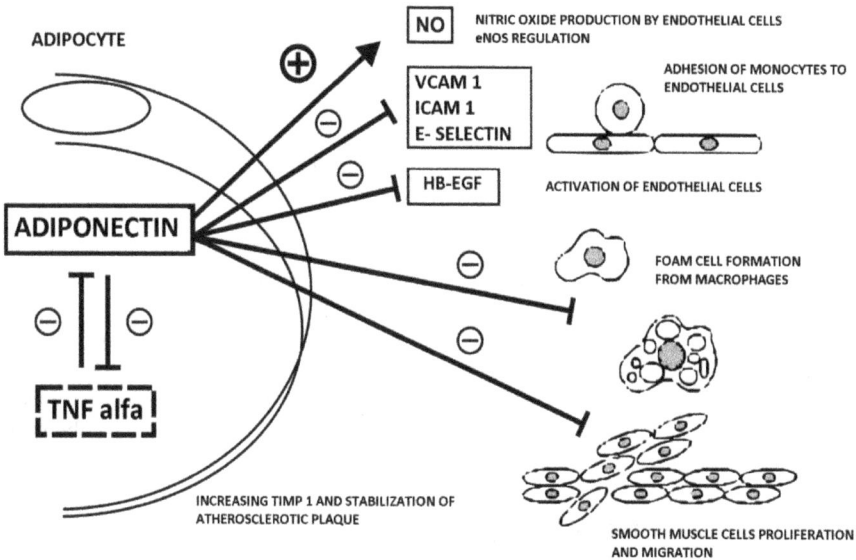

Figure 2. Some of protective actions of adiponectin (adapted from [2]).

A very important finding was observed in recent work describing the relationship of adiponectin-immune system. Adiponectin is able to bind to a number of target molecules, including the damaged endothelium and the surface of apoptotic cells. The significance of this phenomenon is not entirely clear. The study describes in vitro binding of purified C1q complement to recombinant adiponectin and dependence on calcium and magnesium ions. It was found that this binding stimulates the classical pathway of complement activation. Adiponectin does serve as an antiinflammatory factor, but may also induce biological activity through activation of complement. The authors hypotesize the binding of C1q leads to conformational changes in the adiponectin molecule, which induces the classical pathway of complement activation. Adiponectin may play an important role in immunity by its direct biological effect [7]. There is also evidence of adiponection accumulation on injured vascular arterial wall (but not in healthy one). This may lead to the hypothesis of the "consumption" of circulating adiponectin in patients with ischemic heart disease.

In some recent studies, adiponectin has a positive effect in endothelial homeostasis. It acts as a regulator of the enzyme endothelial nitric oxide synthase (eNOS), which is a key determinant of endothelial function and angiogenesis (the production of NO inhibits the inflammatory response in the arteries), and also promotes phosphorylation of eNOS in endothelial cells, increases its expression and induces NO production after suppression of its activity caused by the effect of oxidized low-density lipoproteins (oxLDL) [8].

Adiponectin promotes cyclooxygenase 2 (COX-2) expression and prostaglandin E2 (PGE2) synthesis in cardiac cells. It also has an antiapoptotic properties in vitro, as in endothelial cells. In the heart tissue adiponectin thus acts as a regulator of cardiac damage through its antiinflammatory effect and as a factor preventing the reconstruction of cardiac tissue. In order to become a useful biomarker of cardiovascular risk, it is necessary to determine which of its isoforms exhibit cardioprotectivity, and to clarify mechanism of their action in various pathophysiological conditions [7].

There is an increasing number of papers on experimental models point to the fact that adiponectin plays an important protective role in the development of insulin resistance and diabetes. Severe insulin resistance was seen in adiponectin-deficient knockout mice (KO-AD) after administration of high fat and/or carbohydrates diets. Administration of adiponectin led to reduced hyperglycemia in the diabetic mice without affecting insulin levels. In another study, increased muscle fatty acid oxidation and reduction of plasma glucose, free fatty acids and triglycerides were observed. Studies on experimental animal models revealed the administration of adiponectin has a beneficial action against the development of obesity and atherosclerosis. It seems that adiponectin acts not only as a factor increasing insulin sensitivity, and the protective effect may result from its ability to suppress production of proinflammatory cytokine [4].

2.1. Adiponectin and its relationship to obesity and metabolic syndrome

There has been a growing evidence of significantly reduced levels of adiponectin in obese individuals compared to subjects with normal body mass index (BMI) [9]. An inverse relationship with BMI was observed in both men and women, as well as negative correlation of adiponectin with visceral fat accumulation. It is obvious that hypoadiponectinemia (levels typically less than 4 mg/l) is associated with the development of insulin resistance and type 2 diabetes mellitus, independently of BMI and metabolic syndrome. Low adiponectinemia are considered the independent risk factor for developing hypertension. Kern et al. measured adiponectin plasma concentrations and mRNA levels in adipose tissue in nondiabetic subjects with varying degree of obesity and IR. They found a strong correlation of these two parameters. The obese individuals had significantly lower plasma adiponectin. When BMI was less than 30 kg/m2, women had twice more the body fat than men, but adiponectin levels were higher on average of 65% than in men (14.2 mg/l vs. 8.6 mg/l). Individuals with the highest levels of mRNA secreted the lowest levels of TNF alpha in adipose tissue. The authors conclude that adiponectin expression in adipose tissue is highest in lean subjects and women, and correlates with higher index of insulin sensitivity and lower TNF alpha expression [9]. Another study found that expression of adiponectin mRNA in adipose tissue may reflect short-term energy changes in some obese subjects. Expression of adiponectin and insulin sensitivity may be influenced by genetic variations in the adiponectin gene in response to acute energy fluctuations [10].

Metabolic syndrome characterized by abdominal obesity, dyslipidemia, hypertension and hyperglycemia, is a general risk for the development of atherosclerotic vascular disease. The study with 661 Japanese individuals investigated possible application of adiponectin as a biomarker of the metabolic syndrome [11]. Its plasma levels negatively correlated with waist circumference, visceral fat, TGL concentration, glucose and fasting insulinemia, systolic and diastolic blood pressure, and positively with HDL cholesterol. With decreasing levels of adiponectin, on the contrary, the number of components of metabolic syndrome increased. Total of 52% men and of 38% women with levels below 4.0 mg/l met criteria for MS. The authors suppose hypoadiponectinemia is closely associated with the clinical phenotype and its measurement could be useful in the MS treatment. Saely et al. observed a group of patients undergoing coronary angiography. Low adiponectin levels were independently associated with both metabolic syndrome and angiographically confirmed coronary atherosclerosis [12]. The highest levels of adiponectin were seen in subjects without MS and heart disease (12.1 ± 8.3 mg/l), whereas the lowest levels in patients with MS and presence of heart disease (6.7 ± 3.8 mg/l). Another study then identified a link between serum adipokines and cholesterol metabolism in individuals with MS. In 58 subjects with impaired glucose tolerance or elevated fasting glucose and signs of MS the markers of cholesterol synthesis were measured (determined by the ratio of non-cholesterol sterols to cholesterol and dietary cholesterol portion), in relation to adipokines and ultrasensitive CRP (hsCRP). It was found that adiponectin, leptin and CRP were associated with cholesterol metabolites (variations) and the high ratio of cholesterol synthesis to its absorption is characterized by high levels of serum leptin and low adiponectin [13].

2.2. Adiponectin and its relationship to heart disease

The hypoadiponectinemia was also found in patients with angiographically documented coronary atherosclerosis or acute coronary syndrome. In men, plasma adiponectin significantly predicted the extent of coronary atherosclerosis [14]. A prospective study of patients with end-stage renal disease showed an inverse relationship between cardiovascular events and adiponectinemia. Higher adiponectin levels represent a low risk of myocardial infarction in healthy men individuals and moderately reduced risk of coronary heart artery disease in diabetic men patients [4]. In contrast, adiponectin concentrations did not correlate significantly with the risk of heart disease in American Indians or the British women. A large prospective study involving British men with heart disease combined with a meta-analysis of seven previously published studies found only a weak association of adiponectin with the disease [4]. This inconsistent data could be due to differences in study populations (ethnicity, gender, type of disease etc.). In any cases, it remains unclear whether hypoadiponectinemia is a reliable indicator of heart disease.

2.3. Adiponection and inflammation

CRP is known to be an independent predictor of future risk for cardiovascular events and risk factor for developing MS. Its positive association with BMI is considered as a useful biomarker for chronic inflammation linked to obesity. Plasma levels of CRP correlated negatively with adiponectin levels [15] which was confirmed by various studies. Since CRP mRNA in humans is expressed in adipose tissue, adiponectin can apparently participate in influencing CRP levels in plasma by regulation of its expression. The regulation of CRP synthesis in the liver is also influenced by proinflammatory adipokines IL 6 and TNF alpha. On the one side, adiponectin expression is regulated by proinflammatory cytokines, on the other side, adiponectin modulates the activity and the production of TNF alpha in different tissues. Several studies have found links between hypoadiponectinemia and elevated serum IL 6. So far, there is no evidence of a link between adiponectin and TNF alpha in plasma in humans. Nishida et al. describe the action of IL 6, adiponectin, CRP and metabolic syndrome in subclinical atherosclerosis [16]. The relative influence of these parameters on a group of healthy subjects was observed. In 714 men and 364 women aged 40 to 59 years, thickness of the intima-media complex (IMT), pulse blood flow velocity and components of MS were measured. IL 6 levels correlated with IMT parameter, while adiponectin correlated negatively with IMT only in men. Individuals with either high IL 6 or CRP, or low levels of adiponectin, had increased IMT in the presence of MS. Increasing number of MS components was expressed more strongly in women than in men. The authors speculate IL 6 and adiponectin are important risk factors for premature arterial alterations in men.

In another study, the relationship of adiponectin to markers of inflammation, atherogenic dyslipidemia and heart disease was investigated in patients with coronary artery disease [17]. Study participants were in a rehabilitation program to reduce the cardiovascular risk factors. After adjusting for age and sex, adiponectin was associated positively with HDL cholesterol and N- terminal propeptide of B natriuretic peptide (NT-proBNP), while the association was negative for triglycerides. In this study, no relationship was found with markers of inflammation. The same results were obtained after next adjustment for other parameters; BMI, alcohol intake, smoking, presence of diabetes and/or hypertension and lipid-lowering therapy, and fasting glucose. The authors conclude serum adiponectin is associated with the presence of the atherogenic dyslipidemia and NT-proBNP levels, but not with markers of systemic inflammation (IL 6, CRP) in patients with manifest coronary heart disease. Atherogenic dyslipidemia may be a link between adiponectin and progression of atherosclerosis. The role of systemic inflammation as part of the adiponectin-atherosclerosis relationship may decrease during the course of the disease, and could be more amplified in the earlier stage of disease development.

3. Adiponectin and gene polymorphisms

As mentioned above, the adiponectin gene is located on chromosome 3q27, containing 3 exons and 2 introns. This region also encopasses the susceptibile loci for type 2 diabetes and metabolic syndrome. The sequence polymorphism was found in the form of several single nucleotide polymorphisms (SNPs) and a number of missense mutations. Sequence analysis of the gene for adiponectin in Japanese and Caucasian populations found more than 10 SNPs, some of which are associated with BMI, metabolic syndrome, insulin sensitivity, hyperglycemia, type 2 diabetes, levels of plasma adiponectin, etc. The results of studies, however, are inconsistent, providing conflicting results. In many cases, the haplotype analysis was performed from a combination of alleles of individual SNPs.

Kondo et al. analyzed a cohort of Japanese patients with type 2 diabetes and nondiabetic controls to detect mutations in the gene for adiponectin [18]. Four missense mutations in the globular domain (I+164T, R+112 C, H+241P, R+221S) were identified. The frequency of one mutation, the substitution of I+164T, was significantly higher in patients than in controls of comparable age and body weight. Mutation carriers had lower adiponectin concentrations in plasma and also showed the presence of a feature characteristic of the metabolic syndrome (hypertension, hyperlipidemia, diabetes, atherosclerosis). Hypoadiponectinemia was already evident at the same time in heterozygotes I+164 T mutation carriers and also in R +112 C, but this was the case of only 3 patients. The authors suggest I+164 T variant is associated with low adiponectin levels in plasma and type 2 diabetes mellitus.

Another study has examined the adiponectin gene locus as a candidate site for coronary artery disease [19]. 383 Japanese patients with angiographically confirmed disease and 318 individuals adjusted for age and BMI were the subjects of this study. Analyses of SNPs were performed using real time polymerase chain reaction (rtPCR) and restriction fragment length polymorphism (RFLP). In patients, the higher incidence of T+164 mutation and lower adiponectin levels in plasma were seen, independently of BMI. Subjects with the mutation showed a clinical phenotype of metabolic syndrome. According to the authors, the I+164T polymorphism is associated with metabolic syndrome and coronary artery disease in Japanese population.

Hara et al. examined the relationship between two SNPs located at exon 2 of adiponectin gene (T+45G and G+276T) and type 2 diabetes in the Japanese population [20]. Subjects with the GG genotype at position +45 or +276 had an increased risk of DM compared to TT genotypes. GG +276 homozygotes showed higher insulin resistance index and the presence of G allele at position 276 was characterized by lower levels of plasma adiponectin in subjects with higher BMI (GG: 10.4 mg/l, TT: 16.6 mg/l). The different results showed the study focused on the relationship between haplotypes of the adiponectin gene with obesity and other signs of metabolic syndrome in nondiabetic Caucasian population [21]. Both polymorphisms, T+45G and G+276T, separately significantly correlated with IR. The common haplotype was also closely associated with a number of components of metabolic syndrome. Homozygotes for middle-risk haplotype TG (i.e. individuals with +45 TT variant and +276 GG variant) had higher body weight, waist circumference, blood pressure, fasting glucose, insulin, cholesterol/ HDLcholesterol ratio and lower adiponectin levels, after adjustment for age, sex and body weight. However, in the second group (614 Caucasian individuals with type 2 DM) the risk haplotype was associated with increased body weight, not with DM. It is hypothesized the variability of the adiponectin gene is connected with obesity and other features of insulin resistance, but the risk haplotype is probably a marker of linkage disequilibrium with a polymorphism yet unidentified that directly affects the plasma levels of adiponectin and insulin sensitivity. Moreover, Fillipi et al. found no association of SNP T+45G with insulin resistance [22]. The T+276 G polymorphism was associated with higher BMI, lower insulin and adiponectin, but, unlike previous study, in the TT genotype. In discussion the authors analyzed possible causes of these results and conclude the same mentioned above. There is the high probability of the existence of further SNPs or gene mutations, which is in linkage disequilibrium with SNP +276 and which determines its effect. Variations in the adiponectin gene and risk for subsequent type 2 diabetes in women has been of interest in the study of Hu et al. [23].

A prospective study focused on the determination of SNPs participation in the development of IR in French population found that variations in the adiponectin gene affects weight gain, body fat distribution and the development and the onset of hyperglycemia, as well as serum adiponectin [24]. At the start of a three-year study, the normoglycemic individuals with no signs of diabetes or impaired glucose tolerance were influenced mainly by two SNPs: G-11391A and T+45G.

An interesting work was published in 2006 in Clinical Chemistry by Hegener et al. [24]; the prospective study monitoring the risk of atherothrombotic disease in individuals with no signs of diabetes. Five SNPs in the gene for adiponectin were investigated in 600 Caucasian men with subsequent aterotrombotic events (myocardial infarction or stroke) and 600 controls. After adjustment for potential risk factors, regression analysis then revealed two variants with a decreased risk of stroke (C-11377G and G-11066A). This study has provided evidence of links of specific adiponectin gene variants with reduced risk of stroke.

3.1. Relationship between G+276T single nucleotide polymorphism of adiponectin gene and markers of insulin resistance in dyslipidemic patients

In many recent studies, the adiponectin gene has been proposed as a potential candidate gene for insulin resistance but only a few of them have confirmed this relationship. Insulin resistance is considered the key factor in the patogenesis of common disorders, such as atherosclerosis, metabolic syndrome and diabetes mellitus. The genetic backround is likely to be polygenic but the genes involved are mostly unknown.

In our work, we have studied the possible relationship between single nucleotide polymorphism G+276T and IR markers, including lipid and lipoprotein profiles and adiponectin plasma levels in 355 dyslipidemic patients and their first-degree relatives.

3.2. Subjects

The group consisted of 355 patients attending Lipid Center of 3rd Medical Clinic, Faculty Hospital Olomouc, and their first-degree relatives. Patients had the first examination between January 2004 and January 2006. All patients were examined by a physician and the family history were collected and medical history with physical examinations were performed. All individuals were tested for secondary hyperlipidemia, especially on the presence of diabetes mellitus, hypothyroidism, hepatic and renal failure and nephrotic syndrome. Violation of the following criteria led to exclusion from the study: hypolipidemic treatment in the previous 6 weeks, the presence of secondary hyperlipidemia, acute infection, acute cardiovascular or cerebrovascular attack within the past 3 months, cardiac disease (NYHA III and IV). Participiants were also divided into three groups. Group G1 included the presence of individuals with clinically manifest atherosclerosis, the group G2 individuals with dyslipidemia defined by Sniderman [25] (apolipoprotein B > 1.2 g/l and/or triglycerides > 1.5 mmol/l) but without clinical signs of the presence of atherosclerosis. Group 3 consisted of healthy individuals with the apolipoprotein B < 1.2 g/l and triglycerides < 1.5 mmol/l. The participants signed informed consent before taking a blood sample for DNA testing. The study was approved by the Ethical Committee of the Faculty of Medicine, Faculty Hospital Olomouc.

3.3. Materials and methods

Venous blood for biochemical tests were collected after 12-hour fasting. Total cholesterol, HDL cholesterol and triglycerides were determined enzymatically using an analyzer Modular SWA (Roche, Switzerland), as well as other routine biochemical analyses. LDL cholesterol was calculated using the Friedewald equation for specimens with TG < 4.5 mmol/l (available for 242 subjects). Concentrations of apolipoproteins AI and B were determined by immunoturbidimetric method, as well as C-reactive protein levels, established by highly sensitive method (all Roche, Switzerland). Insulin was determined by IRMA (Immunotech, France). HOMA parameter (homeostatic model) was calculated from the formula: fasting glucose x fasting insulin / 22.5. C-peptide and proinsulin were determined by commercially available kits (Immunotech, France, DRG Instruments GmbH, Germany, respectively). Serum levels of soluble adhesion molecules ICAM 1 and VCAM 1 were analyzed by immunoenzymatic technique (Immunotech, France). Adiponectin determination was performed by the ELISA method (BioVendor, Czech Republic). The following markers of endothelial dysfunction were examined: plasminogen activator inhibitor-1 (PAI-1) and tissue plasminogen activator (tPA), both determined by ELISA methods (Technoclone, Vienna, Austria). Concentrations of adhesion molecules, insulin, proinsulin, C-peptide and adiponectin were measured on samples frozen at - 80 ° C until analysis.

G+276T adiponectin gene SNP was detected by real time polymerase chain reaction with fluorescent hybridization probes (FRET) on the Light Cycler instrument, v.2.0 (Roche), according Fillipi et al [22]. Genotyping was performed after the isolation of DNA from peripheral blood samples using phenol method [26]. DNA isolates were then stored at - 20 °C until analysis. The primer and probe synthesis was made at the in TibMolbiol (Germany). The sequence of oligonucleotides for the detection of SNP +276 G> T were as follows:

Primers:

5'- GGC CTC TTT CAT CAC AGA CC -3'

5'- AGA TGC AGC AAA GCC AAA GT -3'

Probes:

5'- AAG CTT TGC TTT CTC CCT GTG TCT A--FL

5'- LCRed640- GCC TTA GTT AAT AAT GAA TGC CTT—PH

Individual genotypes were determined by melting curve analysis after the amplification process. The fluorescence signal was converted and delivered to the graph as the dependency of negative fluorescence change with temperature (y axis) on temperature (x axis). As the result, creation of the characteristic peaks representing the melting temperature of the product and allow to distinguish the genotypes GG, GT and TT was performed. Example of analysis is shown in Figure 2.

3.4. Statistical analysis

Quantitative data were expressed as a mean ± standard deviation. Parameters with abnormal distribution were logarithmically transformed before statistical analysis. Differences between genotypes in continuous variables were determined by using ANOVA after adjustment for age, gender and waist circumference (SPSS 12.0 statistical package, SPSS Inc., USA). Furthermore, the calculation of frequency of alleles (G and T) and genotypes (GG, GT and TT) in individual groups and subgroups were performed.

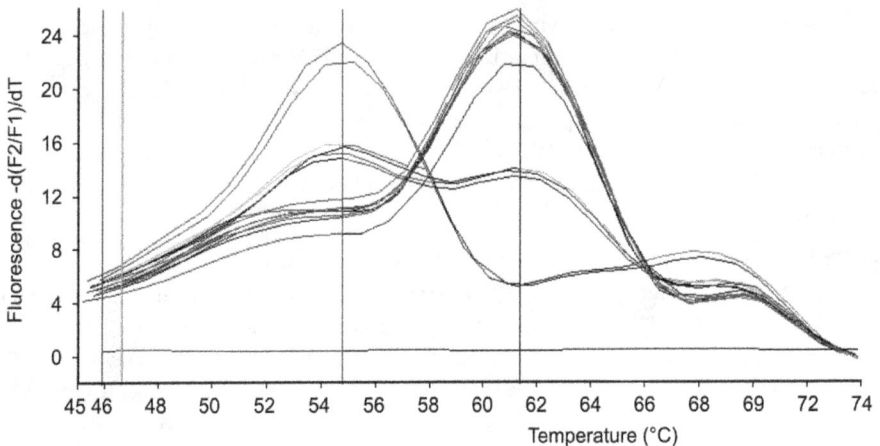

Figure 3. An example of the melting curve analyses for G+276T polymorphism of adiponectine gene. (melting temperature for T and G alelles: $T_{m(T)} = 54.8 \pm 1.5$ °C, $T_{m(G)} = 61.3 \pm 1.5$ °C).

3.5. Results

In Table 1 the clinical and laboratory characteristics of the groups of dyslipidemic patients divided according to genotypes at position +276 of the gene for adiponectin are shown. Table 2 presents the results of laboratory parameters that differed significantly between each of groups determined by genotype at position +276. The data are adjusted for age, gender and waist circumference. The results show that the GG genotype carriers had significantly higher levels of total cholesterol (GG: 6.54 ± 1.74 mmol/l, GT: 6.18 ± 1.45 mmol/l, TT: 6.25 ± 1.64 mmol/l, $p < 0.05$) and LDL cholesterol (GG: 4.12 ± 1.49 mmol/l, GT: 3.78 ± 1.31 mmol/l, TT: 3.70 ± 1.34 mmol/l, $p < 0.05$) than T allele carriers. In heterozygotes, however, the presence of T allele at position +276 was associated with higher concentrations of PAI-1 (GG: 71.50 ± 41.0 μg/l, GT: 81.0 ± 38.7 μg/l, TT: 70.14 ± 44.4 μg/l, $p < 0.05$). We did not find any significant association with other markers of IR, such as BMI, blood glucose, insulin, or serum adiponectin. Table 3 depicts the frequencies of genotypes and alleles at position +276, Table 4 then presents the distribution of genotypes in groups according to triglyceride levels (cut-off value of TGL = 1.5 mmol/l).

	GG	GT	TT
Number	188	144	23
BMI (kg/m²)	26 ± 4	26 ± 4	26 ± 5
Systolic blood pressure (mmHg)	130 ± 18	132 ± 18	129 ± 14
Diastolic blood pressure (mmHg)	80 ± 10	83 ± 9	80 ± 7
Total cholesterol (mmol/l) *	6.47 ± 1.73	6.18 ± 1.44	6.26 ± 1.64
Triglycerides (mmol/l)	2.38 ± 2.97	2.43 ± 2.23	2.30 ± 3.44
HDL cholesterol (mmol/l)	1.47 ± 0.44	1.44 ± 0.44	1.46 ± 0.39
LDL cholesterol (mmol/l) *	4.07 ± 1.47	3.79 ± 1.31	3.70 ± 1.34
Apolipoprotein AI (g/l)	1.55 ± 0.30	1.57 ± 0.34	1.60 ± 0.29
Apolipoprotein B (g/l)	1.21 ± 0.39	1.15 ± 0.32	1.15 ± 0.32
hsCRP (mg/l)	1.99 ± 1.94	2.20 ± 1.93	2.02 ± 1.69
tPA (μg/l)	4.08 ± 4.81	4.31 ± 4.46	4.03 ± 3.89
PAI-I (μg/l) *	69.7 ± 40.7	79.9 ± 39.0	72.3 ± 44.4
VCAM 1(μg/l)	808 ± 247	823 ± 287	743 ± 184
ICAM 1 (μg/l)	563 ± 140	592 ± 165	585 ± 209
Fasting glucose (mmol/l)	5.09 ± 0.91	5.25 ± 1.23	4.82 ± 0.68
Insulin (U/l)	8.33 ± 5.55	7.99 ± 4.84	7.77 ± 4.54
HOMA IR	1.93 ± 1.45	1.92 ± 1.33	1.84 ± 1.18
C- peptide (μg/l)	2.38 ± 1.25	2.40 ± 1.12	2.41 ± 1.30
Adiponectin (mg/l)	12.9 ± 7.6	13.0 ± 7.0	12.0 ± 5.7

* GG vs. GT+TT, $p < 0.05$

Table 1. Clinical and laboratory characteristics according to adiponectin genotypes at position +276 (G+276T).

	GG	GT	TT	p
Number	188	144	23	
Total cholesterol (mmol/l)	6.54 ± 1.74	6.18 ± 1.45	6.25 ± 1.64	< 0.05
LDL cholesterol (mmol/l)*	4.12 ± 1.49	3.78 ± 1.31	3.70 ± 1.34	< 0.05
PAI-I (μg/l)	71.5 ± 41.0	81.0 ± 38.7	70.14 ± 44.4	< 0.05

*only 242 patients included

Table 2. Laboratory characteristics according to adiponectin genotypes at position +276 (G+276T) with significant differences between groups (GG vs. GT+TT, after adjustment for sex, age and BMI).

APM1 G+276T	Patients (n = 355)
Genotype	
GG	188 (53 %)
GT	144 (41 %)
TT	23 (6 %)
Allele	
G	520 (73 %)
T	190 (27 %)

Table 3. Genotype and allele frequencies for G+276T polymorphism in dyslipidemic patients.

Genotype	GG	GT	TT
TG ≤ 1.5 (n = 225)	119 (53 %)	94 (42 %)	12 (5 %)
TG > 1.5 (n = 148)	85 (57 %)	52 (35 %)	11 (8 %)

Chi-square 1.981, p = 0.37

Table 4. Genotype frequencies for G+276T polymorphism in dyslipidemic patients according to level of triglycerides (mmol/l).

3.6. Discussion

Insulin resistance is considered the key factor in the pathogenesis of complex diseases such as atherosclerosis, metabolic syndrome and diabetes mellitus. Genetic background IR is probably multifactorial but the participating genes are largely unknown.

In this study, the relationship of polymorphism G+276T of adiponectin gene and markers of insulin resistance was investigated. We found an association between genotype GT and one marker of IR, PAI-I. However, we found no association with serum adiponectin, insulin, HOMA and BMI. Our work did not confirm the preliminary findings from 2005, where the relationship between the adhesion molecules ICAM 1 and TT genotype was observed [27].

Possible association between SNPs and dyslipidemic phenotypes defined by Sniderman classification, based on serum TGL and apo B, was not seen. We found no linkage (data not specified), even in a situation where the only criterion was TGL alone. The genotype distribution in this case was comparable in both groups.

As shown in Table 1, GG genotype was associated with higher levels of total cholesterol and LDL cholesterol compared with GT and TT genotypes. This was found in our previous study as well [27].

Table 3 displays the fact that distribution of genotypes at position 276 is comparable with those published in previous works [20, 21, 22].

3.7. Conclusions

In summary, our study found only a weak association of adiponectin gene SNP G+276T with IR markers. The relationship of GG genotype and selected quantitative lipid parameters were confirmed, in accordance with several studies. Based on some recent literature we suggest the gene variant G+276T may be marker of one or more haplotypes containing a causal polymorphism determining IR or diabetes mellitus. Differences among populations on the linkage disequilibrium structure may result in association on the disease haplotype with different SNP alleles in different population. More studies will be necessary to perform for evaluation of the influence of G+276T SNP on insulin resistance.

4. Adiponectin and its relationship to endothelial dysfunction

In vitro experiments revealed the physiological concentrations of adiponectin inhibited TNF alpha induced expression of VCAM 1 and ICAM 1 on the endothelium and exhibited other antiatherogenic effects. In 2008 Vaverková et al. published a study concerning the relationship between adiponectin and serum concentrations of soluble adhesive molecules VCAM 1 and ICAM 1 as well as with markers of insulin resistance and inflammation in patients with cardiovascular disease and in dyslipidemic patients at high risk of cardiovascular disease [28].

The aim of the study was to evaluate the relationship of adiponectin to soluble forms of vascular cell adhesion molecule 1 (VCAM 1) and intercellular cell adhesion molecule 1 (ICAM 1) in patients with cardiovascular disease or dyslipidemia.

The data from experimental research in animals support the hypothesis of antiaterogenic properties of adiponectin. Adiponectin accumulates in the arterial wall of injured arteries [29]. In adenovirus-treated animals the increase of adiponectin significantly reduced progression of atherosclerotic lesions [6]. In vitro experiments revealed the fact that physiological concentrations of adiponectin inhibited TNF alfa induced expression of VCAM 1 and ICAM 1 on the endothelium [29] and exhibited other antiatherogenic effects.

We have investigated the relationship between adiponectin and serum concentrations of VCAM 1 and ICAM 1 as well as with markers of insulin resistance and inflammation in patients with cardiovascular disease and in dyslipidemic patients at high risk of CVD.

4.1. Subjects

264 patients of Lipid Center at Faculty Hospital Olomouc were included in the study. All patients were examined by a physician and the following information were obtained:

medical history, physical examination and NYHA classification. Subjects were tested for secondary hyperlipidemia. Patients were divided into three groups, those with the presence of clinically manifest atherosclerosis (G1), those with dyslipidemia defined according to Sniderman, but without clinically manifest atherosclerosis (G2), and healthy individuals (G3).

4.2. Results

The characteristics of the three subgroups of the studied cohort are shown in Table 5. Participants with CVD (G1) had comparable lipid, lipoprotein and apolipoprotein profile to the dyslipidemic subjects without CVD (G2) but were more insulin resistant. These differences persisted after adjustment for age, sex and BMI. The G1 had also the highest soluble ICAM 1, the difference in VCAM 1 was not statistically significant. Subjects with dyslipidemia (G2) had significantly lower adiponectin levels and higher levels of ICAM 1 compared with G3. Lower adiponectin levels in patients with CVD did not reach statistical significance, possibly due to a small number of patients. Adiponectin correlated with many lipid and nonlipid markers of insulin resistance. Adiponectin did not correlate with ICAM 1, but there was a strong positive association of adiponectin with VCAM 1. While ICAM 1 and VCAM 1 were strongly intercorrelated, they showed different association pattern with other risk factors. ICAM 1 correlated strongly with many markers of insulin resistance and hsCRP, while VCAM 1 were negatively associated with apo AI and apo B, and positively with adiponectin. Association of adiponectin with VCAM 1 was most prominent in group G1 and G2, but was not significant with G3. Results of multiple backward stepwise regression analysis confirmed these observations. Adiponectin levels were independently positively associated with sex (higher in women), HDL cholesterol and VCAM 1, and negatively with hsCRP. In multiple stewise regression analysis with VCAM 1 as the dependent variable, VCAM 1 was independently associated with ICAM 1 ($p < 0.0001$), adiponectin ($p < 0.0001$), HDL cholesterol ($p = 0.0208$) and triglycerides ($p = 0.0091$). On the other hand, ICAM 1 was independentely associated with VCAM 1 ($p < 0.0001$), atherogenic index ($p < 0.0001$), hsCRP ($p = 0.0001$) and HOMA ($p = 0.0307$). (More detailed results are given in lit. [28].)

4.3. Discussion

Our study confirms the previously described correlations of adiponectin with many lipid and nonlipid markers of IR as well as its relationships with HDL cholesterol, sex and hsCRP [30, 31, 32]. The unexpected finding was the significant independent positive association of adiponectin with VCAM 1 but not with ICAM 1 serum concentrations in patients with or at risk for CVD. Their expression results in adhesion of circulating leukocytes to the endothelial cells and their subsequent transendothelial migration- an important step in initiation and progression of atherosclerosis. VCAM 1 and ICAM 1 have different expression pattern and probably different roles in atherogenesis [33]. Soluble forms of these molecules can be measured in peripheral circulation. The origins of circulating soluble cell adhesion molecules are not entirely clear, but they may derive from shedding or proteolytic cleavage from endothelial cell.

	G1 (CVD+, DLP+/-)	G2 (CVD-, DLP+)	G3 (CVD-, DLP-)
Number	29 (M 18/F 11)	173 (M 97/F 76)	62 (M 19/ F 43)
Age (years)	60.0 ± 9.1	44.9 ± 13.8	36.4 ± 14.5
BMI (kg/m²)	27.5 ± 3.7	26.3 ± 5.7	23.6 ± 4.3
Waist (cm)	92.3 ± 13.1	88.4 ± 11.4	77.4 ± 10.8
Systolic blood pressure (mm Hg)	143 ± 15	131 ± 17	120 ± 13
Diastolic blood pressure (mm Hg)	86.3 ± 9	83 ± 8	75.6 ± 9.8
Total cholesterol (mmol/l)	6.8 ± 1.2	6.7 ± 1.4	4.7 ± 0.7
Triglycerides (mmol/l)	3.0 ± 2.1	2.7 ± 2.3	0.9 ± 0.2
AIP: log (TGL/HDLchol)	0.29 ± 0.38	0.24 ± 1.16	-0.2 ± 0.2
HDL cholesterol (mmol/l)	1.32 ± 0.43	1.34 ± 0.37	1.56 ± 0.36
LDL cholesterol (mmol/l)	4.2 ± 1.1	4.3 ± 1.3	2.8 ± 0.6
Apolipoprotein AI (g/l)	1.52 ± 0.28	1.51 ± 0.29	1.60 ± 0.30
Apolipoprotein B (g/l)	1.29 ± 0.3	1.33 ± 0.33	0.84 ± 0.17
hsCRP (mg/l)	3.4 ± 4.9	2.68 ± 3.6	2.69 ± 5.6
VCAM 1(µg/l)	885 ± 261	800 ± 285	860 ± 265
ICAM 1 (µg/l)	673 ± 202	601 ± 164	538 ± 114
Fasting glucose (mmol/l)	5.8 ± 1.8	5.1 ± 0.8	4.8 ± 0.6
Insulin (mIU/l)	8.8 ± 4.8	8.9 ± 5.3	6.6 ± 3.4
HOMA IR	2.3 ± 1.6	2.1 ± 1.3	1.4 ± 0.8
C- peptide (µg/l)	3.2 ± 1.2	2.6 ± 1.0	1.9 ± 0.7
Proinsulin (mIU/l)	17.0 ± 8.0	15.6 ± 9.9	11.3 ± 5.2
Adiponectin (mg/l)	15.5 ± 8.0	12.3 ± 6.6	16.1 ± 6.8

Table 5. The demographic, clinical and laboratory characteristics of the study population

The expression pattern of adhesion molecules may explain why VCAM 1 is a marker of increased risk for future coronary events only in patients with atherosclerosis [34]. Patients with stable CAD have moderately increased and in several studies even normal levels of soluble VCAM 1 in comparison with healthy controls. The highest level of VCAM 1 was noted in patients with acute myocardial infarction [35]. In another study, VCAM 1 was a useful marker for predicting future ischemic events in the 6 months after presentation with unstable angina pectoris or nonQ myocardial infarction [36]. In our cohort, levels of VCAM 1 in the CVD patients were not significantly higher than in controls. This is in agreement with several other works.

4.4. Conclusions

Many studies, including experiments in vitro, animal models and studies in human, have shown that adiponectin has antiatherogenic and antiinflammatory properties. Low adiponectin levels were found in patients with CAD independently of other risk factors. Therefore, the finding of positive and independent association of adiponectin with the marker of endothelial dysfunction VCAM 1 was suprising. This positive association was present both in patients with CVD and dyslipidemic subjects without CVD, but it was not significant in healthy subjects without dyslipidemia. We hypothesize that adiponectin, which accumulates in the arterial wall only in place of endothelial injury and atherosclerotic plaques (that is the same places where VCAM 1 is expressed) may be involved in shedding of ectodomains of VCAM 1 from endothelial surface. This may represent a mechanism by which VCAM 1 effects on the cell surface can be downregulated. In this way, adiponectin could protect vascular wall from adhesion of leukocytes and thus from progression of atherosclerosis.

5. Adiponectin and dyslipidemia: Relationship of adiponectin, fibroblast growth factor 21 and adipocyte fatty acid binding protein levels to dyslipidemic phenotypes – Pilot study

5.1. Background

Adipose tissue is an important place of many metabolic and inflammatory processes. Adipokines are considered to be the mediators of these pathways.

Adiponectin (ADP, AdipoQ, apM1, GBP28) is a "favourable" adipokine of fat tissue circulating at relatively high concentrations in human plasma. Adiponectin has the protective effects in early stages and during progression of atherosclerosis probably by its antiinflammatory and antiatherogenic actions.

Fibroblast growth factor 21 (FGF 21) is also a "favourable" cytokine of adipose tissue considered as a new metabolic regulator of non insulin dependent glucose transport in cells [37]. Systematic administration of FGF 21 decreases plasma levels both of glucose and triglycerides, and leads to improving of lipoprotein profiles in genetic compromised FGF transgenic mice and primates [38]. Increased levels of FGF 21 and a negative correlation with HDL and adiponectin were found in patients with metabolic syndrome [39].

Adipocyte fatty acid binding protein (A-FABP) is a „unfavourable" adipokine, probably a new marker and/or predictor of metabolic syndrome [40]. A-FABP is a dominant cytoplasmic protein of mature adipocytes and a regulator of lipid and glucose metabolism, present also in macrophages of fat tissue. Its expression is induced by oxidated LDL [41]. Higher levels of A-FABP are associated with increased fasting glucose, triglycerides, insulin BMI and waist circumference, and decreased HDL in patients with metabolic syndrome. Inhibition of A-FABP action is associated with reversion of atherosclerosis (improving of diabetic and lipoprotein parameters).

The aim of our study was to evaluate the relationship between adiponectin, FGF 21 and A-FABP levels and dyslipidemic phenotypes defined on the basis of concentrations of triglycerides and apolipoprotein B [25].

5.2. Subjects, material and methods

119 pacients of Lipid Center at Faculty Hospital Olomouc were included on the pilot scheme. Routine serum biochemical parameters were analyzed on Modular SWA (Roche, Switzerland) in the day of blood collection. Levels of ADP, FGF 21 and A-FABP were determinated by imunochemical Elisa methods (BioVendor, Czech Republic). The analytical characteristics from data sheets were verificated according to laboratory protocol for all procedures.

119 individuals were divided into four dyslipidemic phenotypes (DLP) according to Sniderman classification- see Table 6.

	TGL (mmol/l)	Apo B (g/l)
DLP1	< 1.5	< 1.2
DLP2	≥ 1.5	< 1.2
DLP3	< 1.5	≥ 1.2
DLP4	≥ 1.5	≥ 1.2

Table 6. Classification of dyslipidemic phenotypes

5.3. Results

Basic clinical characteristics are shown in Table 7. Concentrations of adipokines and other biochemical parameters are given in Table 8.

	Number (n)	F/M	Age (y)	Waist (cm)	SBP (mg Hg)	DBP (mm Hg)	Smoking (n)	Manifestation of ATS
DLP1	32	16/16	41 ± 10.0	85 ± 9.3	129 ± 12	77 ± 9	4	3
DLP2	38	20/18	47.1 ± 10.1	96 ± 12	130 ± 19	78 ± 11	7	3
DLP3	13	3/10	47.8 ± 10.5	88 ± 8.0	125 ± 18	75 ± 7	2	0
DLP4	36	22/15	49.9 ± 10.7	92 ± 9.0	126 ± 15	75 ± 9	9	4

Table 7. Basic clinical characteristics of DLP groups

	ADP (mg/l)	FGF 21 (ng/l)	A-FABP (ug/l)	CHOL (mmol/l)	TGL (mmol/l)	HDLchol (mmol/l)	LDLchol (mmol/l)	Apo AI (g/l)	Apo B (g/l)	GLU (mmol/l)	BMI (kg/m²)
DLP1	10.6 ± 6.0	186 ± 100	22.5 ± 10.3	5.68 ± 0.8	1.04 ± 0.26	1.74 ± 0.41	3.48 ± 0.75	1.74 ± 0.41	0.91 ± 0.16	5.16 ± 0.66	25.9 ± 5.2
DLP2	8.0 ± 5.1	333 ± 360*	33.9 ± 29.0*	6.23 ± 1.92	5.29 ± 8.0	1.22 ± 0.31*	3.14 ± 0.93	1.51 ± 0.30	0.98 ± 0.14	5.57 ± 1.19*	28.3 ± 4.7*
DLP3	8.6 ± 4.9	165 ± 104	14.4 ± 4.6	7.47 ± 1.14	1.11 ± 0.2	1.45 ± 0.33	5.52 ± 1.27	1.51 ± 0.25	1.41 ± 0.25	5.03 ± 0.52	25.2 ± 3.1
DLP4	9.0 ± 5.9	384 ± 347**	29.2 ± 18.4**	8.43 ± 2.0	3.5 ± 2.2	1.27 ± 0.45**	5.59 ± 2.0	1.50 ± 0.43	1.62 ± 0.39	5.36 ± 1.27***	27.2 ± 5.0**

Differences between groups were analyzed with ANOVA. Parameters with skewed distribution (TGL, ADP, FGF 21, A-FABP) were log transformed to normalize their distributions before statistical analyses.

* DLP2 vs. DLP1 and DLP3, $p < 0.01$, ** DLP4 vs. DLP1 and DLP3, $p < 0.01$, *** DLP4 vs. DLP1 and DLP3, $p < 0.05$

Table 8. Adipokines and other biochemical parameters in connection with DLP

The highest levels of ADP were observed in DLP1 (no significance). Suprisingly, there was seen no negative association between adiponectin levels and DLP2 (DLP4). FGF 21 and A-FABP were significantly increased in the groups with the most important atherogenic potential (DLP2, DLP4). These two parameters correlated with higher levels of triglycerides, fasting glucose, BMI and lower HDL cholesterol, both in DLP2 and DLP4.

5.4. Conclusions

No association was found between ADP levels and other adipokines in DLP groups in our study. There was the correlation between FGF 21 and A-FABP in groups with TGL > 1.5 mmol/l. Increased levels of both parameters were associated with increased glucose, BMI and decreased HDL cholesterol levels (in accordance with lit.[40]). The increase of FGF 21 concentrations are probably due to the compensatory response to higher A-FABP that is considered the predictor of metabolic syndrome. In individuals with MS, the determination of A-FABP could be considered as a parameter with the independent metabolic effects [42]. The clinical potential especially of A-FABP in diagnostics and prediction of metabolic syndrome should be continue to observe.

6. Adiponectin in members of families with familial combined hyperlipidemia

Familial combined hyperlipidemia (FCH) is the most common genetic hyperlipidemia which affects 1.0% to 2.0% of the population. The lipids and lipoprotein levels are, however, only moderately elevated and do not fully explain the increased risk of cardiovascular disease. The aim of the study of Karásek et al. [43] was to evaluate plasma levels of adiponectin in asymptomatic, nonsmoking members of families with FCH. We also investigated the association between adiponectin and selected risk factors of atherosclerosis and markers of insulin resistance and chronic inflammation. Furthermore, we investigated the relationship between adiponectin and the intima-media thickness of the CCA (IMT), a recognized morphologic marker of early atherosclerosis.

6.1. Subjects and methods

The study was carried out with 82 members of 29 FCH families. A family with FCH was defined by a proband exhibiting plasma cholesterol and triglycerides concentrations above 90th percentile, adjusted for age and sex, based on data from the Czech population. At least one first-degree relative of the proband should have plasma cholesterol and/or triglycerides above 90th percentile, adjusted for age and sex, or level of apo B more than 1.25 g/l. Secondary hyperlipidemia was excluded by additional testing. Other exclusion criteria were a history of clinically manifest atherosclerosis, heart failure, cerebrovascular ischemic disease, peripheral vascular disease, smoking, hypolipidemic therapy in the previous 8 weeks, hormone therapy with estrogens and acute infection or trauma. Members of FCH were divided into 2 groups: HL (hyperlipidemic members of FCH families, i.e. probands and their hyperlipidemic first-degree relatives) and NL (normolipidemic first-degree relatives). The control groups, C-HL and C-NL, were sex and age matched to groups. Control groups consisted of healthy individuals with a negative family history of hyperlipidemia and early manifestation of atherosclerosis. Nobody was treated for hypertension.

Laboratory parameters were analyzed by routine methods described above. Ultrasound scanning was performed with a 10 MHz linear array transducer (Hewlett-Packard, Image Point, M2410A). All measurements were performed with the subjects in a supine position. Three video records were made of common carotid artery (CCA). IMT measurements were processed off-line using software Image-Pro Plus (v. 4.0, Media-Cybernetics, Silver Spring). The average of the IMT of 3 frozen images of both sides was chosen as the outcome variable. Subjects with an atherosclerotic plaque in the evaluated region were not included in the study. The measurement of IMT was made without knowledge of laboratory results.

6.2. Results

In comparison to sex and age matched controls, HL subjects had significantly higher diastolic blood pressure (DBP), BMI, insulin resistance and elevated levels of C-peptide and proinsulin. They had higher IMT, hsCRP and ICAM 1 as well. By definition, the FCH subjects showed higher plasma cholesterol and triglycerides concentrations compared with controls and normolipidemic relatives. They had a more atherogenic lipid and lipoprotein profile as reflected by increased LDL cholesterol and apo B concentrations. Normolipidemic relatives had significantly higher DBP, TGL and proinsulin concentrations compared with their sex and age matched controls. There was no difference in other measured anthropometric and biochemical parameters.

Compared with healthy controls, HL subjects had lower levels of adiponectin (13.02 ± 4.58 mg/l vs. 16.19 ± 5.39 mg/l, $p < 0.05$). In the NL relatives, there was no significant differences in adiponectin (15.77 ± 2.95 mg/l vs. 16.53 ± 4.26 mg/l). In all FCH families, a significant negative correlation was found between adiponectin and TGL ($r = - 0.35$, $p < 0.01$), proinsulin ($r = - 0.26$, $p < 0.05$), hsCRP ($r = - 0.24$, $p < 0.05$), BMI ($r = - 0.27$, $p < 0.05$) and waist circumference ($r = -0.32$, $p < 0.01$). Levels of adiponectin did not correlate with IMT, in members of FCH families or in controls. By using regression model in HL subjects, levels of adiponectin were predicted by apo B ($p < 0.05$) and hsCRP ($p < 0.05$). (More detailed results are given in lit. [43].)

6.3. Discussion and conclusions

This study reported decreased adiponectin levels in asmyptomatic hyperlipidemic members of FCH families. There was no difference in serum adiponectin levels between their first-degree normolipidemic relatives and healthy controls. A negative correlation between adiponectin and markers of insulin resistance, chronic inflammation and visceral obesity was found in FCH families. The results were consistent with previous findings and support an insulin-sensitizing effect of adiponectin. In hyperlipidemic individuals, the levels of plasma adiponectin were predicted by apolipoprotein B and high sensitive CRP, independent of insuline resistance and visceral obesity. Authors conclude low adiponectin levels are associated with proinflammatory status and insulin resistance, and could partially explain the increased risk of coronary heart disease, even if the lipids and lipoprotein levels are only moderately elevated.

The study did not confirm any correlation between adiponectin levels and IMT, a marker of subclinical atherosclerosis, in FCH subjects. Publications regarding the relationship between these parameters are not entirely consistent. Similar results were observed in other work published by Karásek et al [44] where IMT proved to correlate with age, lipid parameters, markers of insulin resistance and that of visceral obesity and blood pressure. These parameters seem to be risk factors instead of adiponectin. The lack of correlation between adiponectin and IMT does not argue for adiponectin as an independent predictor for next cardiovascular events in clinically asymptomatic, dyslipidemic individuals.

7. Conclusion of chapter

Adiponectin is another promising parameter of the metabolic syndrome, atherosclerosis and associated syndromes. Its effect should be studied in many other situations. Nowadays, its determination in plasma provides valuable information, for example in patients with angiographically documented coronary artery disease, even if not all studies confirm this relationship. We can rely on the fact that its levels show no or very little circadian variability, its concentration is independent of fasting, it has low intraindividual variability, it is present in high concentrations in plasma and its levels can be influenced by diet, lifestyle or medication. Probably the most effective way to increase adiponectin levels in plasma and thus to reduce cardiovascular risk in obese individuals is a reduction in body weight. Beneficial effect of thiazolidinediones are also used to treat patients with type 2 diabetes to increase adiponectin production and plasma levels.

On the other hand, although adiponectin is associated with many of the traditional cardiovascular risk factors and further evidence has shown that hypoadiponectinemia is associated with atherosclerotic cardiovascular events such as myocardial infarction and brain infarction [45, 46], recent epidemiologic studies have shown contradictory results. Some of them revealed that hyperadiponectinemia rather than hypoadiponectinemia is associated with liver cirrhosis, rheumatoid arthritis, inflammatory bowel disease and systemic lupus erythematosus, all of which are conditions predisposed to wasting . Release of adiponectin from fat tissue is increased under conditions of malnutrition and plasma adiponectin concentration rises in the inflammatory state. Therefore, adiponectin can act as a mirror reflecting the degree of systemic wasting, and thus can predict death [47].

We can speculate about the real impact of high adiponectin levels on atherosclerosis: are they protective of harmful? In healthy subjects without clinically important signs of atherosclerosis, adiponectin has the protective effects especially due to its tissue-insulin senzitizing action. However, in individuals with advanced atherosclerosis and/or inflammatory disease, the positive association of adiponectin levels with markers of endothelial dysfunction/hemostasis (VCAM 1, but also with thrombomodulin and von Willebrand factor) could explain the increased total and cardiovascular mortality and the one associated with high adiponectin levels. It could be also the case of other populations, such as elderly people, patients with heart failure, patients with chronic kidney diseases, patients with type 1 diabetes mellitus etc. In recent studies, adiponectin effects should be evaluated in these populations.

Author details

Dalibor Novotný
Department of Clinical Biochemistry, Faculty Hospital Olomouc, Czech Republic

Helena Vaverková and David Karásek
3rd Medical Clinic, Faculty Hospital Olomouc, Czech Republic

Acknowledgement

This work was supported by the grants IGA MZ CR NR/9068-3 and IGA MZ CR NS/10284-3.

8. References

[1] Funahashi T, Nakamura T, Shimomura I, et al (1999) Role of adipocytokines on the pathogenesis of atherosclerosis in visceral obesity. Intern. Med. 38: 202-206.

[2] Shimada K, Miyazaki T, Hiroyuki D (2004) Adiponectin and atherosclerotic disease. Clin. Chim. Acta 344: 1-12.

[3] Yamamouchi T, Kamon J, Ito Y, et al (2003) Cloning of adiponectin receptors that mediate antidiabetic metabolic effects. Nature 423: 762-769.

[4] Ouchi N, Walsh K (2007) Adiponectin as an anti-inflammatory factor. Clin. Chim. Acta 380: 24-30.

[5] Choi KM, Ryu OH, Lee KW, et al (2007): Serum adiponectin, interleukin-10 levels and inflammatory markers in the metabolic syndrome. Diabetes Res. Clin. Pract. 75: 235-240.

[6] Okamoto Z, Kihara S, Ouchi N, et al (2002) Adiponectin reduces atherosclerosis in apolipoprotein E-deficient mouse. Circulation 106: 2767-2770.

[7] Peake PW, Shen Y, Walther A, Charlesworth JA (2008) Adiponectin binds C1q and activates the classical pathway of complement. Biochem. Biophys. Res. Commun. doi: 10.1016/j.bbrc.2007.12.161.

[8] Motoshima H, Wu X, Mahadev K, Goldstein BJ (2004) Adiponectin supresses proliferation and superoxide generation and enhances eNOS activity in endothelial cells treated with oxidized LDL. Biochem. Biophys. Res. Commun. 315: 264-271.

[9] Kern PA, Di Gregorio GB, Lu T, Rassouli N, Ranganathan G (2003) Adiponectin expression from human adipose tissue. Relation to obesity, insulin resistance and tumor necrosis alfa expression. Diabetes 52: 1779-1785.

[10] Liu YM, Lacorte JM, Viguerie N, et al (2003) Adiponectin gene expression in subcutaneous adipose tissue of obese women in response to short-term low calorie diet and refeeding. J. Clin. Endocrinol. Metab. 88: 5881-5886.

[11] Ryo M, Nakamura T, Kihara S, et al (2004) Adiponectin as a biomarker of the metabolic syndrome. Circ. J. 68: 975-981.

[12] Saely ChH, Risch L, Hoefle G, et al (2007) Low serum adiponectin is independently associated with both the metabolic syndrome and angiographically determined coronary atherosclerosis. Clin. Chim. Acta 383: 97-102.

[13] Hallikainen M, Kolehmainen M, Schwab U, et al (2007): Serum adipokines are associated with cholesterol metabolism in the metabolic syndrome. Clin. Chim. Acta 383: 126-132.

[14] von Eynatten M, Schneider JG, Humpert PM, et al (2006) Serum adiponectin levels are an independent predictor of the extent of coronary artery disease in men. J. Am. Coll. Cardiol. 47: 2124-2126.

[15] Ouchi N, Kihara S, Funahashi T, et al (2003) Reciprocial association of C-reactive protein with adiponectin in blood stream and adipose tissue. Circulation 107: 671-674.

[16] Nishida M, Moriyama T, Ishii K, et al (2007) Effects of IL-6, adiponectin, CRP and metabolic syndrome on subclinical atherosclerosis. Clin. Chim. Acta 384: 99-104.

[17] von Eynatten M, Hamann A, Twardella D, et al (2006) Relathionship of adiponectin with markers of systemic inflammation, atherogenic dyslipidemia, and heart failure in patients with coronary heart disease. Clin. Chem. 52: 853-859.

[18] Kondo H, Shinomura I, Matsukawa Y, et al (2002) Association of adiponectin mutation with type 2 diabetes. Diabetes 51: 2325-2328.

[19] Ohashi K, Ouchi N, Kihara S, et al (2004) Adiponectin I164T mutation is associated with the metabolic syndrome and coronary artery disease. J. Am. Coll. Cardiol. 43: 1195-2000.

[20] Hara K, Boutin P, Mori Y, et al (2002) Genetic variation in the gene encoding adiponectin is associated with an increased risk of type 2 diabetes in the Japanese population. Diabetes 51: 536-540 .

[21] Menzaghi C, Ercolino T, Di Paola R, et al (2002) A haplotype locus at the adiponectin locus is associated with obesity and other features of the insulin resistance syndrome. Diabetes 51: 2306-2312 .

[22] Filippi E, Sentinelli F, Trischitta V, et al (2003) Association of the human adiponectin gene and insulin resistance. E. J. Hum. Genet. doi:10.1038/sj.ejhg.5201120.

[23] Hu FB, Doria A, Li T, et al (2004) Genetic variation at the adiponectin locus and risk of type 2 diabetes in women. Diabetes 53: 209-213.

[24] Hegener HH, Lee IM, Cook NR, et al (2006) Association of adiponectin gene variations with risk of incident myocardial infarction and ischemic stroke: a nested case-control study. Clin. Chem. 52: 2021-2027.

[25] Sniderman AD (2004) Applying apo B to the diagnosis and therapy of the atherogenic dyslipoproteinemias: a clinical diagnostics algorithm. Curr. Opin. Lipidol. 15: 433-438.

[26] John SW, Weitzner G, Rozen R, Scriver CR (1991) A rapid procedure for extracting genomic DNA from leukocytes. Nucleic Acids Res. 9: 408.

[27] Novotny D, Vaverkova H, Karasek D, Halenka M (2005) Relationship between +276 G-T single nucleotide polymorphism (SNP) of adiponectin gene and markers of insulin resistance in dyslipidemic patients. 75th EAS Congress, Prague, Supplement of book of abstracts: 7.

[28] Vaverková H, Karásek D, Novotný D, et al (2008) Positive association of adiponectin with soluble vascular cell adhesion molecule sVCAM-1 levels in patients with vascular disease or dyslipidemia. Atherosclerosis 197(2): 725-731.

[29] Okamoto Z, Arita Z, Nishida M, et al (2000) An adipocyte-derived plasma protein, adiponectin, adheres to injured vascular walls. Horm. Metab. Res. 32: 47-50.

[30] Pischon T, Rimm EB (2006) Adiponectin: a promising marker for cardiovascular disease. Clin. Chem. 52: 797-799.

[31] Matsushita K, Yatsuya H, Tamakoshi K, et al (2006) Inverse association between adiponectin and C-reactive protein in substantially healthy Japanese men. Atherosclerosis 188: 184-189.

[32] Kazumi T, Kawaguchi A, Hirano T, Yoshino G (2004) Serum adiponectin is associated with high-density lipoprotein cholesterol, triglycerides, and low-density lipoprotein particle size in young healthy men. Metabolism 53: 589-593.

[33] Blankenberg S, Barbaux S, Tiret L (2003) Adhesion molecules and atherosclerosis. Atherosclerosis 170: 191-203.

[34] The AtheroGene Investigators (2001) Circulating cell adhesion molecules and death in patients with coronary artery disease. Circulation 104: 1336-1342.

[35] Guray U, Erbay AR, Guray Z, et al (2004) Levels of soluble adhesion molecules in various clinical presentations of coronary atherosclerosis. Int. J. Cardiol. 96: 235-240.

[36] Jager A, van Hinsbergh VWM, Kostense PJ, et al (2000) Increased levels of soluble vascular cell adhesion molecule 1 are associated with risk of cardiovascular mortality in type 2 diabetes, The Hoorn Study. Diabetes 49: 485-491.

[37] Spranger J, Kroke A, Mohling M, et al (2003) Adiponectin and protection against type 2 diabetes mellitus. Lancet 361: 226-228.

[38] Kharitonenkov A, Shiyanova TL, Koester A, et al (2005) FGF-21 as a novel metabolic regulator. J. Clin. Invest. 115(6): 1627-1635.

[39] Kharitonenkov A, Wroblewski V, Koester A, et al (2007) The metabolic state of diabetic monkeys is regulated by fibroblast growth factor 21. Endocrinology 148(2): 774-781.

[40] Zhang X, Yeung DCY, Karpisek M, et al (2008) Serum FGF21 levels are increased in obesity and are independently associated with metabolic syndrome in humans. Diabetes 57: 1246-1253.

[41] Stejskal D, Karpisek M (2006) Adipocyte fatty acid binding protein in a Caucasian population: a new marker of metabolic syndrome? Eur. J. Clin. Invest. 36(9): 621-625.

[42] Xu A, Wang Y, Xu JY, et al (2006) Adipocyte fatty acid binding protein is a plasma biomarker closely associated with obesity and metabolic syndrome. Clin.Chem. 52: 405-413.

[43] Novotný D, Karásek D, Vaverková H, Jackuliaková D, Malina P The relationship of adiponectin, fibroblast growth factor 21 and adipocyte fatty acid binding protein levels to dyslipidemic phenotypes- pilot study (2011). Clin. Chem. Lab. Med. 49: S394.

[44] Karásek D, Vaverková H, Halenka M, Jackuliakova D, Fryšák Z, Novotný D (2010) Adiponectin in members of families with familial combined hyperlipidemia. The Endocrinologist 20(3): 117-121.

[45] Karásek D, Vaverková H, Halenka M, Jackuliakova D, Frysak Z, Novotny D (2011) Total adiponectin levels in dyslipidemic individuals: relationship to metabolic parameters and intima-media thickness. Biomed. Pap. Med. Fac. Univ. Palacky. Olomouc 155(1): 55-62.

[46] Chen MP, Tsai JC, Chung FM, et al (2005) Hypoadiponectinemia is associated with ischemic cerebrovascular disease. Arterioscler Thromb Vasc Biol 25: 821-826.

[47] Wang WH, Yu WH, Dong XQ, et al (2011) Plasma adiponectin as an independent predictor of early death after acute intracerebral hemorrhagie. Clin. Chim. Acta 412: 1626-1631.

[48] Jernas M, Olsson B, Sjöholm K, et al (2009) Changes in adipose tissue gene expression and plasma levels of adipokines and acute-phase proteins in patients with critical illness. Metabolism 58:102-108.

Lipoproteins and Hemostasis

Chapter 16

Lipoprotein (a) – An Overview

Anna Gries

Additional information is available at the end of the chapter

1. Introduction

Lipoprotein (a) [Lp(a)], first described in 1963 is an inherited cholesterol-rich particle found in a density range of 1.055-1.120 g/ml. The suggestion that Lp(a) might be a risk factor for cardiovascular diseases was first made by Dahlen et al. [1] who found out that individuals with angina pectoris exhibit an "extra pre-β-band" in lipid electrophoresis. In whites the concentration of Lp(a) in plasma varies from undetectable up to 200 mg/dl in different individuals but seems to be rather constant in the same person [2]. Chemical and physicochemical properties of Lp(a) in comparison with LDL are summarized in table 1.

Plasma Lp(a) concentrations above 30 mg/dl, as measured in about 20 percent of white people, are associated with an approx. two-fold relative risk of coronary atherosclerosis [3] rising to the range of five-fold when LDL and Lp(a) are both elevated [4]. Intrestingly, blacks with high levels of Lp(a) do not experience greatly increased atherosclerotic progression and mortality. In those cases it is assumed that the atherogenicity of Lp(a) must be decreased or counterbalanced by other factors [5].

Till now the site and mechanism of Lp(a) synthesis are quite unclear. Measurements of serum Lp(a) levels of patients suffering from liver disease or from cholestasis who showed significantly lower concentrations than healthy controls gave indications that Lp(a) might be synthesized by the liver [6]. On the other hand there are studies which suggest that apo-a is associated with the postprandial d < 1.006 lipoproteins induced by fat feeding [7] but it is not yet clear however whether apo-a determined in this fraction is really of intestinal origin or whether it originates from free apo-a in serum which might bind to freshly secreted chylomicrons [8]. Because of the chemical similarities between Lp(a) and LDL it is possible that Lp(a) is formed during the metabolic catabolism of chylomicrons, VLDL or LDL. As Lp(a) levels stay nearly constant within one individual and as lipid-rich diet as well as fasting have no influence on Lp(a) concentrations it is assumed that Lp(a) exhibits a metabolic behaviour completely different from other apo-B containing lipoproteins. Turnover studies in vivo performed with labelled VLDL confirmed these assumptions. Nearly all the activity of labelled VLDL could be detected in LDL whereas only trace amounts could be found in Lp(a) [9] confirming the hypothesis that unlike LDL, Lp(a) probably has no triglyceride-rich lipoproteins as precursors but seems to be secreted directly by the liver [10]. On the other hand the site of catabolism of Lp(a) in humans is unknown so far although the kidney is favourized to be implicated [11].

Despite extensive work on Lp(a) its possible physiological function remains unclear till now.

	LDL	Lp(a)
Hydr. Density [g/ml]	1.034	1.085
Mol. Wt. [x 10^6]	2.4	5.5
Diameter [Å]	210	250
E. Mobility	β	pre-β1
Chem. Composition [%]		
Free cholesterol	11	10
Cholesterolester	40	30
Triglycerides	4	4
Phospholipids	21	20
Protein	22	28
Carbohydrates	2	8

Table 1. Chemical and physicochemical properties of LDL and Lp(a)

2. Structural arrangement and catabolism of Lp(a)

The major protein component of this LDL-like particle is apolipoprotein B (apo-B-100) which carries an additional protein called apolipoprotein-a [apo-a] linked to apo-B-100 via disulphide bridges (Fig.1) the lipid moiety however being almost indistinguishable from that of LDL [12]. Human apo-a itself consists of multiple so-called kringle repeats, sequences consisting of 80-90 amino acids arranged in a tripleloop tertiary structure and tandemly arrayed resembling kringles IV and V of plasminogen and a protease domain [13]. Copy number variants in the LPA gene on chromosome 6 coding for apo-a are responsible for a variation of plasma Lp(a) levels of up to 1000-fold among individuals. The most influential is the kringle IV-2 size polymorphism [14] while kringle IV types 1 and 3-10 as well as kringle V occur only once in Lp(a) [15]. The number of kringle IV type 2 structure repeats results in a large number of different sized isoforms of apo-a and correlates inversely with the plasma concentration of Lp(a) [16]. Although the exact mechanism responsible for this inverse correlation has not been elucidated so far an isoform dependent retention and degradation in the endoplasmatic reticulum has been implicated [17].

Contradictory results have been reported about the clearance of Lp(a) and till now it remains unclear whether Lp(a) binds to the B/E receptor via apo-B like LDL or whether it is catabolised independently of the LDL-receptor mediated pathway. Whereas in one study using fibroblasts from normal subjects and from subjects with autosomal dominant hypercholesterolemia the conclusion was reached that Lp(a) enters fibroblasts independently of the LDL-receptor [18] others concluded that Lp(a) is also bound to the LDL-receptor, internalized and degraded but with a degradation capacity of only 25% of that of LDL [19]. Binding studies of native and reduced Lp(a) with different monoclonal antibodies against apolipoprotein B revealed that there was no antibody that failed to react with native Lp(a) but some of the antibodies recognized apoB of Lp(a) to a lesser degree than that of LDL. This favoured the idea that certain regions on apo-B of Lp(a) could be different from those on LDL and led to the assumption that certain domains close to the binding domain of Lp(a) to the B/E-receptor could be covered by apo-a or that apo-a causes conformational changes in the binding region of apo-B thereby constricting the binding of Lp(a) to the LDL-receptor [20] being in agreement with the fact that normal unreduced Lp(a) seemed to be taken up by fibroblasts through B/E-receptor-mediated endocytosis but showed poorer specificity for the receptor than LDL [21].

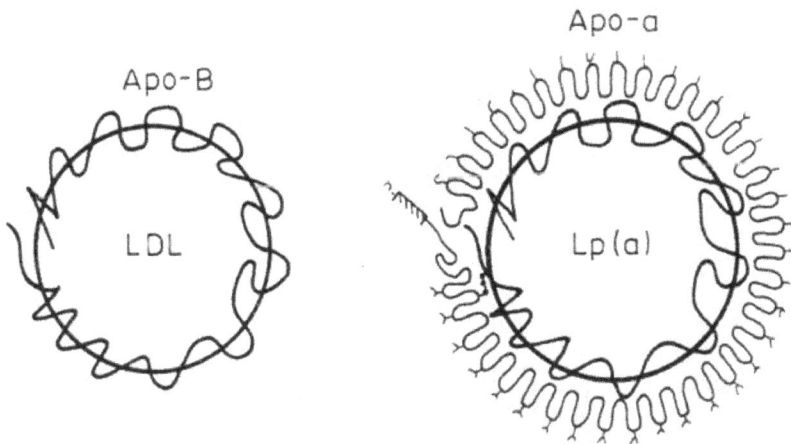

Figure 1. Schematic model of Lp(a) in comparison to LDL

3. Free apolipoprotein-a in human serum

In the beginning of "Lp(a)-research" this lipoprotein was believed to represent a genetically polymorphic form of LDL [22]. According to this assumption apo-a should distribute uniformly between all apo-B-containing lipoproteins. Investigation of this problem in more detail revealed that Lp(a) forms a particular lipoprotein class found primarily in the HDL-2 density region [23] but can also be detected in LDL class (d = 1.019-1.063 g/ml) [24] and even in chylomicrons induced by fat feeding [25]. The fact however that a portion of the Lp(a)-specific antigen can be found in the d > 1.21 g/ml lipoprotein free bottom fraction after ultracentrifugation of plasma [26] led to further investigation of this phenomenon. Apo-a is virtually absent in the VLDL fraction (d < 1.006 g/ml) of freshly drawn fasting human sera while 95% of total Lp(a) can be obtained in the d > 1.006 g/ml bottom fraction. Approximately 5% of the total serum Lp(a) are found in the d > 1.125 g/ml bottom fraction after ultracentrifugation as well as with polyanionic precipitation agents irrespective of the Lp(a) concentration in serum [8]. Due to the lack of Sudan Black staining this bottom Lp(a) is considered as a lipid free "apo-a protein" raising the question whether or not free apo-a can reassociate with LDL to form "native Lp(a)".

4. Lp(a) and platelet aggregation

One of the physiological roles of platelets involves binding to subendothelial tissue after vascular injury [27]. The adherence of platelets to the exposed connective tissue, preferably collagen, leads to aggregation followed by the release of ADP, 5-hydroxytryptamine and Ca^{2+} from their dense granules, causing passing platelets to adhere to the primary clot [28].

There is little doubt that lipoproteins interfere with platelets in vivo being reflected by the fact that platelets from hyperlipoproteinemic patients are hyperreactive [29]. This is confirmed by the fact that incubation of platelets with physiological concentrations of atherogenic apoB-containing lipoproteins such as LDL or VLDL results in enhanced platelet aggregability [30] while antiatherogenic lipoproteins such as HDL exert the opposite effect [31]. Concerning Lp(a) it is generally accepted that elevated plasma concentrations of this lipoprotein are connected with premature atherosclerosis [32] but much uncertainty remains about the influence of Lp(a) on platelet activation, a phenomenon that is believed to be involved not only in long-term processes of plaque formation but also in acute events such as stroke and myocardial infarction [33]. Moreover a two-fold increase in the risk of coronary heart disease (CHD) and ischaemic stroke could be demonstrated especially in subjects with small apolipoprotein(a) phenotypes [34] and prospective findings in the Bruneck study have revealed a significant association specifically between small apolipoprotein(a) phenotypes and advanced atherosclerotic disease involving a component of plaque thrombosis [35]. Indeed, Lp(a) is a "sticky" lipoprotein that self-aggregates, attaches to all sorts of surfaces [36], and precipitates not only in vitro but possibly in vivo. Moreover, Lp(a) binds to proteoglycans and glycosaminoglycans [37] and it has high affinity for fibronectin [38], tetranectin [39], collagen [40], and other connective-tissue structures [41]. Therefore the influence of Lp(a) on platelet aggregation induced with various triggers was investigated measuring serotonin release and thromboxane A_2 formation during collagen-triggered aggregation as well as adhesion of platelets to collagen in flowing blood under the influence of Lp(a). As Lp(a) represents an LDL-like particle an elevated platelet reactivity was expected under the influence of this lipoprotein similar to that described for LDL [42].

Unlike LDL, Lp(a) revealed no proaggregatory effects on platelets, contrary collagen-induced platelet aggregation was inhibited by up to 54% and the aggregation rate was attenuated by 47% compared with platelets incubated with Tyrode's solution (Fig. 2), being accompanied by a significant reduction of serotonin release and TXA$_2$ formation. Furthermore Lp(a) significantly reduced platelet adhesion to collagen by about 20% and the size of platelet aggregates up to 63% especially at high shear rates (Fig. 3) suggesting that Lp(a) exerts antiaggregatory effects at least under well-defined in vitro conditions [43]. If these observations are relevant for the in vivo situation, a variety of potential platelet-collagen binding sites such as GPIa/IIa or GPIV could be covered by Lp(a) the more that binding of Lp(a) to platelets could be demonstrated [44]. As there is conflicting evidence on the role of Lp(a) in thrombosis in vivo and in vitro work has been done to elucidate the mechanisms whereby Lp(a) is influencing platelet aggregation and a variety of mechanisms is suggested how Lp(a) interferes with platelet aggregation and hence fibrin bound clot formation. Lp(a) binds to resting, non-stimulated platelets on the IIb subunit of the fibrinogen (IIb/IIIa) receptor via binding sites distinct from the arginyl-glycyl-aspartyl (RGD) epitope of apo-a [45]. By this way the RGD binding site of Lp(a) could be exposed via conformational change induced by platelet agonist stimulation leading to binding of the RGD epitope of apo-a to the RGD binding site on the IIb protein of the fibrinogen (IIb/IIIa) receptor of the platelet [46] thereby reducing fibrinogen binding to the platelet [47]. Low concentrations of Lp(a) (1-25 mg/100 ml washed platelets) increase intracellular levels of c-AMP of in vitro resting platelets leading to an antiaggregatory condition [48] while at higher in vitro levels of Lp(a) (50-100 mg/100 ml washed platelets) resting platelet intracellular c-AMP levels return to normal [49] which cannot explain the reported progressive Lp(a)-mediated decrease in collagen-induced aggregation [43, 50]. Further investigations strongly support an apo-a mediated, Lp(a) induced reduction of collagen and ADP-stimulated platelet aggregation via diminished production of thromboxane A$_2$ [43, 51]. Concerning the in vivo situation only one study has been published to date looking at adult human type 2 diabetics all of whom where obese (BMI >30). In this in vivo study of human type 2 diabetics there is a positive correlation between fasting serum concentrations of Lp(a) and bleeding time, a strong correlate of in vivo platelet aggregation [52] favouring the inhibitory effect of Lp(a) on platelet aggregation. On the other hand there are studies reporting an apparent proaggregatory action of Lp(a) possibly mediated by the apo-a subunit. While no effect of recombinant apo-a [r-apo-a] derivatives on primary ADP-induced platelet aggregation was observed weak platelet responses stimulated by the thrombin receptor-activating peptide SFLLRN were significantly enhanced by the r-apo-a derivatives accompanied by a significant enhancement of [^{14}C]serotonin release of the dense granules [53]. Further investigations showed that r-apo-a isoforms and Lp(a) do not cause platelet aggregation by themselves but preincubation of platelets with r-apo-a derivatives promotes an aggregation response to otherwise subaggregant doses of thrombin receptor activation peptide (TRAP) and arachidonic acid while inversely platelet stimulation with arachidonic acid enhanced platelet binding of apo-a [54]. In both studies it turned out that the size of r-apo-a determined by the number of KIV type 2 modules seems not to play a crucial role in its proaggregant effect.

Summarizing, in vitro studies indicate that Lp(a) induced decreases, increases or no change at all in platelet aggregation [43, 45, 50, 51, 53, 54]. In all cases the mechanisms involved are quite unclear and only speculative. A recent work strongly supports the evidence to suggest that Lp(a) binds to platelets via its arginyl-glycyl-aspartyl (RGD) epitope of the apo(a) but not via apo(a)'s lysine binding region in both strong and weak agonist-stimulated platelets and inhibits the binding of fibrinogen thus reducing aggregation [55]. On the other side there are in vivo studies published quite recently suggesting that Lp(a) concentrations greater than 30 mg/dl are a frequent and independent risk factor for venous thrombosis [56] and that high levels of Lp(a) could be a more frequently thrombophilic risk factor in young women [57]. To date disagreement exists about increased arterial thrombosis due to elevated blood levels of Lp(a). The fact that this procedure is a result of collagen-exposed platelets in case of plaque rupture followed by clot formation argues against the proaggregatory nature of Lp(a) and maybe procedures others than platelet activation account for the atherogenic nature of Lp(a).

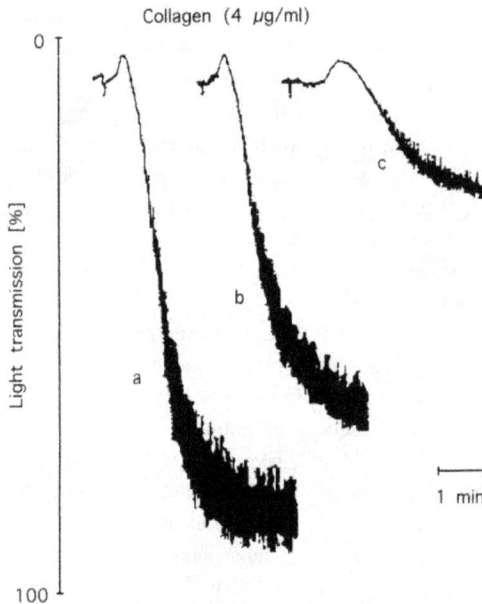

Figure 2. Aggregation curves showing the influence of lipoproteins on collagen-induced platelet aggregation. Gel filtered platelets (200 µl; 2x108/ml) were incubated for 30 min at 37°C with a) LDL 5 mg/ml, b) Tyrodes's buffer or c) Lp(a) 0.5 mg/ml. Aggregation was triggered with 10 µl collagen (final concentration 4 µg/ml).

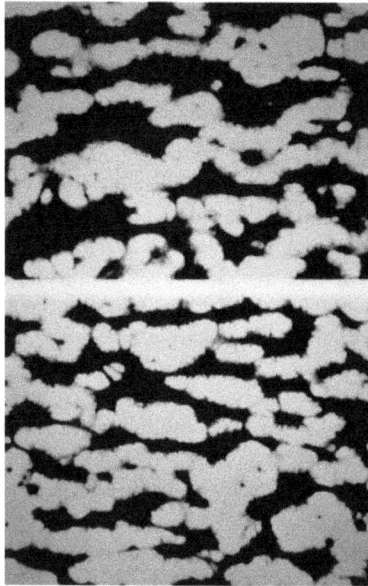

Figure 3. Aggregate formation of fibrillar collagen at a shear rate of 1600/s for a control (top) and under addition of 1 mg/ml Lp(a) (bottom). Aggregates are shown in black.

5. Lp(a) and plasminogen

The mechanism by which Lp(a) accelerates atherosclerosis could not yet been clarified. One possible explanation leads via the connection of Lp(a) to the fibrinolytic system as in 1987 it was found out that Lp(a) and plasminogen are immunochemically related [58] leading to speculations whether Lp(a) might interfere with fibrinolysis. Through partial amino acid sequencing it could be shown for the first time that apo-a has a striking homology of about 70% to plasminogen, the precursor of the proteolytic enzyme plasmin which dissolves fibrin clots [58]. This could be confirmed in our own studies demonstrating that polyclonal antisera from rabbit, sheep and horse as well as three monoclonal antibodies from mouse raised against apo-a reacted with plasminogen on immunoblots and similar to plasminogen, Lp(a) bound selectively but with somewhat lower affinity to lysine-Sepharose [59]. Plasminogen, a protein of 791 amino acids and a molecular weight of about 92 000 D is a plasma serine protease zymogen that consists of five cysteine-rich sequences of 80-114 amino acids each, called kringles, followed by a trypsin like protease domain [60]. The highly glycosylated apo-a exists in various polymorphic forms with molecular weights higher, lower or equal to apoprotein B ($M_r \approx 550\ 000$ D) [61] which are covalently linked to apoprotein B via disulfide bridges [62]. It contains a hydrophobic signal sequence for secretion followed by up to >50 copies of kringle IV of plasminogen predicting the risk for coronary heart disease in the way that apo-a alleles with a low kringle IV copy number (<22) and high Lp(a) plasma concentration are significantly more frequent in the CHD group (p<0.001) [63]. Additionally one kringle V as well as protease domains of plasminogen are found in apo-a [58]. Later on cDNA sequencing revealed that human apolipoprotein(a) is homologous to plasminogen but despite the fact that apo-a contains a protease domain it does not act fibrinolytically like plasminogen because the arginine at the cleavage site for tissue plasminogen activator in plasminogen is changed to serine in apo-a [64].

Nevertheless Lp(a) might interfere with the fibrinolytic system in different ways due to its similarity to plasminogen as it may inhibit the binding of plasminogen to its receptor on endothelial cells thereby preventing generation of plasmin and increasing the thrombotic risk [65, 66]. Furthermore it could be demonstrated that Lp(a) accumulates in atherosclerotic lesions maybe via adherence to fibrinogen or fibrin incorporated in atherosclerotic plaques thereby inhibiting fibrinolysis [66]. Another mechanism by which Lp(a) is thought to attenuate fibrinolysis involves direct competition with plasminogen for fibrinogen or fibrin binding sites thus reducing the efficiency of plasminogen activation [67]. Fibrinolysis is initiated by binding of plasminogen to lysine residues on fibrin thereby initiating activation of plasmin and amplifying fibrinolytic processes [68]. Like plasminogen Lp(a) also binds to lysine residues [69] but without catalytical activity leading to interference with or inhibition of fibrinolysis resulting in hypofibrinolysis and accumulation of cholesterol included in the LDL-like component of Lp(a) [66]. The fact that low molecular weight isoforms of apo-a are associated with greater inhibition of fibrinolysis [70, 71] confirms the hypothesis that subjects with small apo-a phenotypes have a two-fold risk of CHD and stroke compared with those with larger isoforms of apo-a [34]. In contrast Knapp et al. [72] observed that the rate of plasmin formation was inversely related to Lp(a) but inhibition of plasmin generation increased with the size of apo-a using a standardized in vitro fibrinolysis model. From the fact that the inhibitory effect of free apo-a was much stronger than that of the complete Lp(a) particle they conclude that the apo-a component is responsible for the observed reduction of plasmin formation maybe due to the availability of additional lysine binding sites in the unbound apo-a which was formerly reported by Scanu et al. [73]. On the other hand there are also data showing that the plasma concentration of Lp(a) is inversely related to plasmin formation but that this relationship is not influenced by the size of apo-a isoforms [74]. Above all there are other reports explaining the inhibitory effect of Lp(a) on fibrinolysis not only by competition of Lp(a) with plasminogen for the binding sites on fibrin, endothelial cells and monocytes but also by reduction of tissue plasminogen activator or streptokinase-induced fibrinolytic activity [75, 76, 77].

A novel contribution to the understanding of Lp(a)/apo-a-mediated inhibition of plasminogen activation comes from results showing the ability of the apo(a) component of Lp(a) to inhibit the key positive feedback step of plasmin-mediated conversion of Glu-plasminogen to Lys-plasminogen an essential step for fibrin clot lysis [78]. Interestingly, with the exception of the smallest naturally-occurring isoform of apo(a), isoform size was found not to contribute to the inhibitory capacity of apo(a).

In summary, the proposed mechanisms modulating the antifibrinolytic effects of elevated Lp(a) levels in vitro are manifold and emphasize the prothrombotic effects of this lipoprotein particle. The in vivo situation however seems to be much more complex the more that there is a strong positive correlation reported between bleeding time and fasting serum concentrations of Lp(a) [52].

6. Lp(a) and lipid lowering drugs

High levels of Lp(a) are strongly associated with atherosclerosis as revealed by numerous studies [4, 79, 80, 81, 82]. As plasma Lp(a) concentrations above 30 mg/dl, as measured in about 20 percent of white people, are associated with an approx. two-fold relative risk of coronary atherosclerosis [3] rising to the range of five-fold when LDL and Lp(a) are both elevated [4] reduction of plasma Lp(a) concentration is recommended. Dietary interventions do not seem to be effective in lowering Lp(a) plasma levels [9, 83] or even lead to an increase of Lp(a) in plasma, alone [84] or at least when combined with exercise 85]. The same phenomenon could be observed in case of exercise where cross-sectional data suggest that a lifestyle of moderate to intense exercise training does not exert a significant impact on the Lp(a) level [86, 87]. Therefore pharmacological reduction of plasma levels of Lp(a) would be desirable.

Innumerable investigations however indicate that the plasma concentration of Lp(a) is resistant to drug therapy in most cases. As Lp(a) resembles LDL especially with regard to the lipid content (Tab.1) medications reducing LDL-cholesterol should be suitable for lowering Lp(a) as well. Bile acid resins such as cholestyramine which actually cause a significant reduction of LDL-cholesterol as well as of apo-B have no effect on Lp(a) levels [88, 89]. Therapies with bezafibrate or clofibrate [90, 91] showed that there is no role for fibrates in the treatment of elevated Lp(a) concentrations and estrogens also do not seem to significantly affect Lp(a) [92, 93].

Stanozolol, an anabolic steroid used in the treatment of postmenopausal osteoporosis, showed a significant reduction of Lp(a) by about 65% after six weeks therapy but five weeks after the drug was discontinued Lp(a) was near pretreatment levels [94]. Although drastic reductions of Lp(a) up to 40-50% are reported in another study [95] these compounds seem to be unsuitable for the routine treatment due to their harmful side effects [96].

Statins, also known as HMG-CoA-reductase inhibitors are another group of lipid lowering drugs which could be interesting with regard to Lp(a). These drugs have proven to be extremely effective in lowering plasma LDL and apo-B levels presumably through inhibition of intracellular cholesterol synthesis concomitant with an increase of the LDL receptors in the liver [97]. Although Lp(a) and LDL are very similar especially concerning the content of cholesterol, inhibitors of HMG-CoA-reductase, the regulating enzyme of cholesterol biosynthesis, show no influence [98, 99], only modest reduction of about 10% [100, 101] or even an increase of serum Lp(a) levels [102]. Altogether the limited magnitude of decrease of Lp(a) by HMG-CoA-reductase inhibitors confirms the assumption that the LDL-receptor does not seem to play a major role in Lp(a) clearance from plasma [103].

Nicotinic acid, also known as niacin has been shown to lower not only plasma total cholesterol, LDL-cholesterol and triglycerides thereby increasing HDL-cholesterol [104] but also Lp(a) in a dose-dependent manner up to 40% [105]. A more pronounced effect could be observed in a combination therapy with niacin and neomycin showing a reduction of LDL-cholesterol by 48% and of Lp(a) by 45% respectively [106]. In a recently published study niacin was applied in combination with omega-3-fatty acids and the Mediterranean diet. The average reduction of Lp(a) after 12 weeks combination therapy was reported to be about 23%. Additionally a significant association with increasing baseline levels of Lp(a) was observed [107].

Diets rich in fish oils containing considerable amounts of omega-3 polyunsaturated fatty acids are recommended to have beneficial effects on plasma lipids thereby lowering the risk of vascular complications [108, 109]. In a study investigating the influence of dietary fish oils on plasma Lp(a) levels a decrease of triglycerides could be observed after six weeks dietary supplementation while total cholesterol, LDL- and HDL-cholesterol as well as Lp(a) remained unchanged [110]. Furthermore collagen- and thrombin-stimulated platelet aggregation and TXB_2-formation in platelets decreased by approx. 45% irrespective of the plasma concentration of Lp(a) [111]. This is in agreement with many other studies showing that fish oils only seem to be able to reduce Lp(a) in combination with other therapies [107] or moderate exercise [112] but not when used alone [113, 114, 115].

Summarizing it can be shown that increased Lp(a) levels are minimally if at all influenced by drug treatment or drugs reducing Lp(a) to a greater extent like nicotinic acid are not widely used due to undesirable side effects. From previous turnover studies it could be demonstrated that plasma Lp(a) levels correlate with its rate of biosynthesis rather than with the fractional catabolic rate [116, 117] and therefore attempts to reduce Lp(a) should focus on an interference with apo-a biosynthesis. This is supported by the fact that adenovirus-mediated apo-a-antisense RNA expression efficiently inhibits apo-a synthesis in vitro in stably transfected liver cells but also in vivo in transduced mice expressing recombinant human apo-a [118]. In a recently published study it was found that patients suffering from biliary obstructions have very low plasma Lp(a) levels that rise substantially after surgical intervention. Consistent with this, common bile duct ligation in mice transgenic for human apo-a lowered plasma concentrations and hepatic expression of apo-a. Treatment of transgenic mice with cholic acid led to farnesoid X receptor (FXR) activation followed by markedly reduced plasma concentrations and hepatic expression of human apo-a [119]. From that it is concluded that transcription of the apo-a gene is under strong control of the farnesoid X receptor which may have important implications in the development of Lp(a)-lowering medications.

7. Conclusion

High levels of Lp(a) are strongly associated with atherosclerosis. About 10-15% of the white population exhibit plasma Lp(a) concentrations above the atherogenic cut-off value of approx. 30 mg/dl. Therefore the European Atherosclerosis Society recommended screening for Lp(a) in a consensus report, in which the desirable cut-off was set at less than 50 mg/dl [82]. On the other hand it is very well known that Lp(a) is an inherited atherogenic plasma component determined to more than 90% by genetic factors a fact that aggravates the influence on plasma levels of this lipoprotein. So far there are only speculations about the mechanism by which Lp(a) accelerates atherosclerosis and the exact mechanism could not yet be clarified. Its prothrombotic effects may be ascribed to impaired fibrinolysis by inhibition of plasminogen activation rather than to amplification of platelet aggregation which is shown to be reduced by Lp(a) in most cases. At present dietary interventions or drug therapies seem to be only minimal if at all successful concerning reduction of plasma Lp(a). Up to now it was assumed that the atherogenicity of high Lp(a) levels in blacks must be decreased by other factors [5]. However data published recently show that associations between Lp(a) levels and cardiovascular disease are at least as strong in blacks compared with whites [120] and emphasize the recommendation that factors such as total cholesterol, LDL-cholesterol, smoking, diabetes mellitus or overweight that can still increase the atherosclerotic risk of Lp(a) should be kept under observation.

Author details

Anna Gries
Institute of Physiology, Medical University of Graz, Austria

8. References

[1] Dahlen G, Ericson C, Furberg C, Lundqvist K, Svärdsudd K (1972) Studies on an extra pre-beta lipoprotein fraction. Acta Med. Scand. Suppl. 531: 1-29.

[2] Albers JJ, Cagana VG, Warnick GR, Hazzard WR (1975) Lp(a) lipoprotein-relationship to sinking pre-beta lipoprotein, Hyperlipoproteinemia and apolipoprotein B. Metabolism 24: 1047-1052.

[3] Armstrong VW, Cremer P, Eberle E (1986) The association between serum Lp(a) concentrations and angiographically assessed coronary atherosclerosis. Atherosclerosis 62: 249-257.

[4] Kostner GM, Avogaro P, Cazzolato G, Marth E, Bittolo-Bon G, Quinci GB (1981) Lipoprotein Lp(a) and the risk for myocardial infarction. Atherosclerosis 38: 51-61.

[5] Guyton JR, Dahlen GH, Patsch W, Kautz JA, Gotto AM (1985) Relationship of plasma lipoprotein Lp(a) levels to race and to apolipoprotein B. Arteriosclerosis 5: 265-272.

[6] Kostner GM (1976) Lp(a) lipoproteins and the genetic polymorphisms of lipoprotein B. From: Low Density Lipoproteins, eds. Day CE, Levy RS. Plenum Press, New York 229-269 p.

[7] Bersot TP, Innerarity TL, Pitas RE, Rall jr. SC, Weisgraber KH, Mahley RW (1986) Fat feeding in humans induces lipoproteins of density less than 1.006 that are enriched in apolipoprotein(a) and that cause lipid accumulation in macrophages. J. Clin. Invest. 77: 622-630.

[8] Gries A, Nimpf J, Nimpf M, Wurm H, Kostner GM (1987) Free and apo-B associated Lp(a)-specific protein in human serum. Clin. Chim. Acta 164: 93-100.

[9] Krempler F, Kostner G, Bolzano K, Sandhofer F (1978) Studies on the metabolism of the lipoprotein Lp(a) in man. Atherosclerosis 30: 57-65.

[10] Krempler F, Kostner G, Bolzano K, Sandhofer F (1979) Lipoprotein(a) is not a metabolic product of other lipoproteins containing apolipoprotein B. Biochim. Biophys. Acta 575: 63-70.

[11] Kronenberg F, Trenkwalder E, Lingenhel A, Friedrich G, Lhotta K, Schober M, Moes N, König P, Utermann G, Dieplinger H (1997) Renovascular arteriovenous differences in Lp[a] plasma concentrations suggest removal of Lp[a] from the renal circulation. J. Lipid Res. 38: 1755-1763

[12] Fless GM, ZumMallen ME, Scanu AM (1985) Isolation of apolipoprotein(a) from lipoprotein(a). J. Lipid Res. 26: 1224-1229.

[13] McLean JW, Tomlinson JE, Kuang WJ, Eaton DL, Chen EY, Fless GM, Scanu AM, Lawn RM (1987) cDNA sequence of human apolipoprotein(a) is homologous to plasminogen. Nature 330: 132-137.

[14] Kraft HG, Kochl S, Menzel H, Sandhofer C, Utermann G (1992) The apolipoprotein(a) gene: a transcribed hypervariable locus controlling plasma lipoprotein(a) concentration. Hum. Genet. 90: 220-230.

[15] Haibach C, Kraft HG, Kochl S, Abe A, Utermann G (1998) The number of kringle IV repeats 3-10 is invariable in the human apo(a) gene. Gene 208: 253-258.

[16] Lackner C, Cohen JC, Hobbs HH (1993) Molecular definition of the extreme size polymorphism in apolipoprotein(a). Hum. Mol.Genet. 2: 933-940.

[17] Brunner C, Lobentanz EM, Pethö-Schramm A, Ernst A, Kang C, Dieplinger H, Müller HJ, Utermann G (1996) The number of identical kringle IV repeats in apolipoprotein(a) affects its processing and secretion by HepG2 cells. J. Biol. Chem. 271: 32403-32410.

[18] Maartman-Moe K, Berg K (1981) Lp(a) lipoprotein enters cultured fibroblasts independently of the plasma membrane low density lipoprotein receptor. Clin. Genet. 20: 352-362.

[19] Armstrong VW, Walli AK, Seidel D (1985) Isolation, characterization and uptake in human fibroblasts of an apo(a)-free lipoprotein obtained on reduction of lipoprotein(a). J. Lipid Res. 26: 1314-1323.

[20] Gries A, Fievet C, Marcovina S, Nimpf J, Wurm H, Mezdour H, Fruchart JC, Kostner GM (1988) Interaction of LDL, Lp[a], and reduced Lp[a] with monoclonal antibodies against apoB. J. Lipid Res. 29: 1-8.

[21] Krempler F, Kostner GM, Roscher A, Haslauer F, Bolzano K, Sandhofer F (1983) Studies on the role of specific cell surface receptors in the removal of lipoprotein(a) in man. J. Clin. Invest. 71: 1431-1441.

[22] Berg K (1963) A new serum type system in man – the Lp system. Acta Pathol. Microbiol. Scand. 59: 369-382.

[23] Harvie NR, Schultz JS (1970) Studies of Lpa lipoprotein as a quantitative genetic trait. Proc. Natl. Acad. Sci. USA 66: 99-103.

[24] Fless GM, Rolih CA, Scanu AM (1984) Heterogeneity of human plasma lipoprotein a. J. Biol. Chem. 259: 11470-11478.

[25] Bersot TB, Innerarity TL, Mahley RW (1984) Fat feeding in humans induces lipoproteins of density less than 1.006 that are enriched in apolipoprotein a and that cause lipid accumulation in macrophages. Arteriosclerosis 4: 536a.

[26] Parra MG (1976) Isolation of human serum lipoproteins by precipitation and column chromatography methods. Thesis at the University of Marburg/Lahn. Marburg; Mauersperger Press 48-55 p.

[27] Rossi EC (1972) The function of platelets in hemostasis. Med. Clin. North. Am. 56: 25-38.

[28] Jaffe R, Dykin D (1974) Evidence for a structural requirement for the aggregation of platelets by collagen. J. Clin. Invest. 53: 875-883.

[29] Bruckdorfer KR (1989) The effect of plasma lipoproteins on platelet responsiveness and on platelet and vascular prostanoid synthesis. Prostaglandins Leukot. Essent. Fatty Acids 38: 247-254.

[30] Aviram M, Brook JG (1987) Platelet activation by plasma lipoproteins. Prog. Cardiovasc. Dis. 30: 61-72.

[31] Aviram M, Brook JG (1983) Platelet interaction with high- and low-density lipoproteins. Atherosclerosis 46: 259-268.

[32] Cushing GL, Gaubatz JW, Nava ML, Burdick BJ, Bocan TMA, Guyton JR, Weilbaecher D, DeBakey ME, Lawrie GM, Morrisett JD (1989) Quantitation and localization of apolipoprotein(a) and B in coronary artery bypass vein grafts resected at reoperation. Arteriosclerosis 9: 593-603.

[33] Ross R (1993) The pathogenesis of atherosclerosis: a perspective for the 1990s. Nature 362: 801-809.

[34] Erqou S, Thompson A, Di AE, Saleheen D, Kaptoge S, Marcovina S, Danesh J (2010) Apolipoprotein(a) isoforms and the risk of vascular disease: a systemic review of 40 studies involving 58,000 participants. J. Am. Coll. Cardiol. 55: 2160-2167.

[35] Kronenberg F, Kronenberg MF, Kiechl S, Trenkwalder E, Santer P, Oberhollenzer F, Egger G, Utermann G, Willeit J (1999) Role of lipoprotein(a) and apolipoprotein(a) phenotype in atherogenesis: prospective results from the Bruneck study. Circulation 100: 1154-1160.

[36] Zioncheck TF, Powell LM, Rice GC, Eaton DL, Lawn RM (1991) Interaction of recombinant apolipoprotein(a) and lipoprotein(a) with macrophages. J. Clin. Invest. 87: 767-771.

[37] Bihari-Varga M, Gruber E, Rotheneder M, Zechner R, Kostner GM (1988) Interaction of lipoprotein(a) and low-density lipoprotein with glycosaminoglycans from human aorta. Arteriosclerosis 8: 851-857.

[38] Salonen E, Jauhiainen M, Zardi L, Vaheri A, Ehnholm C (1991) Lipoprotein(a) binds to fibronectin and has serine protease activity capable of cleaving it. EMBO J. 8: 4035-4040.

[39] Kluft C, Jie AFH, Los P, DeWit E, Havekes L (1989) Functional analogy between lipoprotein(a) and plasminogen in the binding to the kringle 4 binding protein tetranectin. Biochem. Biophys. Res. Comm. 161: 427-433.

[40] McConathy WJ, Trieu VN (1991) Lp(a) interactions. Prog. Lipid Res. 30: 195-203.

[41] Kostner GM, Grillhofer H (1991) Lipoprotein(a) mediates high affinity low density lipoprotein association to receptor negative fibroblasts. J. Biol. Chem. 266: 21287-21292.

[42] Surya II, Akkerman JWN (1993) The influence of lipoproteins on blood platelets. Am. Heart J. 125: 272-274.

[43] Gries A, Gries M, Wurm H, Kenner T, Ijsseldijk M, Sixma JJ, Kostner GM (1996) Lipoprotein(a) inhibits collagen-induced aggregation of thrombocytes. Arterioscler. Thromb. Vasc. Biol. 16: 648-655.

[44] Ezratty A, Simon DI, Loscalzo J (1993) Lipoprotein(a) binds to human platelets and attenuates plasminogen binding and activation. Biochemistry 32: 4628-4633.

[45] Malle E, Ibovnik A, Steinmetz G, Kostner GM, Sattler W (1994) Identification of glycoprotein IIb as the lipoprotein(a)-binding protein on platelets: lipoprotein(a) binding is independent of an arginyl-glycyl-aspartate tripeptide located in apolipoprotein(a). Arterioscler. Thromb. 14: 345-352.

[46] Shattil SJ, Hoxie JA, Cunningham M, Brass LF (1985) Changes in the platelet membrane glycoprotein IIb/IIIa complex during platelet activation. J. Biol. Chem. 260: 11107-11114.

[47] Hu DD, White CA, Panzer-Knodle S, Page JD, Nicholson N, Smith JW (1999) A new model of dual interacting ligand binding sites on integrin alphaIIbbeta3. J. Biol. Chem. 274: 4633-4639.

[48] Farndale RW, Winkler AB, Martin BR, Barnes MJ (1992) Inhibition of human platelet adenylate cyclase by collagen fibres. Effect of collagen is additive with that of adrenaline, but interactive with that of thrombin. Biochem. J. 282: 25-32.

[49] Barre DE (2003) Apolipoprotein(a) mediates the lipoprotein(a)-induced biphasic shift in human platelet cyclic AMP. Thromb. Res. 112: 321-324.

[50] Barre DE (2004) Apoprotein(a) antagonises the GPIIB/IIIA receptor on collagen and ADP-stimulated human platelets. Front. Biosci. 9: 404-410.

[51] Barre DE (1998) Lipoprotein(a) reduces platelet aggregation via apo(a)-mediated decreases in thromboxane A2 production. Platelets 9: 93-96.

[52] Barre DE, Griscti O, Mizier-Barre KA, Hafez K (2005) Flaxseed oil and lipoprotein(a) significantly increase bleeding time in type 2 diabetes patients in Cape Breton, Nova Scotia, Canada. J. Oleo. Sci. 54: 347-354.

[53] Rand ML, Sangrar W, Hancock MA, Taylor DM, Marcovina SM, Packham MA, Koschinsky ML (1998) Apolipoprotein(a) enhances platelet responses to the thrombin receptor-acitvating peptide SFLLRN. Arterioscler. Thromb. Vasc. Biol. 18: 1393-1399.

[54] Martínez C, Rivera J, Loyau S, Corral J, Gonzalez-Conejero R, Lozano ML, Vicente V, Anglés-Cano E (2001) Binding of recombinant apolipoprotein(a) to human platelets and effect on platelet aggregation. Thromb. Haemost. 85: 686-693.

[55] Barre DE (2007) Arginyl-glycyl-aspartyl (RGD) epitope of human apolipoprotein (a) inhibits platelet aggregation by antagonizing the IIb subunit of the fibrinogen (GPIIb/IIIa) receptor. Thromb. Res. 119: 601-607.

[56] Von Depka M, Nowka-Göttl U, Eisert R, Dieterich C, Barthels M, Scharrer I, Ganser A, Ehrenforth S (2000) Increased lipoprotein (a) levels as an independent risk factor for venous thromboembolism. Blood 96: 3364-3368.

[57] Casals F, Escolar G, Deulofeu R, Casals E (2007) Elevated lipoprotein (a) [Lp(a)] levels: a biological marker of venous thromboembolic risk frequently found in young females. Thromb. Res. 119 (Suppl. 1): S 100.

[58] Eaton DL, Fless GL, Kohr WJ, McLean JW, Xu QT, Miller CG, Lawn RM, Scanu AM (1987) Partial amino acid sequence of apolipoprotein(a) shows that it is homologous to plasminogen. Proc. Natl. Acad. Sci. USA 84: 3224-3228 (1987)

[59] Karàdi I, Kostner GM, Gries A, Nimpf J, Romics L, Malle E (1988) Lipoprotein (a) and plasminogen are immunochemically related. Biochim. Biophys. Acta 960: 91-97.

[60] Brown MS, Goldstein JL (1987) Teaching old dogmas new tricks. Nature 330: 113-114.

[61] Seman LJ, Breckenridge WC (1986) Isolation and partial characterization of apolipoprotein (a) from human lipoprotein (a). Biochem Cell Biol. 64: 999-1009.

[62] Mondola P, Reichl D (1982) Apolipoprotein B of lipoprotein(a) of human plasma. Biochem. J. 208: 393-398.

[63] Kraft HG, Lingenhel A, Köchl S, Hoppichler F, Kronenberg F, Abe A, Mühlberger V, Schönitzer D, Utermann G (1996) Apolipoprotein(a) kringle IV repeat number predicts risk for coronary heart disease. Arterioscler. Thromb. Vasc. Biol. 16: 713-719.

[64] McLean JW, Tomlinson JE, Kuang WJ, Eaton DL, Chen EY, Fless GM, Scanu AM, Lawn RM (1987) cDNA sequence of human apolipoprotein(a) is homologous to plasminogen. Nature 12: 132-137.

[65] Miles LA, Fless GM, Levin EG, Scanu AM, Plow EF (1989) A potential basis for the thrombotic risks associated with lipoprotein(a). Nature 339: 301-303.

[66] Hajjar KA, Gavish D, Breslow JL, Nachman RL (1989) Lipoprotein(a) modulation of endothelial cell surface fibrinolysis and its potential role in atherosclerosis. Nature 339: 303-305.

[67] Rahman MN, Petrounevitch V, Jia Z, Koschinsky ML (2001) Antifibrinolytic effect of single apo(a) kringle domains: relationship to fibrinogen binding. Prot. Eng. 14: 427-438.

[68] Lijnen HR, Bachmann F, Collen D, Ellis V, Pannekoek H, Rijken DC, Thorsen S (1994) Mechanism of plasminogen activation. J. Intern. Med. 236: 415-424.

[69] Harpel PC, Gordon BR, Parker TS (1989) Plasmin catalyzes binding of lipoprotein(a) to immobilized fibrinogen and fibrin. Proc. Natl. Acad. Sci. USA 56: 3847-3851.

[70] Hervio L, Chapman MJ, Thillet J, Loyau S, Angles-Cano E (1993) Does apolipoprotein(a) heterogeneity influence lipoprotein(a) effects on fibrinolysis? Blood 82: 392-397.

[71] Undas A, Stepien E, Tracz W, Szczeklik A (2006) Lipoprotein(a) as a modifier of fibrin clot permeability and susceptibility to lysis. J. Thromb. Haemost. 4: 973-975.

[72] Knapp JP, Herrmann W (2004) In vitro inhibition of fibrinolysis by apolipoprotein (a) and lipoprotein (a) is size- and concentration-dependent. Clin. Chem. Lab. Med. 42: 1013-1019.

[73] Scanu AM, Miles LA, Fless GM, Pfaffinger D, Eisenbart J, Jackson E, Hoover-Plow JL, Brunck T, Plow EF (1993) Rhesus monkey lipoprotein (a) binds to lysine Sepharose and U937 monocytoid cells less efficiently than human lipoprotein(a). Evidence for the dominant role of kringle 4(37). J. Clin. Invest. 91: 283-291.

[74] Testa R, Marcovina SM (1999) The rate of plasmin formation after in vitro clotting is inversely related to lipoprotein(a) plasma levels. Int. J. Lab. Res. 29: 128-132.

[75] Edelberg JM, Gonzalez-Gronow M, Pizzo SV (1989) Lipoprotein(a) inhibits streptokinase-mediated activation of human plasminogen. Biochemistry 28: 2370-2374.

[76] Edelberg JM, Gonzalez-Gronow M, Pizzo SV (1990) Lipoprotein(a) inhibition of plasminogen activation by tissue-type plasminogen activator. Thromb. Res. 57: 155-162.

[77] Donders SHJ, Lustermans FATh, van Wersch JWJ (1993) On lipoprotein(a) and the coagulation/fibrinolysis balance in the acute phase of deep venous thrombosis. Fibrinolysis 7: 83-86.

[78] Feric NT, Boffa MB, Johnston SM, Koschinsky ML (2008) Apolipoprotein(a) inhibits the conversion of Glu-plasminogen to Lys-plasminogen: a novel mechanism for lipoprotein(a)-mediated inhibition of plasminogen activation. J. Thromb. Haemost. 6: 2113-2120.

[79] Költringer P, Jürgens G (1985) A dominant role of lipoprotein(a) in the investigation and evaluation of parameters indicating the development of cervical atherosclerosis. Atherosclerosis 58: 187-198.

[80] Murai AT, Miyahara T, Fujimoto N, Matsuda M, Kameyama M (1986) Lp(a) as a risk factor for coronary heart disease and cerebral infarction. Atherosclerosis 59: 199-204.

[81] Rhoads GG, Dahlen G, Berg K, Morton NE, Dannenberg AL (1986) Lp(a) lipoprotein as a risk factor for myocardial infarction. J. Am. Med. Ass. 356: 2540-2544.

[82] Nordestgaard BG, Chapman MJ, Ray K, Borén J, Andreotti F, Watts GF, Ginsberg H, Amarenco P, Catapano A, Descamps OS, Fisher E, Kovanen PT, Kuivenhoven JA, Lesnik P, Masana L, Reiner Z, Taskinen MR, Tokgözoglu L, Tygjaerg-Hansen A (2010) Lipoprotein(a) as a cardiovascular risk factor: current status. Eur. Heart J. 31: 2844-2853.

[83] Mackinnon LT, Hubinger L, Lepre F (1997) Effects of physical activity and diet on lipoprotein(a). Med. Sci. Sports Exerc. 29: 1429-1436.

[84] Randall OS, Feseha HB, Illoh K, Xu S, Ketete M, Kwagyan J, Tilghman C, Wrenn M (2004) Response of lipoprotein(a) levels to therapeutic life-style change in obese African-Americans. Atherosclerosis 172: 155-160.

[85] Ahmadi N, Eshaghian S, Huizenga R, Sosnin K, Ebrahimi R, Siegel R (2011) Effects of intense exercise and moderate caloric restriction on cardiovascular risk factors and inflammation. Am. J. Med. 124: 978-982.

[86] Hubinger L, Mackinnon LT, Lepre F (1995) Lipoprotein(a) [Lp(a)] levels in middle-aged male runners and sedentary controls. Med. Sci. Sports Exerc. 27: 490-496.

[87] Mackinnon LT, Hubinger LM (1999) Effects of exercise on lipoprotein(a). Sports Med. 28: 11-24.

[88] Vessby B, Kostner G, Lithell H, Thomis J (1982) Diverging effects of cholestyramine on apolipoprotein B and lipoprotein Lp(a). Atherosclerosis 44: 61-71.

[89] Dobs AS, Prasad M, Goldberg A, Guccione M, Hoover DR (1995) Changes in serum lipoprotein(a) in hyperlipidemic subjects undergoing long-term treatment with lipid-lowering drugs. Cardiovasc. Drugs Ther. 9: 677-684.

[90] Kostner G, Klein G, Krempler F (1984) Can serum Lp(a) concentration be lowered by drugs and/or diet? In: Carlson LA and Olsson AG (eds.): Treatment of Hyperlipoproteinemia, Raven Press, New York, 151-156 p.

[91] Neele DM, Kaptain A, Huisman H, de Wit EC, Princen HM (1998) No effect of fibrates on synthesis of apolipoprotein(a) in primary cultures of cynomolgus monkey and human hepatocytes: apolipoprotein A-I synthesis increased. Biochem. Biophys. Res. Commun. 244: 374-378.

[92] Christodoulakos GE, Lambrinoudaki IV, Panoulis CP, Papadias CA, Kouskouni EE, Creatsas GC (2004) Effect of hormone replacement therapy, tibolone and raloxifene on serum lipids, apolipoprotein A1, apolipoprotein B and lipoprotein(a) in Greek postmenopausal women. Gynecol. Endocrinol. 18: 244-257.

[93] Persson L, Henriksson P, Westerlund E, Hovatta O, Angelin B, Rudling M (2012) Endogenous estrogens lower plasma PCSK9 and LDL cholesterol but not Lp(a) or bile acid synthesis in women. Arterioscler. Thromb. Vasc. Biol. 32: 810-814.

[94] Albers JJ, Taggart HM, Applebaum-Bowden D, Haffner S, Chesnut CH, Hazzard WR (1984) Reduction of lecithin-cholesterol acyl-transferase, apolipoprotein D and the Lp(a) lipoprotein with the anabolic steroid stanozolol. Biochim. Biophys. Acta 795: 293-296.

[95] Hartgens F, Rietjens G, Keizer HA, Kuipers H, Wolffenbuttel BH (2004) Effects of androgenic-anabolic steroids on apolipoproteins and lipoprotein(a). Brit. J. Sports Med. 38: 253-259.

[96] Kostner KM, Kostner GM (2005) Therapy of hyper-Lp(a). Handb. Exp. Pharmacol. 170: 519-536.

[97] Ma PTS, Gil G, Sudhof JC, Bilheimer DW, Goldstein JL, Brown MS (1986) Mevinolin, an inhibitor of cholesterol synthesis, induces mRNA for low density lipoprotein receptor in liver of hamsters and rabbits. Proc. Nat. Acad. Sci. USA 83: 8370-8374.

[98] Thiery J, Armstrong VW, Schleef J, Creutzfeld C, Creutzfeld W, Seidel D (1988) Serum lipoprotein Lp(a) concentrations are not influenced by an HMG-CoA reductase inhibitor. Klin. Wochenschr. 66: 462-463.

[99] Kostner GM, Gavish D, Leopold B, Bolzano D, Weintraub MS, Breslow JL (1989) HMG-CoA reductase inhibitors lower LDL cholesterol without reducing Lp(a) levels. Circulation 80: 1313-1319.

[100] Joy MS, Dornbrook-Lavender KA, Chin H, Hogan SL, Denu-Ciocca C (2008) Effects of atorvastatin on Lp(a) and lipoprotein profiles in hemodialysis patients. Ann. Pharmacother. 42: 9-15.

[101] Horimoto M, Hasegawa A, Takenaka T, Fujiwara M, Inoue H, Igarashi K (2003) Long-term administration of pravastatin reduces serum lipoprotein(a) levels. Int. J. Clin. Pharmacol. Ther. 41: 524-530.

[102] Choi SH, Chae A, Miller E, Messig M, Ntanios F, DeMaria AN, Nissen SE, Witztum JL, Tsimikas S (2008) Relationship between biomarkers of oxidized low-density lipoprotein, statin therapy, quantitative coronary angiography, and atheroma: volume observations from the REVERSAL (Reversal of atherosclerosis with aggressive lipid lowering) study. J. Am. Coll. Cardiol. 52: 24-32.

[103] Hobbs HH, White Al (1999) Lipoprotein(a): intrigues and insights. Curr. Opin. Lipidol. 10: 225-236.

[104] Digby JE, Lee JM, Choudhury RP (2009) Nicotinic acid and the prevention of coronary artery disease. Curr. Opin. Lipidol. 20: 321-326.

[105] Linke A, Sonnabend M, Fasshauer M, Höllriegel R, Schuler G, Niebauer J, Stumvoll M, Blüher M (2009) Effects of extended-release niacin on lipid profile and adipocyte biology in patients with impaired glucose tolerance. Atherosclerosis 205: 207-213.

[106] Gurakar A, Hoeg JM, Kostner G, Papadopoulos NM, Brewer jr. HB (1985) Levels of lipoprotein Lp(a) decline with neomycin and niacin treatment. Atherosclerosis 57: 293-301.

[107] Helmbold AF, Slim JN, Morgan J, Castillo-Rojas LM, Shry EA, Slim AM (2010) The effects of extended release niacin in combination with omega 3 fatty acid supplements in the treatment of elevated lipoprotein (a). Cholesterol 2010: 306147.

[108] Harris WS (1989) Fish oils and plasma lipid and lipoprotein metabolism in humans: a critical review. J. Lipid Res. 30: 785-807.

[109] Wei MY, Jacobson TA (2011) Effects of eicosapentaenoic acid versus docosahexaenoic acid on serum lipids: a systematic review and meta-analysis. Curr. Atheroscler. Rep. 13: 474-483.

[110] Gries A, Malle E, Wurm H, Kostner GM (1990) Influence of dietary fish oils on plasma Lp(a) levels. Thromb. Res. 58: 667-668.

[111] Malle E, Sattler W, Prenner E, Leis HJ, Hermetter A, Gries A, Kostner GM (1991) Effects of dietary fish oil supplementation on platelet aggregability and platelet membrane fluidity in normolipemic subjects with and without high plasma Lp(a) concentrations. Atherosclerosis 88: 193-201.

[112] Herrmann W, Biermann J, Kostner GM (1995) Comparison of effects of N-3 to N-6 fatty acids on serum levels of lipoprotein(a) in patients with coronary artery disease. Am. J. Cardiol. 76: 459-462.

[113] Marckmann P, Bladbjerg EM, Jespersen J (1997) Dietary fish oil (4 g daily) and cardiovascular risk markers in healthy men. Arterioscler. Thromb. Vasc. Biol. 17: 3384-3391.

[114] Beavers KM, Beavers DP, Bowden RG, Wilson RL, Gentile M (2009) Effect of over-the-counter fish-oil administration on plasma Lp(a) levels in an end-stage renal disease population. J. Ren. Nutr. 19: 443-449.

[115] Kooshki A, Taleban FA, Tabibi H, Hedayati M (2011) Effects of omega-3 fatty acids on serum lipids, lipoprotein (a), and hematologic factors in hemodialysis patients. Ren. Fail. 33: 892-898.

[116] Krempler F, Kostner GM, Bolzano K, Sandhofer F (1980) Turnover of lipoprotein (a) in man. J. Clin. Invest. 65: 1483-1490.

[117] Rader DJ, Cain W, Ikewaki K, Talley G, Zech LA, Usher D, Brewer HB Jr. (1994) The inverse association of plasma lipoprotein(a) concentrations with apolipoprotein(a) isoform size is not due to differences in Lp(a) catabolism but to differences in production rate. J. Clin. Invest. 93: 2758-2763.

[118] Frank S, Gauster M, Strauss J, Hrzenjak A, Kostner GM (2001) Adenovirus-mediated apo(a)-antisense-RNA expression efficiently inhibits apo(a) synthesis in vitro and in vivo. Gene Therapy 8: 425-430.

[119] Chennamsetty I, Claudel T, Kostner KM, Baghdasaryan A, Kratky D, Levak-Frank S, Frank S, Gonzalez FJ, Trauner M, Kostner GM (2011) Farnesoid X receptor represses hepatic human APOA gene expression. J. Clin. Invest. 121: 3724-3734.

[120] Virani SS, Brautbar A, Davis BC, Nambi V, Hoogeveen RC, Sharrett AR, Coresh J, Mosley TH, Morrisett JD, Catellier DJ, Folsom AR, Boerwinkle E, Ballantyne CM (2012) Associations between lipoprotein(a) levels and cardiovascular outcomes in black and white subjects: the Atherosclerosis Risk in Communities (ARIC) Study. Circulation 125: 241-249.

Chapter 17

An Apolipoprotein CIII-Derived Peptide, Hatktak, Activates Macromolecular Activators of Phagocytosis from Platelets (MAPPs)

Haruhiko Sakamoto, Masaki Ueno, Wu Bin, Yumiko Nagai, Kouichi Matsumoto, Takao Yamanaka and Sumiko Tanaka

Additional information is available at the end of the chapter

1. Introduction

When thrombi are formed, infiltration of leukocytes including neutrophils and macrophages follows in and around the thrombi. On the other hand, in inflammatory lesions where infiltration of leukocytes is observed, thrombi are often observed. When thrombi are formed, activated platelets release a lot of substances while adhering to subendothelial connective tissues, recruiting other platelets and aggregating together. In these regions with thrombus formation or platelet activation, it is possible that platelets affect leukocytic function via the action of the released products, since the substances released from platelets include factors that activate [1-14] and suppress [15-22] neutrophilic functions.

In experiments using human platelets and neutrophils, we found that platelets release several neutrophilic phagocytosis activators. Among substances released from activated platelets, ATP and ADP have been reported to activate iC3b receptor-mediated phagocytosis [1, 2, 23], and some prostaglandins including PGE2, PGF2α and thromboxane B2 and macromolecular activators of phagocytosis from platelets (MAPPs) activate Fcγ receptor-mediated phagocytosis.

MAPPs have two subsets, l-MAPP (3×10^5 Da) and s-MAPP (1.5×10^5 Da) [24, 25]. Platelets stored in the form of platelet-rich plasma lose the capacity to release MAPPs but recover it if incubated with the plasma-derived precursors of MAPPs (precursors of l-MAPP and s-MAPP, 3×10^5 Da and 1.5×10^5 Da, respectively) and thrombin in the presence of Ca^{++} [26]. It was suggested that the loss of platelet ability to release MAPPs is due to escape of the precursors and thrombin during storage with CPD (citrate-phosphate-dextrose) solution. It is possible to produce MAPPs using stored platelet-derived lysate by stimulation with thrombin or trypsin in the presence of the precursors [27]. It has also been suggested that the precursors of l-MAPP and s-MAPP are polymerized transferrins, probably of tetrameric and dimeric forms, respectively [28].

As for the production of MAPPs in platelets, it was suggested that GP Ibα-bound thrombin reacts with a high-molecular-weight substance (HMW activator) to release a low-molecular-weight substance (LMW activator) [29], which can produce MAPPs from the precursors directly.

In an ultracentrifugation study of the platelet lysate, the HMW activator activity was observed in the HDL fraction. Anti-apolipoprotein A1 antibody abolished the HMW activator function from the HDL fraction of the platelet lysate. These findings suggest that the HMW activator belongs to HDL. In an affinity chromatography study of the protein obtained from the HDL-rich fraction of the platelet lysate using an anti-apolipoprotein CIII (Apo CIII) column, it was suggested that LMW activator is derived from Apo CIII. In fact, it was observed that the commercially available Apo CIII could produce LMW activator by the activity of thrombin [30].

The purpose of this study was to determine the structure of the LMW activator.

2. Materials and methods

This study was approved by the local institutional review board of our university hospital. The procedures followed were in accordance with the Helsinki Declaration of 1975, as revised in 1983.

2.1. Phagocytosis experiment

The phagocytosis experiments were performed according to a method previously described [30]. Briefly, neutrophils were separated from heparinized venous blood from healthy volunteers by centrifugation on MonoPoly resolving medium (ICN Biochemicals Japan, Tokyo, Japan). After washing of neutrophils in phosphate-buffered saline (PBS) (Sigma, St. Louis, MO), one thousand neutrophils in 5 μl of PBS were attached to the surface of a Terasaki microplate well (Nunc, Roskilde, Denmark) by centrifugation at 160 g for 2 minutes. Then, stimulation was performed with 10 μl of the test material in PBS supplemented with 1% bovine serum albumin (Sigma) (PBS-BSA) at 37 °C for 15 minutes. After washing, the neutrophils were incubated with 5,000 sheep red blood cells (SRBCs) in 5 μl of RPMI 1640 medium supplemented with 1% BSA and a one-hundredth volume of anti-SRBC rabbit IgG at 37 °C for 25 minutes under 5% CO_2. After lysing the unphagocytosed SRBCs using a Tris-buffered ammonium chloride solution [31] and fixation with 1% glutaraldehyde (Sigma) in PBS, the phagocytosed SRBCs were counted under a light microscope.

All phagocytosis experiments were performed in triplicate. Mean numbers of ingested SRBCs per neutrophil in a well were calculated. Then, means of three wells were obtained. The results are expressed as phagocytic indices normalized by the phagocytic activity of the PBS-BSA control taken as 100.

2.2. Production of MAPPs in vitro

2.2.1. Preparation of the precursors of MAPPs using holo-transferrin

Precursors of MAPPs were produced artificially using holo-transferrin (Sigma) according to a method described previously [28, 30]. Briefly, the peak fraction of holo-transferrin in Superdex 200 (GE Healthcare UK Ltd., Buckinghamshire, England) gel filtration was incubated on ice for 15 min with 0.2% glutaraldehyde (Sigma) to induce polymerization [32], and subjected to filtration again. The fractions obtained at the same elution volumes as the conventional l-MAPP (tetramer transferrin-rich fraction) and s-MAPP (dimer transferrin-rich fraction) were adjusted to an optical density of 0.04 at 280 nm with PBS and used as precursors of l-MAPP and s-MAPP, respectively.

2.2.2. Preparation of LMW activators

Apo CIII (Chemicon International Inc., Temecula, CA) was adjusted to a concentration of 10 μg/ml in PBS and incubated with 0.1 unit/ml human thrombin (thrombin) (Sigma) or 1 unit/ml bovine trypsin (trypsin) (Sigma) at 37 °C for 30 minutes, and then applied on a PD10 column (GE Healthcare UK Ltd.) to obtain the low-molecular-weight fraction, which was serially diluted x10 and used as the LMW activator.

Another Apo CIII (10 μg/ml) was incubated with 0.1 unit/ml thrombin or 1 unit/ml trypsin and gel-filtered through a Superdex peptide column (GE Healthcare UK Ltd.) (elution buffer, PBS; elution speed, 0.5 ml/minute; fraction volume, 1 ml each) to obtain LMW activator fraction.

Samples of the low-molecular-weight fraction of variously prepared platelet-related materials including platelet release products and platelet lysate (see below for details) were obtained using PD-10 columns.

Some peptides associated with Apo CIII were synthesized artificially, and the activity as an LMW activator was examined.

2.2.3. Preparation of MAPPs using LMW activators and precursors of MAPPs

To form MAPPs, 1 ml of appropriately diluted LMW activator derived from Apo CIII, platelet release products or artificially produced peptides was incubated with 10 μl of precursor of l-MAPP or s-MAPP at 37 °C for 30 minutes.

2.3. Preparation of platelet release products using platelets from platelet-rich plasma

To prepare fresh and stored platelets, heparinized venous blood was mixed with 13 vol% citrate-phosphate-dextrose (CPD) solution and centrifuged at 60 g for 20 minutes at 20 °C. The platelet-rich plasma, which was contaminated by less than one erythrocyte per 1,000 platelets, was transferred into another plastic tube and stored at 20 °C in a water bath with agitation once a second for 0, 72 or 120 hours.

These platelets were washed twice in PBS supplemented with 6.7 mM EDTA and twice in PBS, and adjusted to a concentration of $4 \times 10^5/\mu l$. The platelet suspension was stimulated with 0.1 unit/ml thrombin at 37 °C for one minute in the presence or absence of 4 mEq/l Ca^{++}, then cooled immediately on an ice-water bath and centrifuged at 960 g for 15 minutes at 4 °C to obtain the platelet release products in the supernatant.

2.4. Preparation of platelet lysate

Platelet suspensions obtained from platelet-rich plasma as described above were frozen at -15 °C, thawed and centrifuged at 1,500 g for 30 minutes. The supernatant was used as the platelet lysate.

2.5. Peptide synthesis

Peptides such as S1-K21, H18-R40, H18-K24, H18-K21 and T22-K24 of Apo CIII were synthesized by Sawady Technology (Tokyo, Japan). Other peptides that contain part or all of H18-K24 (HATKTAK) of Apo CIII were produced by Thermo Fisher Scientific GmbH (Ulm, Germany).

2.6. Inhibition of the LMW activator activity by anti-HATKTAK rabbit IgG antibody

Antibody against HATKTAK was raised in a rabbit by intracutaneous injection of keyhole-limpet-conjugated CHATKTAK (6 injections at 2-week intervals) by Sigma Aldrich Japan (Ishikari city, Japan). The IgG antibody was refined using a protein A column. The control IgG antibody was obtained from the same rabbit at day 0 of immunization.

By indirect enzyme-linked immunosorbent assay (ELISA) [33], specificity of the anti-HATKTAK rabbit antibody was examined. 100 μl of antigens involving Apo CIII and Apo CIII-related peptides (1 $\mu g/ml$ in PBS) was immobilized to each well of a polystyrene microplate by incubation at 4 °C overnight. 1000-fold-diluted anti-HATKTAK rabbit antibody or 1000-fold-diluted anti-Apo CIII gout IgG (Gene Tex, Inc., San Antonio, TX) was used as the primary antibody. Peroxidase-conjugated anti-rabbit IgG (Fab') or anti-goat IgG (Fab') (Histofine simple stain MAX-PO(R) or -PO(G), respectively, Nichirei Bioscience, Tokyo, Japan) was used as the secondary antibody. Substrate solution consisted of 10 ml of 0.5 M citrate buffer, 10 ml of 0.3% hydrogen peroxide and 10 mg of orthophenylene diamine (Wako, Osaka, Japan). 2 M sulfuric acid was used as a stopping solution. Absorbance was determined at 450 nm in a model 550 microplate reader (Bio-Rad, Tokyo, Japan).

To examine the effect of anti-HATKTAK antibody on an LMW activator, a $\times 10^2$-diluted low-molecular-weight fraction of the platelet release products prepared from fresh platelet- rich plasma and a $\times 10^6$-diluted one from platelet-rich plasma stored for 120 hours were prepared. Then, they were incubated with a one-hundredth volume of anti-HATKTAK IgG or control IgG at room temperature for 30 min, applied to a protein A column to remove the immune complexes and the residual antibodies, and used as the source of the LMW activator.

An Apolipoprotein CIII-Derived Peptide, Hatktak, Activates Macromolecular Activators of
Phagocytosis from Platelets (MAPPs)

363

2.7. Ion exchange chromatography

Cation exchange chromatography was performed using a MONO S HR5/5 column (GE Healthcare UK Ltd.). Samples were dissolved in 20 mM Tris-HCl, pH 8.0 (buffer A). The adsorbed materials were recovered by elution with a linear increase in buffer B (buffer A + 1 M NaCl).

2.8. Assessment of peptide concentrations using the o-phthalaldehyde (OPA) fluorescent method

Peptide concentrations were assessed according to the OPA fluorescent method with intact protein [34]. Fluorescence was measured with an excitation wavelength of 340 nm and an emission wavelength of 450 nm using a spectrofluorometer (FP-6300, JASCO, Tokyo, Japan).

2.9. Mass spectrometry

Matrix-associated laser desorption/ionization time-of-flight mass spectrometry was performed with a Voyager System 4314 (Applied Biosystems, Foster City, CA) in the reflector time-of-flight configuration at an acceleration voltage of 25 kV with delayed ion extraction. Samples were diluted 1:1 with a freshly prepared matrix solution consisting of 10 mg/ml α-cyano-4-hydroxycinnamic acid (Sigma) in 20% acetonitrile with 0.1% trifluoroacetic acid. Aliquots of 1 μl were deposited on a metallic sample holder and analyzed immediately after drying in a stream of air. Mass scale calibration was performed externally.

2.10. Immunohistochemistry

Coagulation of the fresh peripheral blood taken without anticoagulants was induced in a glass tube for 3 hours. This coagulum was formalin-fixed, and paraffin sections of 3 μm thickness were prepared. After deparaffinization and hydration, antigen retrieval was performed by heating in 10 mM citrate buffer (pH 6.0) for 40 minutes. After rinsing in PBS and incubation with 2% BSA in PBS at room temperature for 60 minutes, incubation with a mixture of rabbit anti-HATKTAK IgG (x1,000 diluted) and murine anti-CD 61 monoclonal IgG (Dako Japan, Tokyo, Japan) (x100 diluted) at 4 °C overnight was performed followed by incubation with a mixture of Alexa Fluor 594-conjugated anti-rabbit IgG (x200 diluted) and Alexa Fluor 488-conjugated anti-murine antibody (x200 diluted) (Molecular Probes, Eugene, OR, USA). The fluorescent signals were viewed under a confocal microscope (Bio-Rad Radiance 2100). As a control experiment, an identical immunohistochemical procedure with omission of the primary antibodies was performed.

2.11. Assay of concentrations of Apo CIII, apolipoprotein AI (Apo AI) and apolipoprotein B100

Concentrations of Apo CIII, Apo AI and Apo B100 in the platelet lysate were assayed by ELISA using commercially available assay kits for Apo CIII (AssayPro, St. Charles, MO), Apo AI (Mabtech AB, Nacka Strand, Sweden) and Apo B100 (Mabtech AB), respectively.

A — Phagocytic Index / thrombin / trypsin / Precursor of s-MAPP / Dilution (10^n)

B — Phagocytic Index / NaCl

C — Phagocytic Index / N / Fraction Number / post

D — Phagocytic Index

E — Phagocytic Index / Elution Volume (ml)

F — Molecular Weight / Cytochrome c 12,384 / S1-K21 2,456 / H18-R40 2,365 / H18-K24 756 / H18-K21 455 / T22-K24 318 / K 146 / G 75 / LMW activator / Elution volume

G — H2N-SEAEDASLLSFMQGYMKHATKTAKDALSSVQESQVAQQARGWVTDGFSSLKDYWSTVKDKFSEFWDLDPEVRPTSAVAA-COOH 5 10 15 20 25 30 35 40 45 50 55 60 65 70 75

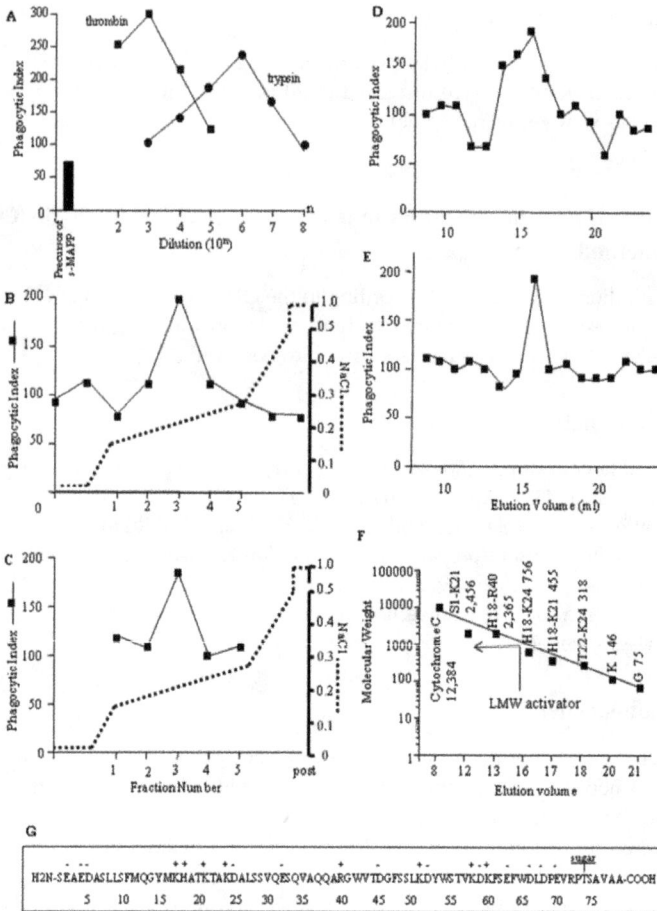

Figure 1. Characterization of the LMW activator produced from Apo CIII. **A.** MAPP formation using Apo CIII. To prepare LMW activator, 10 µg/ml Apo CIII was treated with 0.1 unit/ml thrombin or 1 unit/ml trypsin, and then diluted serially x10. The activity of each diluted sample as LMW activator was examined. **B.** MONO S cation exchange chromatography of thrombin-treated Apo CIII (10 µg). MAPPs were generated using each fraction at a dilution of x10^2 and one of the precursors of MAPPs. **C.** MONO S cation exchange chromatography of trypsin-treated apolipoprotein CIII (10 µg/ml). MAPPs were generated using each fraction at a dilution of x10^4 and one of the precursor of MAPPs. **D.** Superdex peptide gel filtration of thrombin-treated Apo CIII (10 µg/ml). Formation of MAPPs was achieved using each fraction at a dilution of x10^2 and one of the precursors of MAPPs. **E.** Superdex peptide gel filtration of trypsin-treated Apo C-III (10 µg/ml). Formation of MAPPs was achieved using each fraction at a dilution of x10^4 and one of the precursors of MAPPs. **F.** Molecular weight determination of the LMW activator with comparisons with those of Apo CIII-derived peptides. **G.** Amino acid sequence of Apo CIII, distributions of basic (+) and acidic (-) amino acids and sugar binding amino acid (T74) in Apo CIII. The amino acid sequence is cited from the database of GenPex (NCBI Protein DataBase, http://www.ncbi.nlm.nih.gov/protein). A, B, C, D and E were performed using the precursor of s-MAPP.

2.12. Statistics

In this study, the differences were analyzed using Mann-Whitney test, Kruskal-Wallis test, paired t test or Wilcoxon test. A P value of less than 0.05 was considered significant.

3. Results

3.1. LMW activator is considered to be HATKTAK

As it was shown in a previous study that both thrombin and trypsin could produce a substance with LMW activator activity from platelet lysate [27] and that incubation of Apo CIII with thrombin resulted in production of LMW activator, Apo CIII was treated with thrombin or trypsin and the LMW activator activities produced were compared to determine whether LMW activators produced by these enzymes are the same substance. Both thrombin and trypsin could produce LMW activator activity, but 1 unit/ml trypsin could produce LMW activator 100 to 1,000 times more effectively than 0.1 unit/ml thrombin (Figure 1A).

The LMW activator activities prepared using thrombin and trypsin appeared at the same NaCl concentration of the MONO S cation exchange chromatography (0.2 M NaCl, pH 8.0) (Figure 1B and Figure 1C, respectively), and at the same elution volume (16 ml) of the Superdex peptide gel filtration (Figure 1D and Figure 1E, respectively), which corresponds to the approximate molecular size of 800 Da (Figure 1F). These findings suggest that the two LMW activators are the same substance.

The fact that LMW activator can be produced by trypsin digestion of Apo CIII suggests that the C terminal of LMW activator consists of K or R [35]. LMW activator is suggested to be one of the basic peptides that would appear after trypsin digestion of Apo CIII (S1-K24, H18-R40, H18-K24, H18-K21, T22-K24) (see Figure 1G for the sequence of Apo CIII, distribution of basic and anionic amino acids and the sugar binding amino acid). These peptides were artificially produced and filtered on the Superdex peptide column to use as molecular size markers. The optical peaks of OPA fluorescence and LMW activator activity by H18-K24 (HATKTAK) appeared at the same elution volume of Superdex peptide gel filtration corresponding to Apo CIII-derived LMW activator (Figure 1D, 1E, 2A and 2B). In MONO S cation exchange chromatography, the action of HATKTAK as LMW activator was recovered at the fraction with the same NaCl concentration (0.2 M NaCl, pH 8.0) (Figure 2C) as the Apo CIII-derived LMW activator (Figure 1B and 1C).

3.2. Only HATKTAK among examined peptides showed MAPP-forming activity

Among the peptides used for calibration on Superdex peptide gel filtration (Figure 1F), H18-R40 and H18-K21 were separated in the fractions near to HATKTAK (LMW activator). The activity of these peptides as LMW activators were examined. Only HATKTAK showed LMW activator activity with a peak at 1 nM (Figure 2D). Then, some Apo CIII-derived peptides that contain part or all of H18-K24 were examined for their LMW activator activity. All of them were used at a concentration of 1 nM. Only HATKTAK showed MAPP-forming activity (Table 1).

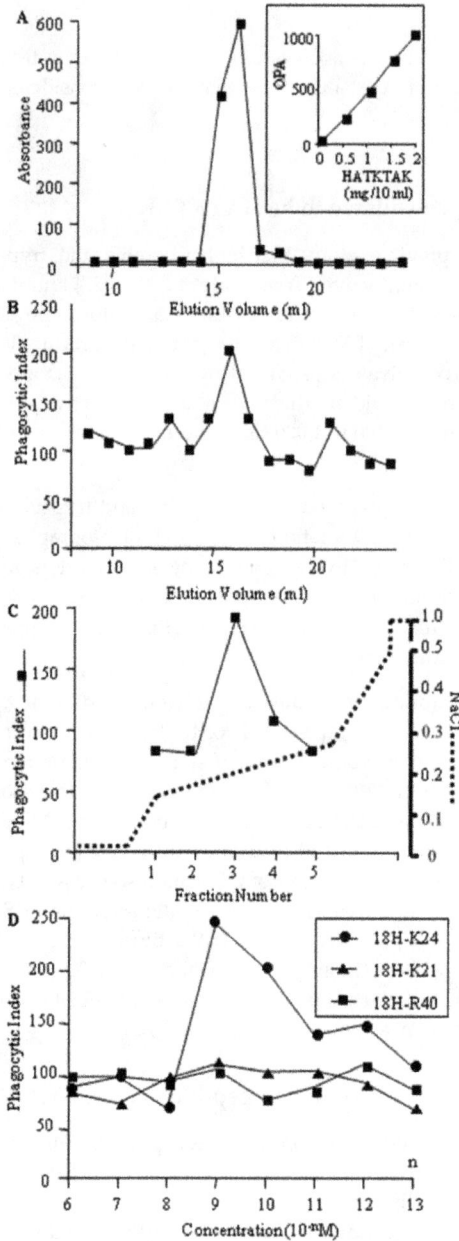

Figure 2. HATKTAK functions as the LMW activator. **A.** Superdex peptide gel filtration of HATKTAK (200 µg, 0.2 ml). The peptide concentration of each fraction was measured by the OPA fluorescent

method. The inset shows the standard graph for the assay of HATKTAK. **B.** MAPP formation using fractions of Superdex peptide gel filtration of HATKTAK (200 µg, 0.2 ml) at a dilution of x10^6 by incubation with one of the precursors of MAPPs. **C.** MONO S cation exchange chromatography of 1 µg/ml HATKTAK. MAPPs were generated by incubation of each fraction from the chromatography at a dilution of x10^4 and one of the precursors of MAPPs. **D.** MAPP formation using peptides containing part of the sequence of apolipoprotein CIII. Each peptide was serially diluted x10 and incubated with one of the precursors of MAPPs. B, C and D were performed using the precursor of s-MAPP.

Peptides	Precursor of l-MAPP	Precursor of s-MAPP
HATK	-	-
HATKT	-	-
HATKTA	-	-
HATKTAK	+	+
HATKTAKD	-	-
K	-	-
AK	-	-
TAK	-	-
ATKTAK	-	-
KHATKTAK	-	-

Table 1. MAPP formation from precursors of MAPPs with peptides derived from K16 to D25 of Apo CIII (KHATKTAKD)

3.3. LMW activator activity in platelet release products and platelet lysate

In the next part of the study, the LMW activator activities in variously prepared samples including platelet release products and platelet lysate were compared to determine whether platelets release LMW activator. Platelet release products were prepared using platelets from fresh and stored platelet-rich plasma. The reason why we examined platelet release products prepared from stored platelet-rich plasma along with those prepared from the fresh equivalent is that platelet releasate from stored platelet-rich plasma was expected to contain a higher concentration of LMW activator because platelets lose the precursors of MAPPs during storage [26]. From all of these samples, the low-molecular-weight fraction was separated using PD10 columns, and the activity corresponding to LMW activator was compared by the largest dilution from the original sample for which a phagocytic index higher than 150 was recorded (effective dilution).

1 ng/ml HATKTAK and platelet release products from fresh platelets stimulated with 4 mEq/l Ca^{++} and 0.1 unit/ml thrombin showed an effective dilution of 10^2 to 10^3, whereas fresh platelets stimulated only with 4 mEq/l Ca^{++} or 0.1 unit/ml thrombin for 1 minute showed far lower effective dilution, and platelet release products prepared from platelets after storage for 72 hours and 120 hours in the form of platelet-rich plasma showed extremely high effective dilution (10^3 to 10^5, platelets from platelet-rich plasma stored for 72 hours; 10^7 to 10^8, those stored for 120 hours). None of the platelet lysate produced from platelets in fresh and stored platelet-rich plasma showed LMW activator activity (Figure 3).

To confirm whether these LMW activator activities were by the same substance, Superdex peptide gel filtrations of the low-molecular-weight fractions of platelet release products prepared from fresh platelets (A), and those stored for 72 hours (B) and 120 hours (C), with stimulation with 0.1 unit/ml thrombin in the presence of 4 mEq/l Ca^{++} were performed. Fractions of the gel filtrations were diluted $x10^2$ (A), $x10^4$ (B) and $x10^6$ (C). All of these samples showed LMW activator activity at the fraction corresponding to HATKTAK (compare Figure 4A, 4B and 4C with Figure 2B).

3.4. Inhibition of LMW activator activity by anti-HATKTAK antibody

Then, to obtain evidence of the existence of HATKTAK in the platelet release products, actions of an anti-HATKTAK rabbit antibody against LMW activator activity were examined.

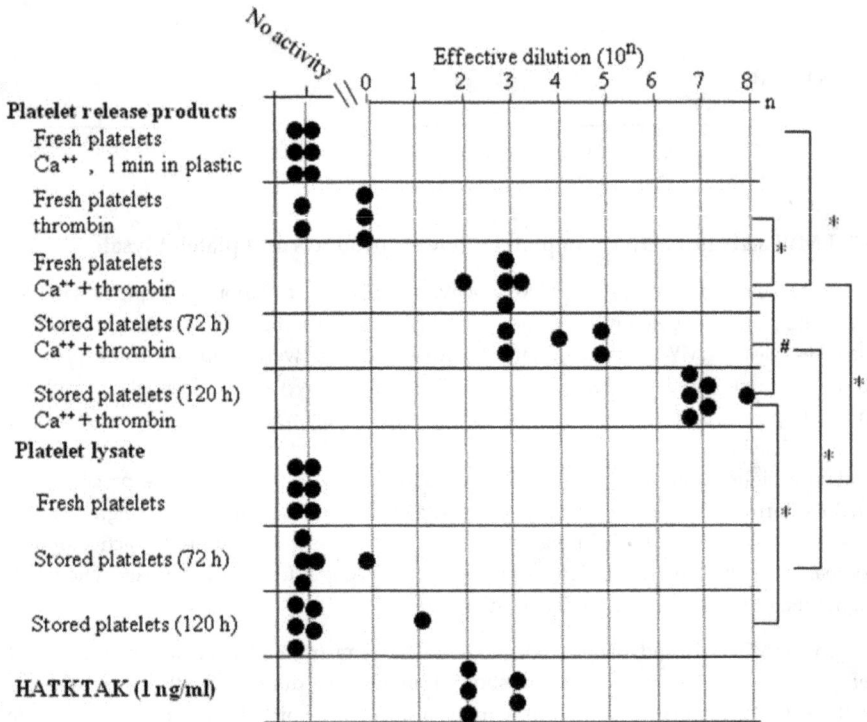

Figure 3. Comparisons of effective dilution of LMW activator in platelet release products and platelet lysate prepared using platelets that were obtained from platelet-rich plasma stored for 0 hours, 72 hours and 120 hours. A black circle means a case with the effective dilution of LMW activator indicated by the horizontal axis. *, $P<0.01$ with the Mann-Whitney test, and #, $P<0.01$ with the Kruskal-Wallis test. All experiments were performed using the precursor of s-MAPP.

An Apolipoprotein CIII-Derived Peptide, Hatktak, Activates Macromolecular Activators of
Phagocytosis from Platelets (MAPPs)

369

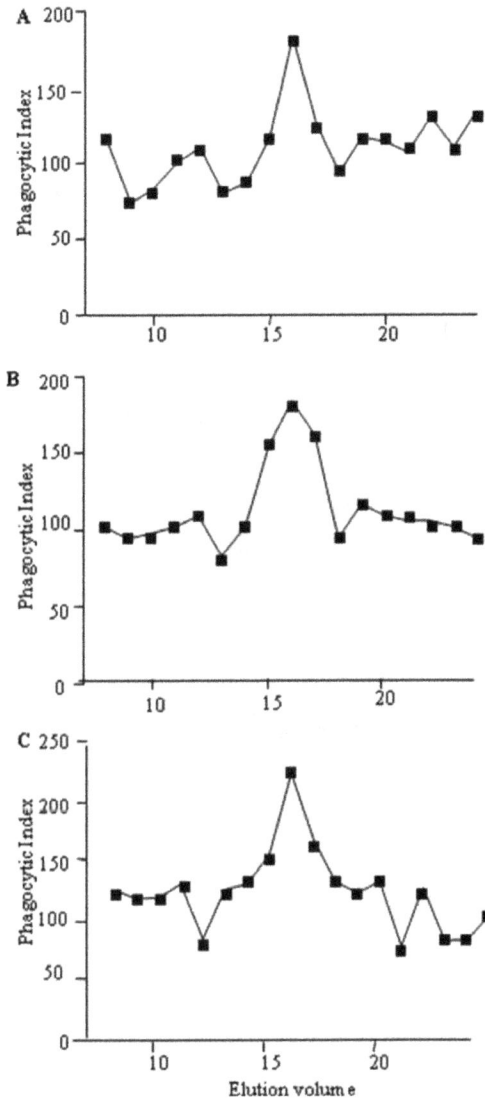

Figure 4. Comparisons of molecular size of LMW activator in platelet release products prepared using
(A) platelet release products using fresh platelets stimulated with 0.1 unit/ml thrombin in the presence
of Ca^{++}; (B) platelet release products using platelets stored for 72 hours stimulated with 0.1 unit/ml
thrombin in the presence of Ca^{++}; (C) platelet release products using platelets stored for 120 hours
stimulated with 0.1 unit/ml thrombin in the presence of Ca^{++}. The LMW activator activity of each
fraction was examined at dilutions of x10^2 (A), x10^3 (B) and x10^6 (C). All experiments were performed
using the precursor of s-MAPP.

ELISA revealed that anti-HATKTAK reacted strongly to H18-K24 (HATKTAK), K17-K24 and S1-K24 of Apo CIII, but very weakly to Apo CIII and H18-D25 of Apo CIII, whereas anti-Apo CIII goat antibody showed a positive reaction only to Apo CIII (Figure 5A and 5B, respectively).

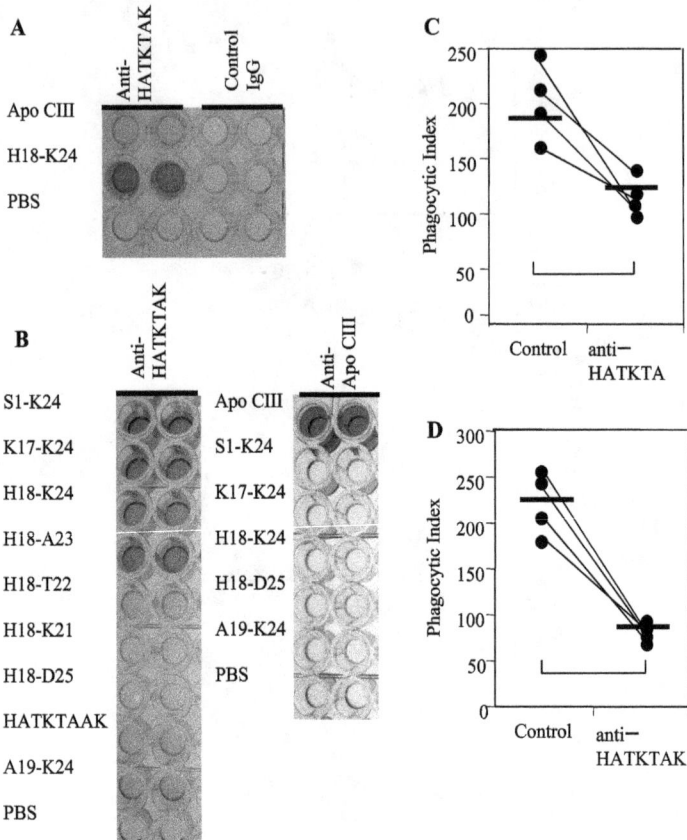

Figure 5. Effects of anti-HATKTAK antibody on LMW activator activity. A, ELISA using anti-HATKTAK rabbit IgG and control rabbit IgG against Apo CIII and H18-K24 of Apo CIII. B, ELISA using anti-HATKTAK rabbit IgG and anti-Apo CIII gout IgG against peptide associated with Apo CIII, HATKTAAK and Apo CIII. C, Effects of control rabbit IgG and anti-HATKTAK rabbit IgG on LMW activator activity in the low-molecular-weight fraction of platelet release products prepared from platelets of fresh platelet-rich plasma. D, Effects of control rabbit IgG and anti-HATKTAK rabbit IgG on LMW activator activity in the low-molecular-weight fraction of platelet release products prepared using platelets of platelet-rich plasma stored for 120 hours. In C and D, figures reveal the result of experiments using the precursor of s-MAPP. The bars in the figure show the average. *, $P<0.01$; **, $P<0.05$ by paired t test.

Both LMW activators derived from fresh platelet and platelets stored for 120 hours were inhibited by the anti-HATKTAK IgG antibody (Figure 5C and 5D, respectively).

3.5. Evidence of existence of HATKTAK by mass spectrometry

In mass spectrometry, HATKTAK showed m/z 756 (Figure 6A). Although it was impossible to show the existence of a substance with m/z 756 in thrombin-digested Apo CIII (10 μg/ml) and in platelet release products released from fresh platelets and those stored for 72 hours (data not shown), a substance with m/z 756 was detected in the trypsin-treated Apo CIII (10 μg/ml) (Figure 6B) and platelet release products prepared from platelets stored for 120 hours (Figure 6C).

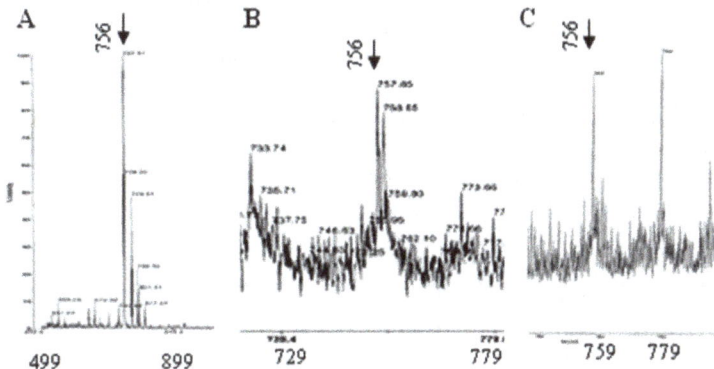

Figure 6. Mass spectrometry of HATKTAK, trypsin-digested Apo CIII and platelet release products from platelets of platelet-rich plasma stored for 120 hours. **A**, 1 mg/ml HATKTAK in PBS; **B**, 10 μg/ml Apo CIII digested with 1 unit/ml trypsin; and **C**, platelet release products using platelets of platelet-rich plasma stored for 120 hours. Platelets were stimulated with 0.1 unit/ml thrombin in the presence of Ca++.

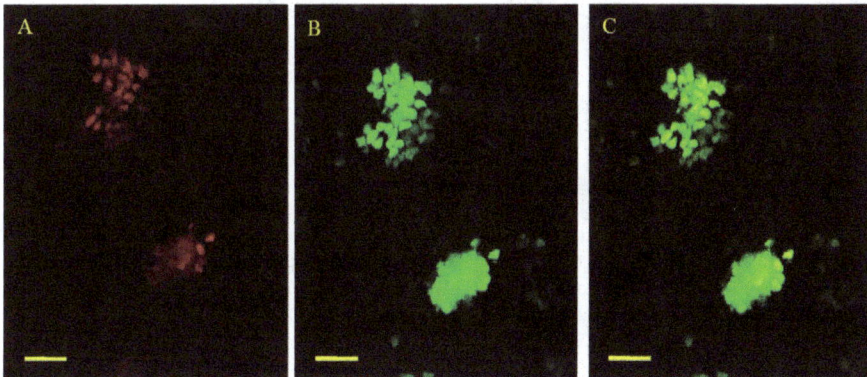

Figure 7. Confocal microscopic images of double staining for the anti-HATKTAK antibody (visualized as red in A), and the anti-CD61 antibody (visualized as green in B). The colocalization of two antibodies is indicated by the conversion of green and red to yellow (C). Scale bar indicate 20 μm.

3.6. Immunohistochemical evidence of existence of HATKTAK in activated platelets

To show the existence of HATKTAK in activated platelets, an immunohistochemical study of the blood coagula was performed. Platelets in the coagula showed positive reactions with both anti-HATKTAK and anti-CD 61 antibodies simultaneously (Figure 7).

3.7. Increases in concentrations of Apo CIII, Apo AI and Apo B100 in the platelet lysate

As the LMW activator activity in the products released from platelets increased markedly if stored platelets were used, concentrations of Apo CIII, Apo AI and Apo B100 in the platelet lysate prepared from platelet-rich plasma stored for 120 hours at 20 °C were compared with those prepared from fresh equivalent plasma to determine whether lipoproteins are internalized by platelets during storage of platelet-rich plasma. All of Apo CIII, Apo AI and Apo B100 in the lysate from 120 hour-stored platelets were high (20 ng/ml, 18 ng/ml and 252 ng/ml in average, respectively), whereas those from platelet-rich plasma before storage were very low (1.4 ng/ml, 0.6 ng/ml and 130 ng/ml on average, respectively) (Figure 8).

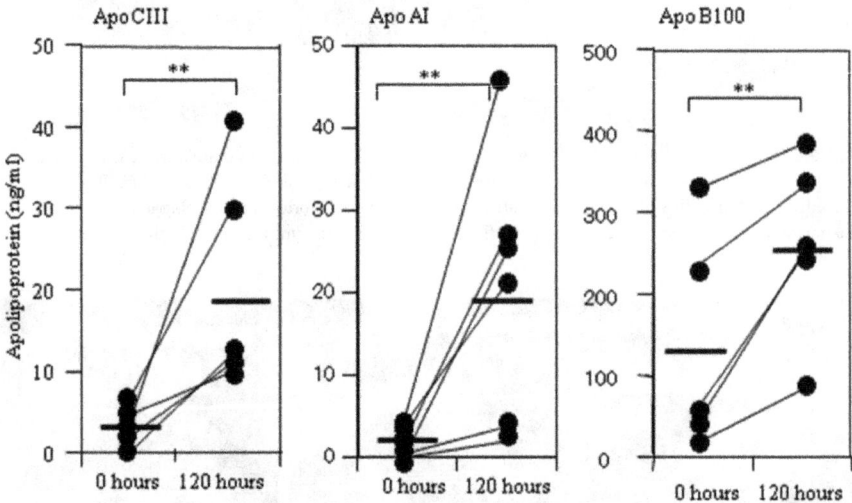

Figure 8. Concentrations of Apo CIII, Apo AI and Apo B100 in the platelet lysate. Concentrations of Apo CIII, Apo AI and Apo B100 in the platelet lysate, which were prepared from platelet-rich plasma before (0 hours) and after storage for 120 hours (120 hours), were determined using commercially available ELISA kits. The bars in the figure show the average. **, P<0.05 by Wilcoxon test.

4. Discussion

Since we found MAPPs in platelet release products, we have investigated their structure and the mechanisms of production and release by platelets. When fresh platelets are stimulated by thrombin, collagen or centrifugation in the presence of Ca^{++}, MAPPs are released. When platelets are stored longer than 72 hours in the form of platelet-rich plasma, they lose the capacity to release MAPPs. It is speculated that this is because platelets lose their contents of thrombin and precursors of MAPPs during storage in the form of platelet-rich plasma. They recover the MAPP-releasing function by washing with PBS and incubation with plasma-derived precursors of MAPPs and thrombin in the presence of Ca^{++} [26]. This suggests that some platelet factor other than thrombin and the precursors of MAPPs is necessary for MAPP production. To analyze this factor, the action of the platelet lysate obtained by freeze-thaw of stored platelets on MAPP formation was investigated. By incubation of the platelet lysate with thrombin and the plasma-derived precursors, it was possible to produce MAPPs. When producing MAPPs using plasma-derived precursors and the platelet lysate, we found that trypsin instead of thrombin can produce MAPPs [27].

After we found that the plasma precursors of MAPPs are dimer and tetramer transferrins [28], we produced precursors of MAPPs by glutaraldehyde treatment of commercially available holo-transferrin and used them to produce MAPPs in vitro.

In the former study [30], we found that Apo CIII associated with the high-density lipoprotein in platelet lysate can be the source of LMW activator. In this study, we compared the LMW activators produced from Apo CIII by the actions of thrombin and trypsin. It was revealed that both LMW activators appeared in the fractions of the same elution volume of Superdex peptide gel filtration and in the fractions with the same NaCl concentration of MONO S cation exchange chromatography. These findings suggest that the LMW activators produced by thrombin and trypsin are the same substance. The fact that trypsin can form LMW activator suggests that LMW activator is a peptide with a C-terminal amino acid of lysine or arginine and that the N-terminal is an amino acid next to lysine or arginine in the amino acid sequence of Apo CIII. It can be asserted that the action of thrombin in LMW activator formation occurs by its trypsin-like activity, although the activity of thrombin to release LMW activator from Apo CIII is far lower than that of trypsin. Cation exchange chromatography in this study revealed that the LMW activator is cationic, suggesting that it is a peptide rich in basic amino acids. In Apo CIII, K17 to K24 is a region rich in basic amino acids, containing four basic amino acids among eight amino acids in total. Several candidate peptides for LMW activator that contain all or part of the K17 to K24 peptide were raised. By comparing the elution volumes of peaks of these peptides with that of LMW activator function on Superdex peptide gel filtrations and by examining MAPP formation using these peptides, it was strongly suggested that HATKTAK is the LMW activator.

As for the reaction of thrombin, the most abundant natural substrate of thrombin is fibrinogen [36], whereas O'Mullan et al. [37] reported that the action of thrombin to various proteins including Apo CIII is more variable and various peptides cleaved from Apo CIII appear by the action of thrombin on Apo CIII. Our study revealed that the trypsin-like thrombin activity that digests Apo CIII to release HATKTAK is weak, but does in fact exist. We have shown that GP-1bα-bound thrombin functions in MAPP production [29]. GP-Ibα is a high-affinity thrombin receptor on platelets [38-39]. Binding of thrombin with GP-1bα might enhance the trypsin-like activity of thrombin in platelets.

To confirm that LMW activator, HATKTAK, is produced by the platelets, we examined the LMW activator activity in the low-molecular-weight fractions of variously prepared platelet release products. We found that, in the platelet release products prepared using $4\times10^5/\mu l$ fresh platelets (0.1 unit/ml thrombin in the presence of 4 mEq/l Ca^{++}), the LMW activator involved is as much as 1 ng/ml HATKTAK.

By storage of the platelet-rich plasma, release of LMW activator from platelets induced by thrombin in the presence of Ca^{++} increased prominently. It was suggested that the concentration of LMW activator released from platelets stored for 120 hours is 10,000 times as much as that from fresh platelets.

One reason for the tremendous increase in LMW activator, HATKTAK, in the released products during storage is probably the loss of the precursor of MAPPs during storage, as shown in a previous report [26]. It is speculated that, in activated fresh platelets, LMW activator (HATKTAK) is produced as much as in stored platelets, but it decreases markedly because precursors of MAPPs remove it in fresh platelets. Another possible reason is that lipoproteins, apolipoproteins or fragments of apolipoproteins might be transported at high levels into platelets from the plasma during storage. In fact, it was shown that the concentrations of Apo CIII and Apo A1 in the platelet lysate increased markedly after storage of platelets in the form of platelet-rich plasma, but this was still too small to explain the observed increase in the effective dilution by as much as 1,000 times.

Indirect ELISA of the platelet release products using anti-HATKTAK antibody was undertaken to prove the existence of HATKTAK. The results were satisfactory if synthesized pure peptides were used, and it was shown that the anti-HATKTAK rabbit antibody reacted positively to Apo CIII-derived peptides with C-terminal HATKTAK. However, we have not succeeded in establishment of a method to analyze HATKTAK in platelet release products. It is postulated that some substances derived from the platelet release products interfere with the adherence of HATKTAK on the wall of microtiter plate. Therefore, we examined the effect of the antibody on the LMW activator function. It was confirmed that the anti-HATKTAK antibody cancels the activity of the LMW activator in the platelet release products from fresh platelet-rich plasma and that stored for 120 hours. Mass spectrometry study revealed the presence of a substance corresponding to HATKTAK (m/z 756) in the platelet release products from platelets stored for 120 hours. Immunohistochemistry of the blood coagula revealed the existence of platelets with double-positive reaction to anti-HATKTAK and anti-CD61 antibodies. These findings strongly suggest that the LMW activator is HATKTAK and is produced and released by platelets.

A schematic illustration of the probable mechanism of production and release of MAPPs and HATKTAK by platelets is depicted in Figure 9.

At present, the mechanism of how MAPP contributes to neutrophilic phagocytosis enhancement after binding to neutrophils is not known. Because MAPPs possess transferrin molecules and anti-transferrin receptor antibody inhibits the action of MAPPs [28], it is suggested that the action of MAPPs occurs via the transferrin receptor. As shown previously, treatment of neutrophils with MAPP does not result in an increase in the number of Fc receptors on neutrophils [40]. This suggests that the effect of MAPP is to strengthen the affinity of Fc receptors with the Fc portion of IgG and to internalize more foreign materials inside the cell. The necessity of HATKTAK for MAPP function suggests that the transferrin receptor might have a site for binding to both transferrin and HATKTAK, and the fact that MAPPs consist of multimers (dimer and tetramer) of transferrin suggests that transferrin receptors must be fixed at an appropriate distance when MAPPs induce enhancement of phagocytosis via the Fc receptors. The mechanisms of how the transferrin receptor, which is stimulated by HATKTAK and transferrin, transfers the information to Fc receptors remain to be elucidated.

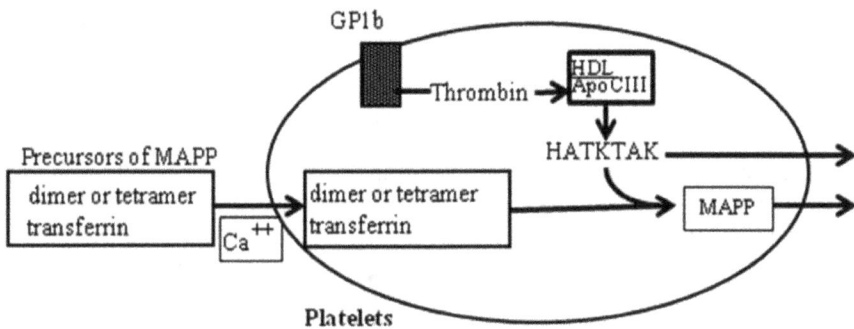

Figure 9. A schematic illustration of the mechanism of production of MAPPs and HATKTAK in platelets and their subsequent release.

In conclusion, we could show that HATKTAK is the LMW activator, which is produced in activated platelets and activates MAPPs, and that the residual HATKTAK is released with other platelet-related substances including MAPPs.

Author details

Haruhiko Sakamoto[1,*], Masaki Ueno[1], Wu Bin[1,2], Yumiko Nagai[3], Kouichi Matsumoto[1],
Takao Yamanaka[1,4] and Sumiko Tanaka[1]
[1]Inflammation Pathology, Department of Pathology and Host Defense, Faculty of Medicine,
Kagawa University, Kagawa, Japan
[2]Department of Gynecology and Obstetrics, Second Affiliated Hospital, China Medical University,
San Hao Road, Shen Yan, China
[3]Division of Research Instrument and Equipment, Life Science Research Center,
Institute of Research Promotion, Kagawa University, Kagawa, Japan
[4]Internal Medicine, Minami-Okayama Medical Center,
Incorporated Administrative Agency National Hospital Organization, Okayama, Japan

---

* Corresponding Author

Acknowledgement

The authors wish to thank staff and students of the Faculty of Medicine, Kagawa University, for their cooperation in providing their blood for this study.

5. References

[1] Sakamoto H, Firkin FC, Chesterman C (1984)   Stimulation of leukocyte phagocytic activity by the platelet release reaction. Pathology. 16: 126-130.

[2] Sakamoto H, Firkin F (1984)   Characterization of leukocyte phagocytic stimulatory material released by activated human platelets. Br. j. haematol. 57: 47-60.

[3] Ward P, Cunningham TW, McCulloch KK, Phan SH, Powell J, Johnson KJ (1988) Platelet enhancement of O2- response in stimulated human neutrophils. Identification of platelet factor as adenine nucleotide. Lab. invest. 8: 37-47.

[4] Spisani S, Giuliani AL, Cavalletti T, Zaccarini M, Milani L, Gavioli R, Tranello (1992) Modulation of neutrophil function by activated platelet release factor. Inflammation. 16: 147-158.

[5] Ruf A, Schlenk RF, Maras A, Morgenstern E, Patscheke H (1992) Contact-induced neutrophil activation by platelets in human cell suspensions and whole blood. Blood. 80: 1238-1246.

[6] Tsuji T, Nagata K, Koike J, Todoroki N, Irimura T (1994) Induction of superoxide anion production from monocytes and neutrophils by activated platelets through the P-selectin-sialyl Lewis X interaction. J. leuk. biol. 56: 583-587.

[7] Carulli G, Barsotti G, Cupisiti A (1995) Platelet-neutrophil interactions in uremic patients: effects on neutrophil superoxide anion production and chemiluminescence. Nephron. 69: 248-252.

[8] Zalavary S, Grenegard M, Stendahl O, Bengstsson T (1996) Platelets enhance Fc(gamma) receptor-mediated phagocytosis and respiratory burst in neutrophils: the role of purinergic modulation and actin polymerization. J. leuk. biol. 60: 58-68.

[9] Piccardoni P, Evangelis Vt, A. Piccoli, de Gaetano G, Walz A, Celetti C (1996) Thrombin-activated human platelets release two NAP-2 variants that stimulate polymorphonuclear leukocytes. Thromb. res. 76: 780-785.

[10] Neumann F, Max JN, Gawaz M, Brand K, Ott I, Rokitta C, Sticherling C, Meinl C, May A, Schomig A (1997) Induction of cytokine expression in leukocytes by binding of thrombin-stimulated platelets. Circulation. 95: 2387-2394.

[11] Peters MJ, Dixon G, Kotowicz KT, Hatch DJ, Heyderman RS, Klein NJ (1999) Circulating platelet-neutrophil complexes represent a subpopulation of activated neutrophils primed for adhesion, phagocytosis and intracellular killing. Br. j. haematol. 106: 391-399.

[12] Petersen F, Bock L, Flad HD, Brandt E (1999) Platelet factor 4-incduced neutrophil-endothelial cell interaction: involvement of mechanisms and functional cosequences different from those elicited by interleukin-8. Blood. 94: 4020-4028.

[13] Kirton CM, Nash GB (2000) Activated platelets adherent to an intact endothelial cell monolayer bind flowing neutrophils and enable them to transfer to the endothelial surface. J. lab. clin. med. 136: 303-313.

[14] Brandt E, Petersen F Ludwig A, Ehlert JE, Bock L, Flad HD (2000) The beta-thromboglobulins and platelet factor 4: blood platelet-derived CXC chemokines with divergent roles in early neutrophil regulation. J. leuk. biol. 67: 471-478.

[15] MacGarrity ST, Heyers TM, Webster RO (1988) Inhibition of neutrophil functions by platelets and platelet-derived product: Description of multiple inhibitory properties. J. leuk. biol. 44: 93-100.

[16] Dallegri F, Ballestero, Ottonello AL, Patrone F (1989) Platelets as scavengers of neutrophil derived oxidants: A possible defence mechanism at sites of vascular injury. Thromb.haemost. 61: 415-418.

[17] Moon DG, Der Zee HV, Weston LK, Gudewicz PW, Fenton JW, Kaplan JE (1990) Platelet modulation of neutrophil superoxide anion production. Thrombosis. haemost. 63: 91-96.

[18] Naum C, Kaplan SS, Basford RE (1991) Platelet and ATP prime O_2^- generation at high concentration. J. leuk. biol. 49: 83-89.

[19] Bengtsson T, Zalavary S, Stendahl O, Genegard M (1996) Release of oxygen metabolites from chemoattractant-stimulated neutrophils is inhibited by resting platelets: role of extracellular adenosine and actin polymerization. Blood. 1996: 87: 4411-4423.

[20] Jancinova V, Drabikova K, Nosal R, Danihelova E (2000) Platelet-dependent modulation of polymorphonuclear leukocyte chemiluminescence. Platelets. 11: 278-285.

[21] Losche W, Temmler U, Redlich H, Vickers J, Krause S, Spangenberg P (2001) Inhibition of leukocyte chemiluminescence by platelets: role of platelet-bound fibrinogen. Platelets. 12: 15-19.

[22] Reinisch CM, Dunzendorfer S, Pechlaner C, Ricevuti G, Wiedermann CJ (2001) The inhibition of oxygen radical release from human neutrophils by resting platelets is reversed by administration of acetylsalicylic acid or clopidogrel. Free. radical. res. 34: 461-466.

[23] Miyabe K, Sakamoto N, Wu YH, Mori N, Sakamoto H (2004) Effect of platelet release products on neutrophilic phagocytosis and complement receptors. Thromb. res. 114: 29-36.

[24] Sakamoto H, Ooshima A. (1985) Activation of neutrophil phagocytosis of complement coated and IgG coated sheep erythrocytes by platelet release products. Br. j. haematol. 16: 173-181.

[25] Sakamoto H, Yokoya Y, Ooshima A. (1987) In vitro control of neutrophilic phagocytosis of IgG coated SRBC by macromolecules involved in released products from platelets. J. leuk. biol. 41: 55-62.

[26] Sakamoto H, Ogawa Y, Sakamoto N, Oryu M, Shinnou M, Hirao T (1996) Recovery of macromolecular activators of phagocytosis from platelets (MAPP) producing and releasing function in stored human platelets. Int. j. hematol. 63: 145-148.

[27] Ogawa Y, Sakamoto H, Oryu, M Shinnou M, SakamotoN, Wu Y, Khatun R, Nishioka M (2000) Production of macromolecular activators of phagocytosis by lysed platelets. Thromb. res. 97: 297-306.

[28] Sakamoto H, Sakamoto N, Oryu M, Kobayashi K, Ogawa Y, Ueno M, Shinnou M (1997) A novel function of transferrin as a constituent of macromolecular activators of phagocytosis from paltelets and their precursors. Biochem. biophys. res. com. 230: 270-274.

[29] Sakamoto H, Ueno M, Wu Y, Khatun R, Tanaka S, Miyabe K, Ogawa Y, Onodera M (2000) Glycoprotein Ibα-bound thrombin functions as a serine protease to produce macromolecular activators of phagocytosis from platelets. Biochem. biophys. res. com. 270: 377-382.

[30] Sakamoto H, Wu B, Nagai Y, Tanaka S, Onodera M, Ogawa T, Ueno M (2011) Platelet high-density lipoprotein activates transferrin-derived phagoctosis activators, MAPPs, following thrombin digestion. Platelets. 22: 371-379.

[31] Boyle W (1968) An extension of the 51Cr-release assay for the estimation of mouse cytotoxins. Transplant. 62: 761-764.

[32] Avrameas S, Ternyck T (1971) Peroxidase labeled antibody and Fab conjugates with enhanced intracellular penetration. Immunochem. 8: 1175-1179.

[33] Nakamura R, Voller M, Bidwell DE (1986) Chapter 27 Enzyme immune assay: heterogenous and homogeneous system. In: Weir DM, Herzenberg LA, Blackwell C, Herzenberg LA, editors. Handbook of Experimental Immunology. 1 Immunochemistry. Oxford: Blackwell Scientific Publications. pp. 27.1-27.20

[34] Peterson GL (1983) O-phthalaldehyde fluorescent methods for protein quantitation. In: Hirs CHW, Timasheff SN editors. Methods in enzymology 91. New York: Academic Press. pp. 95-98.

[35] Hafon ST, Baird TT and Craik CS. (2004) 452. Trypsin In: Barret AJ, Rawlings ND, Woessner JF, editors. Handbook of Proteolytic Enzymes, Vol 2. Second edition. London: Elsevier Academic Press. pp. 1483-1488.

[36] Brown MA, Sternberg LM, Stenflo J (2004) 510. Thrombin. In: Barret AJ. Rawlings ND, Woessner JF, editors. Handbook of Proteolytic Enzymes, Vol 2. Second edition. London: Elsevier Academic Press. pp. 1667-1672.

[37] O' Mullan P, Draft D, Yi J, Gelfand CA (2009) Thrombin induces broad spectrum proteolysis in human serum samples. Clin. chem. lab. med. 47: 685-693.

[38] Harmon JT, Jamieson GA. (1985) Thrombin binds to a high-affinity approximately 900,000-dalton site on human platelets. Biochem. 24: 58-64.

[39] Jamieson GA (1997) Pathophysiology of platelet thrombin receptors. Thromb. haemost. 78: 242-246.

[40] Sakamoto N, Sakamoto H, Tanaka S, Oryu M and Ogawa Y. (1998) Effects of platelet release products on neutrophilic Fcγ receptors. J. leuk. biol. 64: 631-635.

Permissions

The contributors of this book come from diverse backgrounds, making this book a truly international effort. This book will bring forth new frontiers with its revolutionizing research information and detailed analysis of the nascent developments around the world.

We would like to thank Saša Frank and Gerhard Kostner, for lending their expertise to make the book truly unique. They have played a crucial role in the development of this book. Without their invaluable contribution this book wouldn't have been possible. They have made vital efforts to compile up to date information on the varied aspects of this subject to make this book a valuable addition to the collection of many professionals and students.

This book was conceptualized with the vision of imparting up-to-date information and advanced data in this field. To ensure the same, a matchless editorial board was set up. Every individual on the board went through rigorous rounds of assessment to prove their worth. After which they invested a large part of their time researching and compiling the most relevant data for our readers. Conferences and sessions were held from time to time between the editorial board and the contributing authors to present the data in the most comprehensible form. The editorial team has worked tirelessly to provide valuable and valid information to help people across the globe.

Every chapter published in this book has been scrutinized by our experts. Their significance has been extensively debated. The topics covered herein carry significant findings which will fuel the growth of the discipline. They may even be implemented as practical applications or may be referred to as a beginning point for another development. Chapters in this book were first published by InTech; hereby published with permission under the Creative Commons Attribution License or equivalent.

The editorial board has been involved in producing this book since its inception. They have spent rigorous hours researching and exploring the diverse topics which have resulted in the successful publishing of this book. They have passed on their knowledge of decades through this book. To expedite this challenging task, the publisher supported the team at every step. A small team of assistant editors was also appointed to further simplify the editing procedure and attain best results for the readers.

Our editorial team has been hand-picked from every corner of the world. Their multi-ethnicity adds dynamic inputs to the discussions which result in innovative outcomes. These outcomes are then further discussed with the researchers and contributors who give their valuable feedback and opinion regarding the same. The feedback is then collaborated with the researches and they are edited in a comprehensive manner to aid the understanding of the subject.

Apart from the editorial board, the designing team has also invested a significant amount of their time in understanding the subject and creating the most relevant covers. They scrutinized every image to scout for the most suitable representation of the subject and create an appropriate cover for the book.

The publishing team has been involved in this book since its early stages. They were actively engaged in every process, be it collecting the data, connecting with the contributors or procuring relevant information. The team has been an ardent support to the editorial, designing and production team. Their endless efforts to recruit the best for this project, has resulted in the accomplishment of this book. They are a veteran in the field of academics and their pool of knowledge is as vast as their experience in printing. Their expertise and guidance has proved useful at every step. Their uncompromising quality standards have made this book an exceptional effort. Their encouragement from time to time has been an inspiration for everyone.

The publisher and the editorial board hope that this book will prove to be a valuable piece of knowledge for researchers, students, practitioners and scholars across the globe.

List of Contributors

Vikram Jairam
Yale University School of Medicine, USA

Koji Uchida
Nagoya University, Japan

Vasanthy Narayanaswami
California State University Long Beach, USA

Mohammad Z. Ashraf and Swati Srivastava
Genomics Group, Defence Institute of Physiology & Allied Sciences, India

Andriy L. Zagayko, Anna B. Kravchenko, Mykhaylo V. Voloshchenko and Oxana A. Krasilnikova
Biochemistry Department, National University of Pharmacy, Kharkiv, Ukraine

Amany M. M. Basuny
Department of Fats & Oils, Food Technology Research Institute, Agriculture Research Centre, Giza, Egypt

Ivana Pejin-Grubiša
Department of Human Genetics and Prenatal Diagnostics, Zvezdara University Medical Center, Belgrade, Serbia

Isaac Karimi
Division of Biochemistry, Physiology and Pharmacology, Department of Basic Veterinary Sciences, School of Veterinary Medicine, Razi University, Kermanshah, Iran

Masashi Shiomi, Tomonari Koike and Tatsuro Ishida
Kobe University Graduate School of Medicine, Japan

Etsuro Matsubara
Department of Neurology, Institute of Brain Science, Hirosaki Graduate School of Medicine, Japan

Sanja Stankovic, Milika Asanin and Nada Majkic-Singh
Clinical Center of Serbia, University of Belgrade, Serbia

Armando Sena, Carlos Capela, Camila Nóbrega and Elisa Campos
Centro de Estudos de Doenças Crónicas (CEDOC), Faculdade de Ciências Médicas, UNL, Campo Mártires da Pátria, Lisboa, Portugal

Armando Sena, Carlos Capela, Camila Nóbrega and Rui Pedrosa
Serviço de Neurologia, Hospital dos Capuchos, Centro Hospitalar de Lisboa-Central, Lisboa, Portugal

Armando Sena, Camila Nóbrega and Véronique Férret-Sena
Interdisciplinary Centre of Research Egas Moniz (CRiEM), Cooperativa Egas Moniz, Monte da Caparica, Portugal

Caryl J. Antalis and Kimberly K. Buhman
Indiana University School of Medicine & Purdue University, USA

Vladana Vukojević
Department of Clinical Neuroscience, Karolinska Institute, Stockholm, Sweden

Ludmilla A. Morozova-Roche
Department of Medical Biochemistry and Biophysics, Umeå University, Umeå, Sweden

Adebowale Bernard Saba and Temitayo Ajibade
Department of Veterinary Physiology, Biochemistry and Pharmacology, University of Ibadan, Ibadan, Nigeria

Yoshio Aizawa, Hiroshi Abe, Kai Yoshizawa, Haruya Ishiguro and Yuta Aida
Jikei University Katsushika Medical Center, Division of Gastroenterology and Hepatology, Internal Medicine, Aoto, Katsushika-ku, Tokyo, Japan

Noritomo Shimada
Shinmatsudo Chuo General Hospital, Department of Gastroenterology and Hepatology, Shin-Matsudo, Matsudo City, Chiba, Japan

Akihito Tsubota
Jikei University Kashiwa Hospital, Institute of Clinical Medicine and Research (ICMR), Kashiwa City, Chiba, Japan

Dalibor Novotný
Department of Clinical Biochemistry, Faculty Hospital Olomouc, Czech Republic

Helena Vaverková and David Karásek
3rd Medical Clinic, Faculty Hospital Olomouc, Czech Republic

Anna Gries
Institute of Physiology, Medical University of Graz, Austria

Haruhiko Sakamoto
Inflammation Pathology, Department of Pathology and Host Defense, Faculty of Medicine, Kagawa University, Kagawa, Japan

Masaki Ueno
Inflammation Pathology, Department of Pathology and Host Defense, Faculty of Medicine, Kagawa University, Kagawa, Japan

Wu Bin
Inflammation Pathology, Department of Pathology and Host Defense, Faculty of Medicine, Kagawa University, Kagawa, Japan
Department of Gynecology and Obstetrics, Second Affiliated Hospital, China Medical University, San Hao Road, Shen Yan, China

Yumiko Nagai
Division of Research Instrument and Equipment, Life Science Research Center, Institute of Research Promotion, Kagawa University, Kagawa, Japan Kouichi Matsumoto

Takao Yamanaka
Inflammation Pathology, Department of Pathology and Host Defense, Faculty of Medicine, Kagawa University, Kagawa, Japan Internal Medicine, Minami-Okayama Medical Center, Incorporated Administrative Agency National Hospital Organization, Okayama, Japan

Sumiko Tanaka
Inflammation Pathology, Department of Pathology and Host Defense, Faculty of Medicine, Kagawa University, Kagawa, Japan